A+C

Architecture+Conception

A+C 建筑+方案

凤凰空间·上海　编

江苏人民出版社

序言 Preface

作为中国主导性景观建筑出版机构之一，《国际新景观》杂志社与世界建筑协会网强强联手，共同推出本书。书中汇集了世界各地优秀建筑师的各类建筑设计作品，与时同时，嵌入一定的中国图书特色元素，为世界建筑设计界提供了又一精彩的参考读本。

世界建筑协会，成立于2008年3月1日。该协会由包括Hans Hollein、Mario Botta以及Fumihiko Maki在内的200位名誉成员组成。世界建筑协会建立了世界上首个互动式资料库，为世界各地的建筑师提供一个展示自己作品的平台，以获取全世界建筑界的认可。通过建议、讨论、修改、评价和共享他们的建筑作品，这一展示平台的独特功能为它的所有成员将来新的建筑思想的诞生奠定了一定的基础。

设立世界建筑协会奖的目的在于突出和展示卓越的建筑作品，以避免这些优秀的建筑作品被国际建筑界所忽视，同时，激发整个国际建筑界就当代建筑这一话题展开讨论，并通过讨论提出一些振奋人心的见解。

《国际景新观》杂志社不仅为国际景观设计界的互动与交流做出了卓著的贡献，对国际建筑界也贡献颇多，曾陆续出版若干册优秀建筑图书。

在此，笔者谨代表世界建筑协会期待着《国际新景观》及其读者们对世界建筑事业的更多关注与支持。

International New Landscape collaborates with Online World Architecture Community initiatively as a leading architectural publisher of China to publish this Book. Embracing Chinese culture is sine qua non for the Community's prospect of connecting architects from all countries in a common platform.

World Architecture Community launched on March 1, with 200 honorary members including Hans Hollein, Mario Botta, and Fumihiko Maki, is probably the first interactive database created to provide an opportunity for all local practices to become internationally recognized. Unique features of this portal enable all members to contribute in the making of the future of architectural thought by submitting, discussing, editing, rating, sharing their work.

The aim of the WA Community Awards is to highlight and publish remarkable projects that might otherwise remain unnoticed by the international public but have the potential to inspire exciting questions about contemporary architectural discourse.

The mission of International New Landscape is especially illustrating in this respect: focusing on landscape not only as an addition of architecture but as the very essence of it.

We expect that you and your readers play active roles in developing World Architecture Community.

Dr. Suha Özkan，美国建筑师协会　名誉理事

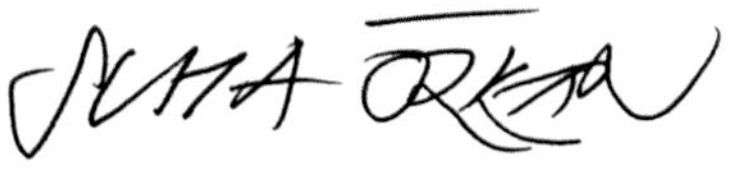

CONTENTS 目录

Project Team: Ângela Frias and Gonçalo Dias, architects

500 Cubic Meters — The Perfect Embodiment of the Architectural Art

500立方米——建筑艺术的完美体现

Given the program of creating a new and renewed area for cultural development, within the city of Beijing, the concept brings forward the idea of individualizing every kind of art and every kind of artist, whatever their craft. Thus, the buildings that are exactly 500 cubic metres in space are slightly separated between them. An objective approach to a bigger urban scheme is also the fact that this separation creates visual and light connections that allow the public to briefly interact with the dwelling of the artists and their work.

More spaces are created beside the courtyards of the houses, for exhibitions that are more public, vaster. These can bring together several kinds of work in a single space.

The concept of the individual is subverted in the interior of the houses. Each has the same arrangement of rooms, with small alterations on the length of the rooms themselves, always reflecting the size of the shell. They are presented in three different kinds of sizes, although maintaining the same cubic area. Each of the sizes also varies in three different ways. This is presented in both the size of the rooms and the layout and size of the windows. The ground floor, access level is a presentation room, accessible through a ramp; this is used as a showroom to each of the artist's work. The ramp exists because the houses are one metre bellow, so they can use their terrace and maintain the 10-metre rule for the façade.

The first floor is a workshop, a place for the development of the art itself, where the person will be able to create freely, in an open space, and then quickly interact with the rest of the ground floor scheme, as well. This open space is only interrupted for the volume of the bathroom and the closet space.

The second floor is a subtle arrangement of one volume that creates a room structure separated from the circulation and living room, by sliding panels. The rest of the second floor is an open space as well, that allows the artist to arrange his own personal objects.

All the volumes have a private swimming pool on the rooftop. Someone for afar will look at this scenery as a big lake, interrupted here and there by the spaces between the houses and the concrete roof slabs.

From the inside these swimming pools will provide lighting as well since their floor is partly transparent and someone inside the bedroom can enjoys the calming effect of waving water.

Different arts, different artists, different houses. We make distinguished facades to announce the several artists, each house, and each art. Meanwhile the interior is equal for all houses.

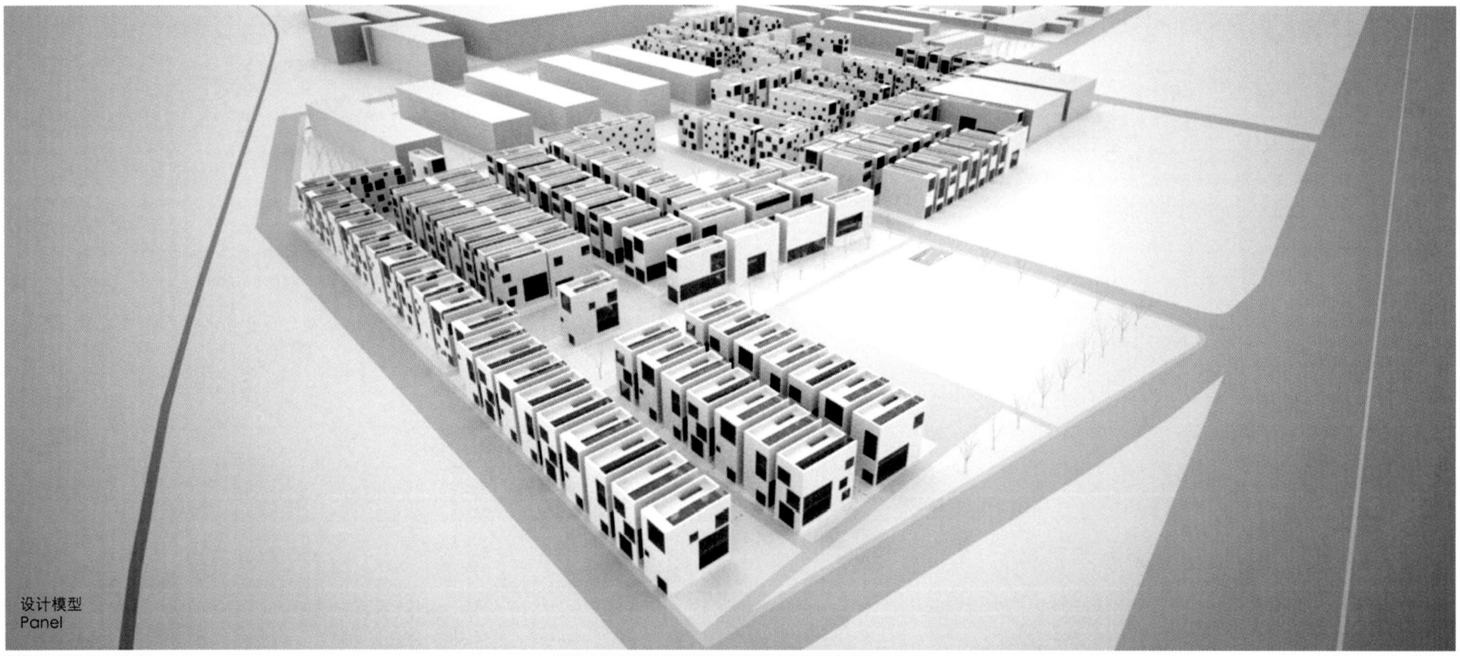

设计模型
Panel

总体景观
Global Views

总体景观
Global Views

A

500m³ .ARTiculate
用语

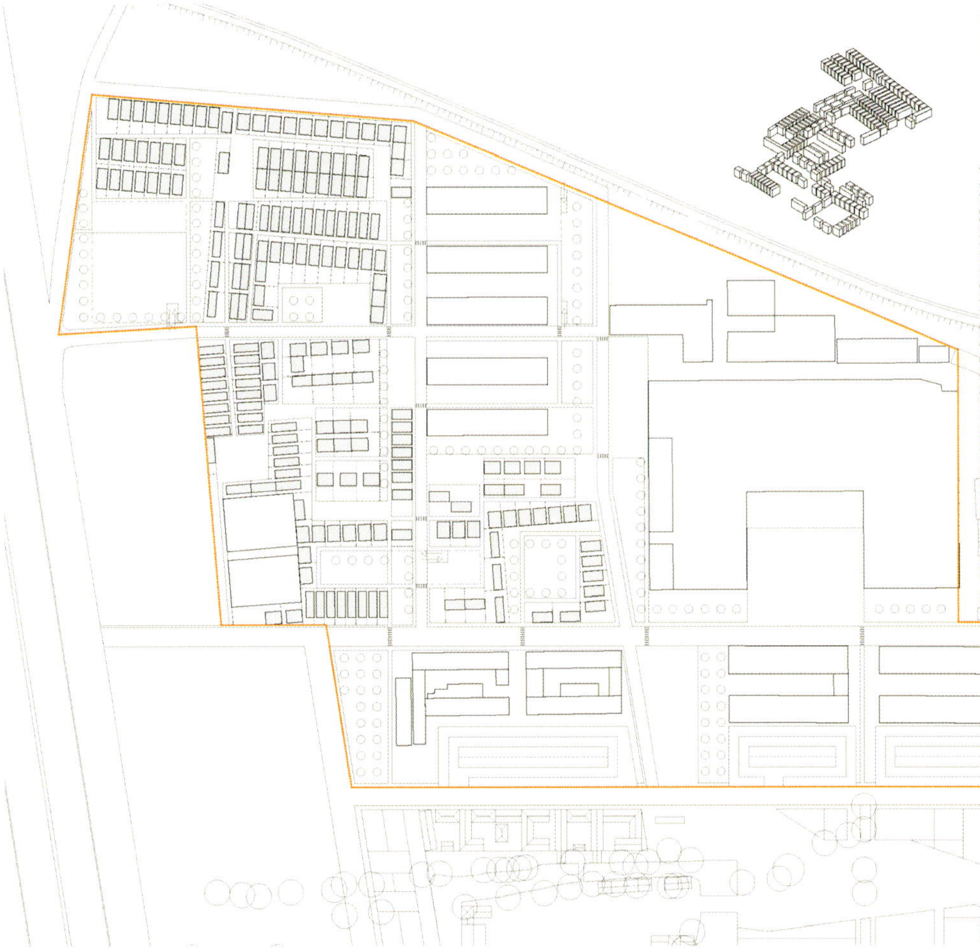

500 cubic metres

The concept,
Given the program of creating a new and renewed area for cultural development, within the city of Beijing, the concept brings forward the idea of individualizing every kind of art and every kind of artist, whatever their craft. Thus, the buildings that are exactly 500 cubic metres in space are slightly separated between them. An objective approach to a bigger urban scheme is also the fact that this separation creates visual and light connections that allow the public to briefly interact with the dwelling of the artists and their work.
More spaces are created beside the courtyards of the houses, for exhibitions that are more public, vaster. These can bring together several kinds of work in a single space.

The houses,
The concept of the individual is subverted in the interior of the houses. Each has the same arrangement of rooms, with small alterations on the length of the rooms themselves, always reflecting the size of the shell. They are presented in three different kinds of sizes, although maintaining the same cubic area. Each of the sizes also varies in three different ways. This is presented in both the size of the rooms and the layout and size of the windows. The ground floor, access level is a presentation room, accessible through a ramp, this is used as a showroom to each of the artist's work. The ramp exists because the houses are one metre bellow, so they can use their terrace and maintain the 10-metre rule for the façade.
The first floor, is a workshop, a place for the development of the art itself, where the person will be able to create freely, in an open space, and then quickly interact with the rest of the ground floor scheme, as well. This open space is only interrupted for the volume of the bathroom and the closet space.
The second floor is a subtle arrangement of one volume that creates a room structure separated from the circulation and living room, by sliding panels. The rest of the second floor is an open space as well, that allows the artist to arrange his own personal objects.

All the volumes have a private swimming pool on the rooftop. Someone for afar will look at this scenery as a big lake, interrupted here and there by the spaces between the houses and the concrete roof slabs.
From the inside these swimming pools will provide lighting as well since their floor is partly transparent and someone inside the bedroom can enjoys the calming effect of waving water.

500 立方米

概念

根据北京市新建和改造文化开发区的计划，此概念提倡艺术个性化和艺术家个性化的设想，无论他们属于哪一行业。因此空间达到500立方米的建筑物间隔会很小，这一间隔会形成某种直观而微妙的联系，使人们能短暂地与艺术家的住所及其作品进行互动。这也是处理较大城市规划的一种客观方法。
除了房屋的庭院，还能创造更多空间，用于更加公开、规模更大的的展览会。这样就能将多种作品集中在一个地方。

房屋

房屋的内部结构颠覆了个性化的理念。每座房子的房间结构都大同小异，仅在房间长度上有所差别，这是由房屋框架大小不一所致。尽管都保持了500立方米的空间要求，但在尺寸上却分为三种比例，每种比例又分为三种类型，体现在房间大小和窗户大小及其布局上。底层入口处为陈列室，可通过坡道进入，用于陈列艺术家的每件作品。鉴于房屋要低出1米，因而专门修建了这一坡道，这样就能利用房屋平台，遵循房屋正面高度10米的设计规则。
一楼工作室是创作艺术作品的地方，艺术家可在此处的开阔空间中自由创作，利用底层设施也很方便。除浴室和厕所外，此开阔空间再无其它阻碍。
二楼结构设计微妙，以滑门将过道和客厅隔开，剩余部分则为开阔空间，用于摆放艺术家的个人物品。

所有房屋的顶楼均建有私人泳池，隔空远眺就如一池被房屋间距和混凝土屋面板隔开的湖水一般。
屋内，由于地板采用半透明材质，这些泳池还可起到照明的作用，呆卧室内的人可以体味到池水波纹带来的宁静。

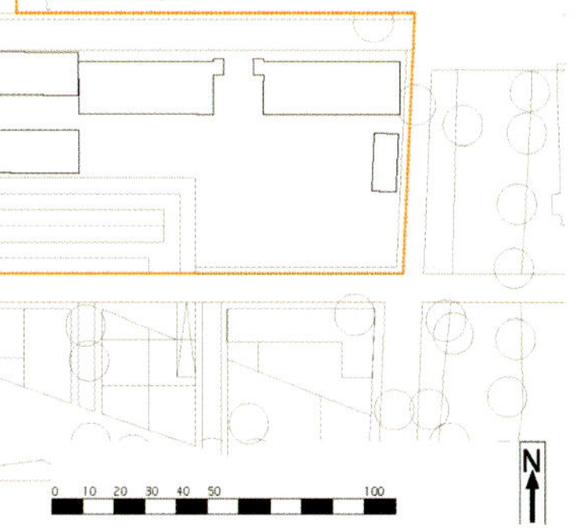

B

500m³ .ARTiculate
用语

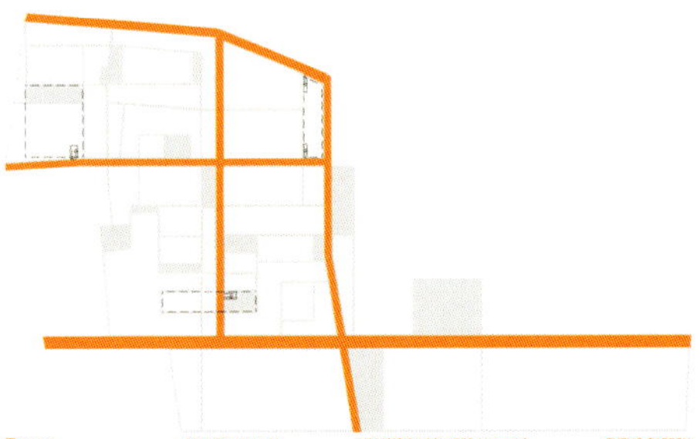

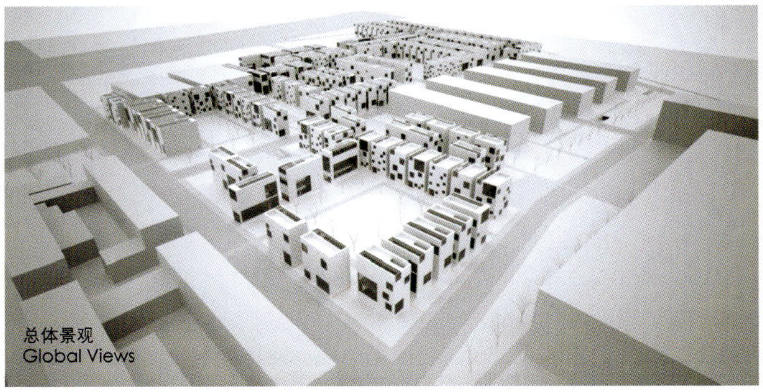

总体景观
Global Views

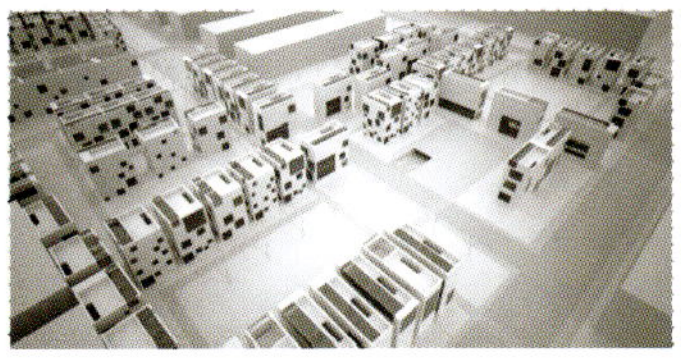

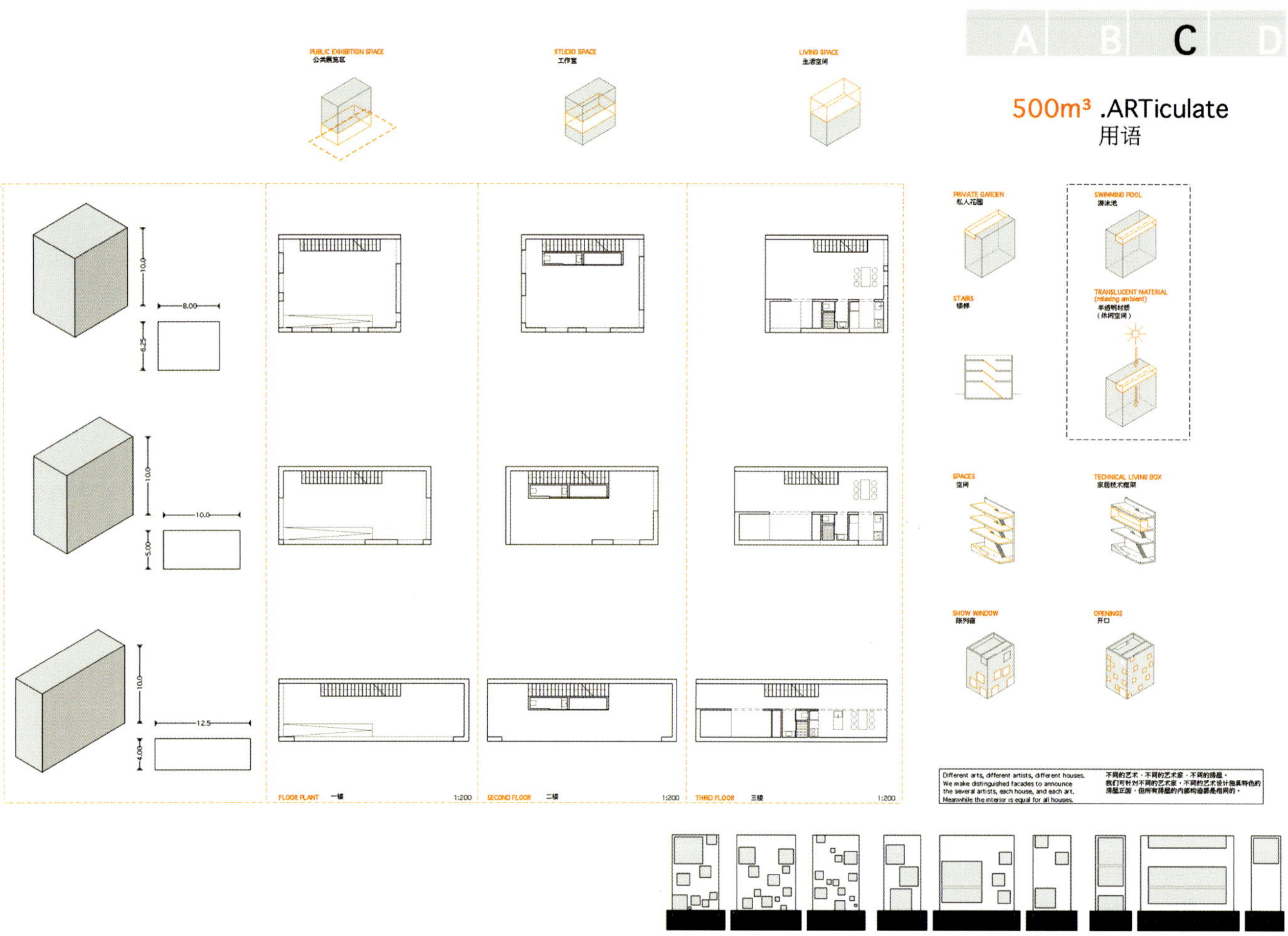

设计模型
Panel

设计模型
Panel

室内布局
Interior

室内布局
Interior

室内布局
Interior

A House in Jeddah

吉达住宅

The house is to be built on a 28m by 30m site in Jeddah. The design is an attempt to exploit some of the spatial, form vocabulary rooted in local traditional architecture, and use it in a modern context. Paradigms like "Roshan" (bay-window) and open courtyard are merged with "facade painting", terraces and light-shade articulation.

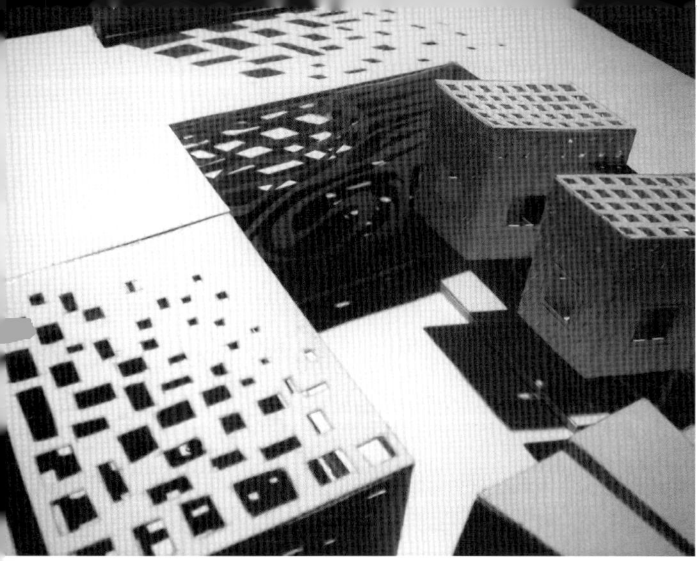

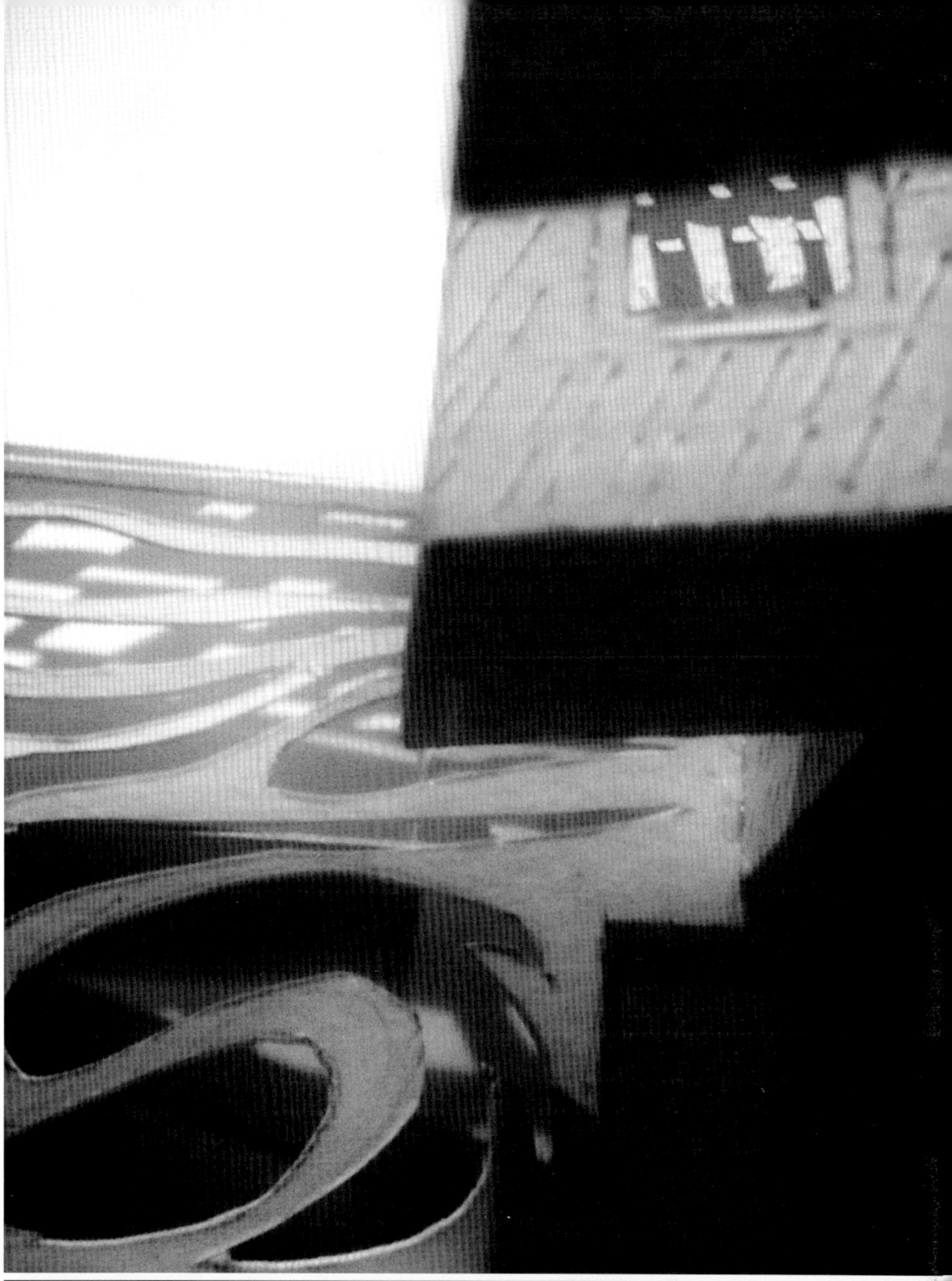

Project Team: Alessandro Calvi Rollino, Micaela Tolio

A Kindergarten pedagogical Building Filled with Naiveness

童趣十足的幼儿园教学楼

The requirement for this project was to build a 6-classroom kindergarten, to be completed in two phases, giving prominence to the existing natural landscape, to the environmental impact, to sustainable building solutions and to pedagogical issues.

The architectural design is strictly influenced by the surrounding landscape; the entire building with its main façade, is oriented toward South because of bioclimatic reasons (radiant heating and solar photovoltaic modules); the space between the façade and an opposite green artificial hill, creates a corridor toward the agricultural landscape. Outdoor space flowing into the school has a psycho-pedagogical function, since children feel comfortable with the view of a familiar landscape: Learning is made easier. The school is a symbolic centre for the life of the children and it becomes a landmark because of the Rainbow Façade, an element for playing and a stimulus for the fantasy.

Materials and colors have some pedagogical purposes with particular interest to visual, acoustic, haptic, olfactory and ecological qualities. The activity rooms are set into a higher glass curving volume; their structure is made of timber while the envelope is made of rough lime plaster and the finishing floor is linoleum; common floor spaces are covered with solid Oak (indoor) and Ipè (outdoor); cladding for the facilities spaces is made of reddish brown copper. Each activity inside the school has an own peculiar space for color and material so that the children easily understand differences and hierarchies among the spaces.

The building develops horizontally throughout the allotment: the high glass volume acts as a plaza for common activities while the activity rooms and the facilities spaces (service department, teacher office, parent room, storage) are set into it.

教学楼整体外观

Cover Image for the Rainbow School

建筑物彩色外观
The Rainbow Facade

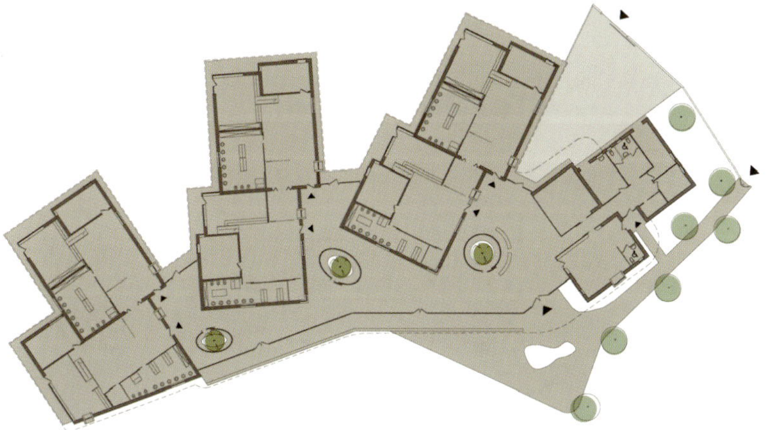

底层平面图
Ground Floor Plan

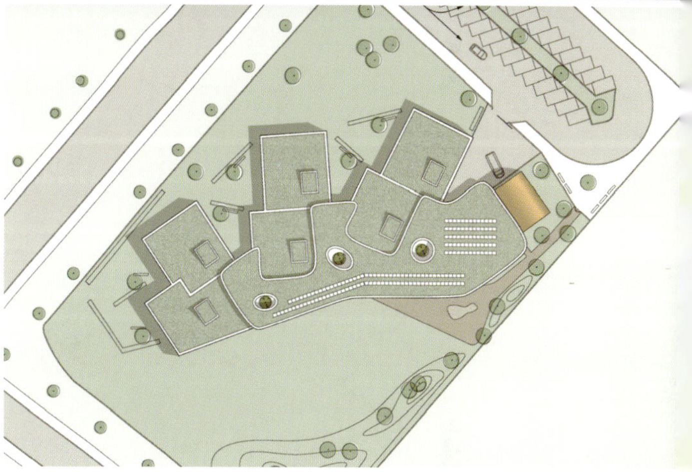

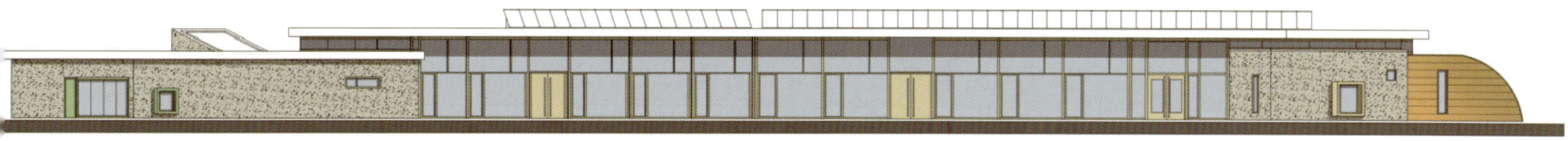

南立面
South Elevation

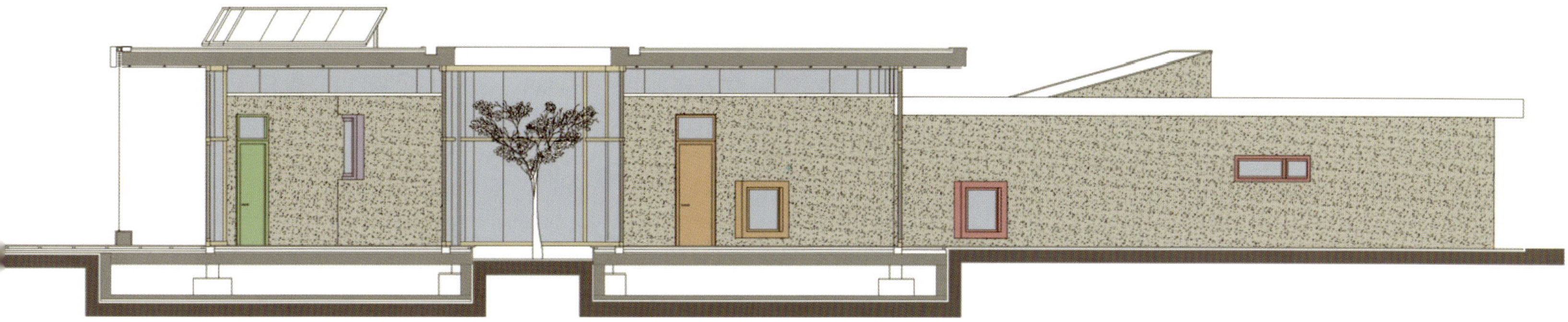

横截面
Cross Section

Project Team: Flavio Galvagni, Brian Gallagher, Yasmin Muhsin Client: World Population

A Novel Carapace House

新颖的甲壳住宅

PROGRAM

The old idea of a "living machine" has been explored with this residential project radically, landing to a solution much depending on mass industry concepts and ship construction processes rather than using their evocative forms to produce a building with traditional materials and methods. Main issues are the pragmatic use of those strategies, the possibility to adapt the house to a variety of radically different environments and domestic scenarios, as well as the need to fulfill high performance requirements.

The program explores the nature of site independent architecture and provides an environmentally friendly-low impact dwelling with high living standards and flexible spaces.

SOLUTION

The C.H. is meant to be a fully autonomous prototype house for a family of four to live in challenging natural environments. The building is raised on legs for lowest rates of footprint usage: the elevation of the body from the ground allows great adaptability to different site conditions and frees an unexpected amount of green space at ground level. The existing topography is left untouched thanks to three mechanical cranes acting as dynamic legs used to support a couple of shipping containers. These elements create the main basic structure of the house capable of pivoting over the legs themselves. Surface walls {blobs} interact with the lower steel structure {boxes} to create the spaces of the house and the partitions within. The blobs are made with composite pre-insulated panels covered with brightly colored glass-fiber layers for full protection. Outer special glass, which dynamically whiten up if exposed to direct sun for long hours, is combined with common glass to create a "thick active wall" used to stop the solar heating during the day and the cold in winter seasons, ensuring comfortable living conditions in any situation. The entrance door to the house is positioned at ground level. The middle level contains two bedrooms, a study/bedroom and a bathroom/sauna. The top floor is where a multipurpose space reveals its ability to transform itself from an open space into a specific space for living, dining, cooking, watching TV, meditating etc., thanks to a deployable system of devices containing the kitchen facilities, the fan/cooling and the LCD TV.

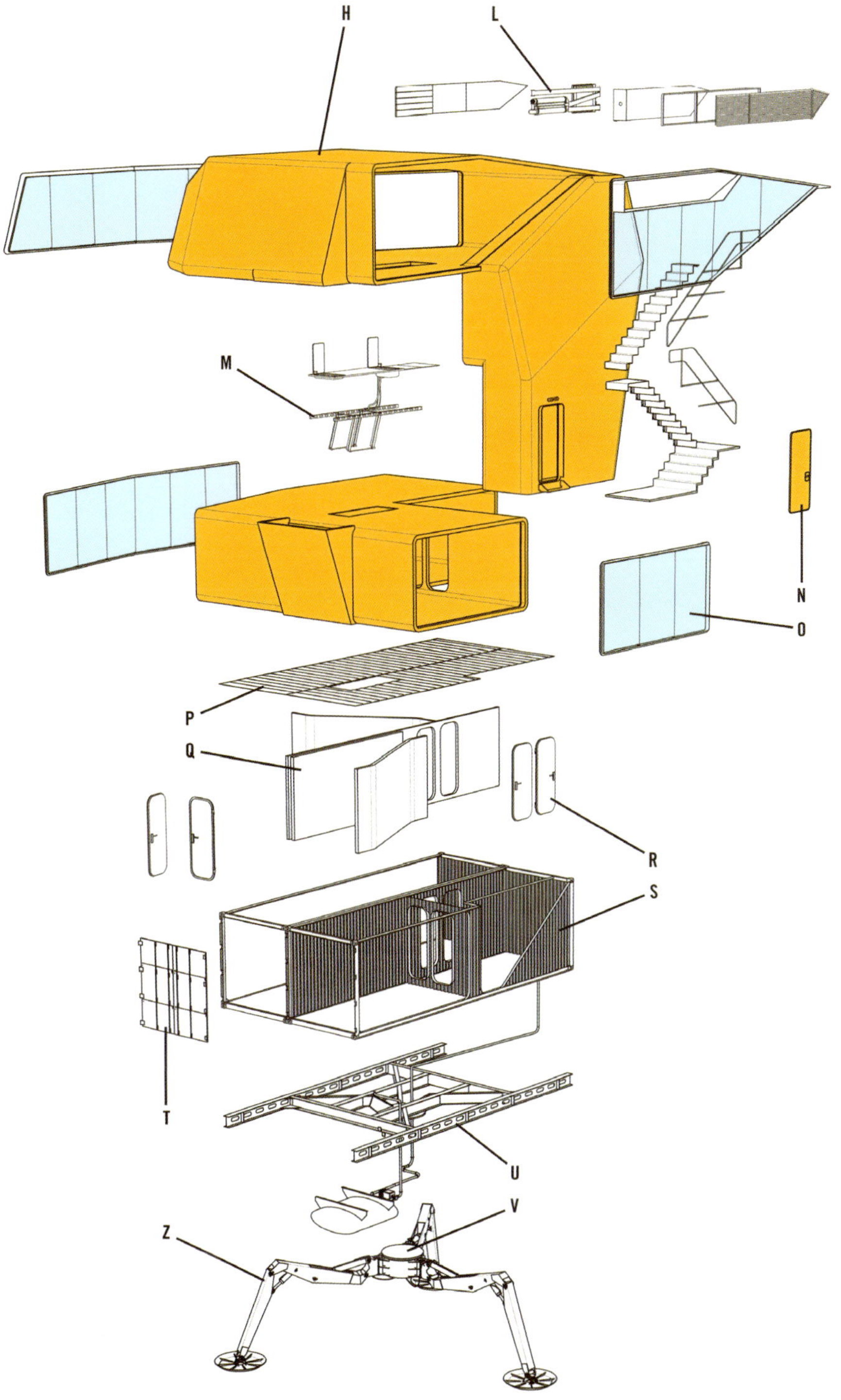
H
L
M
N
O
P
Q
R
S
T
U
V
Z

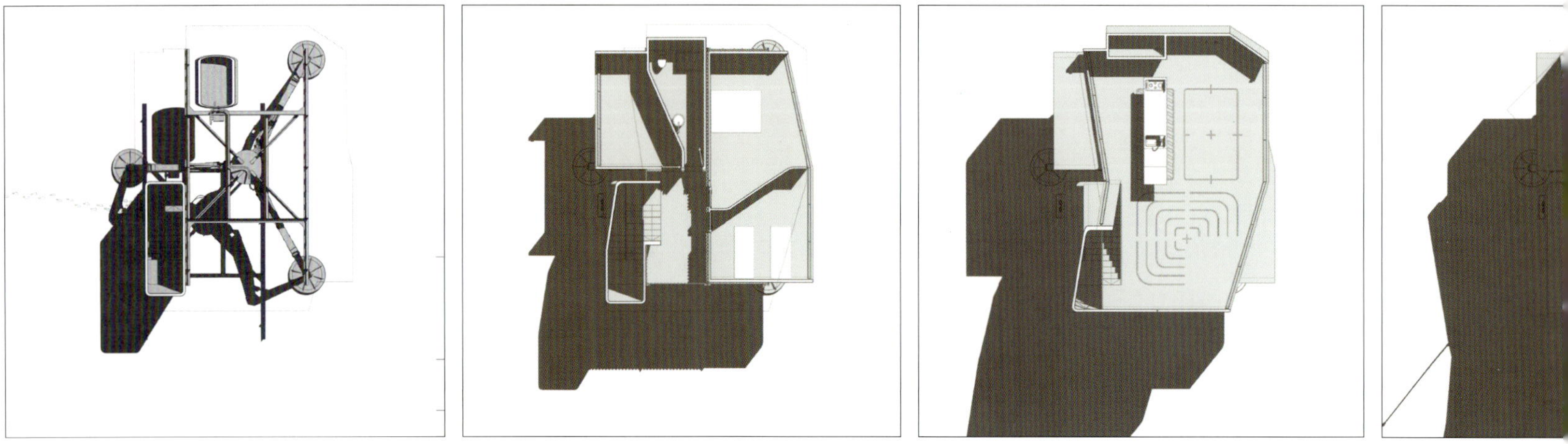

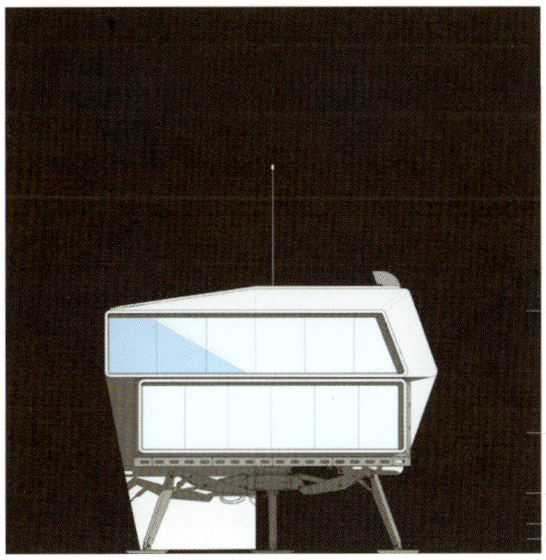

Project Team: Shigeru Kuwahara/Shigeru Kuwahara Architects + Hirofumi Ono/Ono Japan

A Secluded House in the Suburb Tokyo

东京市郊一座清幽的私宅

This residence for a couple and a dog locates in the typical satellite town, 30 minutes train traveling distance from the heart of Tokyo, Japan. The rapid expansion of Tokyo had already involved this area 20 years ago. But there are still some woods secured from the development. The site was a left piece at the edge of urbanization and very calm environment with the rich sun light on the southern slope of a small hill with the view to the green park. However this fragment was an irregularly shaped trapezoid.

Escaping from the difficult condition of the site shape, the plan study was started from the center of the site looking for the gentle connection to those good environments. We placed three certain functions in ideal positions in relation both to the site orientation and to each other, connected those with the uncertain functioned space, and created one homogenous and trifurcated room. Although the space is continuous, the trifurcation system can softly divide functions and generously allow multiple usages. The root part is an ambiguous buffer space, gives free scope of activities among the branches, and works as distributing circulation. Three wings also create the three each characterized gardens in between which provide the situation of soft blending of inside and outside.

The basement floor is an equal level with to the road. We assign the easiest accessible floor to the office space and each of three clusters becomes meeting, work, and archives. The entrance is shared and the stair leads above to the middle of the diverse programs; living, kitchen, dining, library, bath, and bedrooms. The first and second floor is connected through the void which each room faces. By sharing this unfixed and empty space, this entire envelopment can afford multiple activities.

西南视角

Looking from the South-West

南面视角
Looking from the South

西面视角
Looking from the West

局部外观
Exterior Detail

局部外观
Exterior Detail

东面视角
Looking from the East

一楼，LDK空间
1FL, LDK Space

一楼与二楼之间的小空间
Looking up the Void from 1FL

二楼通道
2FL, Bridge

通往二楼的楼梯
Stair Upto 2FL

Project Team: Stefano Dosi, Claudia Cirolli, 3TI Progetti Italia s.r.l., Crea s.r.l.

ACEN — Pilot Project for a Modern Trial Detention Center

现代感化中心实验项目

Concept is principally based on the development of two different paths, each taking advantage of its own relationship-world (physical and conceptual), but inseparably convergent. One is the formal and functional aspect: reality (the "outside" world) jumps into prison with formal "functional voids" and appropriate dosage of borders' transparency (density): infecting and permeating all the trail detention. The other is the introduction and emphasize of a "premium strategy" inside development of trail detention: as the detention positively develops (transition), a wider and quality improved "exposure" to "external world" is guaranteed through different solutions.

To these concepts is also linked the functional "glove" alike configuration, which fits into its palm all the services' structures and among its "fingers" all the penalty sections.

A continuous external embankment surrounds three of the four sides of the squared plan: inside this "natural barrier" stand all the offices (relationship, organization, managing and police&security premises and residences) and the one and only check-point to and from outside world.

The main building develops on three over ground floors and one partially underground floor. Each modern jail "cell" faces onto inner voids (big and bigger), which functions as visual connections and outdoors spaces for leisure and sport activities. The number and dimensions of these voids increase proceeding from entrance (south) to northern boundary. It's also possible to maximize cells' surface and also join two adjacent cells, in order to host 4 persons in 56 sq.m. (with toilet and services like a basic kitchenette and wardrobe). Plants functioning is enhanced by thermal mass studies, optimized porosity configuration of the building and by renewable energy fonts like geo-thermal and solar panels. So comfort and energy saving attitude are both provided in a general "green-building" approach.

概念图
Concept

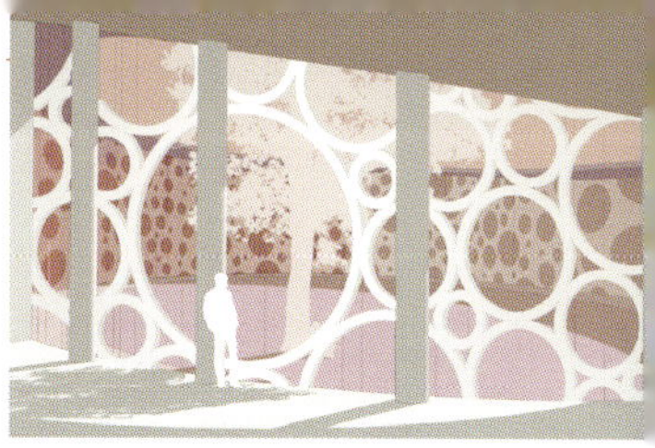

外部透视图和内部设计图
Outer Rendered View & Inner Court View

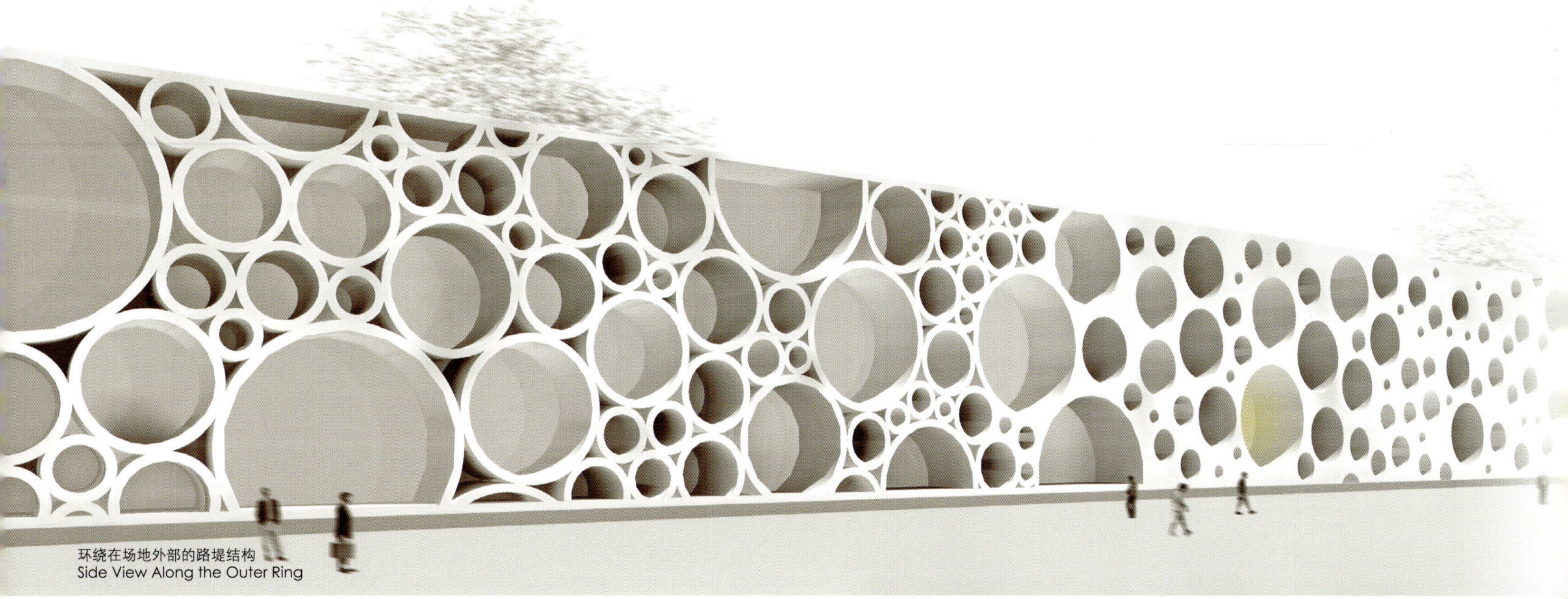

环绕在场地外部的路堤结构
Side View Along the Outer Ring

模型和概念图
Model and Quantity Sketches

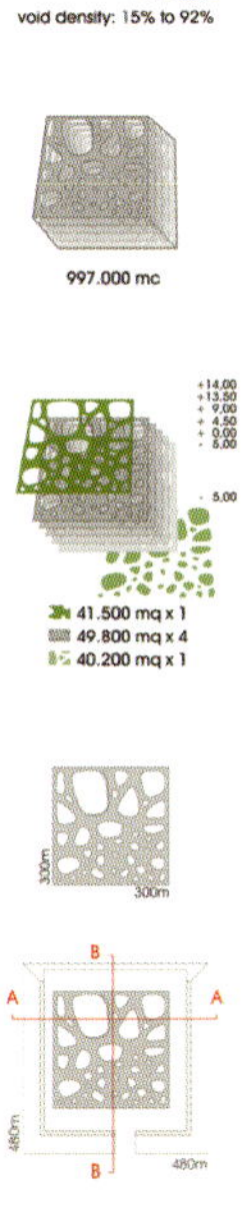

边界设计图
side views (Image: studioDOSI)

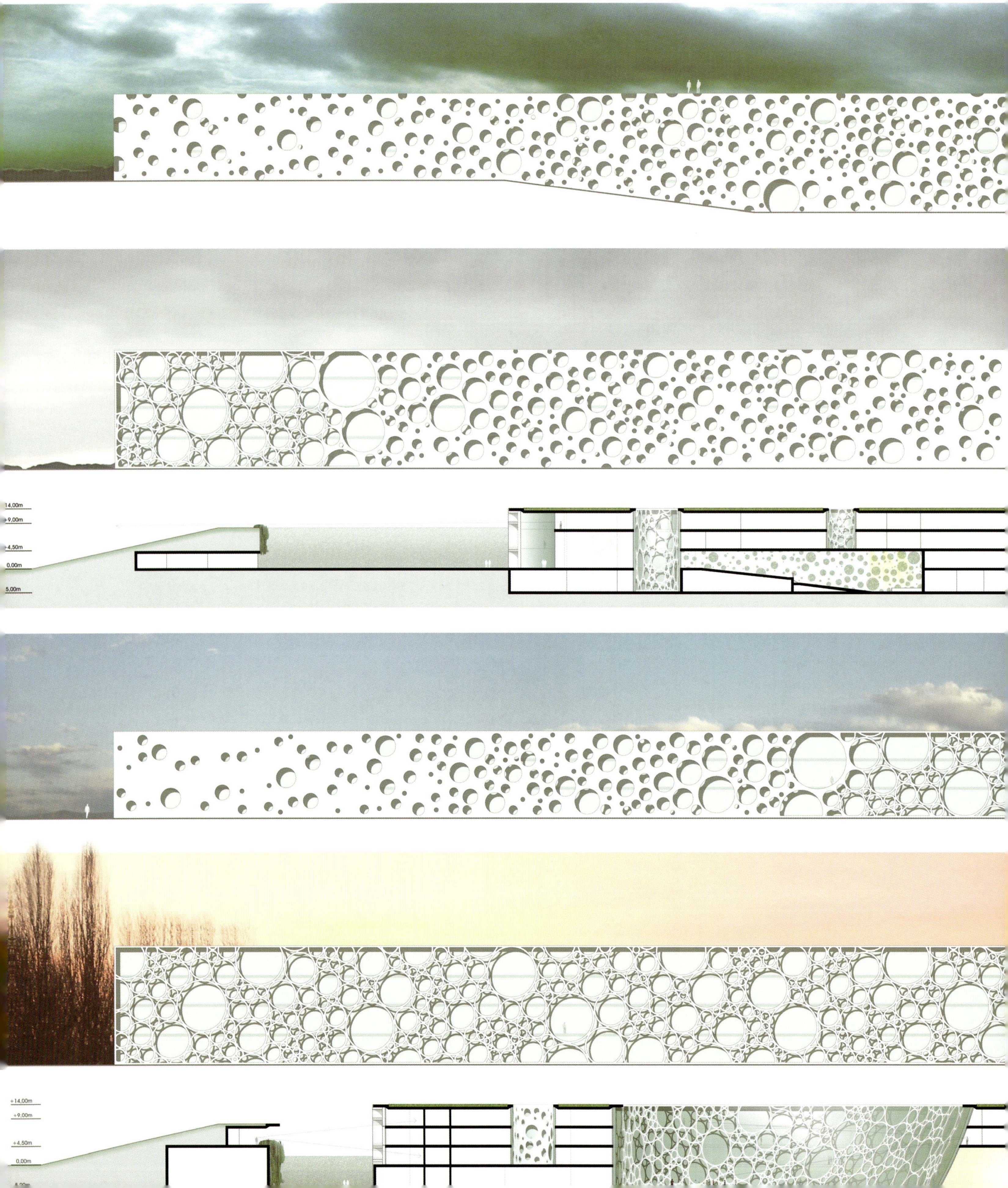

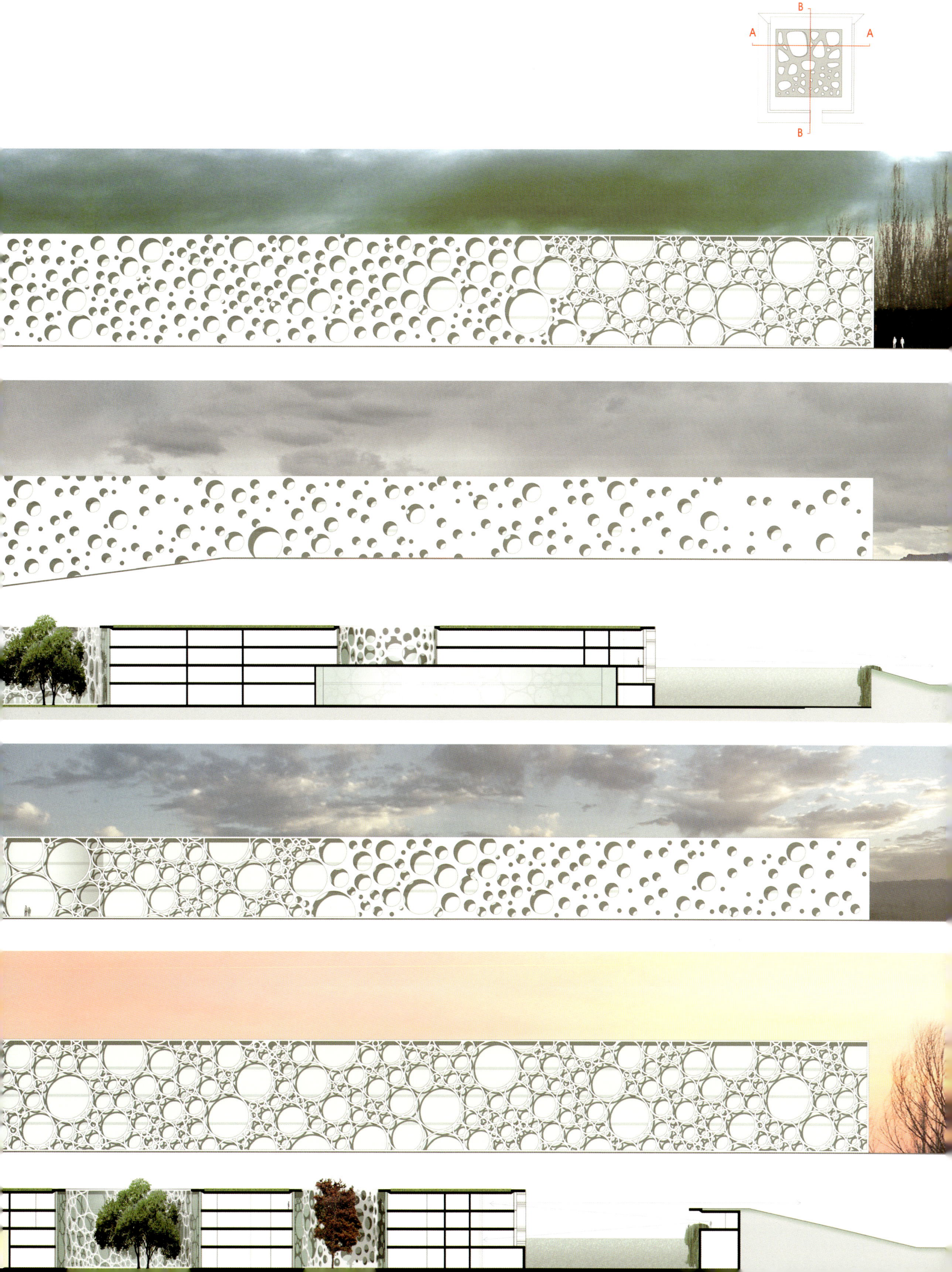
B
A
A
B

Adharshila Vatika
清新的学前教学楼

DESIGN BRIEF

The design required for 5-6 classrooms with capacity of 30 children meant for children age group of 2½-5 year with activity room, Dance & Music room, and Medical room. The reception room should have waiting for around 12-15 people with Administration & Principal room near it. The requirement was to have classrooms well lit up with nature & natural light. Further outer areas will be landscaped & designed for outdoor activities like swimming pool, slides, skating, assembly space & sand pit. The School was to be centrally air-conditioned. The building to have a character of kinder garden school with use of pastel shades (Client had mentioned the colors used in Disneyland).

DESIGN PROCESS

The prior emphasis was kept on the functional requirement and the character (elevation) of the building. The building two storied was placed at middle of the site where all rooms should have natural ventilation. The plan kept as simple as possible & only had a corridor connecting all the spaces which over looked the central courtyard. An environment to be created inside which transforms into children's space, one with which they can identify with, the areas & space they can easily remember and move around. The building plan worked on a hexagonal shape enclosing an open courtyard acts as a secured space for children's to play.

The building longer side facing the main G.T Karnal road has an entrance from front but the main entrance to be used by children's was kept from side road (Side road has less traffic & Children's movement can be monitored properly). The two entrances are kept at side, one lead to the reception area with the administration & principal room and the other to be used by children's which opens on to the staircase well leads them to basement as well as first floor. The swimming pool is kept at the corner in the rear side of the plot.

Big glass windows gave children's to have a good view of open areas.

The building elevation was intended to be derived using basic forms of circles, triangles and square. Further it had to differ from castle shaped or fairly tale exterior. It had to be modern and raw in its outlook.

The use of colors & forms used in exterior was reflected in the interiors too.

The foremost attempt in execution of the building or selection of materials or colors was not to loose the character of the building which it was meant for (an apprehension from client that it should not look like a mall). The landscape was also an important aspect of design with transitional between two spaces achieved by different materials & their colors.

The classrooms has been worked on a theme and has been designed on its function and utility where the children's identify and educated as in one of the classroom has flags painted of different countries and at other has colorful display of alphabets and jungle book characters.

The corridor which connects all the classrooms has interesting elements on the wall and floor which act as a good transitional space, further big windows overlooking the corridor creates an interactive environment among children's.

COLOUR COORDINATION

The colors have played the most important part in defining the character to the building. The selection of colors on the building forms should be eye pleasing, Therefore pastel shades were the obvious choice avoiding bright or dark colors like Red, Green, and Orange.

The form & colors are like a composition with one complimenting each other and not standing apart, it's like a fruit basket where all the fruits of different shapes & colors form a perfect

西南角立面图
Corner Elevation-South East

正立面
Front Elevation

南侧面立面图
Side Elevation-South

从楼梯平台望向庭院
Courtyard View from Landing

庭院景致
Courtyard View

接待处
Reception

走廊
First Floor Corridor

走廊
Ground Floor Corridor

活动室
Activity Room

校长室
Principal Room

composition.
The entrance gate has kept very simple where in longer one rectangular panel has painted in different shades of pink, while the smaller one has different circles painted in pastel shade (distinguishing from normal entrance gates where gate has cutout of Cartoons to get a pre- school look).
The central circular element (front elevation), is in axis of front entrance, cladded with Al. Composite Panel, which neutralizes with "yellow ocre" color on Parapet wall at terrace level on either side. The bands of pastel cream blends with brown color of the corridor walls in front (since "yellow ocre", brown, cream are of same family colors- if used in right proportion always blends well together). The grape paint, dark shade on the front part acts as a frame (always dark). All the classrooms concept has been worked on theme & decisions of colours has been taken on instinct (at site) taking all the three dimensional spaces in consideration. Dark colors were avoided.

CHILDREN'S SAFETY

The entrance of the school is kept from the road side instead from the main G.T.Karnal road at front as the traffic is less and the movement can be monitored well. Similarly all the risers in the staircase for steps as kept 5 inches for children to access easily. Antiskid tiles are used to avoid children's to fall. Further railing is designed keeping children's height in consideration. The swimming pool at the rear side has monitoring camera which can be viewed from the reception & principal room further the depth of the swimming pool is kept 1'-6". Ramp from front leads one to the corridor on ground floor (used for disabled children).

Project Team: Brendan O'Grady Tom Trenolone

Aeroform — Subsidized Urban Living Module

机翼形建筑形态——都市生活空间的补充形式

With an ever increasing global population land is becoming one of our most valuable resources and it comes at a premium. Specifically in dense urban areas, the rising demand and limited supply of land makes home ownership very difficult for most people who want or need to live in the city. We must start to look at new ways of occupying space within the city.

Rather than destroy preexisting space to make room for the new, why not examine ways to occupy residual space within the city? Let's take a common element which is pervasive in every major metropolitan area – the billboard. Now most people perceive the billboard as a simple structure used for the display of two-dimensional graphics. In their minds it is merely a flat plane. What if we start to think about the billboard in a more dimensional way? How can we occupy the space behind or around it?

Aeroform is conceived as a prototype which explores the dimensionality of the billboard and creates a subsidized living module that would allow people with limited means to live in a home in a major metropolitan city where escalating real estate values would normally prohibit such home ownership. This fusion of the billboard and the house creates an opportunity for inhabitants to generate income from the billboard, in turn subsidizing the exorbitant cost of living in a home in a major metropolis. Aeroform essentially pays for itself.

Conceived as a prefabricated, lightweight unit, Aeroform is constructed off-site, shipped to the site, and grafted to the existing structure of a billboard. The outer shell is made of structural insulated panels covered with a layer of carbon fiber in order to reduce the weight of the structure and minimize any

机翼形住宅透视图
Aeroform Perspective View

机翼形住宅透视图
Aeroform Perspective View

structural reinforcement to the existing billboard. The form of the building is directly influenced by its location and program. The shell, which reflects its context of the roadway and the automobile, is streamlined on the exterior for optimal aerodynamic and climatic performance. The name Aeroform is derived from the term aerofoil, which is a structure having a shape that provides stability, or directional control in a flying object. Operable windows at each end of the unit allow for cross ventilation and provide an abundance of natural light within the living spaces. On the interior a programmatic shrink-wrap occurs to optimize the overall volume of enclosed space. This process produces an aerodynamic section that continuously changes from one end of the unit to the other.

Sustainability is not an option but rather it is a necessity for Aeroform. Since land is at a premium the goal of Aeroform is to reduce the building's footprint to a single point that touches the ground. By using an existing billboard's support structure Aeroform has a minimal impact on the environment and its site. It also reduces the amount of raw materials needed to construct the home by re-using the existing steel structure of the billboard.

In some cases there is the opportunity to use the space in-between two or more billboards. In the future several different versions of this prototype will be developed to accommodate various billboard structures such as V-face, double face or triangular configurations. Since the billboard is a universal structure the prototype can easily be adapted for use in any city, anywhere in the world.

后透视图

Aeroform Rear Perspective (Nocturnal Design Lab)

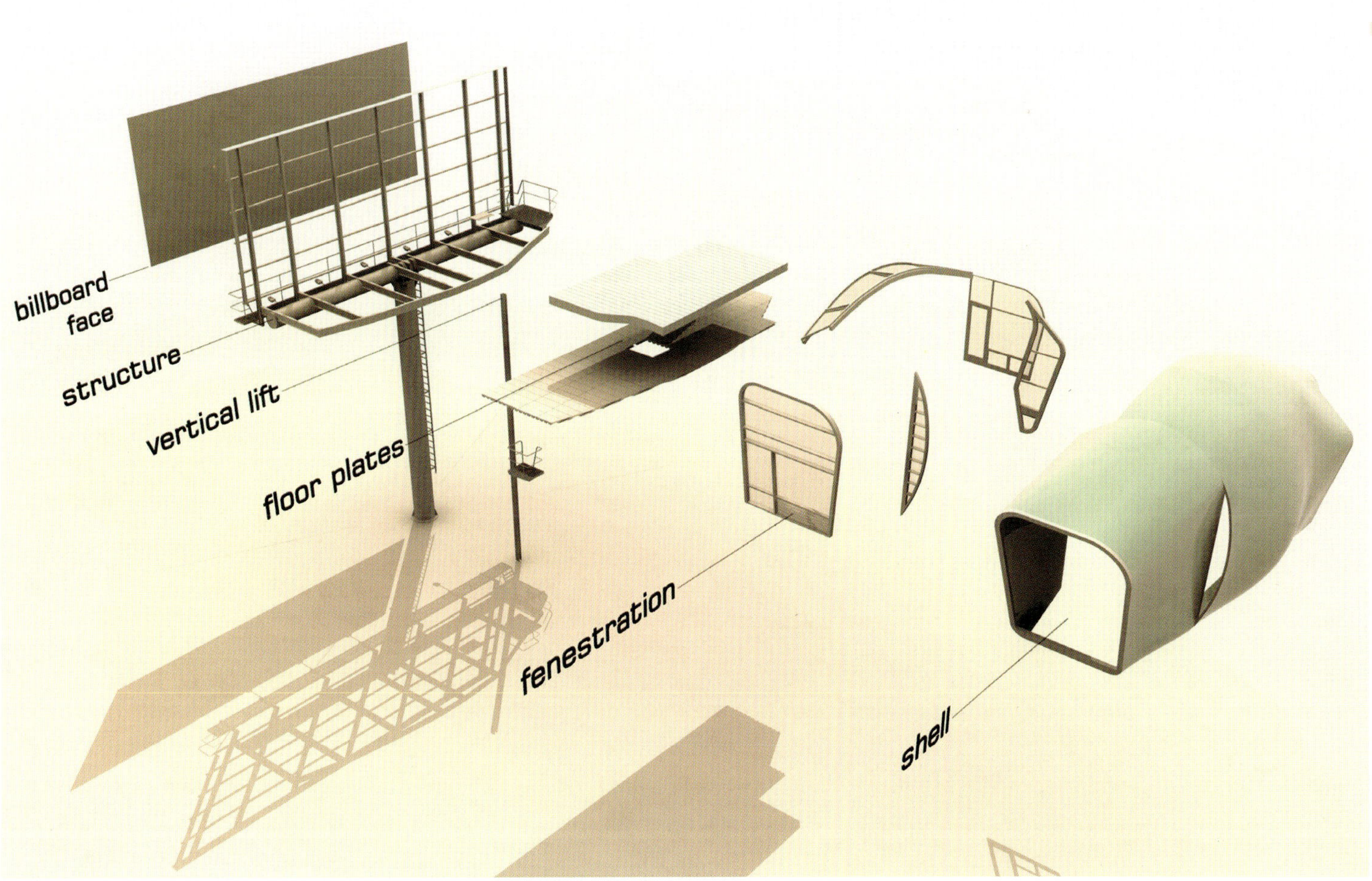

subsidized urban living module

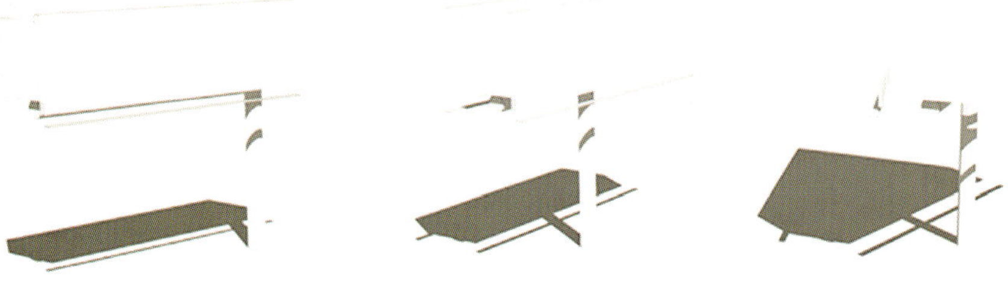

Aeroform is a *prefabricated*, lightweight structure that is **shipped to the site** and **attached** to the base structure of an **existing billboard**. The shell of Aeroform is made of structural insulated panels with an outer layer of carbon fiber in order to reduce the weight of the structure and minimize any structural reinforcement to the existing billboard.

stage 01

- Order and customize your dwelling online.

stage 02

- Living module components are fabricated in an off-site plant on an addembly line similar to the automotive industry.

stage 03

- Components are delivered to the project site by truck.

stage 04

- Module sections are assembled on-site and a crane lifts the module into position.

stage 05

- Module is secured to existing billboard structure, utilities are connected, and the vertical lift is installed.

*The name **aeroform** is derived from the term **aerofoil**, which is a structure having a **shape** that provides **stability**, or **directional control** in a flying object.*

The **form** of the building is directly influenced by its location and program. The **shell**, which reflects its context of the highway and the automobile, is **streamlined** on the exterior for **optimal aerodynamic** and **climatic performance**. On the interior a **programmatic shrink-wrap** occurs to optimize the overall volume of enclosed space. This process produces a section that **continuously changes** from one end of the unit to the other.

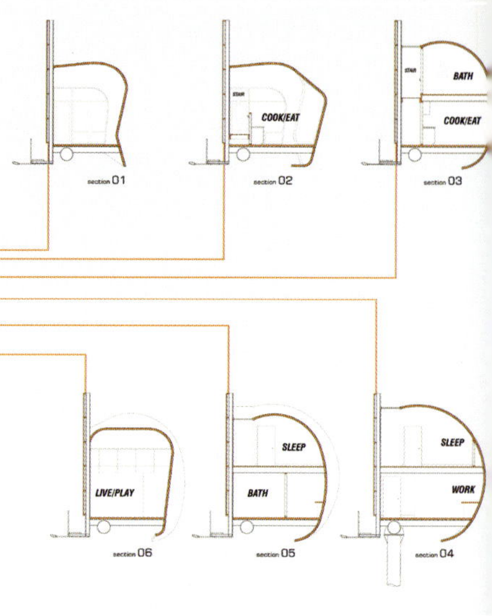

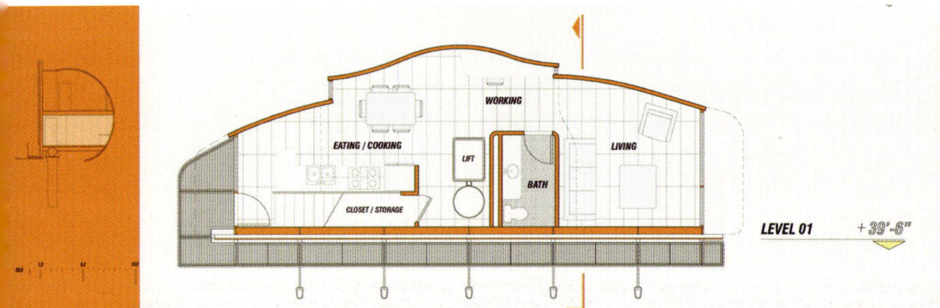

平面图
Aeroform Plans

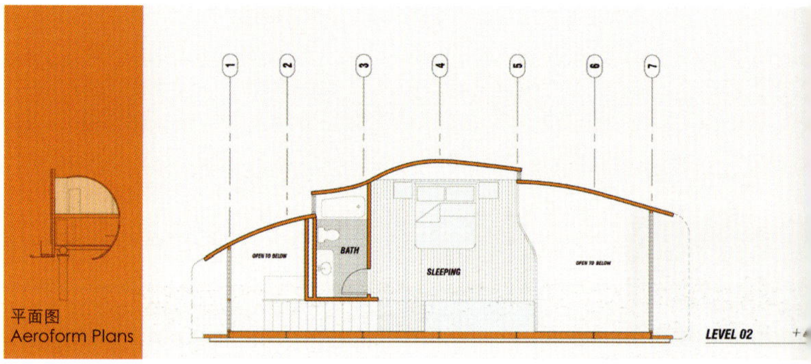

平面图
Aeroform Plans

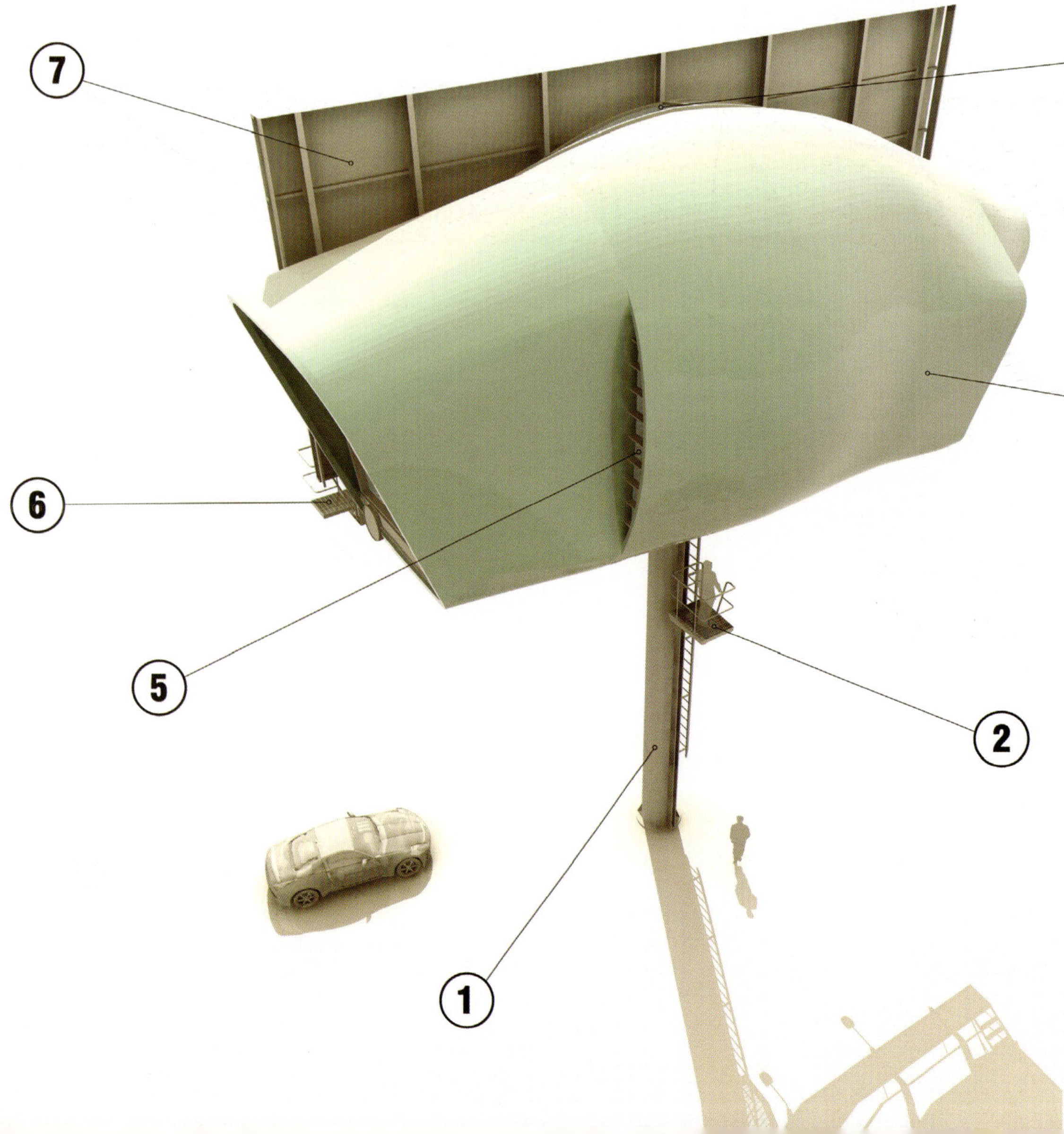

customize your shell skin

solar photovoltaic

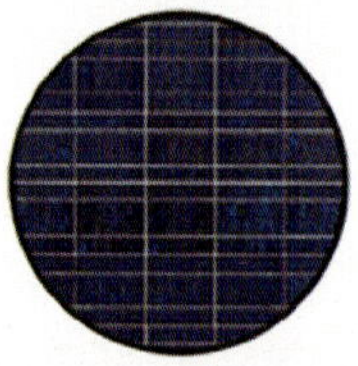

carbon fiber

racing stripe

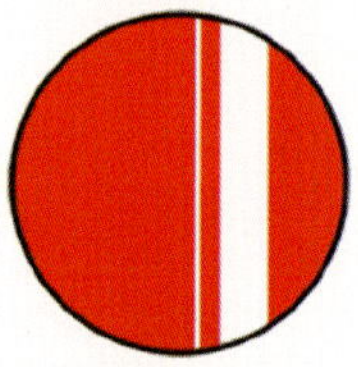

digital camouflage

01. existing structure
02. vertical lift
03. carbon fiber shell
04. skylight
05. windows / air-intake
06. catwalk
07. existing billboard

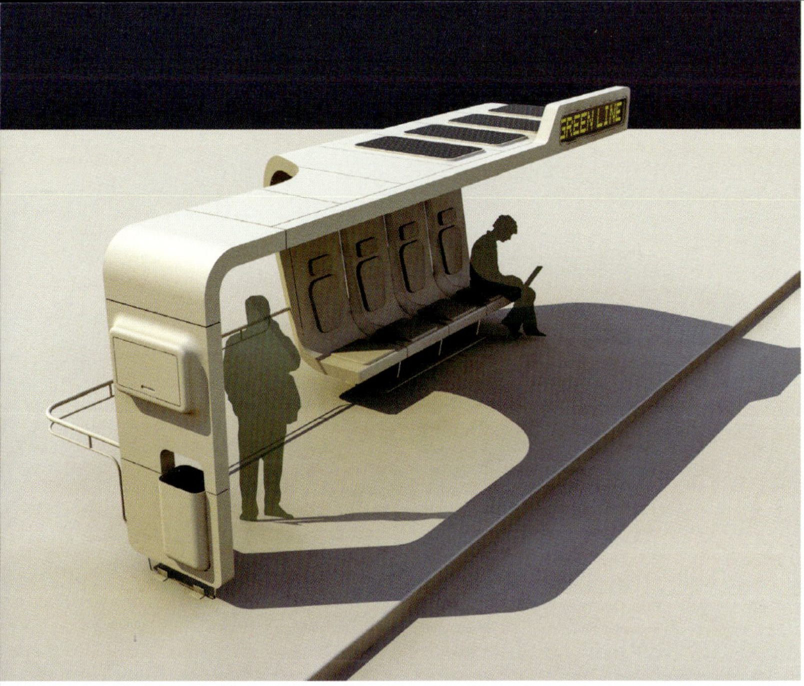
GREEN LINE

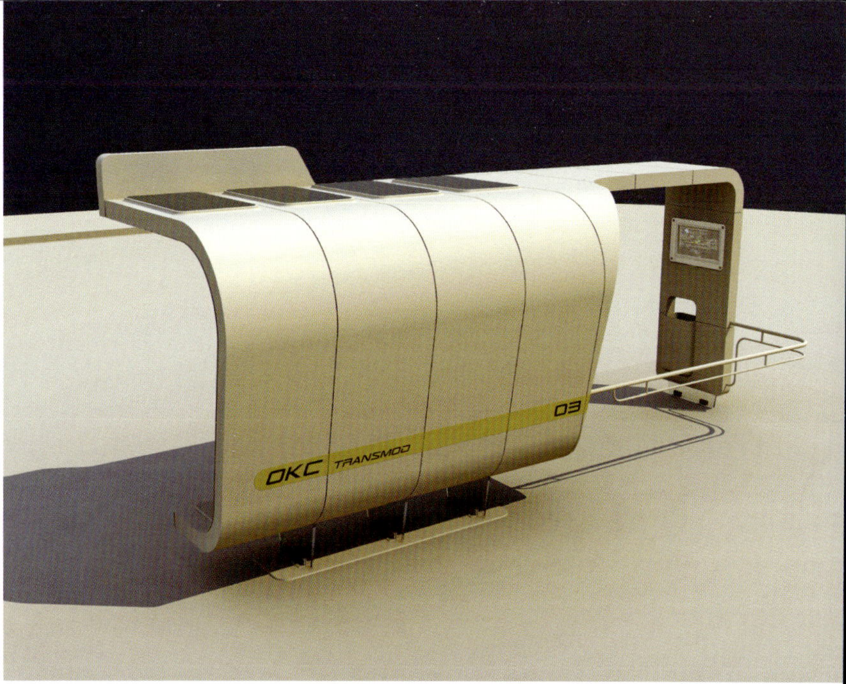
OKC TRANSMOD
03

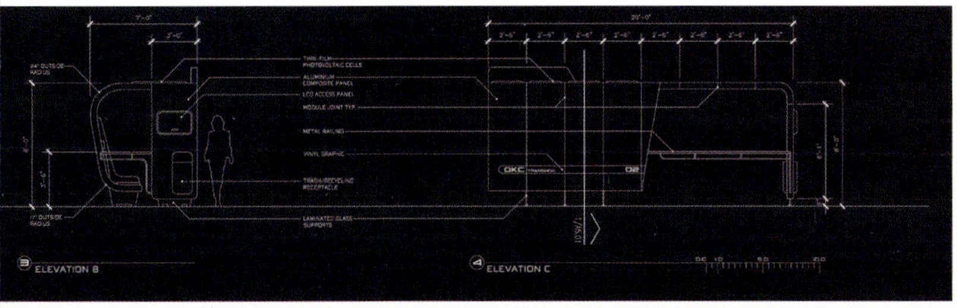
ELEVATION B
ELEVATION C

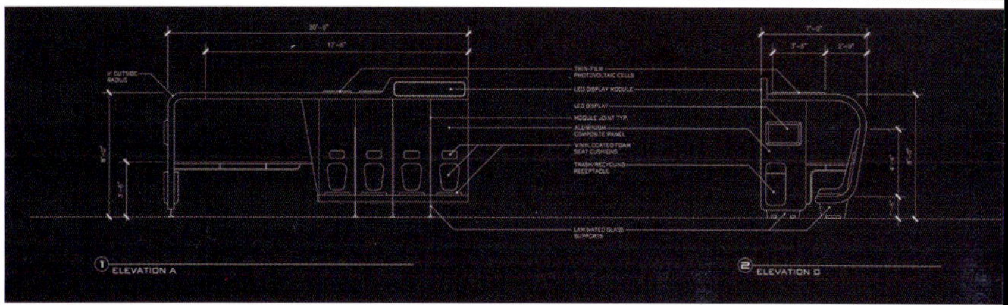
ELEVATION A
ELEVATION D

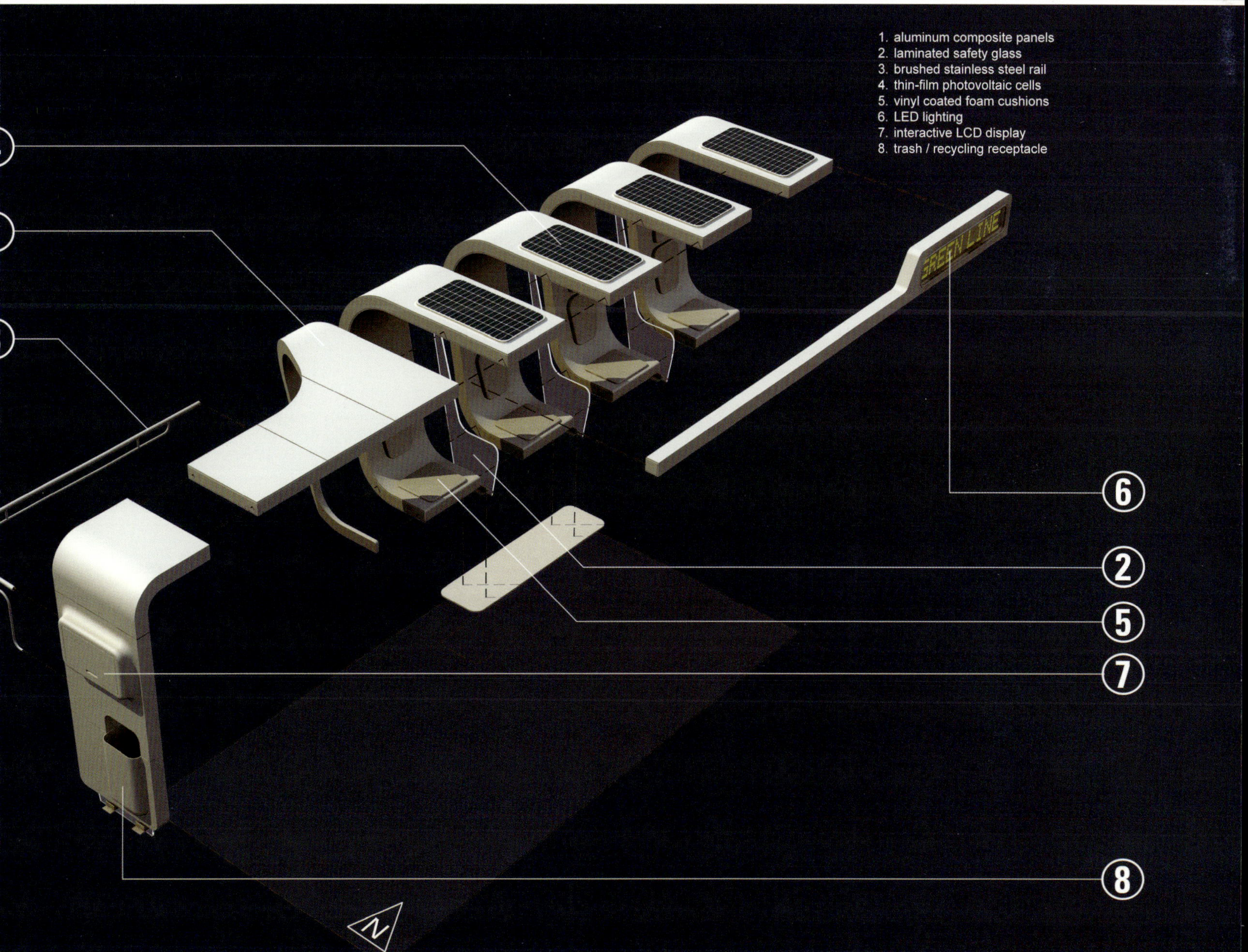
1. aluminum composite panels
2. laminated safety glass
3. brushed stainless steel rail
4. thin-film photovoltaic cells
5. vinyl coated foam cushions
6. LED lighting
7. interactive LCD display
8. trash / recycling receptacle
GREEN LINE
6
2
5
7
8

Project Team: Paul W. Athineos, Principal Architect + Adam J. Wanaselja, Architect in training

Alley Pond Park

艾丽池塘公园

Alley Pond Park which features an 18-hole miniature golf course, a nine-hole kiddie course (both designed by Harris Miniature Golf Courses Inc.) and an 88 tee-stall driving range as well as a pro shop and snack bar housed in an inviting 2,300-square-foot clubhouse. In harmony with the land and designed to maximize sunlight, the low-slung, flat-roofed clubhouse is both inviting and energy efficient. A central glass atrium captures sunlight all day long. A particularly striking feature is the high ceiling in the snack bar, flooded with sunlight by the atrium, and, when the sun goes down, studded with energy-efficient Halogen up lights which create evening drama.

Clad in cool, grey concrete and warm, cedar wood accented with bold geometric windows echoing the building's shape, the clean-cut clubhouse is designed with environmentally sensitive materials such as recycled glass terrazzo countertops and sustainably harvested woods. The stylish, sunny clubhouse includes an outdoor patio where patrons can sit and enjoy scenic views of the garden and kiddie course.

Architect Paul Athineos swung into action after the client asked for an eco-friendly "green" clubhouse. Athineos designed a clubhouse that appeals to families, young people on a date as well as weekend warriors and senior citizens. The clubhouse is a place for golfers to putter around and have fun. The children can play while their parents sit on the patio and enjoy the view.

The steel-frame clubhouse is clad in cool grey concrete and warm cedar wood with geometric windows piercing the façade which allows natural light to flood the interior space. Of course, the windows (both on the façade and in the atrium) open to allow crosswinds to circulate and help minimize the amount of energy used for heating and cooling.

The smooth concrete finish is durable, fire and insect-resistant (may be crushed and used for sub-base material or fill if removed). Cedar wood weathers well in the Northeast. The interior space includes stylish polished concrete floors, light, bright bamboo floors (bamboo is a fast-growing hallow grass) and ceramic tile in the restrooms. Countertops for the snack bar, restrooms and cashier counter consist of recycled glass (see cut sheets in the presentation booklet).

The rectangle-shaped clubhouse consists of a snack bar/lounge, kitchen, restroom, pro shop and staff office all of which is contained within a 2,312-square-foot space. The building consists of two volumes bisected by a gracious entryway flooded with light through a glass atrium that runs the entire length (south to north).

The snack bar/lounge is conveniently located to the right where patrons enter the building. It contains seating which can be easily rearranged to accommodate patrons. An early morning coffee and pastry may be the perfect way to start the day before hitting a bucket of balls. Alternatively, it makes for a great place to have a light lunch and cold drinks when golfers are finished.

Adjacent to the snack bar is the outdoor patio with a glass covered aluminum trellis faced with wood along its exterior perimeter. This canopy will allow visitors to enjoy the outdoors even if it rains. The floor surface is gravel base several inches deep with concrete pavers spaced an inch apart to assist in the overall drainage and water run-off.

The restroom facilities have a generous amount of space which more than adequately satisfies ADA requirements and additionally provides for a make-up counter in the women's room. For easy, convenient access, the restrooms are centrally located in the building yet the entrances are screened by a banquette (high-back bench) and raised planter for privacy.

The pro shop is situated on the northwest corner of the building and has adequate display area for patrons in search of golf gear. Above the pro shop entrance is a skylight that allows natural light to flood the space. There is a generous view to the north overlooking the driving range.

The repair shop is nestled in the south-west corner of the building and provides for direct access from the pro shop as well from an outdoor pass through. If patrons need to use the repair service, they can drop off clubs on the way in and fetch them on the way out after they pay at the cashier's counter. This conveniently located repair shop is another amenity for the patron. Additionally, the awning, which appears as part of the wall when the repair shop is closed for business, serves to protect patrons from the sun.

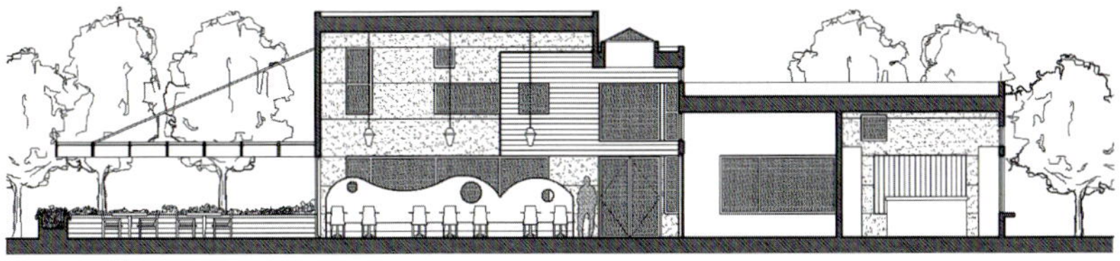

LONGITUDINAL SECTION
SCALE: 3/32" = 1'-0"
(INTERIOR ELEVATION LOOKING SOUTH)

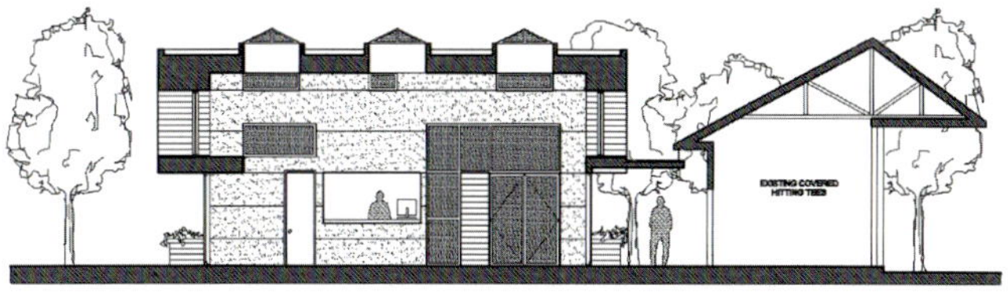

LATITUDINAL SECTION
SCALE: 3/32" = 1'-0"
(INTERIOR ELEVATION LOOKING WEST)

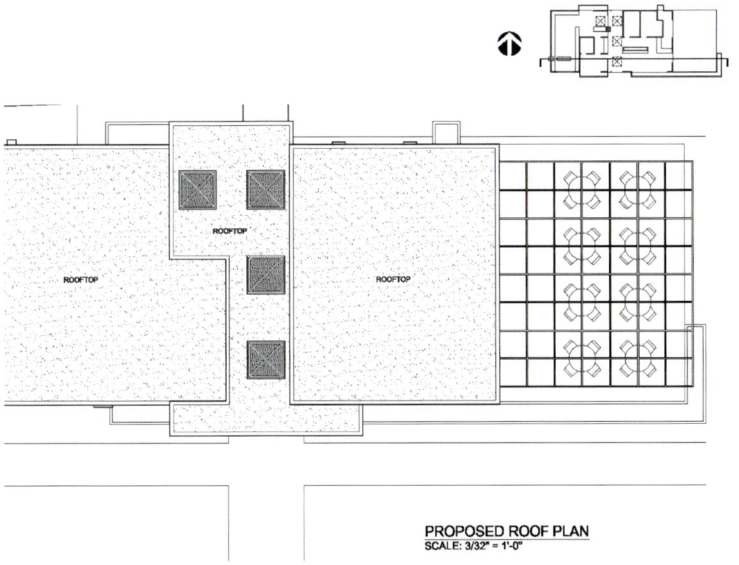

PROPOSED ROOF PLAN
SCALE: 3/32" = 1'-0"

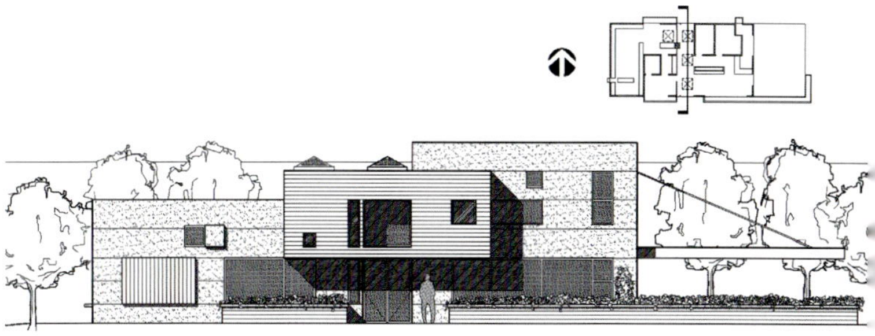

SOUTH FACING ELEVATION
SCALE: 3/32" = 1'-0"
(NORTHERN BLVD. FACING ELEVATION)

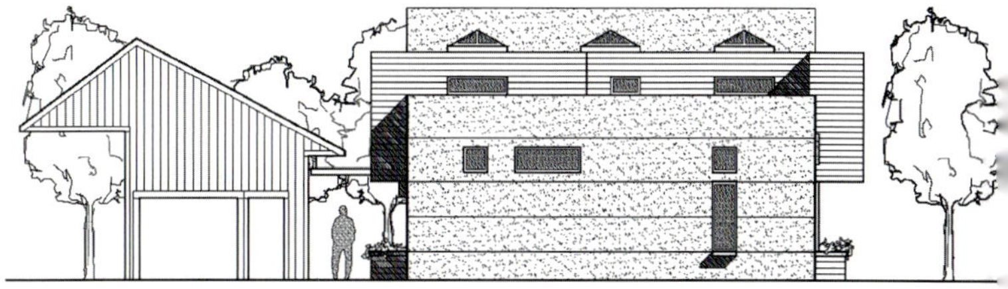

WEST FACING ELEVATION
SCALE: 3/32" = 1'-0"

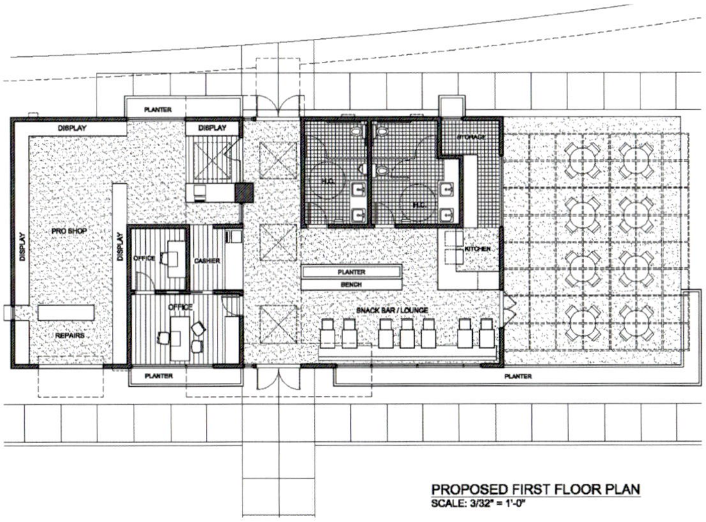

PROPOSED FIRST FLOOR PLAN
SCALE: 3/32" = 1'-0"

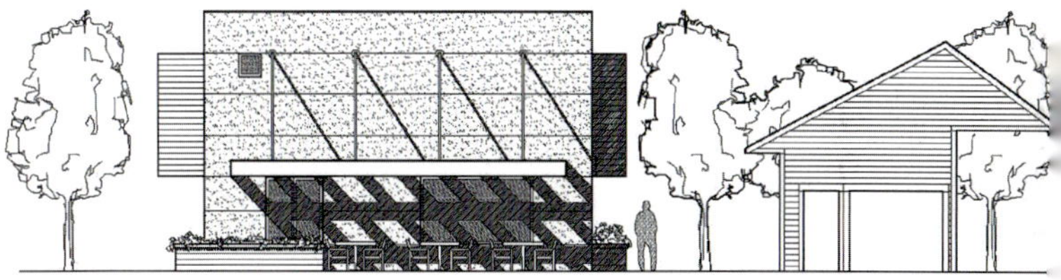

EAST FACING ELEVATION
SCALE: 3/32" = 1'-0"

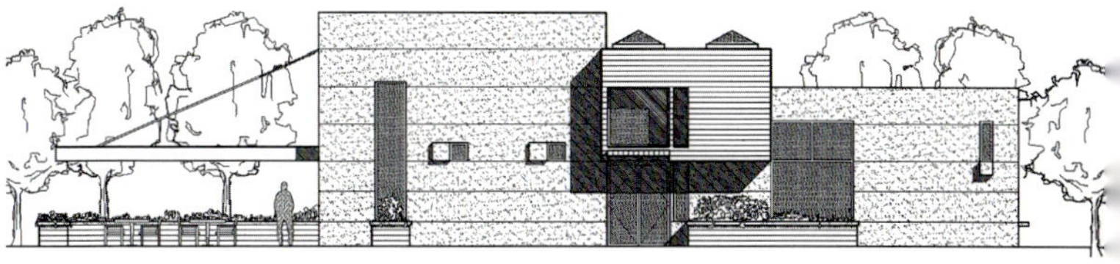

NORTH FACING ELEVATION
SCALE: 3/32" = 1'-0"

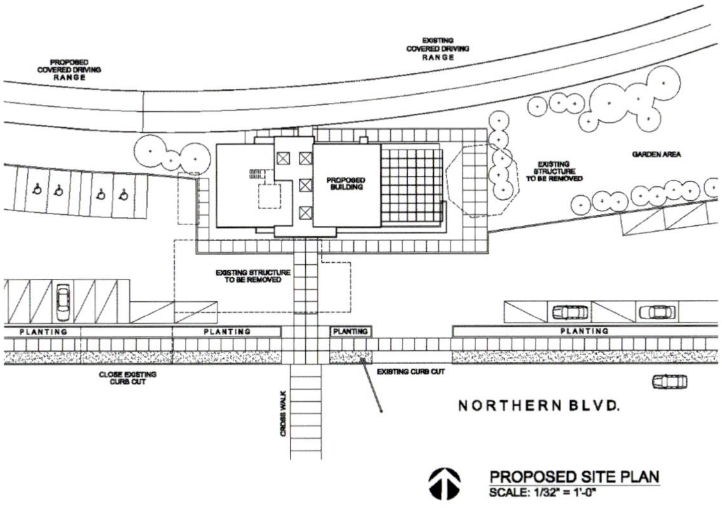

PROPOSED SITE PLAN
SCALE: 1/32" = 1'-0"

Project Team: F. Pagliano Tajani, L. Campagna, A. Senesi

Arezzo Courthouse – The Emblem of Architectural Art

阿雷佐市最高法院办公大楼——建筑艺术的象征

Near the Medici's Fortress, inside an historical park, the new building houses the main courtrooms and the Hall of Justice President headquarter and is linked to a neoclassical building, once a large hospital, to be restored and used for offices and archives.

The design of the New Wing, symbolises its environment: the rich vegetation of the park and the city's medieval structure, which was protected by a belt of conch walls on one side and, on the other, open to the squares, gardens and the territory.

Also the New Wing is enclosed on the North side by a coved wall of black granite slabs whose surface is flamed, as to acquire a very elegant matted dark-grey tone. To the South, an undulating transparent sunscreen façade of stainless steel is shaped according to a geometrically warped geometry which is characteristics of the skin of many living creatures, including leaves. Those complex curvilinear forms can be built using only rectilinear elements.

Thus, silver bioclimatic foliage protects the interior spaces by a luminous shadow and blends with the park without clashing with the neoclassical nearby architectural elements.

In the interior spaces, the astonishing reflection of the mirror-like polished black granite floor is contrasting with the greenish glazed envelop and the horizontal perforated acoustic aluminium grey panels separated by narrow maple fascias. The three level high entrance hall receives light from the roof and is dominated by the vertical elevator case clad alike the floor by shining black granite panels.

建筑总体外观
General View

建筑局部景观
Detail

入口处
Entrance

遮阳板
Brise Soleil

楼梯
Stairs

建筑北面花岗岩石墙
North Granite Wall

建筑局部
Detail

建筑局部夜景
Night View Detail

Office Name: Daniel Ventura Arquitecto

Argerich Building
阿格里奇住宅

"Good architecture starts always with efficient construction.Without construction there is no architecture. Construction embodies material and its use according to its properties. That is to say, stone imposes a different method of construction from iron or concrete.
I believe we can create contemporary architecture with all materials – with any material as long as we use use it correctly according its properties. In areas where we find nothing but stone, we shall build with that stone…" — Aris konstantinidis, Architecture 1964
The five houses building is situated over a lot of 8m front and 35m depth on the neighborhood of Villa del Parque, Buenos Aires city, four of them are 3-story flats and one is a ground floor, this is the one which has zero level and the one which appropriates of its territory and incorporates two patios on its surface.
The construction and structural logics are part of a material work which organizes both: The shape and the space. The economical resources used on the constituents elements of such structures show the multiplicity of senses detached from them. The esthetic structure of the compositive elements is synthesized to see the various formal, functional and semantic answers.
The building is composed by two blocks of 12m height separated by a patio of 6 m. which functions as the common trajectory of all of the houses.
The construction of this gap is stressed by three poplars which sieve the entrance of light that allows, among other things, certain privacy between both blocks' facades. The patio is the meeting place; the light column detaches from it and strengths it. The concrete armed box is the piece to go through and is discovered by the diverse perceptions of the horizontal and vertical trajectory. The new space is incorporated to the subject not as a watcher but as a member implied on the universe of the perceptions.

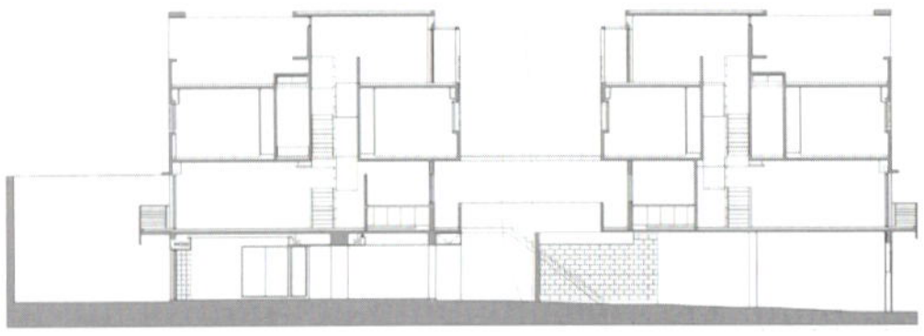

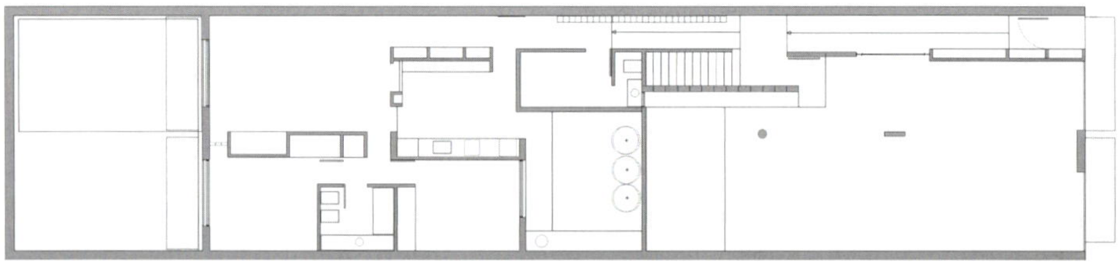

露台
Patio

露台细部
Patio Detail

立面细部
Façade Detail

立面
Façade

通往二层的楼梯
Interior Space from Stairs

入口空间
Interior Space from Access

Asian Culture Complex in GwangJu
光州亚洲文化综合楼

INTERACTIVE CULTURE STAGE
(International Competition, 3th prize)
We propose to constitute a stage that acts as a new container for the culture and everyday events. If new stage for the city is creating an empty open space out of penetrations pathways and plazas, making the new 'city stage' is as same as to conduct an enormous mapping operation with penetrating programs and landscape.
We propose the Asian Culture Complex as the earth in between creation and disappearance. The earth that is transformed into fragmented pieces and become urban landscape filled with memories of the site and historical evolution of various spectrums of relationships.
EXTRUDED PLAZA AS COMPOUND BODY
We propose to make the new plaza as the compound body a system that randomly assembles urban fabric, landscape, paths, open space, built structure, and programs through which the new cultural courage can be generated. Two-dimensional skin of the plaza is extruded to become spatial skin. Not only stopping at changing the urban topography through spatial morphing, the plaza is rearranged by matching with diverse codes required cultural code, artistic code, urban code,....matched with spatial code. The plaza is not a two-dimensional urban surface, but rather the multi-dimensional space with layers and folds of spaces that is matched by several codes. It becomes the framework of earthly generation that accommodates multiple channels of arts.
RECORDING OF GWANGJU - HISTORY AND URBAN FABRIC
We propose to establish the Complex through morphing the earth. The Complex becomes the strata of contours tracing the history of Gwangju and its memories of various events, while preserving the existing urban fabric.
1) Trace of History
2) Urban Landscape of Historical Memories
3) Event Stage for Activities
4) Integration of Stages
CULTURE STAGE
City Stage for historical memories and cultural activities Making a city - scale urban plaza of Culturescape Street Mat City - We propose the process of extruding out the existing city structures and fabrics as a way of recording the history of Gwangju existing bands of streets are extruded to become the volume in order to frame the new city condition. This means that the city's existing fabric with historical value is also the spaces and streets with historical memories and events. Through the reversal of old streets and masses, city-scale urban mat is defined, followed by morphing it into the earthly monument. Urban mat framed out by the streets then forms the landscape slicing through the surrounding urban panorama, hosting various events and programs, while providing needed circulations.
Three-dimensional event plaza embedded in the urban mat during the culture stage (stage 01), is inserted in between streets to become the multi-dimensional skin containing cultural programs and activities. In cohesive with outside landscape elements, it acts as a stage to host diverse events. Outdoor spaces framed out by the streets constitute various themes of programs parks, commercial facilities, cultural and media centers and others.
STREET CELL - HISTORIC RECORDER
Through a process of spatial reversal, the three-dimensional volume of the streets become 'objet' to which its combinations are formed as a historical matrix. Streets are no longer the left-out voids of the city, but to become an active framework containing diverse range of programs. Through each street being accentuated by the city's important historical events over time, programs and boundaries are defined for the Historic Recorder. Reaching beyond the limits of the concept that a gallery describes only the historical events, historical matrix is generated as a symbolic space for the memories and commemoration.
New cultural network is organized at the heart of the cross-road connecting the city of Gwangju and Asia. Cultural network acts as one of apparatus linking functional mechanisms of the surrounding area with the topological context of the site. Through the rearrangement of lines and orientations embedded in the network, the compound body is created out of the existing pathways of the history and the future cultural pathways still to come.
NETWORK FLOW
Hyper-street map, acting as the city plaza and landscape, is organized by the series of network cells. The site is not an isolated property, but rather becomes an open stage. The lower level of the city stage is a network of space freely interacted by the people, while the event stage at the upper level is penetrated by the series of programs to possess ideal urban dynamics.

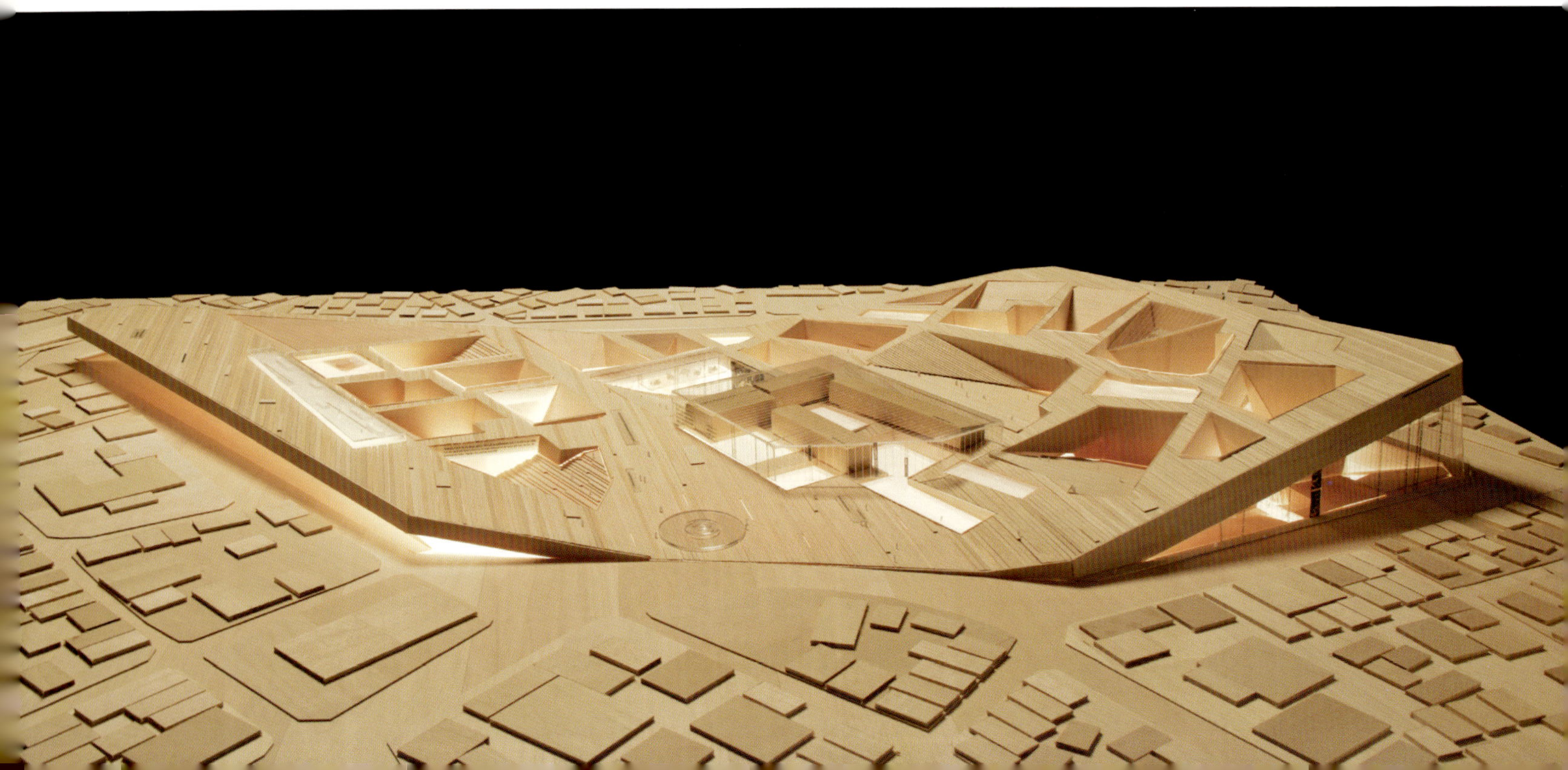

1 Lighting Box 1
2 City Edge Stage
3 Lighting Box 2
4 Lighting Box 3
5 Children's Stage
6 Green Gadern
7 Future Gadern
8 Record Gadern
9 Media Gadern
10 Canvas Stage
11 Tree Box
12 Asian Culture Stage
13 Tree Box
14 Asian Cultureal Network Center
15 Lighting Box 4
16 Green Box
17 Asian Cultureal Network Center
18 Asian Cultureal Network Center
19 Disappear Box
20 Lighting Box 7
21 Creat Zone Stage Box
22 Lighting Box 5
23 Tree Box
24 Media Stage
25 Green Box
26 Lighting Box 6
27 Water Box
28 Green Box
29 Imagination Box
30 Tree Box
31 Water Gadern

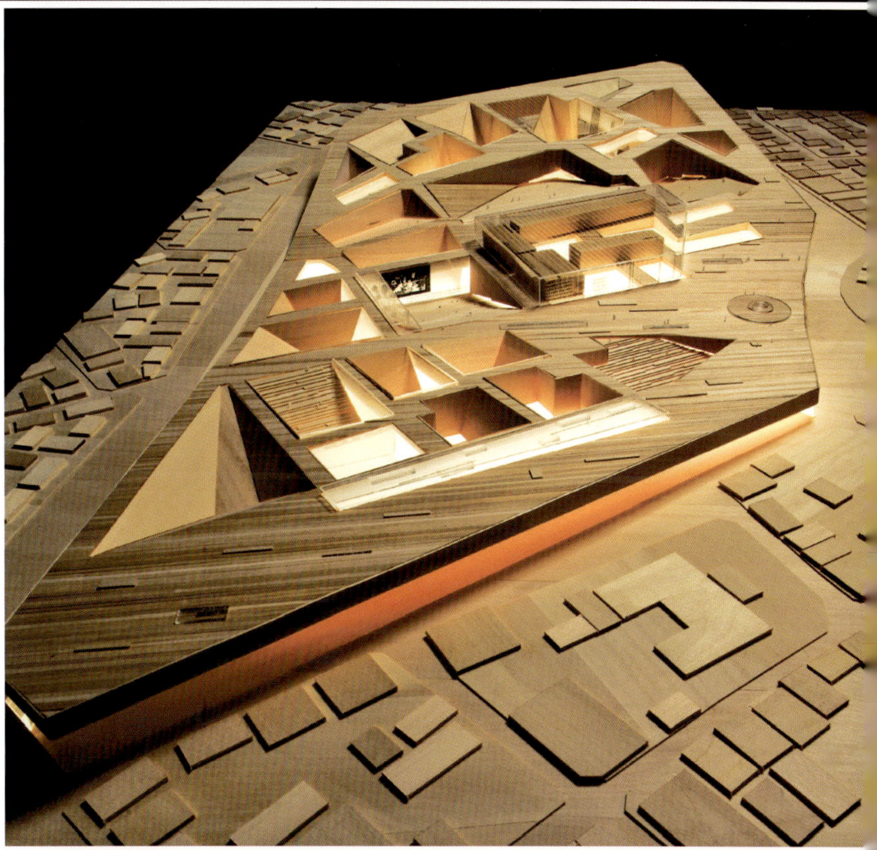

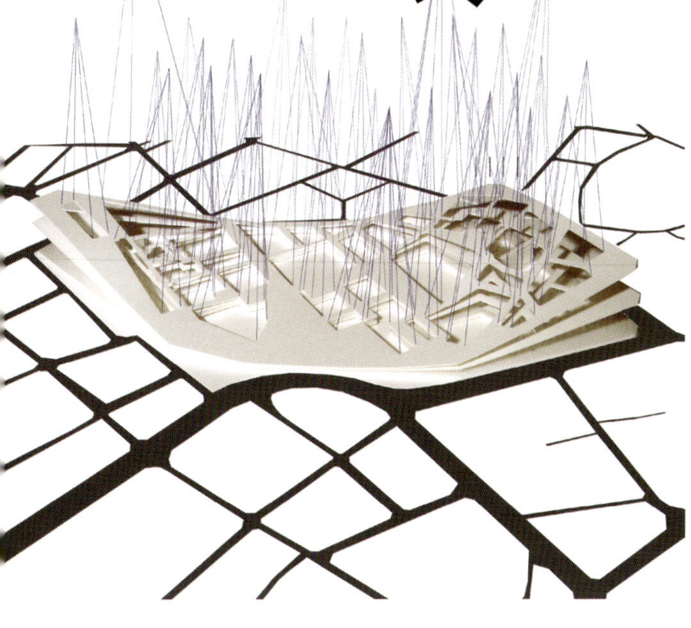

Architectural Design: Cafer Bozkurt Architecture

Atatürk Culture Center and Merinos Park

Atatürk文化中心与Merinos公园

The site of this project is the old Merinos textile factory grounds, one of the landmarks of Turkey's industrialization process during the 20th century. Originally a public building built during Turkey's Early Republican period and opened by M.K. Atatürk on 2 February 1938, the Merinos factory was abandoned after the year 2000 and given to the Bursa Metropolitan Municipality in 2005, to be developed for cultural and creative activities. With the financial support of the Bursa Chamber of Commerce and Industry, the Bursa Metropolitan Municipality has undertaken this project to give the building and its environs new functions that meet the contemporary social and cultural needs of the growing city. This urban renewal project consists of three main parts, the Merinos Park, the Merinos Culture Center, and the Atatürk Opera House and Concert Hall.

Because of strict time constraints, the construction process started along with the design process. Specifically the infrastructural and landscape works started in 2006, as soon as the project was tendered. While the Merinos Culture Center and the Merinos Park have now come to completion, construction of the new Atatürk Opera House and Concert Hall continues as we speak.

MERINOS PARK

Total Land: 270,000 m^2

Total Landscape Area: 210,000 m^2 (42,000 m^2 hardscape, 168,000 m^2 softscape)

The Merinos Park serves the needs of the surrounding area as well as the larger urban area. With its thousands of new and old trees, artificial lake, jogging and cycling paths, seating areas, children's play areas, cafe, snack bars and other facilities, the Merinos Park is an important, centrally located recreational complex for the city of Bursa. This landscaping project started in June 2006 and will be completed June 15th, 2008.

MERINOS CULTURE CENTER

Total Construction Area: 61,800 m^2 (includes the new 9,600 m^2 parking garage)

Ground Floor Area: 36,240 m^2

The former Merinos Factory building, that had lost its practical and technical functionality, constitutes the main site of this re-use project. The newly transformed Merinos Culture Center houses the following cultural facilities: A music conservatory, art galleries, art studios, library, textile and silk museums. This renewal project started in June 2006 and will be completed June 15th, 2008.

ATATÜRK OPERA HOUSE AND CONCERT HALL

Total Construction Area: 44,600 m^2

Ground Floor Area: 5,750 m^2

The Atatürk Opera House and Concert Hall has been designed to integrate both formally and functionally with the new Merinos Culture Center and Park. It will include a concert hall seating 1800 people and designed for opera, ballet and symphony orchestra performances, an auditorium seating 800 for conferences and concerts, a smaller concert hall seating 300 people, a ballroom with a capacity for 1000 people, galleries, a cafeteria and other facilities. When completed it will be a new cultural focal point for Bursa.

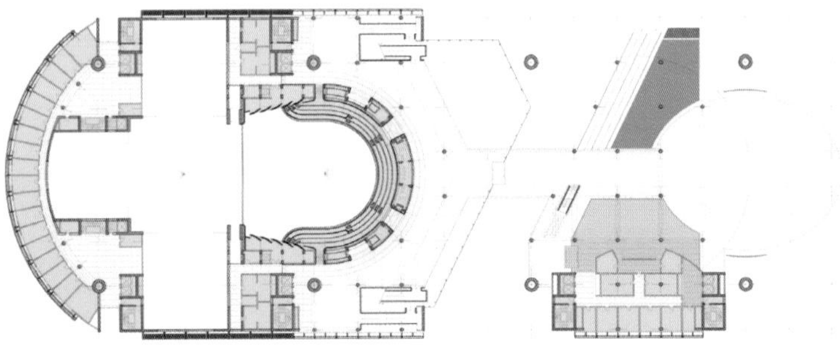

LEVEL +5.40 and +3.60 FLOOR PLAN

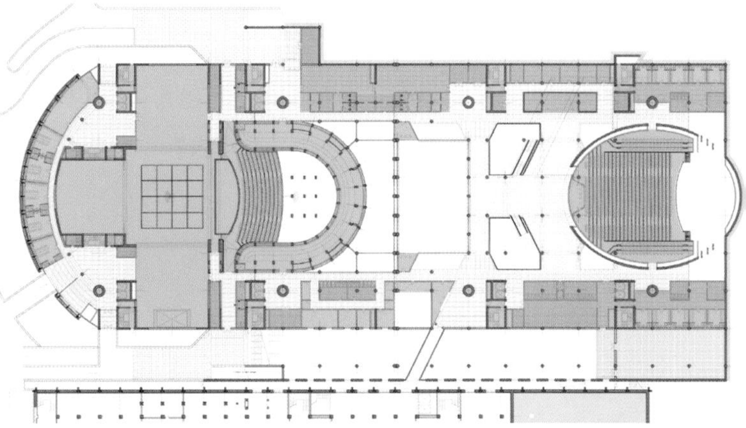

ATATURK OPERA AND CONCERT HALL

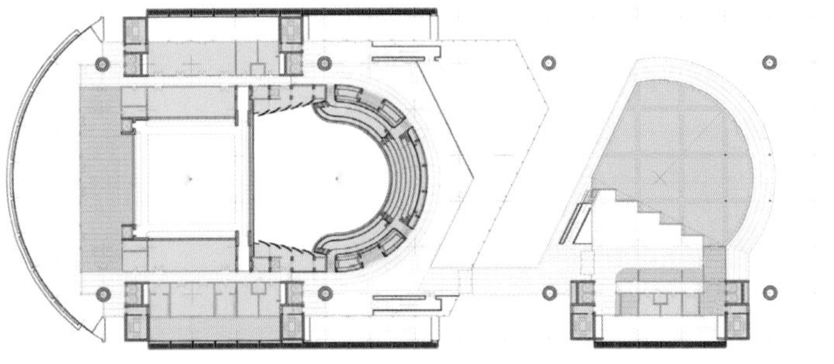

LEVEL +10.80 FLOOR PLAN

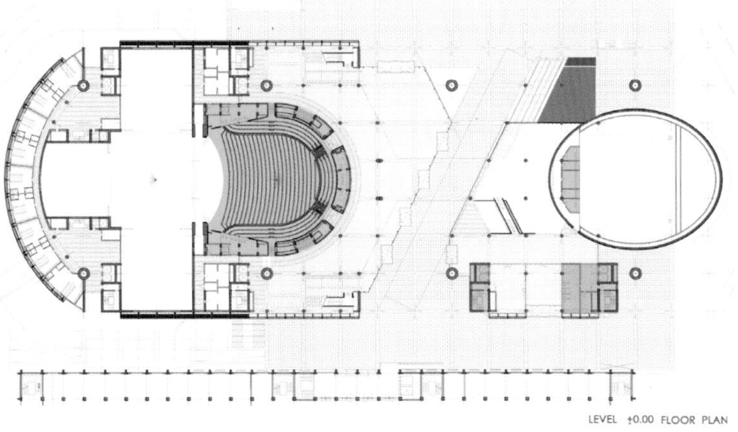

LEVEL ±0.00 FLOOR PLAN

NORTH ELEVATION

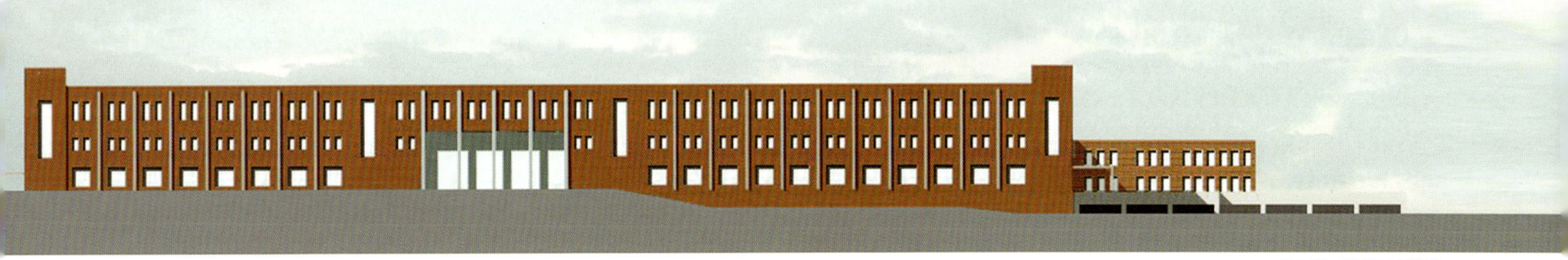

EAST ELEVATIO

WEST ELEVATION

SOUTH ELEVATION

Being situated in front of the east façade of the Merinos Culture Center, the Atatürk Opera House and Concert Hall faces the pedestrian walkway that arrives from the east across the park. Volumetrically the building opens up at this point to create a public plaza underneath its broad metal roof that addresses the surrounding urban fabric and invites citizens to use the Merinos Culture Center and the new Opera House as a whole.

F
08 / 2006

ROOF
06 / 05 / 2008

SOUTH ELEVATION
18 / 05 / 2006

SOUTH ELEVATIC
06 / 05 / 2008

Old Merinos Factory Settlement Plan

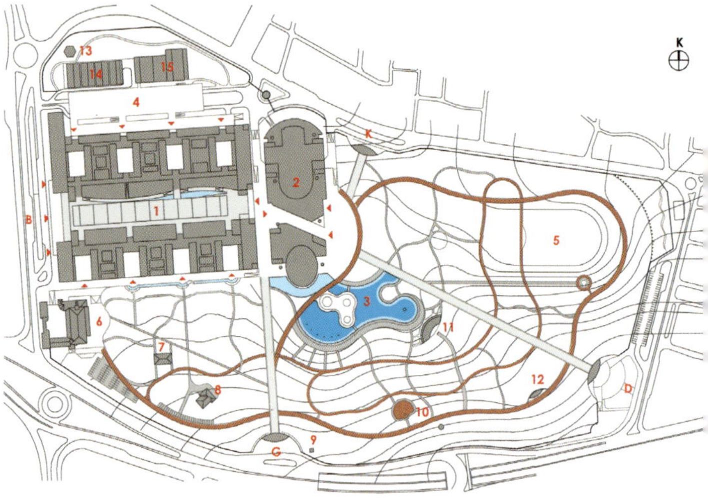

New Merinos Settlement Plan

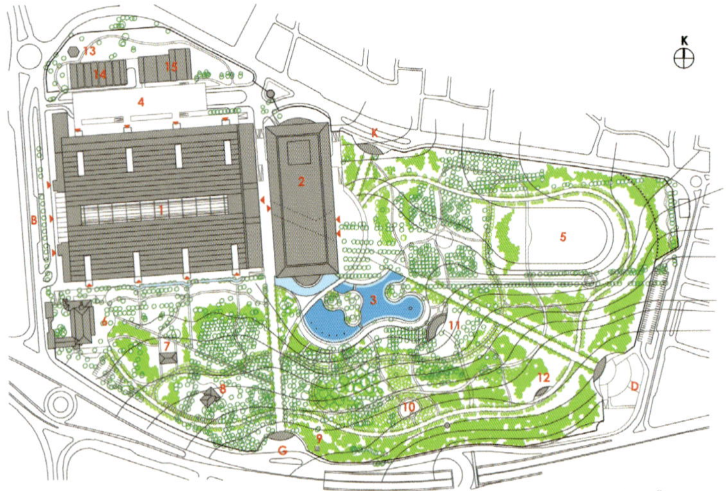

Site Settlement Scheme

1 MERINOS FACTORY - Old textile and thread workshop
2 ATATURK CULTURAL CENTER
3 POND
4 OPEN-AIR EXHIBITION AREA
5 DISASTER MANAGEMENT and OPEN AREA PERFORMING
6 WEDDING SALOON– Old table d'hote building
7 TEAHOUSE – Old tally clerk building
8 RESTAURANT – Old Manager House
9 TOWER – Old Water Tower
10 PLAY GARDEN
11 CAFETERIA
12 BICYCLE RENTAL
13 TOWER – Old Cooling Tower
14 CAFETERIA – Old Storehouse
15 INDUSTRIAL MUSEUM – Old Electric Station
16 WEAVING ENTERPRISE
17 CONFECTION ENTERPRISE
18 WORKER DINING HALL
19 DAY-CARE CENTER
20 GUEST HOUSE
21 EMPLOYEE CLUP HOUSE
22 ENTRANCE
23 MATERIAL WAREHOUSE
D EAST ENTRY
B WEST ENTRY
K NORTH ENTRY
G SOUTH ENTRY

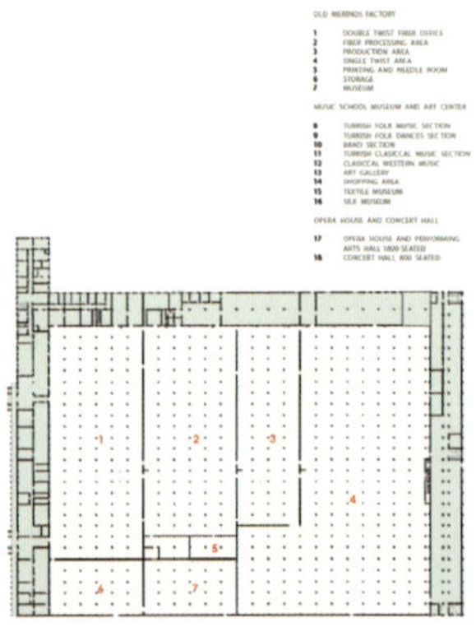

OLD MERINOS FACTORY GROUND FLOOR PLAN

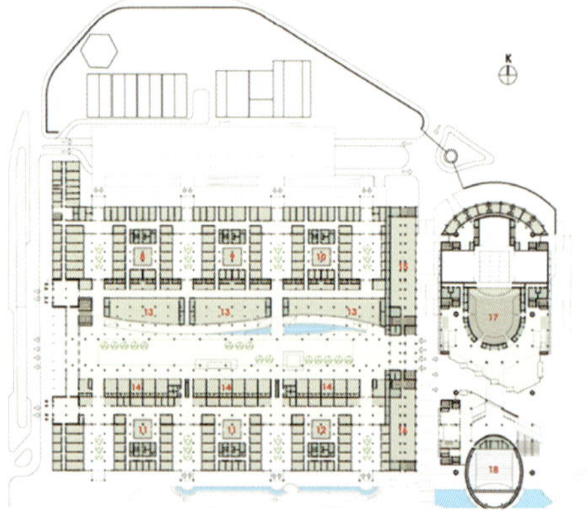

NEW MERINOS FACTORY and ATATÜRK CULTURAL CENTER

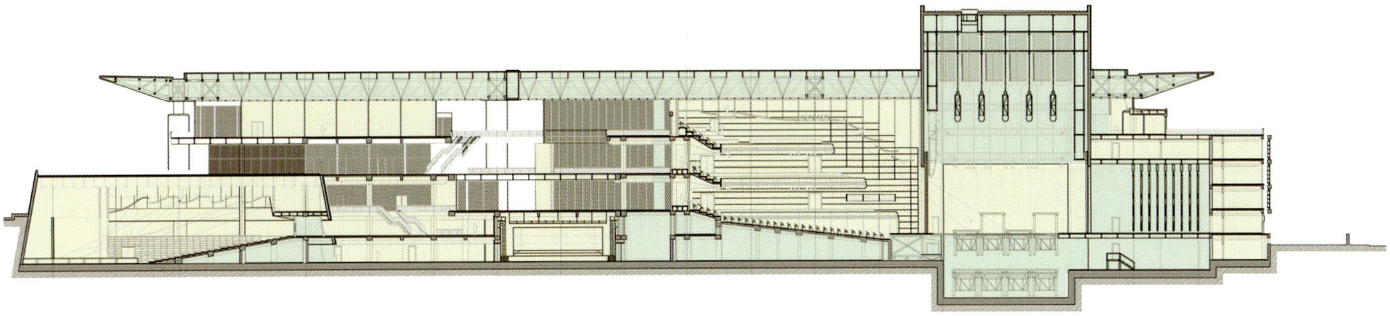

A-A SECTION

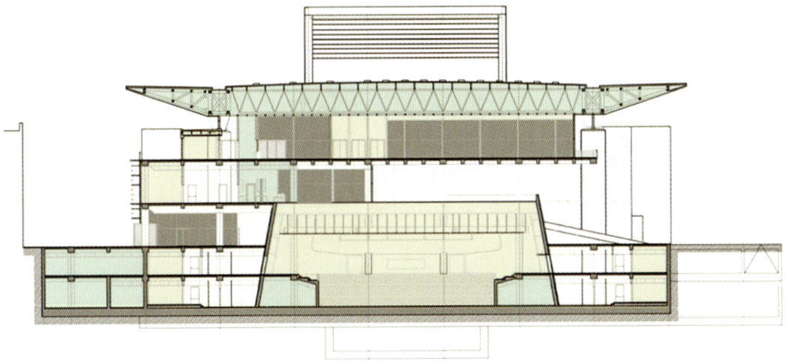

B-B SECTION

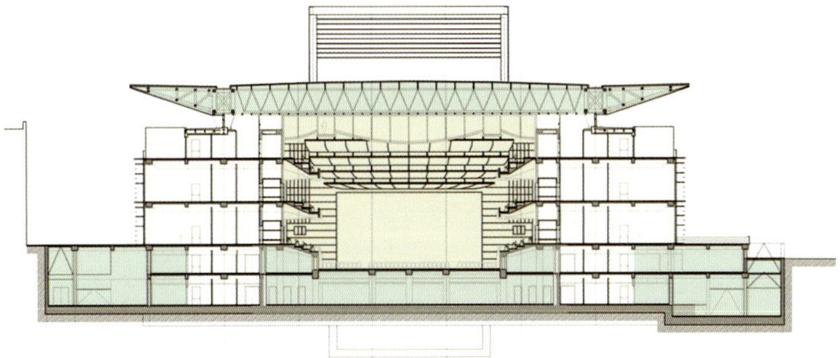

D-D SECTION

ELEVATIONS OF THE NEW ATATURK OPERA AND CONCERT HALL

WEST ELEVATION

EAST ELEVATION

SOUTH ELEVATION

NORTH ELEVATION

Bungalow Olger
木质住宅大楼

The Bungalow Olger is my first personal command I received in a very specific context:
I was making my practice in Costa Rica as a work-site supervisor on an Eco-lodge previously designed by TAG arquitectos: The Celeste Mountain Lodge. The constructing team was mostly composed by local workers, so they literally needed "eyes" to be at the construction on an everyday basis. That is how I received the opportunity to make one year practicing on this 1,000m^2 hotel.
So I started to live at the village, Bijagua, which was situated 3km away from the lodge's lot. The small number of inhabitants, 800 persons, helped me to get to know rapidly the people in town. Only one month being there, I got to meet and become friend of Mr. Olger Arrollo who told me about a vacation house he dreamt to see come true...
He owns a great property on the way to the Volcano Tenorio national park, where you can find the international renowned tourist attraction "Rio Celeste". The man had perfect knowledge of his land so he knew exactly where he wanted his house to be built. The location benefited of an impressive panoramic view on the region. From there you were able to see the Tenorio and Miravalles Volcanoes as for the Lake Nicaragua.
So I started to work on the project during nighttime as I was working every day of the week on the lodge's construction and I always kept my week-ends free for visiting the country.
The owner is a bike seller but he also owns on his lands a forest plantation and a sawmill I made him an all-wood project, except for the concrete tower which contains the sanitary fittings. The house has 120m^2 and has two bedrooms and a unique bathroom upstairs and the living areas (kitchen, terrace, living room, toilet, laundry) on the main floor. The climate in this region is extremely rainy and humid with a short summer from February till May. The interiors are naturally ventilated thanks to the teak louvers on the windows. They are also protected from the rain and the sun by the overhanging zinc roof. The terrace was going to be the most used place of the house so I gave it a central position to increases the interaction between the different living areas and also with the outdoor spaces.
The building's dimensions have been conceived keeping in mind the reduction of material loss (especially for the wood beams) as well as rational use of manufactured products.
It also was the first time I ever drew technical plans so the project's evolution equally followed the lodge's building progress. Every week I learned something new, it could be a new detail, a new aspect to take in count, a new material, a new construction's technique...I finished to draw the plans exactly when the lodge's construction was over. The bungalow's project can be taken as a condensation of the hotel's concept in a wood residential building.

平房景致
2 Views of the Bungalow

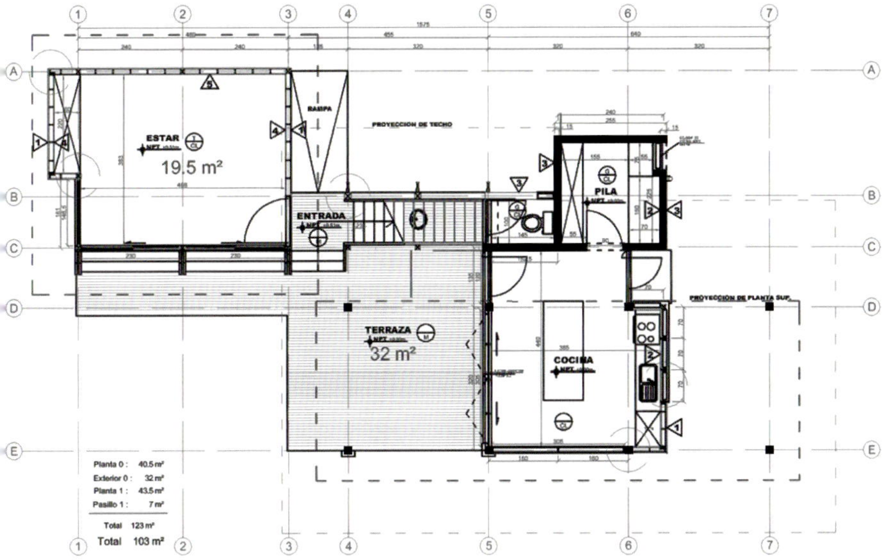

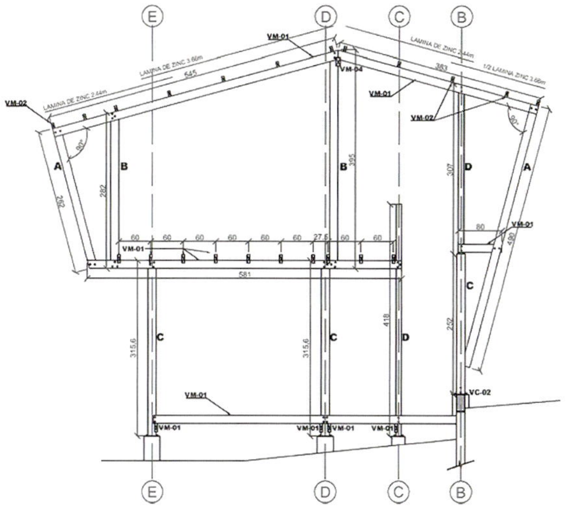

ESQUEMA CERCHA // EJE4-5

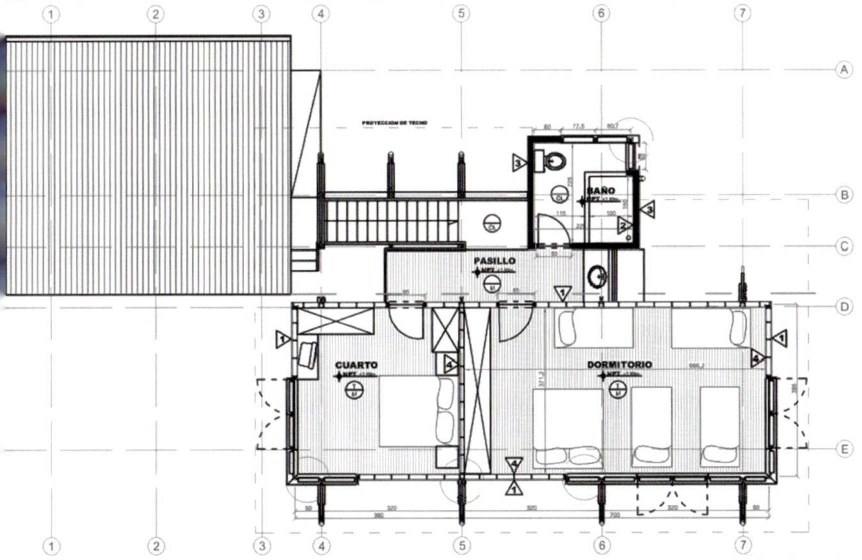

Cash and Carry Food Store

C&C食品店

The cash & Carry Food Store is situated on the north side of Budapest, on a former industrial territory, which, according to the urban development concept, will be used as brown zone investment. The basic area of the building is 4,400m^2 gross. The value of the investment is 3 million EURO.

On the area is going on continuously the demolition, recultivation of the old, strongly polluting leather factory. Just like on the other areas in the neighborhood (DIY shop, car showroom and service, etc.), the city planners assigned business function to this area.

The point of the development is that the new zonings must be up to date, must grow the green areas, and have to assure adequate service and energy-saving operation.

The appearance of the building was developed accordingly to these principles; it is a hall building with reinforced concrete structure, sandwich panel revetment assuring the right thermo-isolation.

Its design is stirring; the fractions are visually diminuating the size of the building.

The materials and colors are changing at each cube, accentuating their importance. The cubic system changes into a transparent glass-surface at the entrance of the building, attracting the attention to the main entrance.

Because it is the first building of the zone, at planning it was principle to give adequate architectural basic conception to the later buildings, predetermining a possible line for the future.

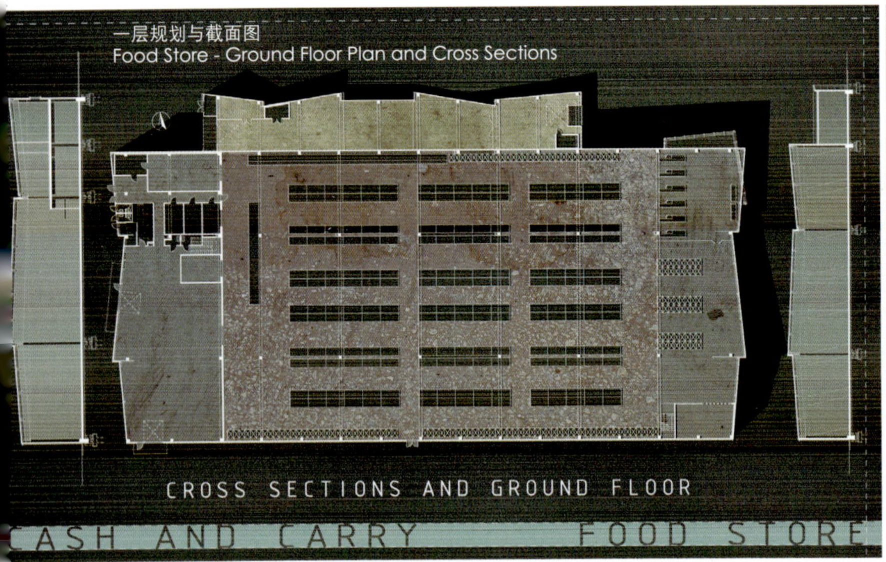

Project Team: Aline Wallach, Arais Reyes, Sandra Ortiz, Veronica Alatorre

CB30

CB30大楼

This building has 3 apartments looking to the street, as well as a two-story house in the back that looks into an indoor patio. The public areas are taken into a single space facing the street or an open interior space. There is a spectacular tree in front of the building, and a view to the Polanco Park. These conditions became the central design objective, as we wanted to flood the interior out to the street and the park.

To this end, the facade responds to the principle of eliminating obstructions between the inside and outside areas, generating openness and continuity. Working with structural geometry, we managed to support the 15 m. span on a 40 cm slab with no columns that could interfere with the view; we then sharpened the floor slabs and walls to the minimum, hiding the edges within the facade.

Cellular Network City — Peculiar Architectural Conception

蜂窝式建筑——奇特的建筑构想

PROPOSAL

This project is aiming to redefine 'urbanism' as a series of cellular space subdivisions and transitions. Like all the ancient and historical cities that emerged from their own spatial ordered qualities, and geometrical distributive hierarchies, am: Pm proposes a networked urbanism configured from cellular volumes, initially starting as a thick multiplicitous urban ground strata for the World's Fair Expo 2010.

The landscape is interconnected with interior and exterior public spaces as well as cultural facilities and domestic spaces. Networking lines act as infrastructural connections between the distributed spatial pockets. Cellular organizations morphs from conventional aggregate spaces to extreme transformed ones, gradually transforming into an open space.

Extreme moments of transformations act as experimental pavilions that transform into radical landscapes and public spaces, and therefore enduring the legacy of the expo, whereas private spaces maintain their conventional appearance and homogenous transformation.

On both urban, building scale and material level, the cellular network organization acts as ground strata and spatial pockets organization, as well as porous occupiable sections, all seeking a coherent yet different oriented urban order, that spans across the two river banks into the two sides of the city, and thus over time influencing its formation.

It is an order that has the capacity to organize differential interactivity, sectional movements, as well as diversity of scales, densities, heights, and programmatic mixities.

The Expo 2010 landscape will start to be generated across both sides of the river banks, and the networked system of spatial cellular organizations and connections is proposed to extend and intensify across the site in the period intervals that follows, producing highly dense sliver organizations providing a horizontal view to the few emerged vertical organizations.

PARAMETRIC DESIGN SYSTEMS

A scripted 3d networking system of line segments is deployed at the site scale, coded with specific rules of growth and expansion configured from straight line segments, to provide gradual spatial transitions and subdivisions. Specific behaviors are assigned to transitional arrays of curved surface module and created within the guidelines of a bounding network, and thus transforming the conventional cube into curved surface modules, which is further applies at different scales expressing interaction and achieving forms of notational complexity that result from simple local order and behavior.

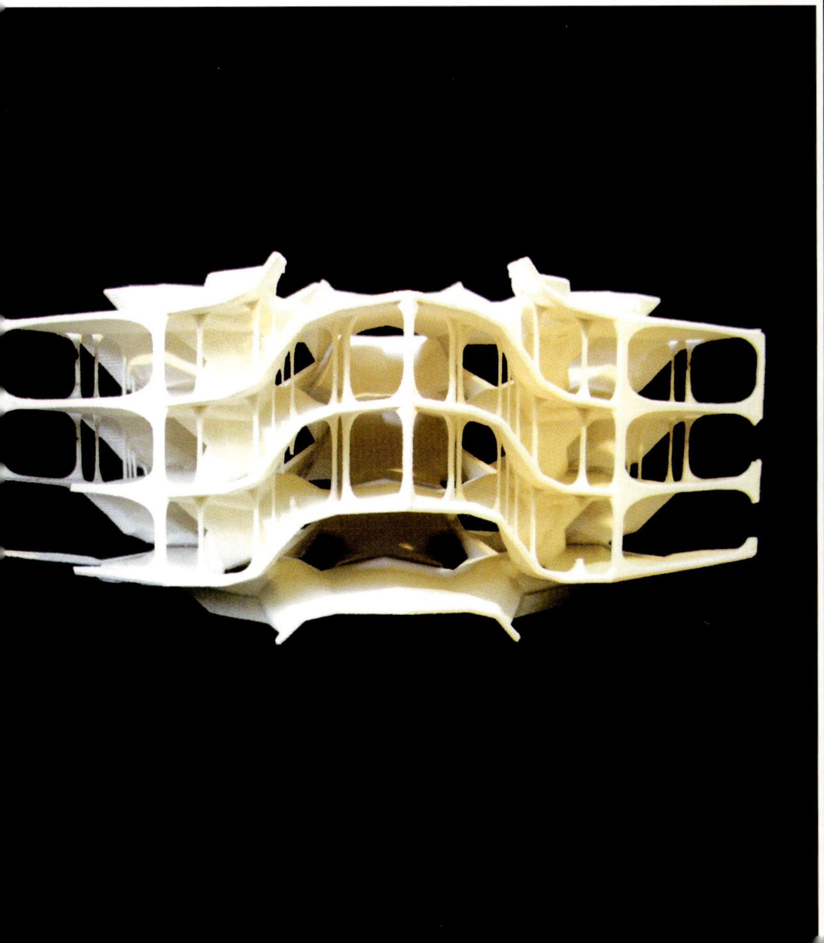

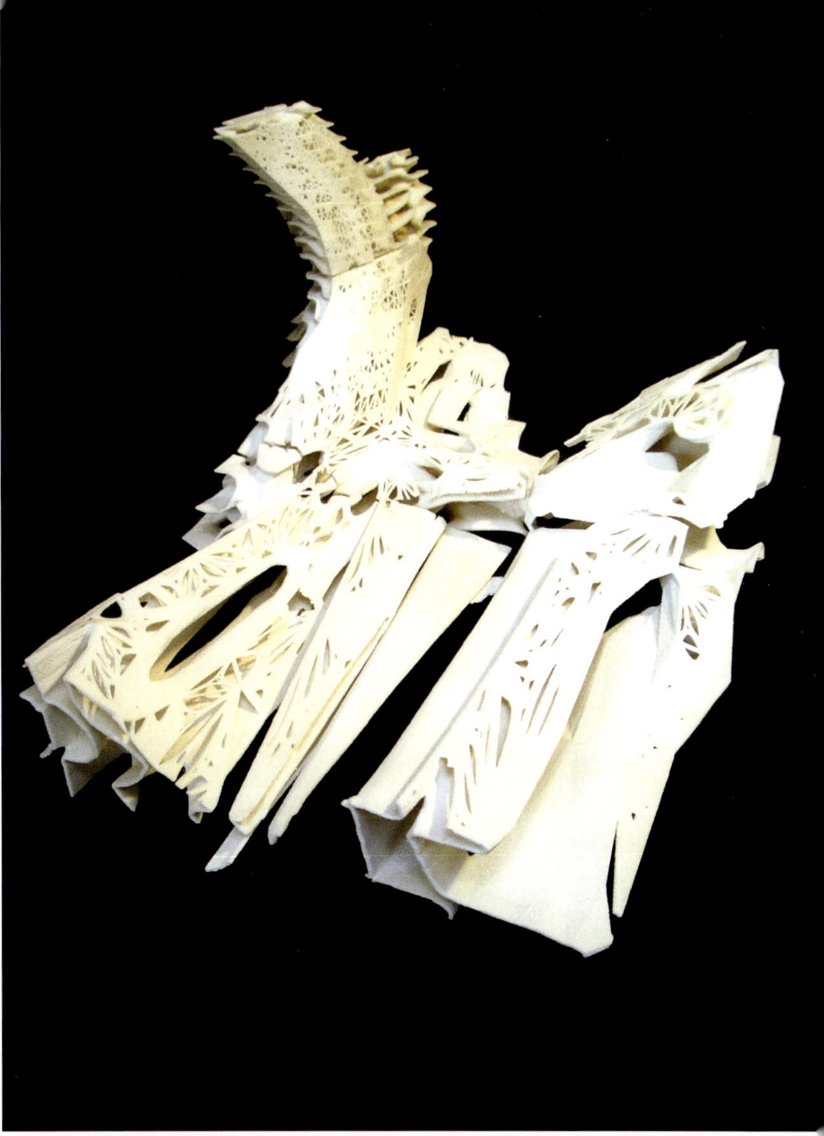

Chief Architect/office Name: Olga Banchikova

Centre for Nanoscience

纳米科学中心

According to the existing governmental program of supporting the development of nanoscience and nanotechnologies in Russia, it is proposed to create a centre on the territory of Moscow State University which would host multiple functions, starting with advanced research, experimental manufacturing and education, and concluding with the functions of informational exchange, show-room for advanced methods and achievements.
The place for such headquarters of Russian nanoscience is chosen in the neighborhood of important city landmarks and good transportation and ecological conditions. The new building is situated on the crossing of Michurinsky Av. and Akademika Hohlova Str., on the territory of the largest and most prominent Russian university, the Moscow State University. Besides, there are two important urban planning axes intersecting at the very site of the new building.
To maintain the transparency and pedestrian penetrability along the first axis, which goes lengthwise through the University building and botanical garden, the Centre building is relatively low and bent in archways along the axis.
Second axis is one of "fan" axes, which pivot on the Main Olympic Stadium and the Sparrow Hills. As an important urban vector it is integrated in the Centre building as the direction of main ramp entrance to it.
Starting with the entrance, the whole Nanoscience Centre building can be divided into five zones according to the levels of admission.
Zone 1 (open admission for students and guests) – main entrance and exhibition space;
Zone 2 (which has a secondary entrance on the ground floor) – educational area;
Zone 3 (accessible both for students, guests and researches) – media library, main auditorium;
Zone 4 (which has a separate entrance on the ground floor; restricted admission) – research offices, labs, workshops;
Zone 5 (accessible only from zone 4 via a lock) – cleanroom ISO 1 with technical floors below and above. Main cleanroom and zone 1 have visual connection, which meets all safety and cleanness requirements and at the same time serves to increase Centre's attractiveness for students, professionals and visitors.
In terms of spatial correlation with surroundings, these zones are orientated correspondingly:

从超大尺度到纳米尺度的顺序，还值得人类继续探索：纳米中心夜景图
The Unbreakable Infinite Sequence: From Mega- To Nano-scale. Still Much to Pursuit. Night View of the Centre.

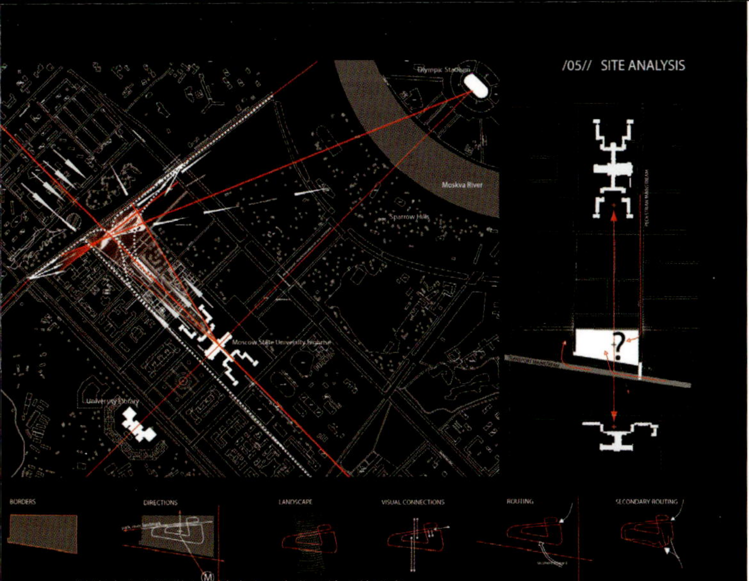

纳米中心在视觉上、技术上、文化场地上的需求
Visual, Technical, Cultural Site Requirements for the Centre of Nanoscience.

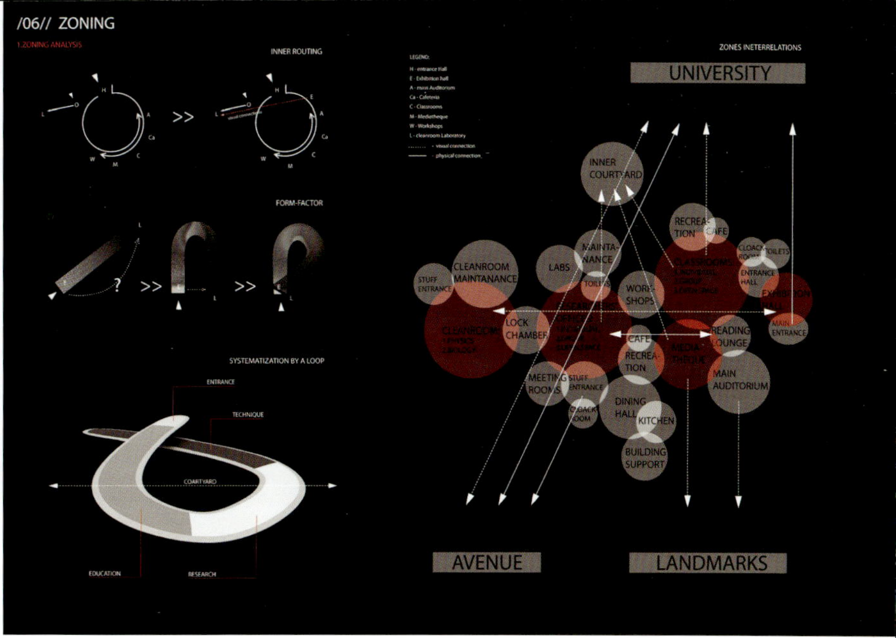

educational – to the University, research offices – face the street, and the public space is directed towards the Moscow River, Sparrow Hills and the Olympic Stadium. Zone 4 is hosted in a higher part of the building, which is north to the other parts and has an inner courtyard space from south. Together with multiple skylights this allows quality illumination. The general shape of the Centre, which fits all the premises and the composed program, possesses characteristics of non-orientable surface, known as Klein bottle. Where an "exterior" length of the surface turns into an "interior", there is the main entrance to the Centre, there's the place of visual connection between Zone 1 and Zone 5. The Klein bottle, proper, doesn't self intersect. But for this, it needs one extra dimension. Metaphorically, this fifth dimension reminds us of new perspectives, which might be revealed in front of humanity via the development of nanoscience.

Hexagonal lattice, quite widespread in nanoworld, e.g. organic molecules, serves as a basis for the design of buildings site and exoskeleton. The site development includes, first, pathways, which are expressed by condensation of hexagonal grid and which "pierce" the Centre building through archways, and secondly, the structure of "extruded" volumes for street furniture and engineering.

The building's exoskeleton geometrically is the vertical honeycomb hexagonal structure, which fills a volume of the aforementioned Klein surface as if

/07// STRUCTURE

3-D LATTICE VIRUALLY EXTRUDED FROM HEXAGONAL GRID TRIMMED BY A KLEIN SURFACE

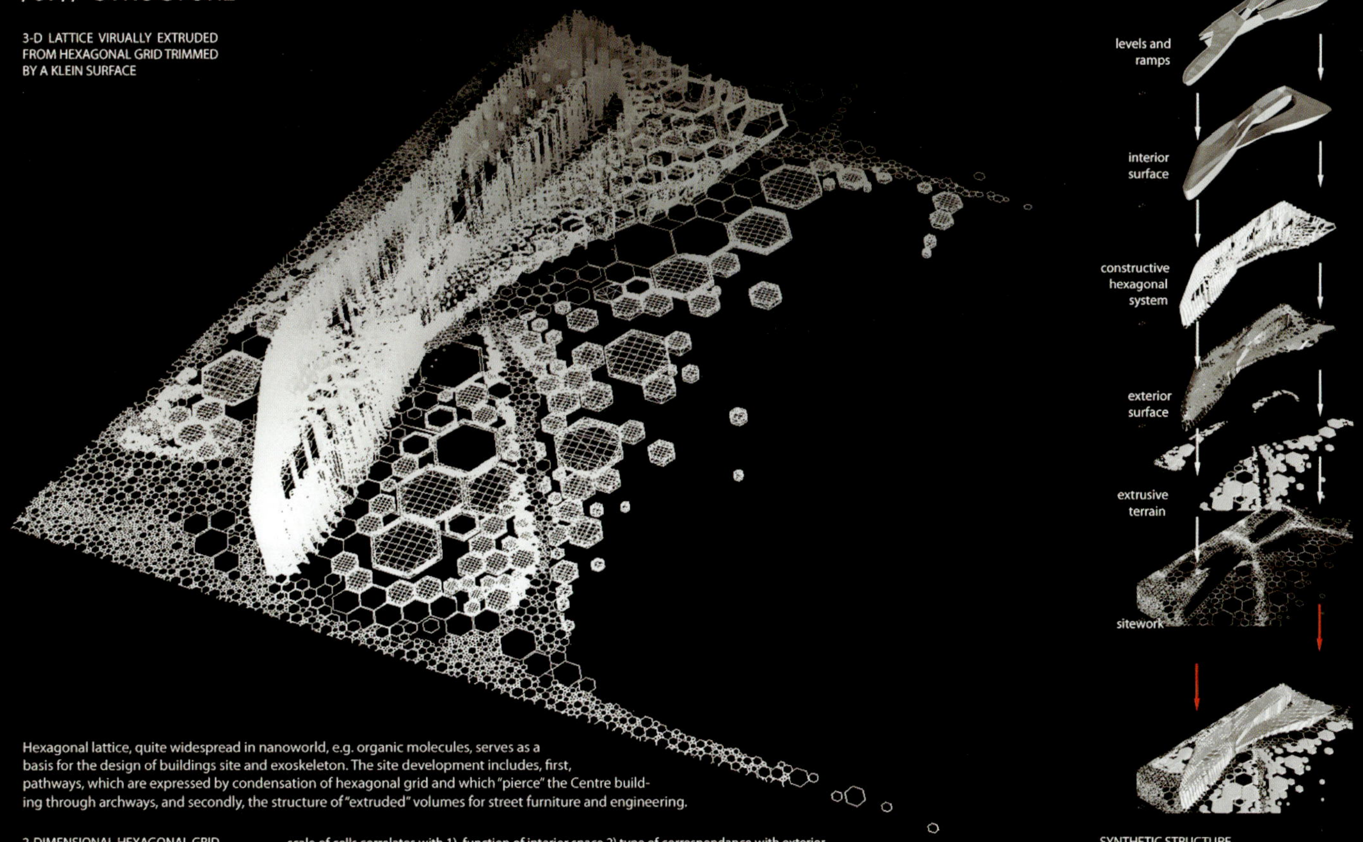

Hexagonal lattice, quite widespread in nanoworld, e.g. organic molecules, serves as a basis for the design of buildings site and exoskeleton. The site development includes, first, pathways, which are expressed by condensation of hexagonal grid and which "pierce" the Centre building through archways, and secondly, the structure of "extruded" volumes for street furniture and engineering.

2-DIMENSIONAL HEXAGONAL GRID   scale of cells correlates with 1) function of interior space 2) type of correspondance with exterior   SYNTHETIC STRUCTURE

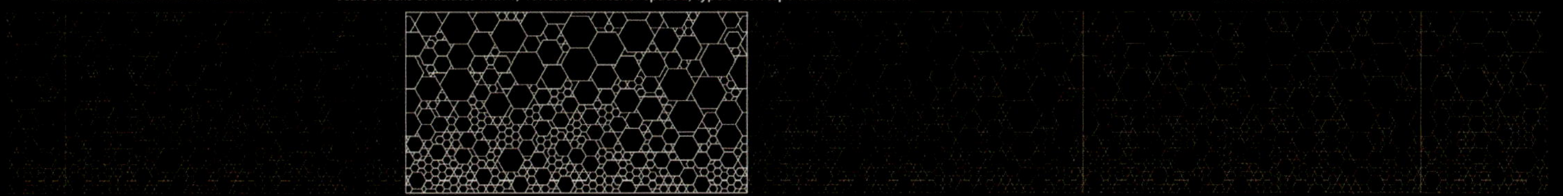

这座大楼是建筑学和科学技术的完美统一
The Centre, Referentially to Architectural or Scientific Examples Might Remind on the Sequence of Gothic Buttresses or a Terrifically Zoomed Particle Structure.

a surface could be solid and have some thickness (which is changing according to constructive and functional requirements, e.g. insolation correction). This concrete honeycomb with nano-dispersed reinforcement constitutes a load bearing and covering three-dimensional structure, which occasionally pierces the inner volume to perform better constructive rigidity and special planning integrity. The size of hexagonal cells AND the tilting of virtual Klein surface, which "cuts"/bounds the lattice, vary accordingly to the function and requirements of inner spaces they refer to. For instance, the biggest openings can be found around the public zone, smaller and more regular – around the offices. Shape of the cutting birecurvate surface forms the slanting auditorium, and in the cleanroom zone the exoskeleton lets lights inside and prevents looks from outside at the ground level. On the other hand, this consecutive scaling of the cells allows the new Centre to answer the surrounding on multiple scale levels, starting with the buildings, landscape objects, up to the scale of human tactile perception (e.g. at the main entrance). Besides, withdrawn from environment, the Centre building can be seen as an object of any size, just like any nanosize object examined without scale affixment.

And, finally, the series of cell sizes metaphorically represents the infinite sequence of fractal scaling of all objects in universe, starting with the mega-scale of galaxies and solar systems through macro-scale and micro-scale, which till recently seemed to mark the boundary of human capacity for miniaturization, concluding (for now) with the nano-world, which is still to be studied.

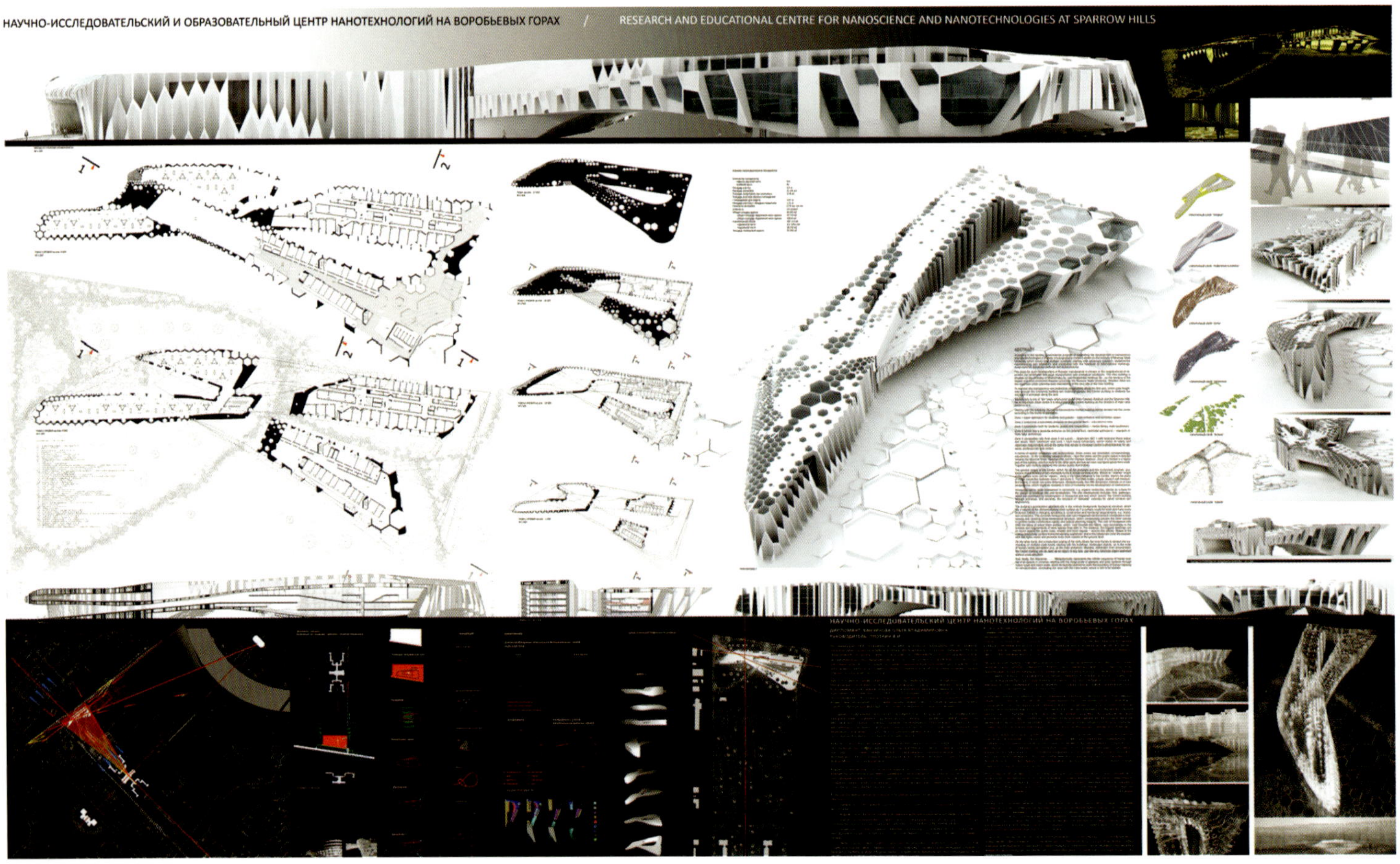

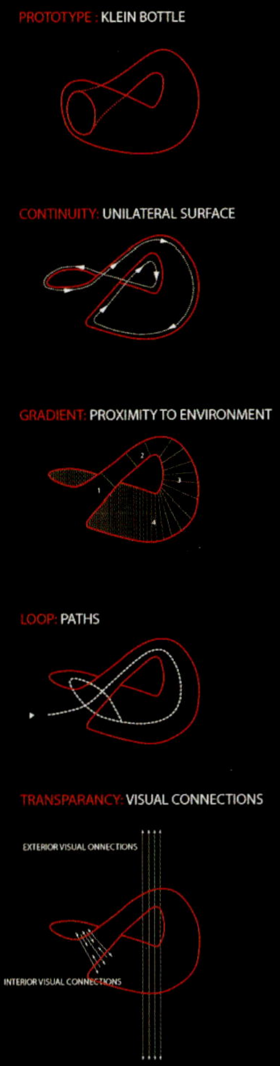

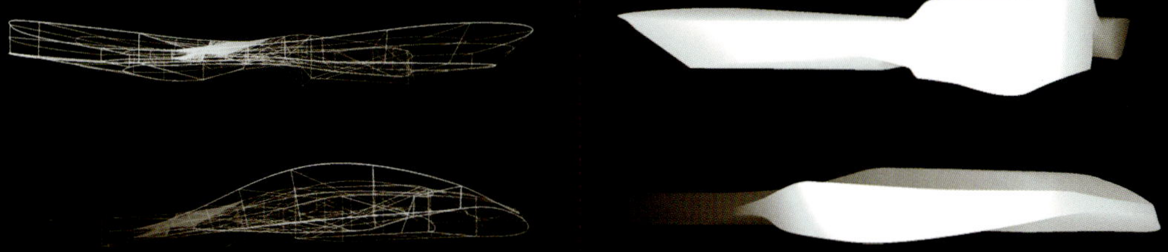

The general shape of the Centre, which fits all the premises and the composed program, possesses characteristics of non-orientable surface, known as Klein bottle. The Klein bottle, proper, doesn't self intersect. But for this, it needs one extra dimension. Metaphorically, this fifth dimension reminds us of new perspectives, which might be revealed in front of humanity via the development of nanoscience.

克来因单侧曲面需要一种特别的尺寸，通过纳米科技的发展有可能将其实现
The Proper Klein Unilateral Surface Requires One Extra Dimension, Where We Might Dream to Step Via Evolution of Nanoscience.

建筑的外形和照明设计非常符合人们的使用需求
The Plicate Building Skin and Light Design are Well Corresponsive to Human Scale.

东立面
Eastern Fasade.

公共区域室内拥有绝佳的视野，可以欣赏到州立大学的摩天大楼
Interior of Public Spaces Provides Great Views on the Skyscraper of State University.

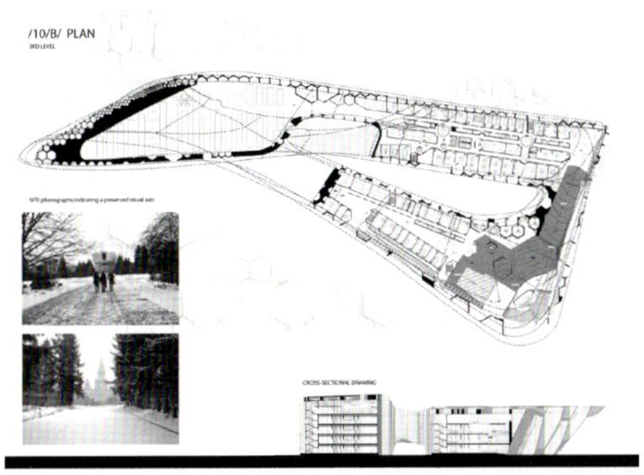

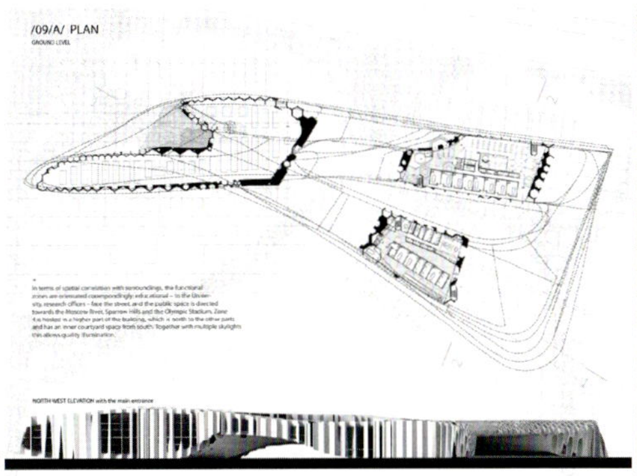

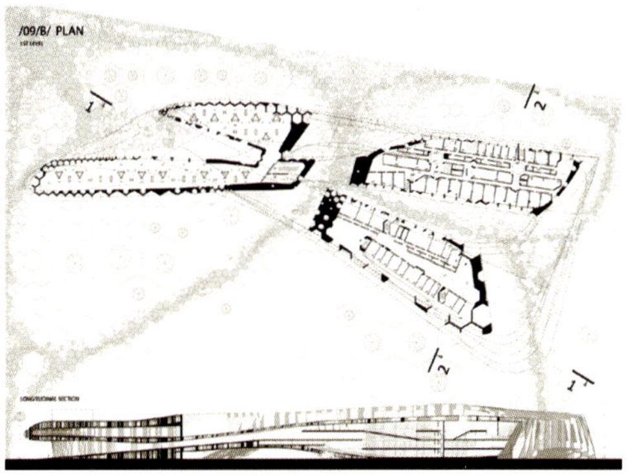

Project Team: Patricio Mora Araya and Daniela García Letelier

Childhood Evocative Mantle

孩童记忆弥漫的建筑覆盖物

CHILDHOOD EVOCATIVE MANTLE is a project that aims to propose a new habitability model able to give an answer to the current life-styles of the older adult. The proposal arises as a qualitative solution for the inhabiting of the older adult that, along with the creation of a set of houses, seeks the endowment of an inhabitable environment, constituting a combo that places itself spatially, formally and functionally under a mantle, which allows having solutions under different circumstances to the main aspects of the life of a current older adult: dwelling-house, garden-nursery, neighborhood-common green areas.

The project places itself in the city of Concepción, Bío-Bío region, Chile; a city with mild-rainy climate and with an eight-month precipitations period yearly. This situation makes it difficult to enjoy outdoor recreation and has an impact on the complete social development of the older adult, where recreation plays a fundamental role for keeping their mental health.

The project Manto Evocador de la Infancia propitiates meetings for people 365 days per year; this condition is complemented with sustainability variables in space design and creation, which is the case of the handling of the greenhouse effect produced under the mantle. This allows controlling the temperature difference of the interior space between 5 and 10°C in comparison with the exterior, which affects by being warmer during the winter and cooler during the summer due to the solar protection that the enveloping surface of deciduous generates over the mantle.

The main variables considered in the design of this project are:

SOCIAL - FINANCIAL

The project is a model that integrates a flower growing system that aims to reinsert the older adult productively. The challenge is to achieve social management to avoid segregation and to strengthen self-confidence by giving them a tool that makes them feel useful again to society. From this point arises "Huerta de Los Ancianos de Tucapel Bajo", transforming this center in a social referent to the city which, in practice, would allow generating extra incomes for this age group that demands a strong financial support due to medical expenses made through this period of life, among other reasons.

FUNCTIONAL

The concept of combo corresponds to a mini-city in which three big areas are distinguished among which the common spaces and flowery nursery stand out. These areas not only give economic support, but also a welcoming and pleasant environment for its residents. The central volume of the combo corresponds to the common center for senior citizens, in charge of satisfying the recreational and union needs of the elderly, which also has spaces for flowery workshops where they can create their own products to be sold later in the flower shop of the combo. On the right area (north) is placed the guided village for senior citizens, which has 24 dwelling-modules conforming a common yard between them. Towards the left side of the club (south) is placed the collective residence for senior citizens, which has 30-double bedrooms conforming one module and that in the center contains common spaces

建筑覆盖物外部
Outside Mantle

建筑覆盖物内部
Inside Matle

as dining rooms, visiting rooms, living rooms, etc.

SUSTAINABILITY

The mantle covers 100% of the combo, making it like the function of a greenhouse that obtains heat energy from the environment, reducing heating costs for the enclosed areas (during the winter) and propitiating the controlled temperature through the deciduous enveloping surface that the mantle has (during the summer), which gives abundant shadow and creates a microclimate favorable to the senior citizens. Ventilation inside the mantle is worked through air-crossed systems with mechanical regulation, set with north-south direction due to the predominant winds in Concepción. It complements with direct extraction chimneys for the wet areas of the houses, which crosses through the mantle allowing the release of contaminated air out of it. Lineup of the houses is continuous in order to generate a faster transfer of heat between mass in contact and the rainwater is collected within the mantle perimeter, which becomes the watering fount for plants, and is also sent to gray waters chambers that distribute them through the residential combo for its usage in toilets of the three areas of the project.

SPATIAL

One of the main characteristics of the elderly is seeing and feeling like a child once again. This involution is produced in many aspects of their lives making the simplest things to stimulate them, somehow constituting a rebirth, a discovering or rediscovering

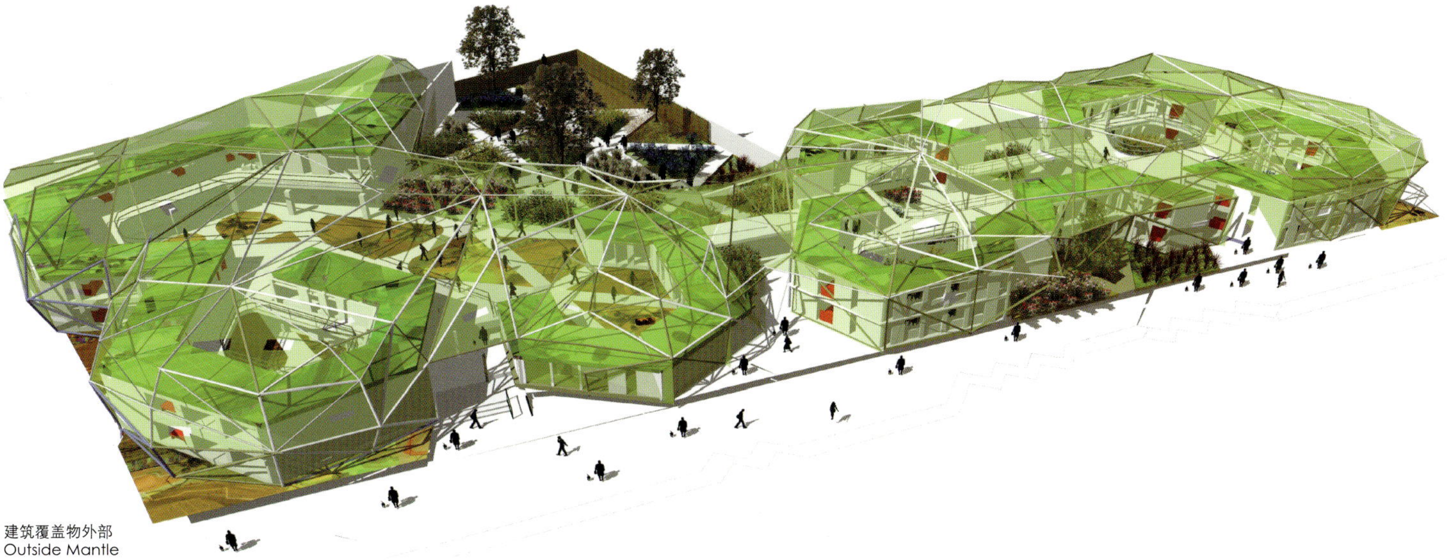

建筑覆盖物外部
Outside Mantle

建筑覆盖物外部
Outside Mantle

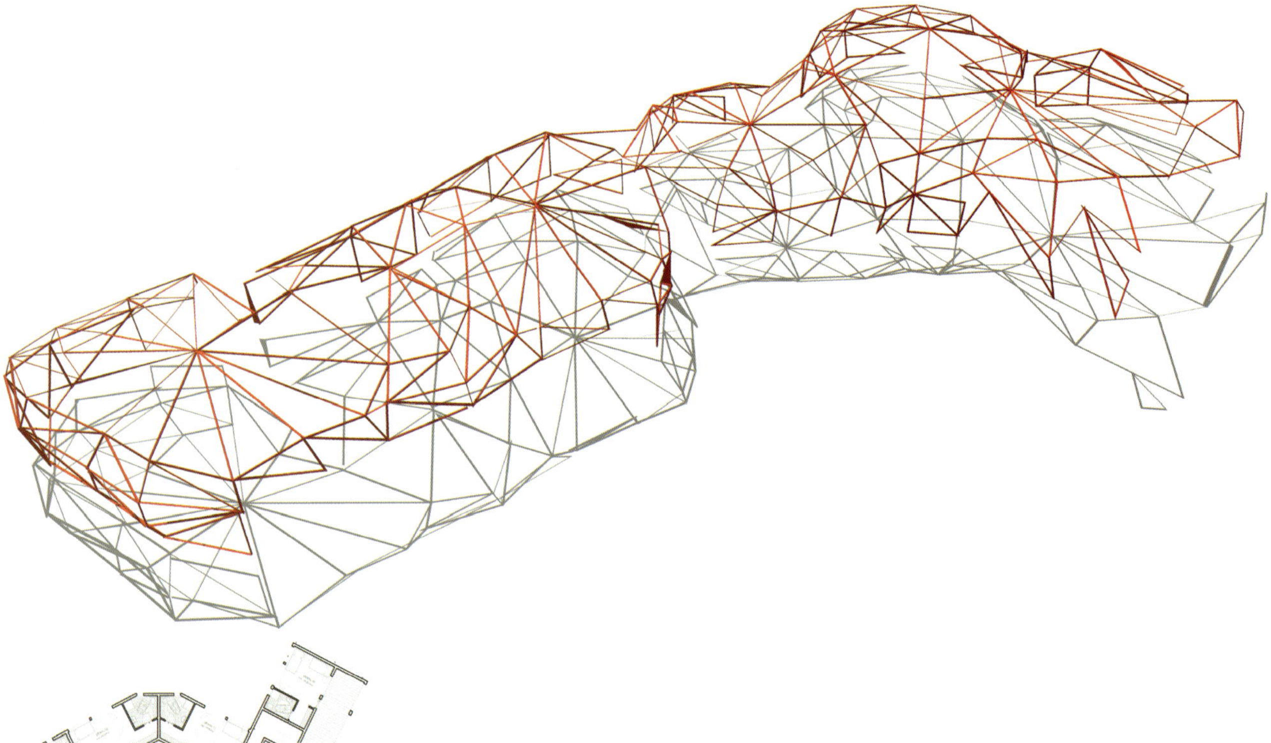

老人指南村
Placed the Guided Village for Senior Citizens

of the world. Physical conditions change, their body posture shrinks making their sight angle to point downwards, approximating to a child's height. From this point emerges the intention of relating this childhood evocative process with what is special about a garden, seeing it as the place that naturally, as an inherent issue to the childhood, reminds us of our happy moments of games, friends and freedom. Considering the effect of emotional resonance for the senior citizen that produces an open space surrounded by nature is why the mantle that allows evoking colors, textures and odors pertaining to flowers and plants is proposed, creating an interior space full of life and harmony, where what is inert is expelled and joy is welcomed, promoting collective participation and projecting a green image in constant change to the saturated gray city.

建筑覆盖物内部
Inside Matle

Continuity of Heritage
历史遗产的延续

"Each generation writes its own story which stays engraved on its architecture."
Lewis Mumford
Tel Aviv, "spring's hill", includes the 2/5 of Israel total population. Born in 1909 on dunes, the city was created by small districts linked up together. Its youthfulness resolutely makes it modern. Laboratory of a new architecture at the beginning, urban theatre where towers and other constructions grows up today, Tel Aviv became a city bursting with energy and optimism. Sir P.Geddes drew its first urban plan and imagined it with public spaces, gardens for everybody. Thanks to that, it is a green city which is breathing and its white part, jewel of the Bauhaus architecture, is registered to the world heritage by UNESCO.
Every year, thousands of immigrants arrive at Tel Aviv and the municipality must manage the increasing of housings capacities because city's limits spread to suburbs. Nowadays, the "tayelet", walk by the sea, is one of the most dynamic place. But at the north entrance things are like blocked. An abandoned peninsula remains. Roads don't serve correctly this zone that is disused since the river got dry. Its strategic position, at the Yarkon River mouth, offers all the qualities required for a pleasant life space in continuity with Geddes mind. Then a global urbanism, which takes care of the existing and opens news fields, is experienced. It spreads on the peninsula and moves forward the sea until it recreates a visual link with Yafo. The impact of history gives a strong symbolic meaning to this place.
A free flow for the development of roads, a protection of the river and an important place for the public spaces are the main points of this urban approach without forgetting rehabilitation and restoration of the trade fair's pavilions and the old port, other articulation of the past.
New equipments are developed to reply to the city requests: housing, stores, offices, gardens, the city continues and regenerates. In direction of a new territory: the sea, towards the Mediterranean, towards Europe...
Geddes Centre results from the continuity of heritage. At the end of the peninsula, near an esplanade and a garden, this vertical public space is the result of the city's image and is also dedicated to it. Like a landmark, it represents the formation of Tel Aviv, by fragments linked up together. Its program touches art, culture and city. Its vocation is to transmit a "heritage". The tower brings us from noise to silence, from contemplation to reflection, from discovery to surprise. Through its visit, we cross exhibitions and projections spaces, we can see study and meditation spaces where we can read, listen to music, communicate with other people by internet... New media technologies help us a lot and offer us a way to learn an infinite number of data: other cultures, other languages, other visions... Finally, its highest terrace is an observatory of the city where we get conscience of our scale and where we break loose with all our usual perceptions.
Aim's project is to revive urbanity's notion and to position Tel Aviv as a collective work. A view of the future has a meaning only if it looks in the past. Geddes Centre, conclusion of the complex, let the opportunity to anyone to express himself, to learn and to understand.
"Culture is what links all knowledge and fertilizes them"
Edgar Morin

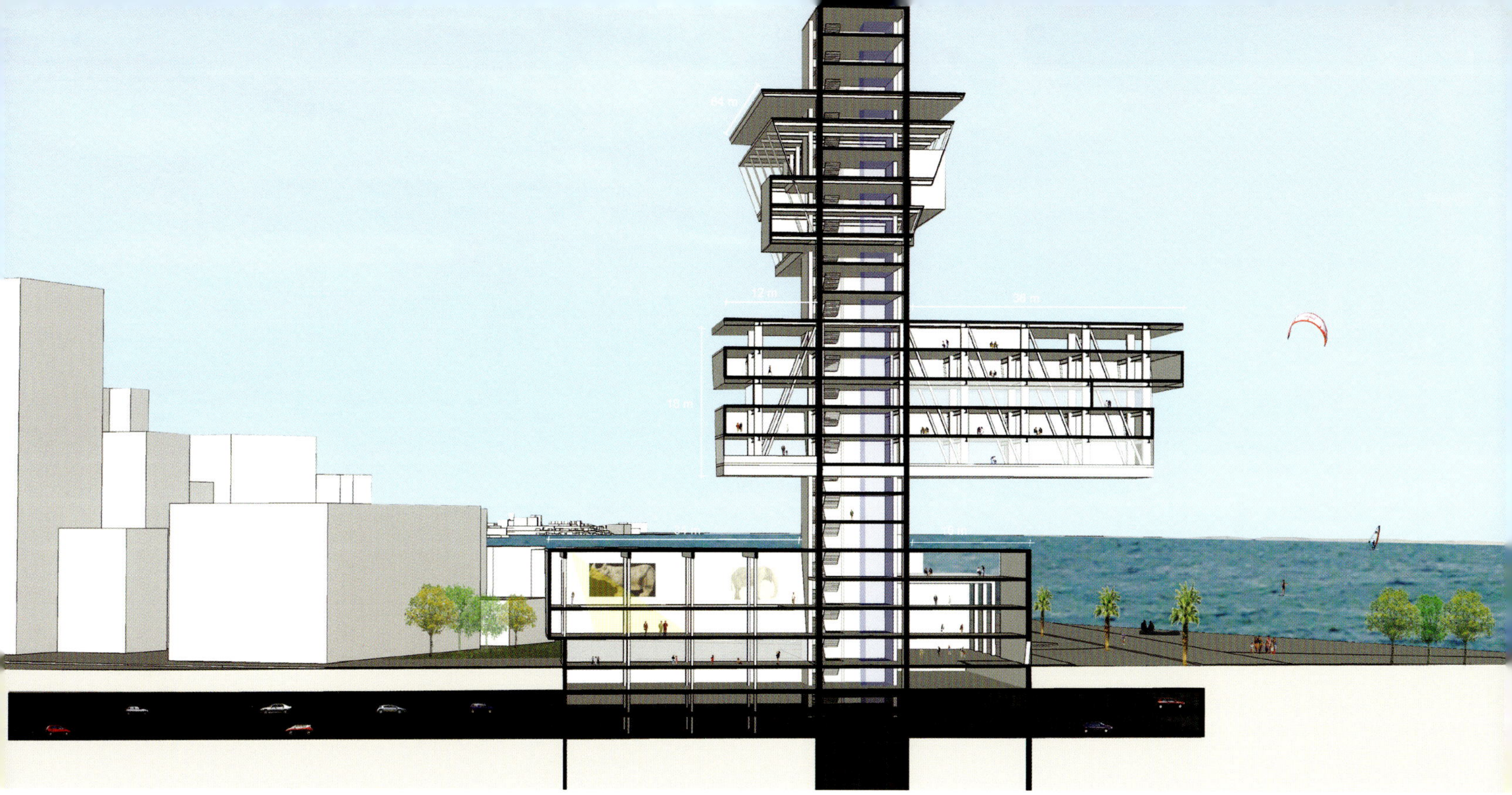

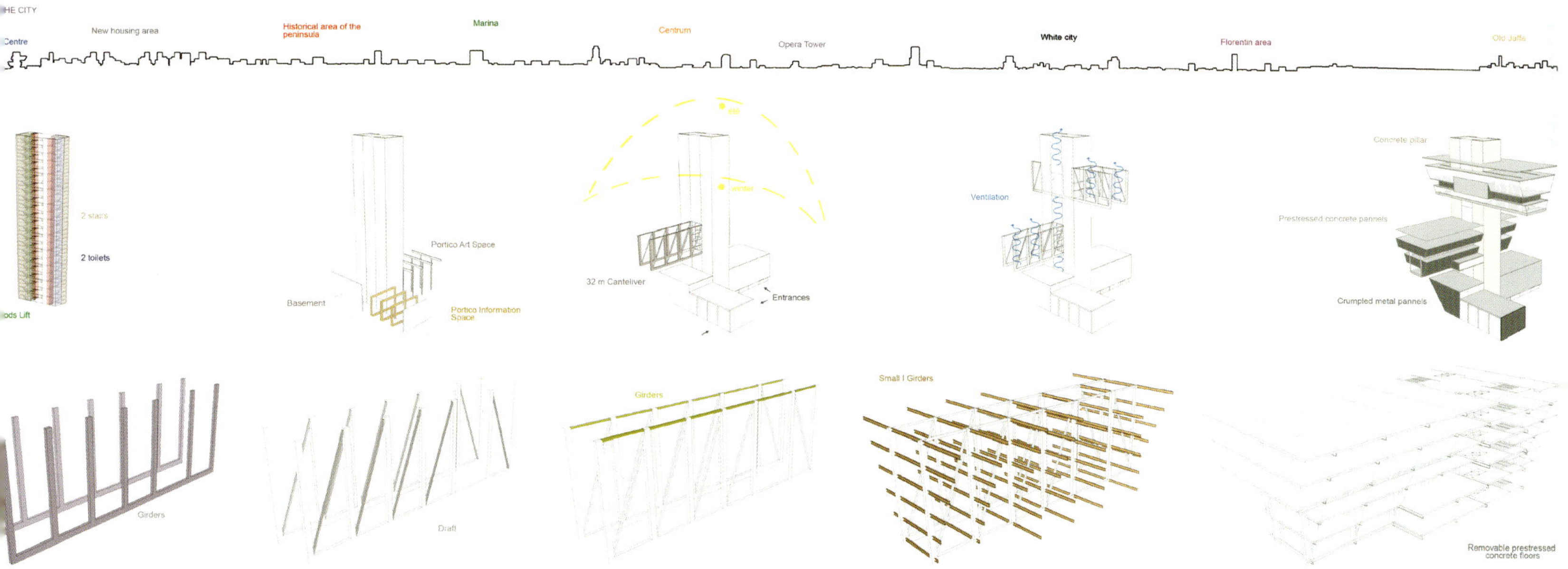

HE CITY
Centre
New housing area
Historical area of the peninsula
Marina
Centrum
Opera Tower
White city
Florentin area
Old Jaffa
2 stairs
2 toilets
ods Lift
Portico Art Space
Basement
Portico Information Space
32 m Canteliver
Entrances
Ventilation
Concrete pillar
Prestressed concrete pannels
Crumpled metal pannels
Girders
Draft
Girders
Small I Girders
Removable prestressed concrete floors

Project Team: Architect Dr. Joseph Cory, Architect Farah Farah, Flavio Adriani

Contour Museum

轮廓博物馆

The Museum of Contemporary Art (won 2nd prize in the competition) that was supposed to erect in the Arab town Um el-Phahem in Israel has minimal footprint in the local and dramatic environment using floating structure frames are floating above the main street. In this way galleries are acting as spatial terraces being part of nature with constant awareness to nature and time. The museum is born from the earth and bridges between the cliff and the grove allowing synergetic dialog between urban and nature, tradition and modernity, old and new, east and west. The museum symbolizes progress, modesty and optimism for local residents, tourists and professional from around the world transforming the city to become a key player in the world of art discussion. The museum has clear and simple appearance but yet striking from different angles. It can be built in phases while each phase has its own

DESIGN PRINCIPLES

PRESERVATION AND CONNECTIVITY USING INNOVATIVE HOVERING

Minimal footprint in the site: doubling open area by using a floating structure

Lower parking place on metal grills allowing small vegetation growth and ongoing water flow

The museums frames floating above the main street

Galleries acting as spatial terraces

BEING PART OF NATURE

Nature enhancement

Public areas landscaping accessible 24 hours a day

Green roofs: soil, local vegetation and rocks as an integral part of the design

Minimal interference: low negative impact on the environment

Preserving the local grove and allowing access by foot

Walkways and bicycle paths around the museum in the green open spaces of the city

Preserving the land and allowing storm water infiltration

Open, shaded and roofed spaces using local vegetation and the building shape for winter and

summer activities
Using light and shadow as a dramatic and dynamic design tools for the interior and the exterior
The museum is created from the rock and bridges between the cliff and the grove
Unique viewpoint from the museum outside and from the city and nature inside
Exterior, yet roofed sculpture garden protecting the sculptures and allowing full year activities
Synergetic dialog between urban and nature, tradition and modernity, old and new, east and west
The museum exterior walls allow projection of video performances enhancing the environment
The landscape architecture has a minimal design approach as of the local identity, allowing contemporary interpretation to local terms such as terraces, Mediterranean groves and plantations
Notion of Belonging by preserving identity values of local materials and vegetations
Connection to the modernism and contemporary art while developing a dialogue between local and modern through architecture representation

SUSTAINABLE ARCHITECTURE

Building green with minimal impact on the site, creating clean energy and using energy efficient systems
Creating acoustic barrier between the local residents and the future road
Letting optimal natural light in and saving on artificial light with Tizana elements and glass walls
Putting photovoltaic cells on top of the technical roof and in the upper parking place
Local material usage
The structure of the "bridge like" museum can be realized using metal or concrete
Inner controlling systems for air, light and volume parameters
Passive (positioning, isolation, shadow, natural ventilation) and active climatic strategy
Accessibility to handicaps

TRADITION AND MODERNISM

Local stone and expertise with modern construction
Modern mashrabia in the well of light that connects the masses

CULTURAL ISRAELI-ARAB ICON

The museum symbolizes progress, modesty and optimism

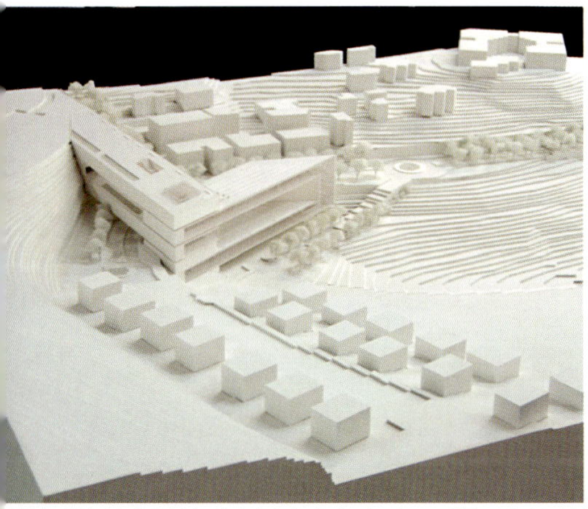
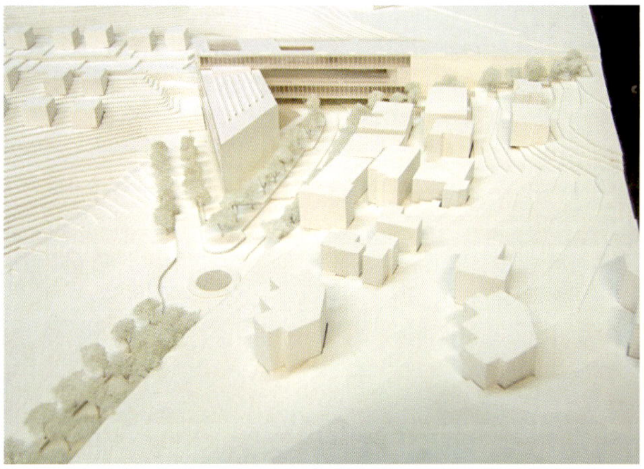
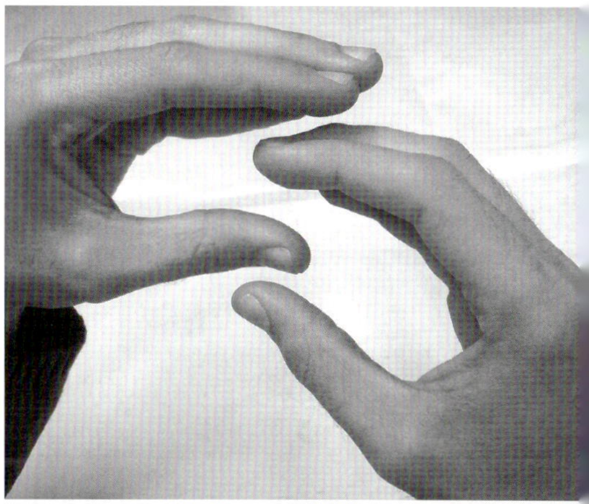

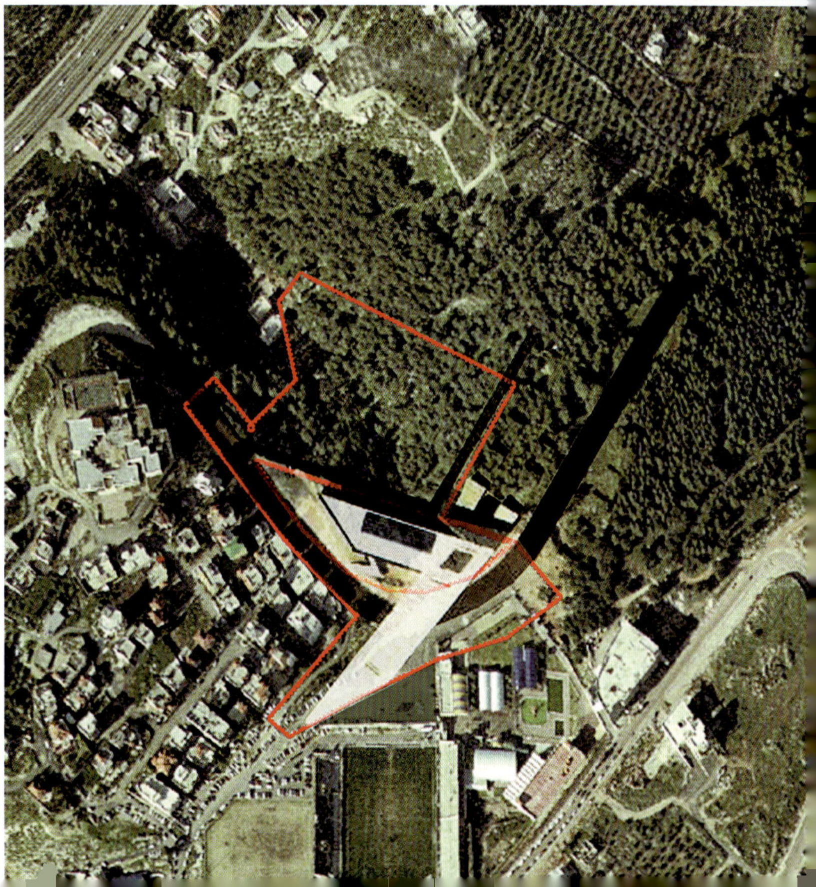

Interior Design: Tabanlıo□lu Architects

Doğan Medya Center Ankara Cube as a Perfect form and More

安卡拉DMC中心

On Ankara-Eski□ehir road, Do□an Medya Center is planned to be a distinctive media figure, housing TV channels, Hurriyet and Milliyet newspapers of the group, creating the basis of interaction between the brands and providing a common source.

The main form of the building is decided to be a cube in respect to the square site; yet starting on the surface the straight cube form deformed and re-structured by again cubic annexes and subtractions that resulted in extra volumes.

The emblematic use of façade creates a visually legible dynamic ambiance with reference to today's fast moving and assertive image of the media. The building is strongly perceived from afar and in diverse prospects at various angles due to perforated shield resembling Braille alphabet at range of sizes.

The separate units stick out on the façade dedicated to each TV channel and newspapers so that special units are obvious from outside as well, and from within the building there are panoramic views of the city. The assemblage of smaller cubes within the structure of the large box can be read as a metaphor for the diversity of the company's operations.

In accordance with topographic directions, slightly angled solutions of the modules on the front elevations and south-north axis enhance the vivacity.

It is a genuine building in terms of transparency; strong in-out correlation enables a well-defined description before getting in the building. Composed harmoniously with the environment; it is open to the cityscape.

The concrete structure reaches up to 7 stories by feasible use of galleries so that extra operative space created by mezzanines, whereas the standard is 4-4 ½ floors. The story height is 4 meters, every two floors a cube is formed; the modular structure provides a flexible basis for planning. Intermediate floors are supported on secondary steel columns and beams.

Main food court is located in the atrium after the main lobby and a terrace-lounge is reserved as the VIP meeting point at the upper floor.

There is an open car park in an L-form parallel to the building besides two underground levels reserved for parking. The first basement is saved for technical facilities and storage requirements are solved in basement floors.

南立面
South Façade

外立面细部
Façade Detail

从室内望出去的细节
Detail From Inside

工作室特写
The Studio

平行的外立面
Double Façade

展览厅
The Gallery

VIP办公区
VIP Office

VIP楼层
The VIP Floor

Technological facilities and healthy infrastructure makes the building user-friendly and provides easy maintenance.
The building is dynamic and interactive with the road in front of it, as an eye-catching "sign". Separate sections like studios and offices of press people preserve their exclusivity and each segment is accentuated in the integral form of the re-assembled blocks of cube.

Dormitory for Staff of NGO Magic Bus — An Expression of Contemporary Sustainable Architecture

NGO Magic Bus员工宿舍楼——当代可持续建筑艺术的精彩演绎

The dormitory for the staff of the NGO Magic Bus will be located in the valley below the Karjat Dam. Sited on the tip of a small plateau it overlooks paddy fields and sleepy villages. Embedded in the topography, the dormitory reveals different elevations depending on the viewpoint. By utilizing the steep slope an intimate low façade faces the entrance court on the forest side. A more prominent vertical façade faces the paddy fields. The interior of the dormitory will reveal itself as one moves through the building. A staircase cutting through the building mass breaks up the passage on the ground floor. This cut in the building frames the exterior beyond, creating interplay between inside and outside. This staircase leads you down to a large double height recreational space. Suspending the two-storey structure over the ridge creates this space below. The top floor is surrounded on all sides by a veranda. At the end of a colonnaded passage a large shaded deck overlooks a most spectacular view over the trees, towards the contours of the Matheran mountain range in the distance.

A 'cage' of bamboo columns wraps around the building. Public circulation and veranda's occupy the interstitial spaces between the external bamboo columns and the internal mass. The large covered veranda's and the relatively narrow width of the building envelope allow for comfortably ventilated and shaded semi-indoor spaces. The bamboo enclosure creates a dialogue between the interior and the dramatically changing landscape. The natural landscape is from a dense bright green jungle-like forest during the monsoon months to a pale brown shrubby wasteland during the dry and hot summer months. The building has to respond to these extreme conditions by allowing enough shade and breeze during the summer and providing a waterproof indoor environment during the stormy monsoons. The screen of columns creates an ever-changing pattern of light and shadow throughout the seasons and times of the day, making the building a 'sensor' of light.

A reduced material palette using bamboo for columns and walls, and natural stone and polished concrete for floors, neutralised by the whitewashed interiors gives the calm rational architectural language a sensual physical presence. The row of bamboo posts at 1.5-meter intervals forms a permeable skin around the building as well as supports the entire building. Bamboos, being the fastest growing grass in the world, are a high yielding renewable material

建筑外观
Perspective Exterior

建筑外部入口远景
Perspective Exterior Entrance

resource. The high strength and low weight factor of bamboo makes it remarkably earthquake and cyclone resistant.

With invaluable inputs from engineer Vilas Gore of GeoScience Services, innovative technologies and methods are used in order to reduce the amount of non-renewable materials throughout the building. The selection of materials and building methods is optimized to decrease the load on the building, which in turn allows the supporting columns to be sleek and elegantly dimensioned.

The columns consist of two bamboos, with a diameter of 90mm, joined together by steel strips. Steel dowels inserted in between the two bamboos avoid shear movements. Trusses made of pre-cast concrete and steel tension cables span between the columns. The floors are made of half cut bamboo, on which a layer of 50 mm concrete is laid, finished with a layer of natural stone. Walls are infill panels made of split bamboo covered with reinforced plastered on the inside.

The construction details emphasize the lightness of the structure. A steel plate, which is embedded in the pre-cast concrete beam, connects to both the lower and upper bamboo twin-column as well as the bamboo tie beams. While this steel plate cuts through the bamboo column to make use of reverse bending, the continuity of the bamboo columns is maintained accentuating the verticality of the structure. Combining natural and local materials with state of the art technologies and materials such as wood, concrete and steel this building will portray a new construction idiom – an expression of contemporary sustainable architecture.

建筑模型
Model

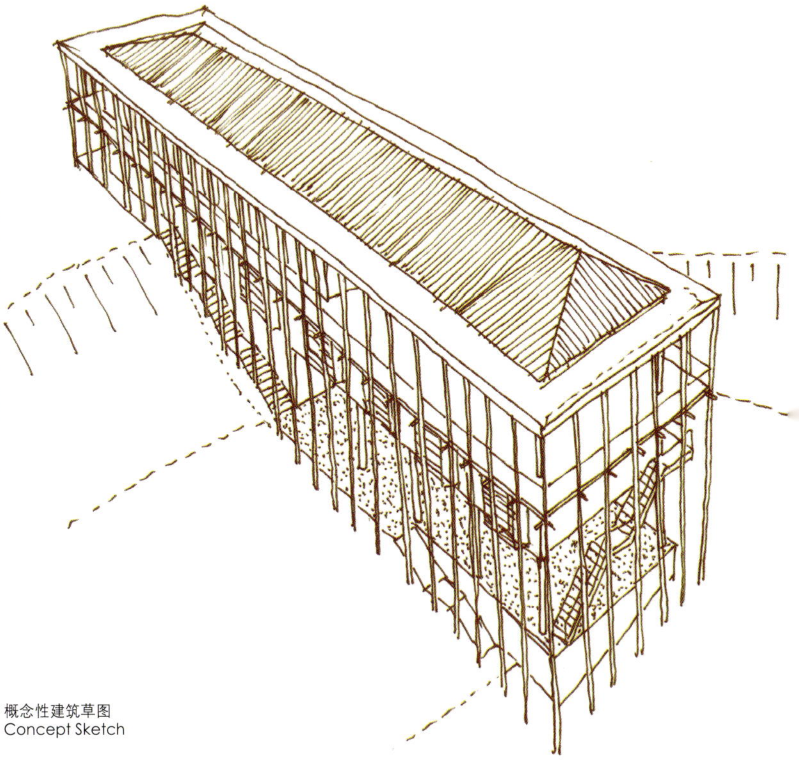

概念性建筑草图
Concept Sketch

建筑总体规划
Area Plan Campus

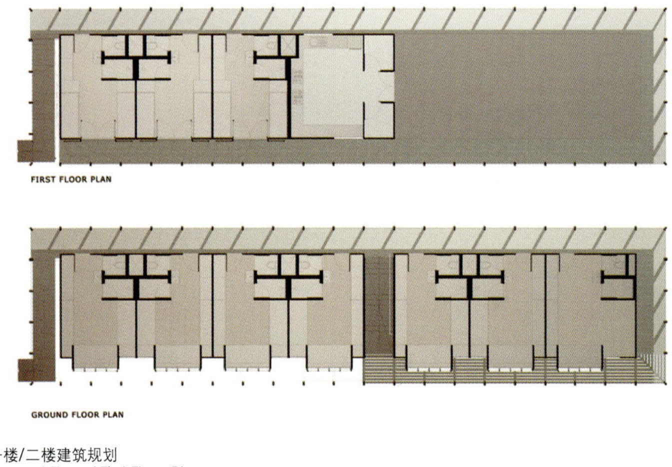

一楼/二楼建筑规划
Ground Floor / First Floor Plan

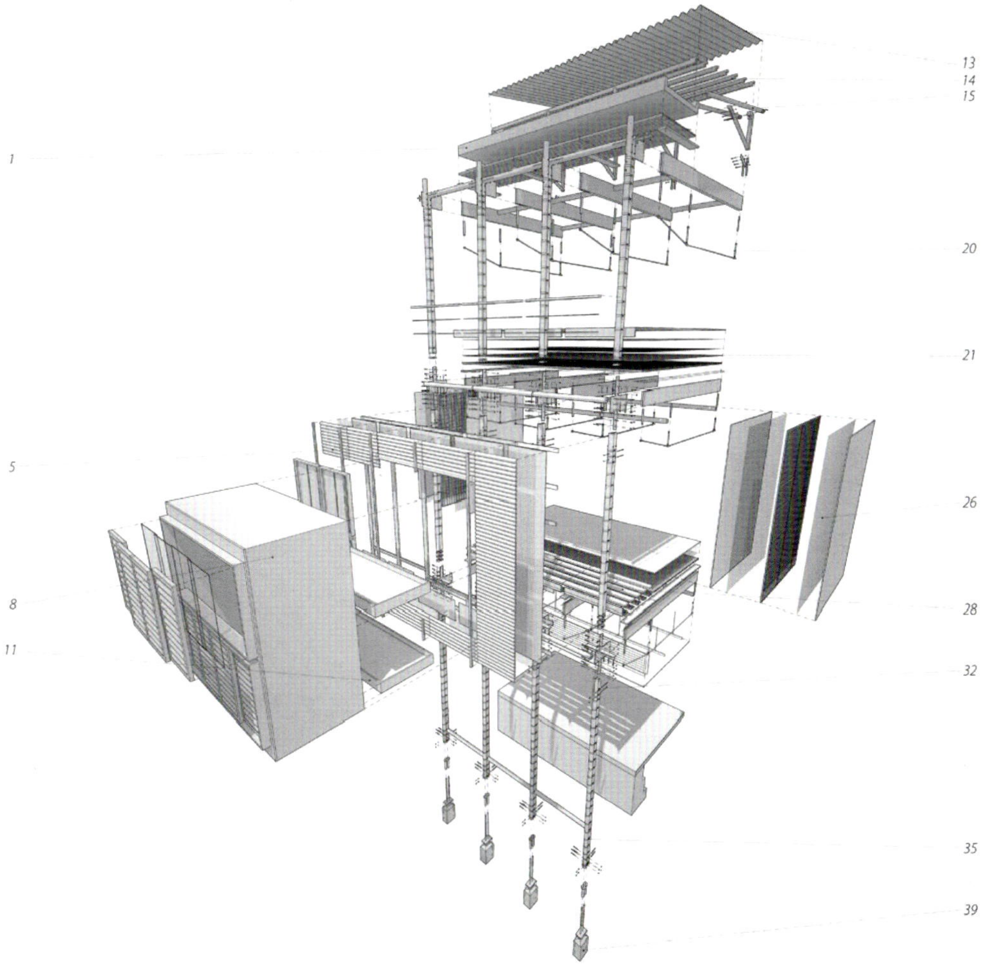

建筑物分解图
Exploded View

局部模型
Sectional Model

内部结构
Perspective Interior

Project Team: arch.Catalin Trandafir, Conf.dr.arch. Cornelia Barbulescu, arch. Irina Filofi

Drobeta Archaeology Museum

迪罗贝塔考古博物馆

WHY?

The need for the promotion of the national cultural heritage and its integration in the international circuit represents a fact in the European and world-wide conjuncture. In order to achieve this desideratum, it's mandatory to revitalize the whole museum complex, the archeological park "Drobeta", from Drobeta Turnu-Severin, which gathers not only multiple elements, remains of the roman civilization, but also a museum, which, at present time, functions as a general museum complex, that includes various specializations. Here, you can admire until present day, aspects of "citadin" roman life and, later on, elements of the roman colonial life in Drobeta. One of these elements is the roman castrum Drobeta, which is known to be the first stone fortress built by the romans in the Dacia province. The castrum had the strategic role of defending the bridge. The uncovering completed in 1950 revealed an important infrastructural part and, in certain areas, even the castrum's superstructure - elements that were subsequently consolidated and integrated in the tourist circuit. In a general planimetry, we can clearly distinguish, not only the elements of the first construction (II nd century), but also the ones belonging to the second one (III rd –IVth century) and also the castrum's perimeter, excluding the eastern part, where it can be estimated, thanks to its symmetrical composition.

The castrum is also equipped with an independent construction, destined for the roman baths. The roman baths of Drobeta, probably the most complex of the kind, built by the romans in the Oltenia's region, are placed in the south-western extremity of the museum complex. The relocation towards the north of the railway (in the 1980's), operation imposed by the realization of the Hydro energetic Dam of the Iron Gates I and II, affected a major part of the roman baths ruins and of the roman palestra, which, combined with the lack of supervising, led to the destruction of the major part of the bath's installations and, subsequently, their elimination from the tourist's circuit.

The key-artifact of the site, The Trajan's Bridge, which had a lot to suffer from the poor management, emphasized with the lack of the heritage conservation regulations and deteriorated by some interventions that were intended to be appropriate, is still holding on, even after nearly 2000 years, preserving in the visitor's conscience the memory of this place. Designed by Apolodor of Damasc, at the specific orders of the Emperor Trajan, the bridge was considered by the romans the 8th wonder of the world. Completed in only 2 years, during the second Dacia invading campaign, between 103 and 105 A.C., the bridge was 1134. 9 meters long, 14.55 meters wide and 18. 60 meters high and had in its structural composition 20 piles of stone, with weaved core from brick with hydraulic roman cement, with an opening of 50 meters. Its superstructure was built exclusively of oak wood. Unfortunately, the various attempts on this place left their marks. The defective restorations combined with the lack of maintenance, had

鸟瞰
Aerial View

affected and still affect parts of the ruins. Even more, the indirect long term effect provoked by these actions (the decreasing number of visitors), forced the administration to seek new solutions. In this direction, a development plan was conceived including a project theme for a new museum corpus, which represents, in this case, the starting point of this project.

THE CONCEPT - "DROBETA" ARCHEOLOGY MUSEUM

The project theme left for the architect to decide, among other things, an essential element of the architectural project - the location. Although some boundaries, dictated by the property documents, existed, these have not confined in any way the possibility of deliverance of the architectural concept, regardless of its type. So, in the attempt of placing the building in this sensitive context, I tried to keep the memory of the place, but, in the same time to define the way of exhibiting for the visitor, using exactly this antic-modern transposable. The symbols of the place became the symbols of the museum. The regulating directions, not only roman, but also of the modern city, have define the shape of the museum. The architectural concept suggests the evocation of these key elements of the site. The roman axis give meaning to the directions - the directions become transposed into language. The relation of the museum with its objects is one of interaction, as have already seen. The imagined museum is created by its objects, but its mission must not stop here, the museum must recreate the objects - and so, became necessary an active and continuous intervention in their protection - the "In situ Museum".

As we already have seen before, the "in situ museum" approaches the most to the concept of the optimal condition in the customer - object - museum scheme. It confers the vestige its well deserved space - it includes it in the attempt to protect it, it attends it in the attempt to exhibit it, it exhibits it in the attempt to render or to explain its value. The object's name, in this case, is Trajan's Bridge. The Trajan's bridge suffers in the attempt to exhibit itself. Its remains, specifically the four ending piers (the portal pier, piers no. one and two, and the abutment pier) no longer have the power of visual effect, necessary for the exhibition of the past, but the bridge pulses at symbolic level, its importance in the antic context, the political and cultural factors assure this object an undying place in the universal history. All these aspects, transposed into a well controlled scenography, will render the Trajan's bridge its long lost value and positions in its relation to the viewer.

As I already stated before, I choose for the "in situ" approach, having as a weight center the vestiges of the Apollodorus of Damascus's bridge. Crossing the bridge to the Romanian side passed through the castrum. The castrum was equipped with an articulation which connected with the bridge's floor, being the only way of reaching Romanian soil. This essential element in the defensive system is resumed in my proposal in an evocative shape, which concurs with the "loisir" needs of the archeological park. In the same way, the object brings on the castrum's guard layout, trying to offer the viewer a vision of this operation. Fulfilling the initial role, it will connect the museum with another proposal of semiotic interpretation - a light variant of a protecting tower of the "pricipalis destra" gate, which, having a vertical component, will serve at the same time as a belvedere point as well as a signal.

In the concept of development and arrangement thought for the city at Danube's shore, the proposed museum occupies, on medium term, the ending position (unfortunately, in the present stage, some areas with industrial character and inappropriate placement represents a setback for the normal course of development). This type of positioning in a track that includes many more other objectives, gives to the museum a privileged position in its relation with the visitor - it can remain in the memory of the viewer as the route subject. This type of relation, as we have already seen, it's not always positive, being able to transform itself in the dusty node of interaction, which we are trying to setback. The Antic Drobeta Harbor, whose remains were completely destroyed, has no way in participating in the museum representation. It remains at the latitude of our interpretations, and we have the obligation to present it, in one way or another. A sort of city-square accentuated by a powerfully anchored in the Danube's waters tries to present the place's memory, in a today Osnabrück shape, which combines itself very well with the utility element.

A really problematic element for the site and archeological park and for the whole concept of city's shore development, since its appearance, is represented by the railway. Obviously, its current position, between the bridge's ruins/Danube and the roman castrum/roman baths, influenced the whole museum organization. The permanent exhibition includes the ruins, crosses the railway and then it takes the shape of an underground corpus, taking advantage of the space offered by the natural amphitheatre. In this way, of adapting

入口
Entrance View

水景
Water View

立面——图拉真桥遗址展示大厅
Side View - Trajans Bridge Ruins Expo Hall

水景
Water View

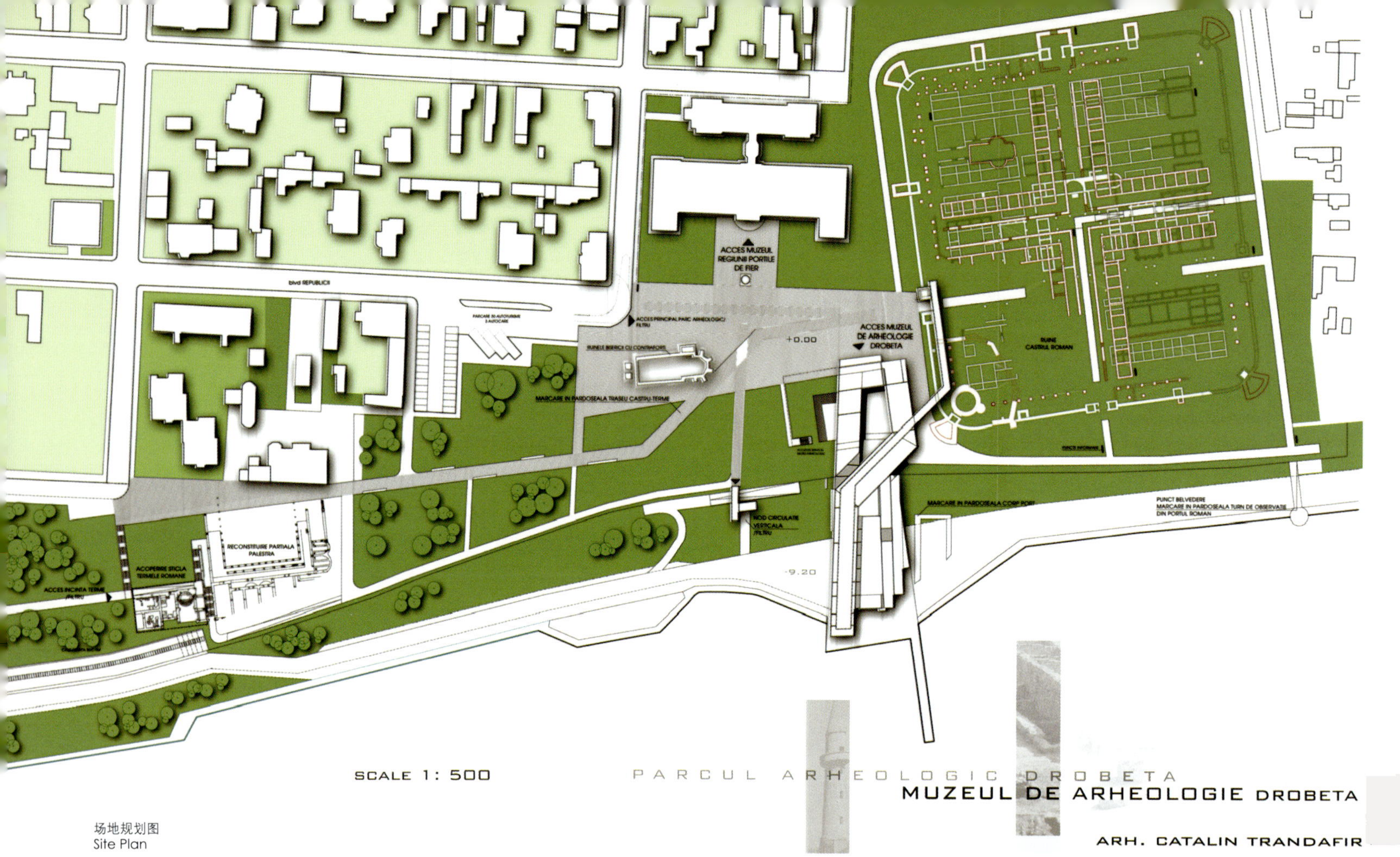

场地规划图
Site Plan

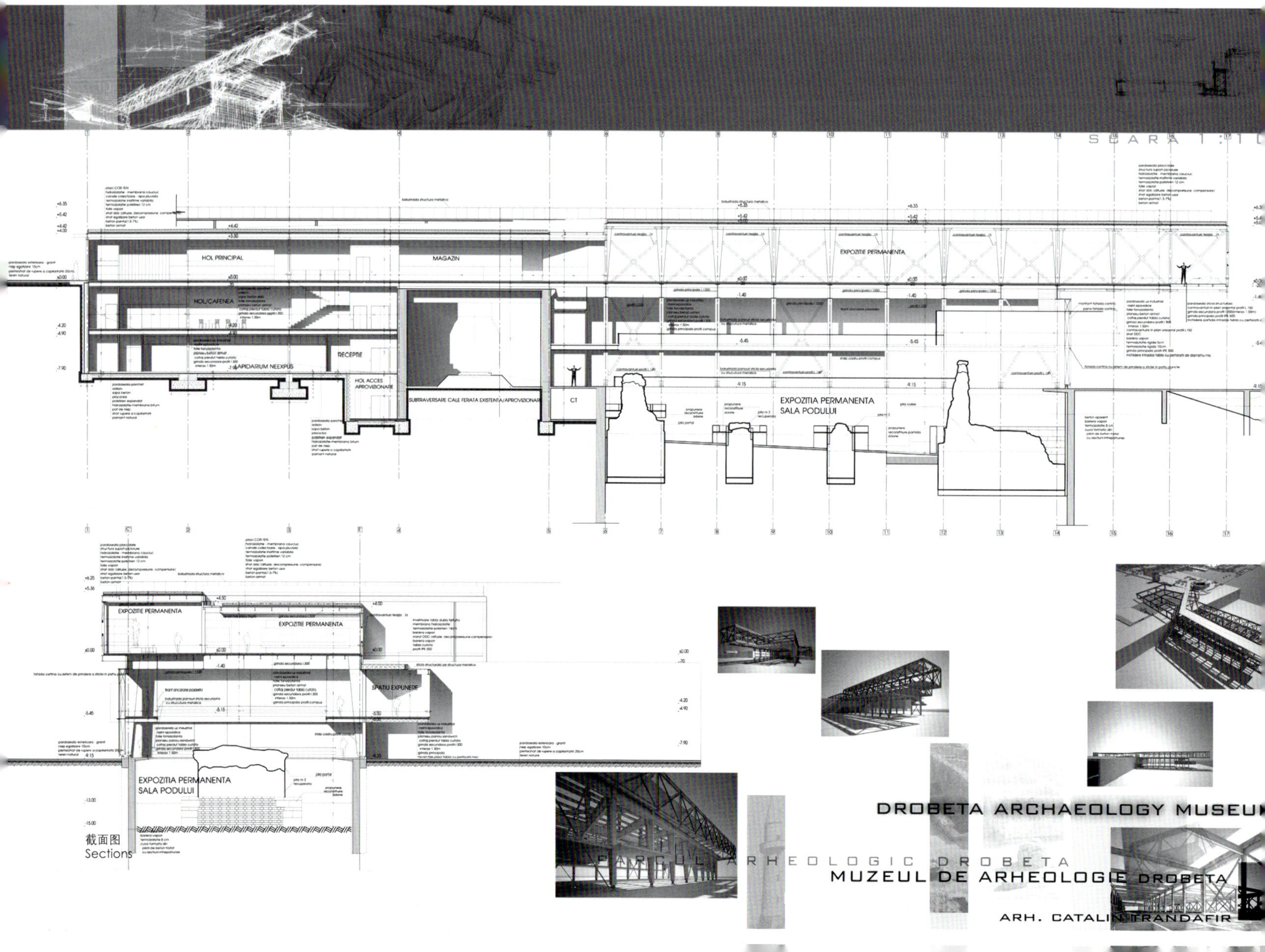

截面图
Sections

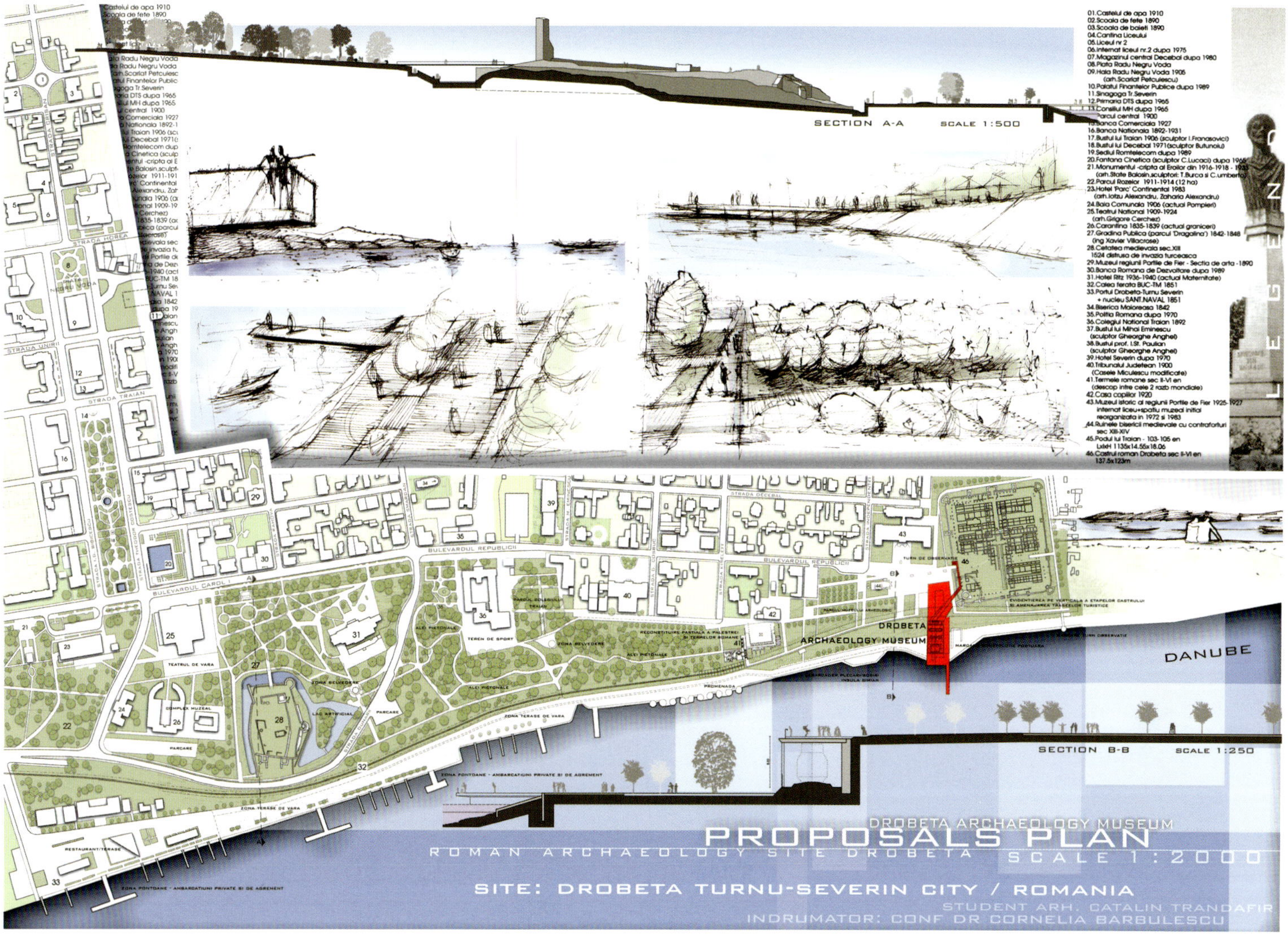

迪罗贝塔水岸研究与设计方案
Study - Conclusion and Proposals for Drobetas Shore

at the natural landscape, I tried to conserve a part of the current character of the archeological park. Giving their underground placement, the corpus that serves as an administrative, research and public equipped area, is organized around 2 inner-yards. These have not only the role of natural illumination and ventilation, but it also separates the corpus from the visiting sites of the archeological park.
We can safely say that all these connections took, in the end, to an image that tries to express itself, a self-characterization in a way adapted to our times - the symbol of the bridge between distant civilizations, a part of a glass bridge that crosses the Danube, but which connect with nothing, at least physically.

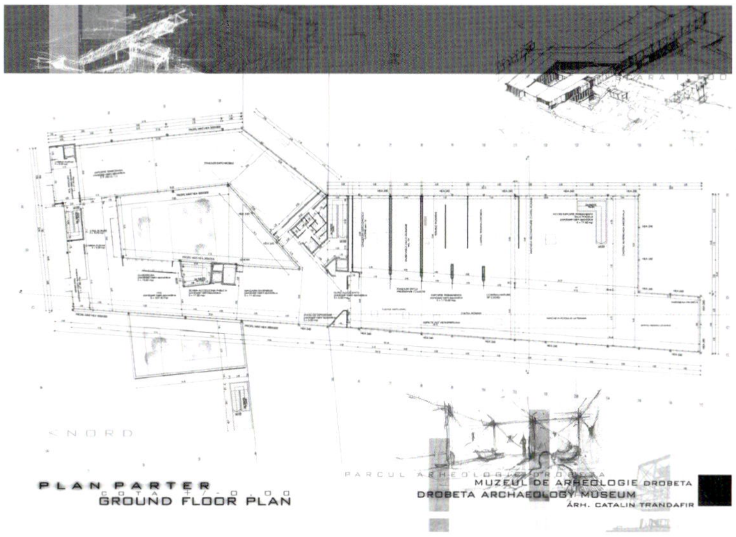

一层规划图
Ground Floor Plan

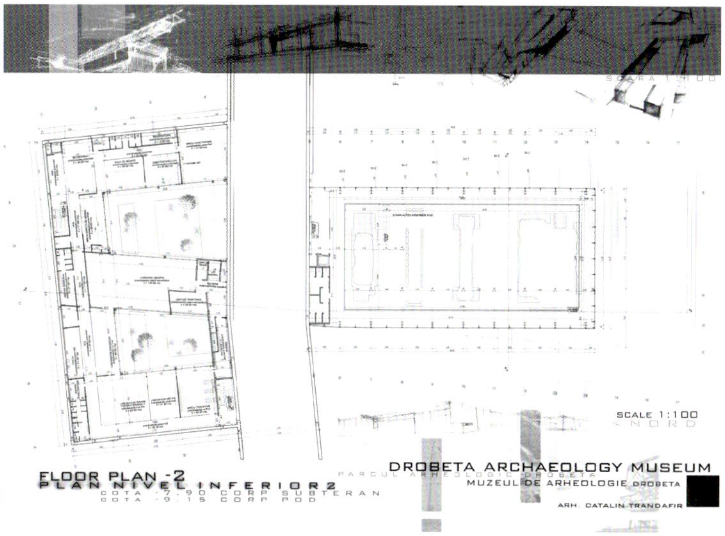

三层规划图
Floor Plan -2

Team: LAN Architecture (lead architect), Franck Boutté (HEQ consultant), Batiserf Ingénierie (structure), Michel Forgue (surveyor), Base (landscape architects), LBE (utilities)

EDF Archives Centre

法国电力公司档案中心

Within the framework of the Meuse and Haute Marne economic support program, EDF has decided to centralize all its intermediary Engineering Production Management archives in Bure.

Until now, these paper copy archives had been stored in nuclear, hydraulic and thermal production units, as well as in engineering units and associated services. As well as being spread over several sites, they were also handled in different ways. It had therefore become necessary to group them to rationalize and optimize the archiving functions: sorting, indexing, management and access.

Over and above the functional aspect needed to manage the archives, this was also a strategic project providing a positive social and environmental impact on the region.

This led us to work closely with local elected representatives on the economic aspects, the creation of jobs in the region as well as in the village itself, and on the architectural characteristics. The building needed to be able to fully integrate into the landscape as well as meet environmental quality standards, an aspect now fundamental to EDF's building strategy.

The objective is to obtain an energy-efficient building through the use of an intelligent morphology, a high performance skin, choice of adapted systems and the production of renewable energy.

The principles underlying our approach were to compress and minimize impacts. This approach results in:

1) Considerable saving in terms of the building's envelope

2) Improved functionality translated by a reduced number of kilometers covered per year

3) A marginal impact on the landscape (with viewpoints at a considerable distance from the building)

4) The possibility of a maximum use of the excavated land around the building's footprint to control water recuperation and treatment on the site

5) An energetically and environmentally extremely high performance building

6) The creation of a symbol representative of the approach taken by the Meuse and Haute Marne economic support program

The question was how to integrate lightness into a massive object? How to avoid interrupting the continuity of the natural landscape?

Firstly, to give the impression of a lightweight building in movement, we proposed incorporating stainless steel studs into the earth-colored concrete cladding. This solution had the effect of blurring the building's limits and reflecting the surrounding colors and changing seasons. The building has a very high performance envelope resulting from the materials employed and the technology used, which ensures that the building has a high level of inertia favoring comfort during the summer and reduces cooling requirements.

The site is isolated and has no sewerage treatment network. Consequently a low-cost natural wastewater system will be installed to meet the building's needs: sewerage, wastewater and rainwater treatment.

The choice made for the ventilation was to use a double flow ventilation system with heat recuperation. This limits energy consumption resulting from heating and ensures the good sanitary quality of the air.

The high performance of the envelope combined with economic ventilation and lighting systems reduces energy requirements. The use of renewable energies and a heat pump will result in a high level of energy autonomy.

建筑与景观融为一体
The Building is Fully Integrated into the Landscape

从道路望向建筑
View From the Departmental Road

中轴景观
Axonometric View

天井
Patio View

建筑冬景
View of the Building in Winter Season

Project Team: Chandu Arisekere, Veer Nanavatty

Eicher-Volvo India Headquarters, Gurgaon

艾彻-沃尔沃印度总部

Eicher House building is conceived in a way that brands the engineering core of Eicher business. The building reflects this emphasis on engineering and expresses its structure on the outside of the building surface. By designing an innovative structure based on the tetrahedron within a cube, the Architects working in close collaboration with Engineers in Pune have devised an economic and bold solution to a multi-storied structure.

The Eicher House is primarily divided into 3 main components I) the base, II) the main office and III) corporate head office.

The base is a linear building sitting on the ground; comprising entrance lobbies, cafeteria, games rooms, service areas, as well as the Research and Development wing. Sitting astride the base are two volumes, one large and the other small. The former comprises the entire office floor space of the Corporate Building, and the latter houses within it the offices of the senior management of the Eicher corporate infrastructure.

Both these volumes are meant to juxtapose each other.

VEGETATION AND SUN PROTECTION

The volumes that don't house buildings on the base, i.e. the open terraces, have been planted over with roof gardens. The main building also has creepers growing down the side of it, to act as sunshades for the building, reducing heat intake. The West and South facades of the buildings have louvres angled to reduce the direct intake of heat into the building. The louvres are angled specifically for the latitude and longitude of the Eicher building.

We have tried to balance here a building which is engineered, as well as climatically responsive.

三维景致
3d View

三维景致
3d View

三维景致
3d View

孩子们的娱乐场地景致
Childrens Playground

艾彻-沃尔沃模型图
Photograph of Model

Project Team Coordinator: Yapi-Endustri Merkezi A.S.

EKOyapı Research Center

EKOyapı科研中心

EKOyapı is a research and education centre to be established as part of the Nature-Environment Science-Society Park Project at Istanbul Technical University Maslak Campus. The project was coordinated by the Construction-Industry Centre, and the design and application were prepared by HAS Architects. All project design and consultancy as well as most of the construction material and fixtures were provided by sponsors with volunteering principles.

When EKOyapı was being planned, the initial discussion was on determining what example "green design" was. The LEED points system widely accepted in the USA, was taken as a guide. According to the description of LEED, green design is all of the design and construction methods to remove or reduce to a minimum the negative effects of buildings on the environment and the users.

LEED System is collected under these 5 main topics.

- Planning for Renewable Sites
- The Protection of Water Resources and Water Efficiency in Use.
- Energy and Atmosphere
- Construction Materials and the Use of Natural Resources.
- Interior Quality

EKOyapı's evaluations have therefore also been grouped under the same topics.

EKOYAPI SITE

LOCATION

EKOyapı, is located near the lake on Maslak Campus of ITU, as part of the Nature Environment Science-Society Park Project. The site was planned to face southwest in order to make the best use of solar energy. With the steepness of the site considered, the project was established on a slope so as to cut down on any excavation. This site and its compact form reduce the building's projection on the ground and make for little interference in the natural environment. Its proximity to a water body is something to be desired in the LEED Evaluation System.

USE OF WATER

ARMATURES WITH SENSORS

In order to keep water use at EKOyapı to a minimum, armatures in washrooms were picked to be the ones with sensors. Flush-basins in the toilet were picked to be the ones that have two levels of flows.

RAIN WATER

Another method to cut down on the use on mains water is to make use of rain water. Rain water collected off the rooftop gets stored in a tank underground and pumped into a reservoir to make future use of it.

WASTE

Waste water will be expelled through natural means in a cesspit. The cesspit will be located at a point in site where it won't leak into the lake. EKOyapı's solid waste will collected based on renewability potential.

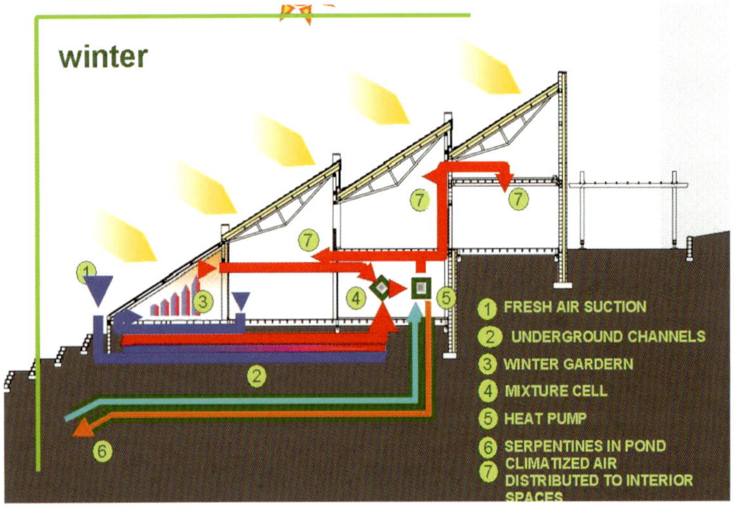

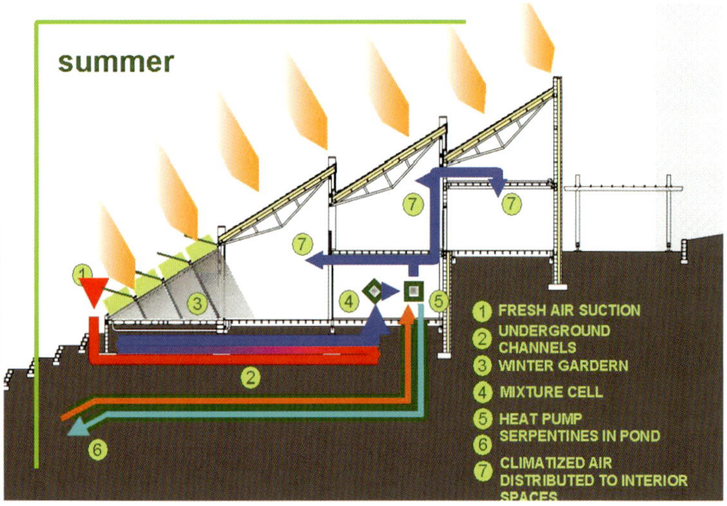

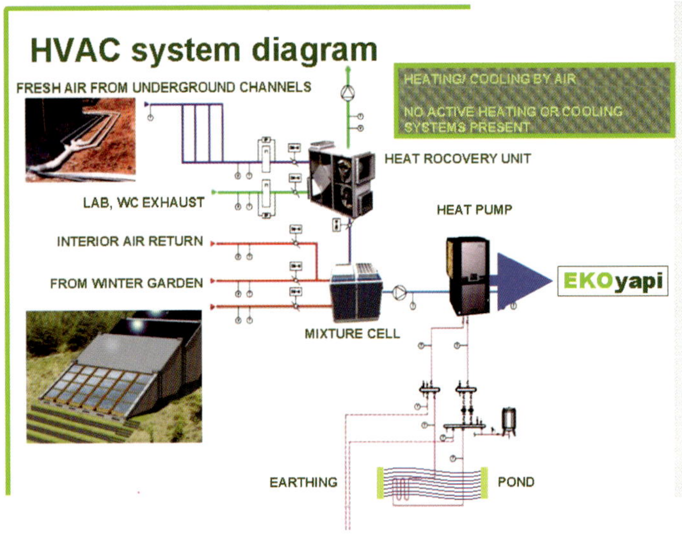

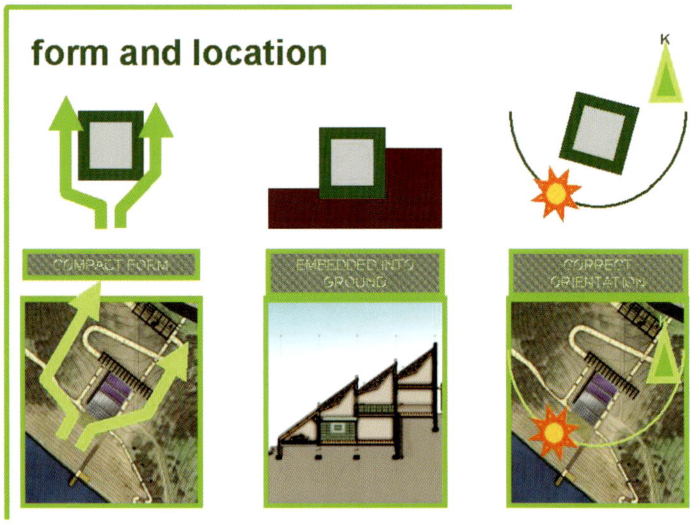

ENERGY AND ATMOSPHERE

The aim when determining EKOyapı's strategies was to keep the dependency on fossil fuels and greenhouse gas emissions to a minimum. Planning is bidirectional on energy sparing and local production.

NO AIR CONDITIONING, NO BOILERS

There will be no boilers or air conditioners for heating or cooling at EKOyapı. In order for climate control to be achieved through alternative means the construction shell was designed as one with super-insulation based on the standards set by German "Passivhaus Institut." External walls were designed to be slightly aggregated and have two layers. Part of the building volume is underground. Heat bridges were avoided in the building shell.

ALTERNATIVE ENERGY PRODUCTION

Two wind turbines placed in suitable places around campus and the photovoltaic (PV) cells on the rooftop covering an area of approximately 25 metres squared will have EKOyapı produce electricity. In other words, the building will produce more electricity than is uses and keep the grid like a warehouse.

ACCOUNTABLE PERFORMANCE

Since EKOyapı is a research centre, steps have been taken to manage, monitor and report on energy systems, not to mention installing them.

CONSTRUCTION MATERIALS AND USE OF NATURAL RESOURCES

The following criteria were kept in mind when picking construction materials for EKOyapı.

- To have low embodied energy values
- Local and readily available
- Produced from renewable or recycled sources
- Long lasting and requires no maintenance

INTERIOR QUALITY

VOLATILE ORGANIC COMPOUNDS

As the building shell of EKOyapı has high insulation, and is impermeable, the inner air is under risk of any toxic gasses to be released. All paint, glue, varnish and insulation materials were picked from a range of Volatile Organic Compounds (VOCs) with low levels. Same rules of measure were in place where composite woodwork is concerned.

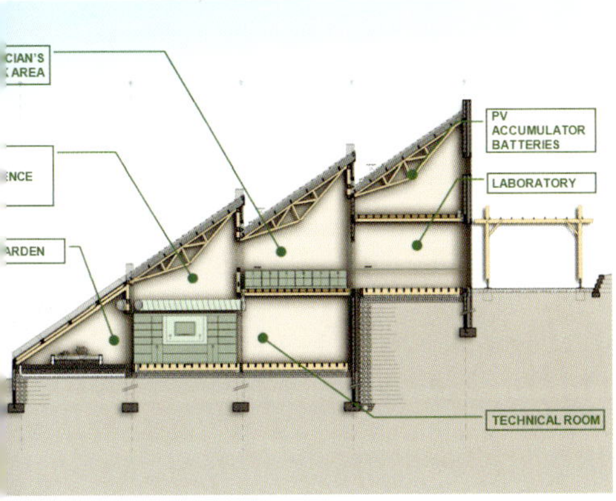

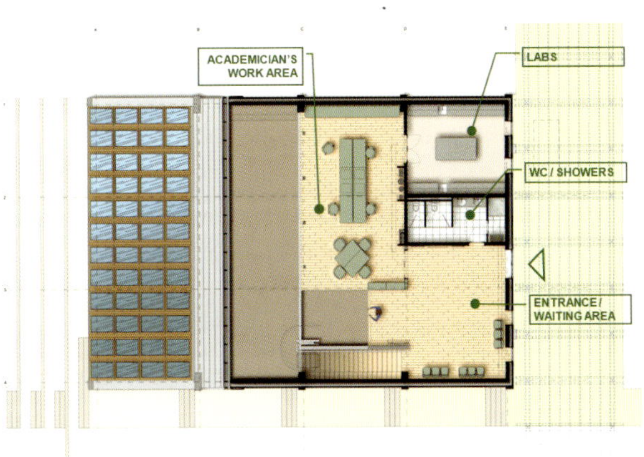

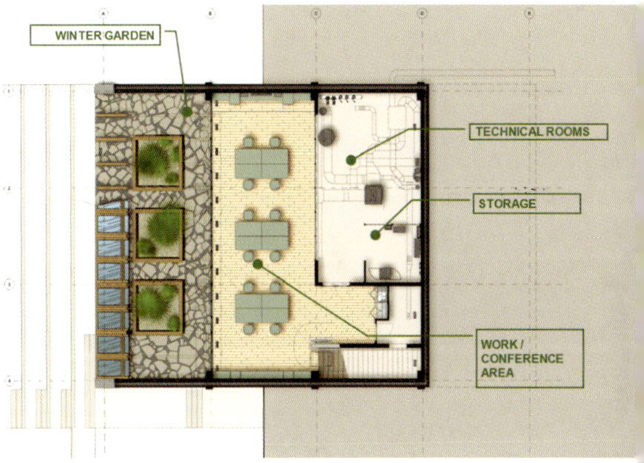

Project Architect: Pedro Barran

ETC n 330 Gutierrez Ruiz, Maronas

n 330 Gutierrez Ruiz, Maronas的全日制学校

This public school creates a continuous environment, which is variable in form and scale and seeks to engage the community. The site is small, so we had to design the school in three levels to allow for a large courtyard. The existing slope let us divide the courtyard in terraces, from an open soccer field to 3 different playgrounds; and it suggested us to work with split levels in the building, which are close to children´s scale. The school building is a coherent system formed by different sectors. There are four classroom clusters (of different sizes, heights and color-coded identities) with subtle formal distortions seeking the best views. These clusters are unified by a triple height atrium, which flows among the volumes and integrate the educational community. Classrooms volumes are transparent to that common space and most are elevated by columns to achieve a free ground floor. This lower floor has subtle differences in levels that suggest different shared facilities (access, atrium/multipurpose area and dining common), but maintain them flexible and allow them to expand or shrink.

从庭院向中庭看
View From Courtyard to Atrium

学校通道
Access to School

外视图
Exterior View

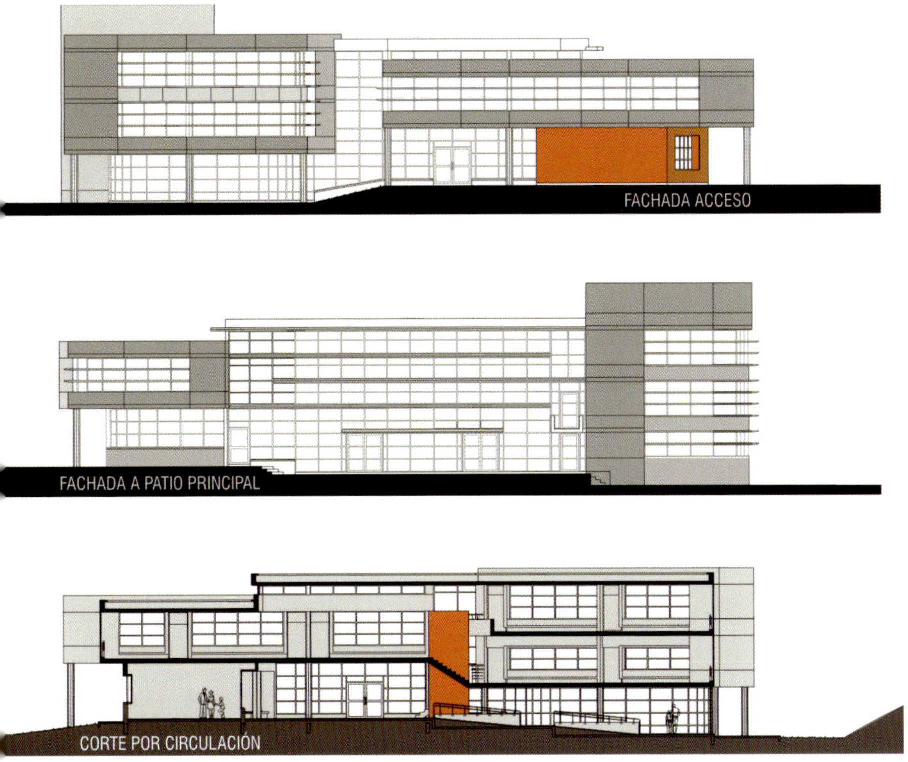

外立面和截面
Facades and Section

从行政管理室看
View From Administration

门厅通道
Access Hall

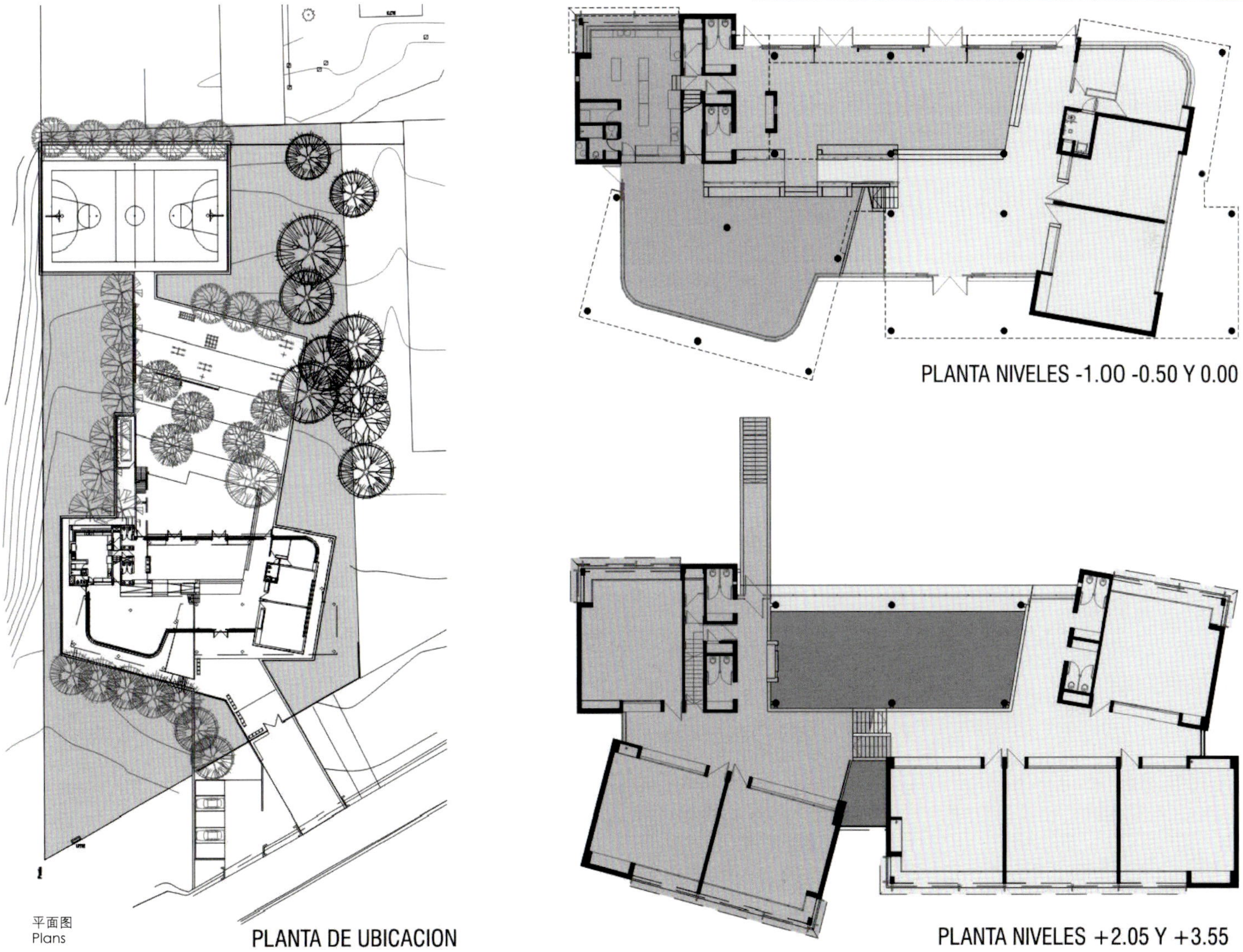

平面图
Plans

Eye Hospital
眼科医院大楼

Bangladesh, the largest delta on earth, has 52 rivers that carry water from the Himalayas in an intricate pattern to the Bay of Bengal. During Monsoon these rivers inundate two third of the country's land depicting water as the major element of the country landscape. When the water recedes, it leaves a fine layer of fertile alluvial soil and the entire landscape transforms into large patches of paddy fields dancing with the wind.

Except the parliament complex most public buildings in Bangladesh are mere buildings, not really public. The effort here is to make the whole approach to offer more space for the public inspired by the context.

Louis I. Kahn and his beaux-arts approach for the overall parliament complex and Sangsad Bhaban has been taken in the scheme as a "basics". The "celestial square" of Sangsad Bhaban has been shifted to the site as a "transferred square" through a "transitional square". Basically the repetition of 11'x11' square grid is forming the whole project as a living transparent and porous urban wall, connecting the "Old Eye Hospital" with a green-courtyard and the Farmgate in distance on the east, the Manik Mia Avenue and the Sangsad Bhaban on the west.

Facing west is the unique challenge for the site, which has been taken care of softly, with poetic intuition, without loosing its visual connections. Bangladesh is a country of six seasons with its unique subtle and substantial variations. The persistent sun, monsoon rain and wind have been willfully taken into account to make the scheme responsive to our climate. The use of thin water layer acting as rain collector is not only functional but Deltaic and connotative. The use of linearity, transparency and porosity in the building is not only for cross ventilation and connection but also to make it a traditional pavilion transformed. The use of shades and layers of green and parasols is not only essential but also sensual.

The main material in this project is concrete. Its level of permeability and sustainability with structural ability in combination is remarkable. Use of exposed concrete as Kahn used in the parliament building in the hot and humid Bangladesh is not only compatible but also vibrant giving it a poetic essence underneath.

建筑侧面图
Lateral View

建筑主立视图
Main Elevation View

建筑室内外观及庭院
Inner Façade and Courtyard View

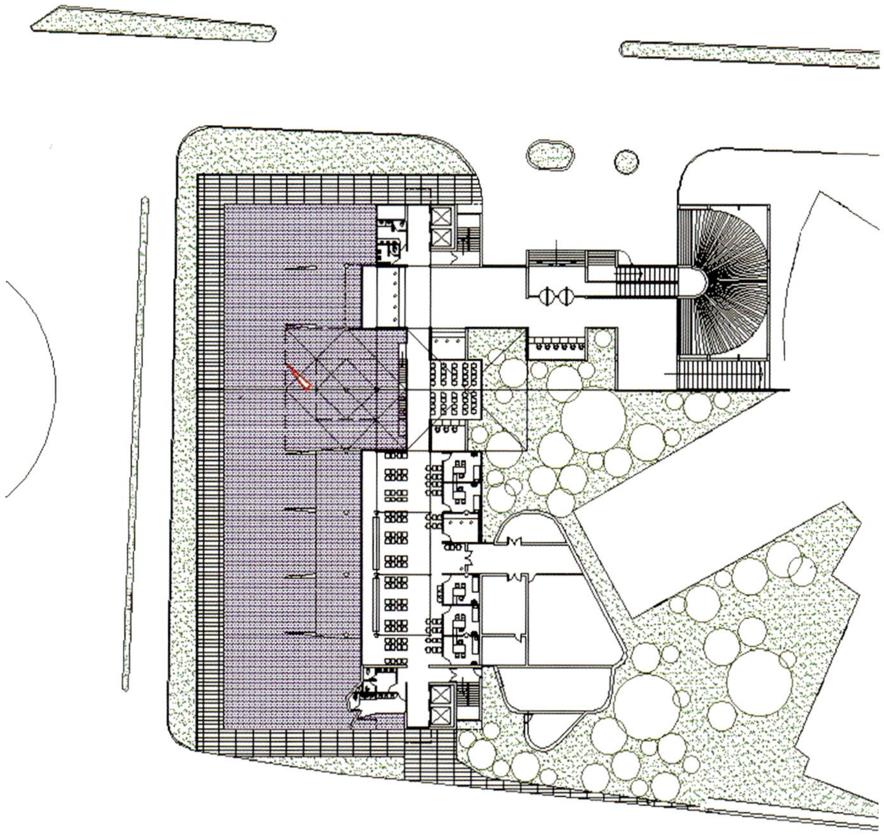

建筑规则
Plan

建筑顶视图
Top View

建筑室内外观局部部图
Inner Façade Detailed View

室内布局
Interior View

Project Team: Mr.Jongsarit Jinachan, Mr. Akaphol Jarnpian

Floating Spaces of the Residence

浮动空间住宅

This small project is a design of a new 300 sq.m. house, situated within the same compound of an existing house. Our client requested for a new modern building of steel structure, but requires that it does not deface their existing 30 years old house. We proposed a semi-detached building which could be connected with the existing house. After a careful consideration, we decided that our new design should not resemble the existing building in order to give possibilities for new creative idea to develop, as requested by the client. As the new building will be situated at the back of the site behind the existing house, we, as an architect, would like to envision a new design that shines its life rather than becoming dormant at the back. Our conceptual approach is to utilize the existing drive-way as the main approach to the new building extension. At night, the light from the 3 meters span cantilevered room on the 2nd and 3rd floor would generate an effect of a bright floating space above ground. At the same time, the living room and the bedroom of this new extension would have an open view of the gardens; as well as full ventilation from the wind direction. Owing to the limitation of available space, a unique manipulation by means of juxtaposition of the structure will be the major focus for our design consideration

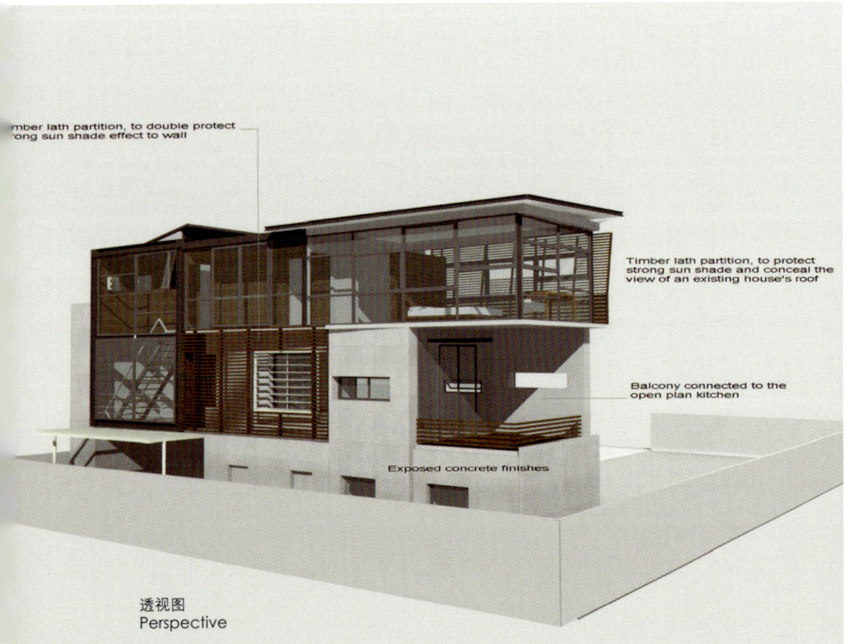

透视图
Perspective

透视图
Perspective

透视图
Perspective

透视图
Perspective

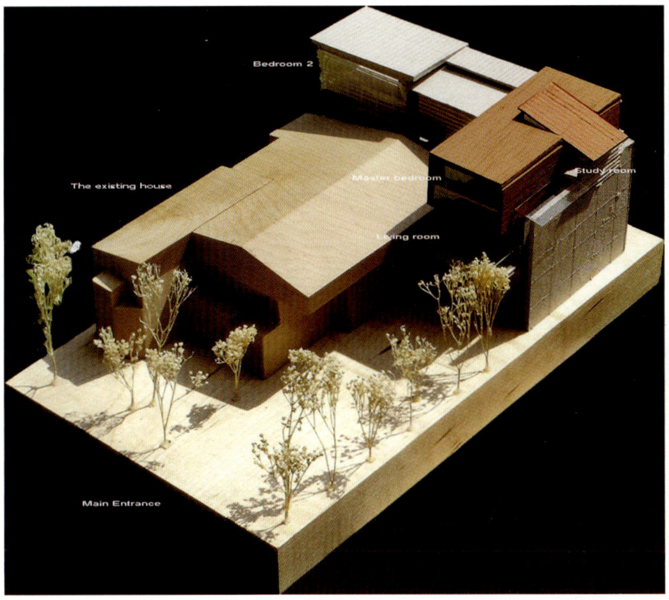

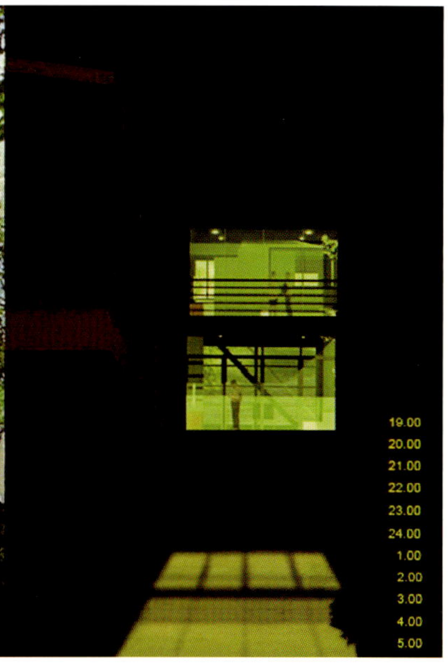

研究模型
Study Model

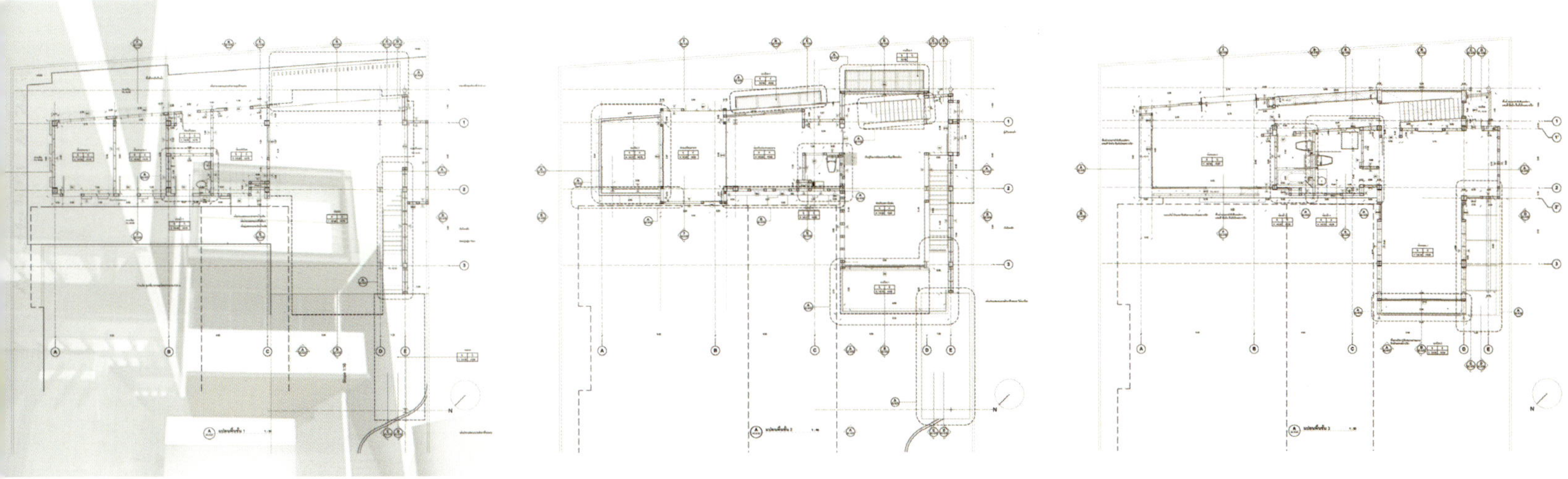

平面图
Lay Out Plan

截面图
Section

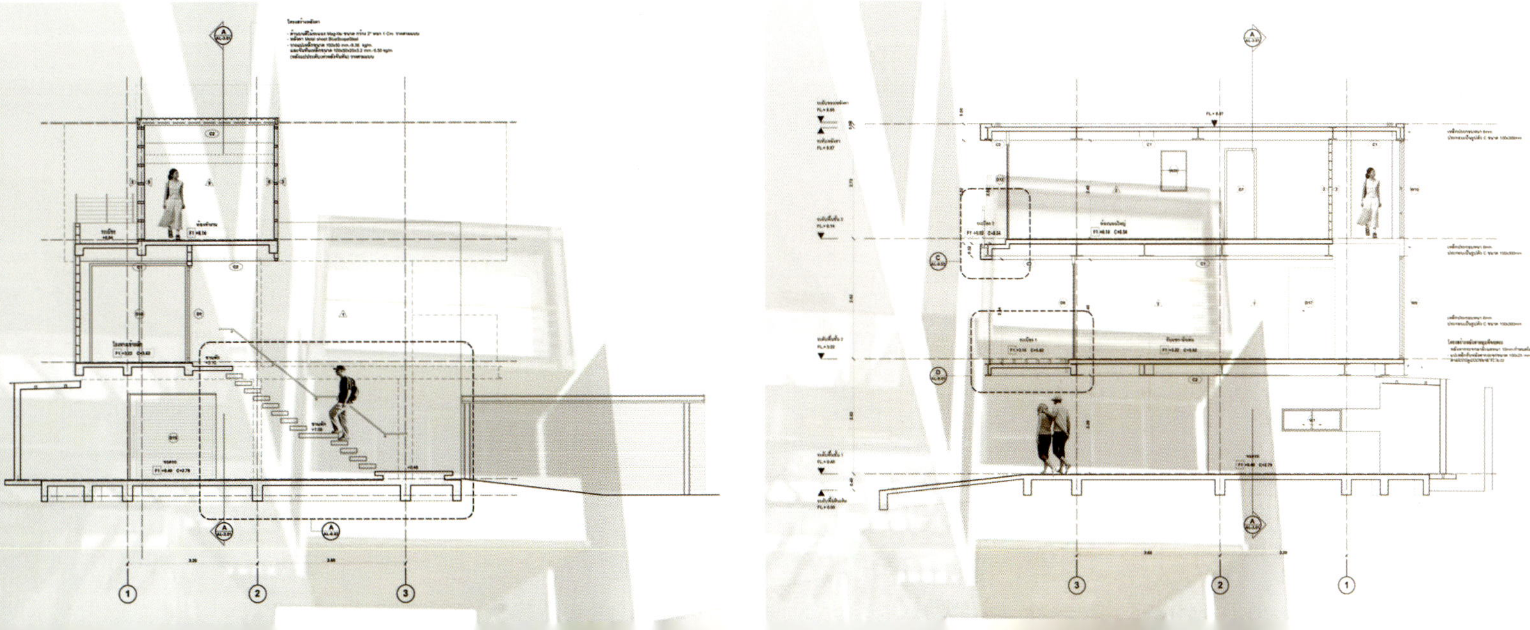

Designer: Ming Tang, LEED AP

Folded Bamboo + Paper House

"竹制和纸制折叠住宅"

After a magnitude 7.9 earthquake struck central China last May, killing 69,000 people, injuring hundreds of thousands and leaving millions homeless, the government is planning an extensive reconstruction project that includes building more than 1.5 million temporary homes, which are expected to last two or three years. Hundreds of Chinese factories that make tents and temporary buildings are maximizing their production for the sake of the earthquake survivors, many of whom lost everything.

The central feature of our project is the development of a temporary shelter for the homeless people, a kinetic structure that exhibits characteristics of umbrella and folded fans, with the potential of arranging themselves into various contexts and dwelling requirements. We named it as Bamboo + paper House, a self reconstructive structure for instant installations, which, according to the changing internal requirements and site topography, can produce potentially infinite scenarios.

BAMBOO HOUSE

As the fastest growing plant, Bamboo reaches full height in one growth spurt of about two months. When bamboo is harvested, the root system is unharmed and healthy, ready to produce more shoots, just like a grass lawn. Thick bamboo poles are 2-3 times stronger than comparable size of wood timber. Bamboo can be harvested in 7 years versus 10-50 years for softwoods and hardwoods, yielding up to 20 times more than wood. It produces greater biomass and 30% more oxygen than a hardwood forest on the same area, while improving watersheds, preventing erosion, restoring soil, providing sweet edible shoots and removing toxins from contaminated soil.

Rather than using the industry mass production to generate uniform dwellings, the Folded Bamboo House uses a simple kinetic structure made by bamboo, a kind of bottom-up assembling of complex adaptive systems that self-regulate, in opposition to top-down overarching principles. The straight bamboo poles are used to create ruled surfaces-helicoid, hyperbolic paraboloid and hyperboloid of revolution. The result of Folded Bamboo House is a reflection of the logic of fold versus unfolded, self-construction versus de-construction, permanent structure versus mobility. The characteristic of the Folded Bamboo House heavily relies on the umbrella structure's open angle and the spatial relations of each rib. The Folded House is transported to site and modified by the social, economic and culture requirements of the user. With this user customized system, the house can react to external stimuli and be transformed with a short time responding to the light, wind or temperature change.

FOLDED PAPER HOUSE

With over 50% of the average landfill mass comprised of paper, and with the growing need to decrease landfill input for economic reasons, the ability to turn this paper into usable, stable, strong building products is a great advantage. Paper is recycled content (post-consumer and pre-consumer). Use of savaged paper reduced

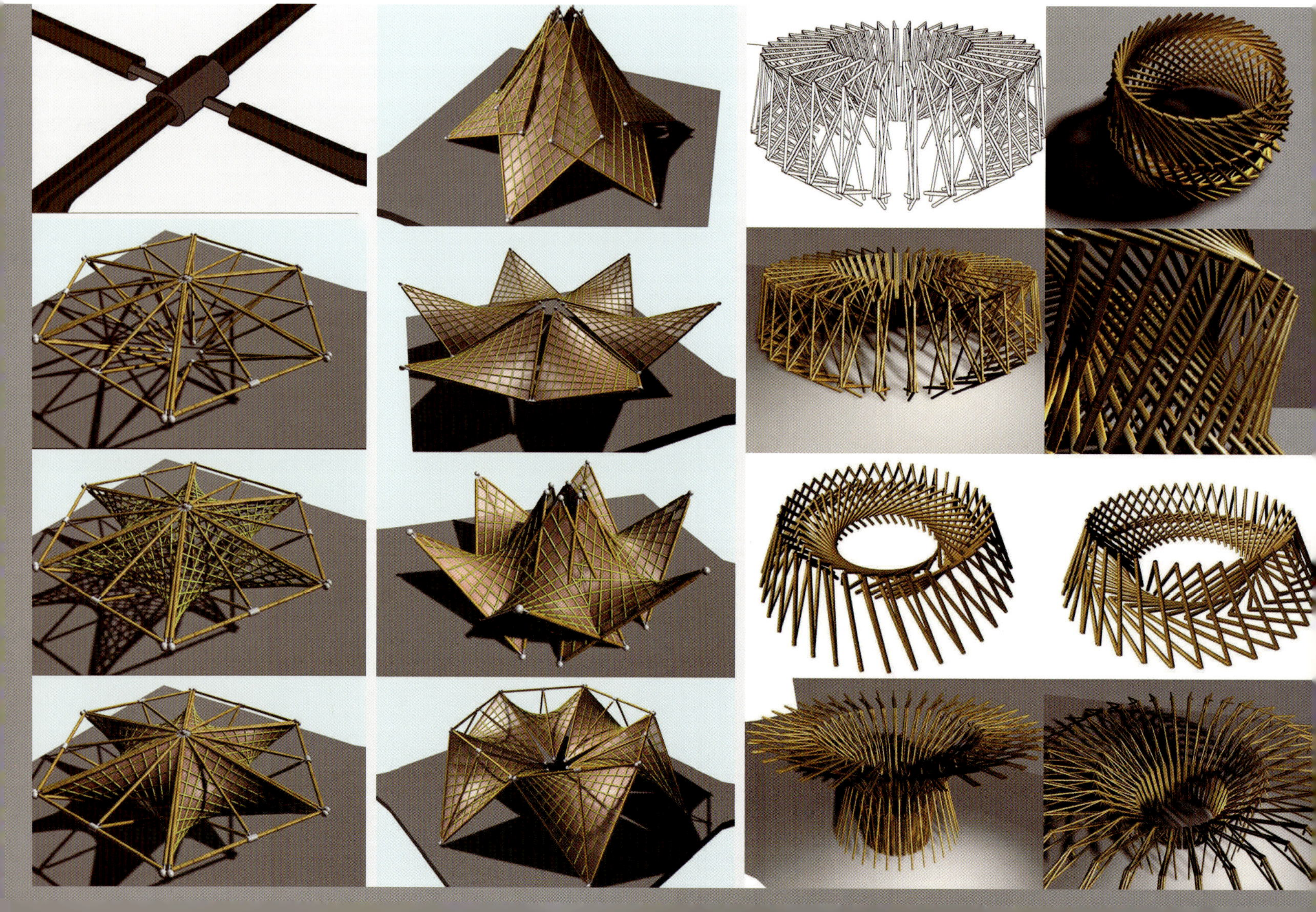

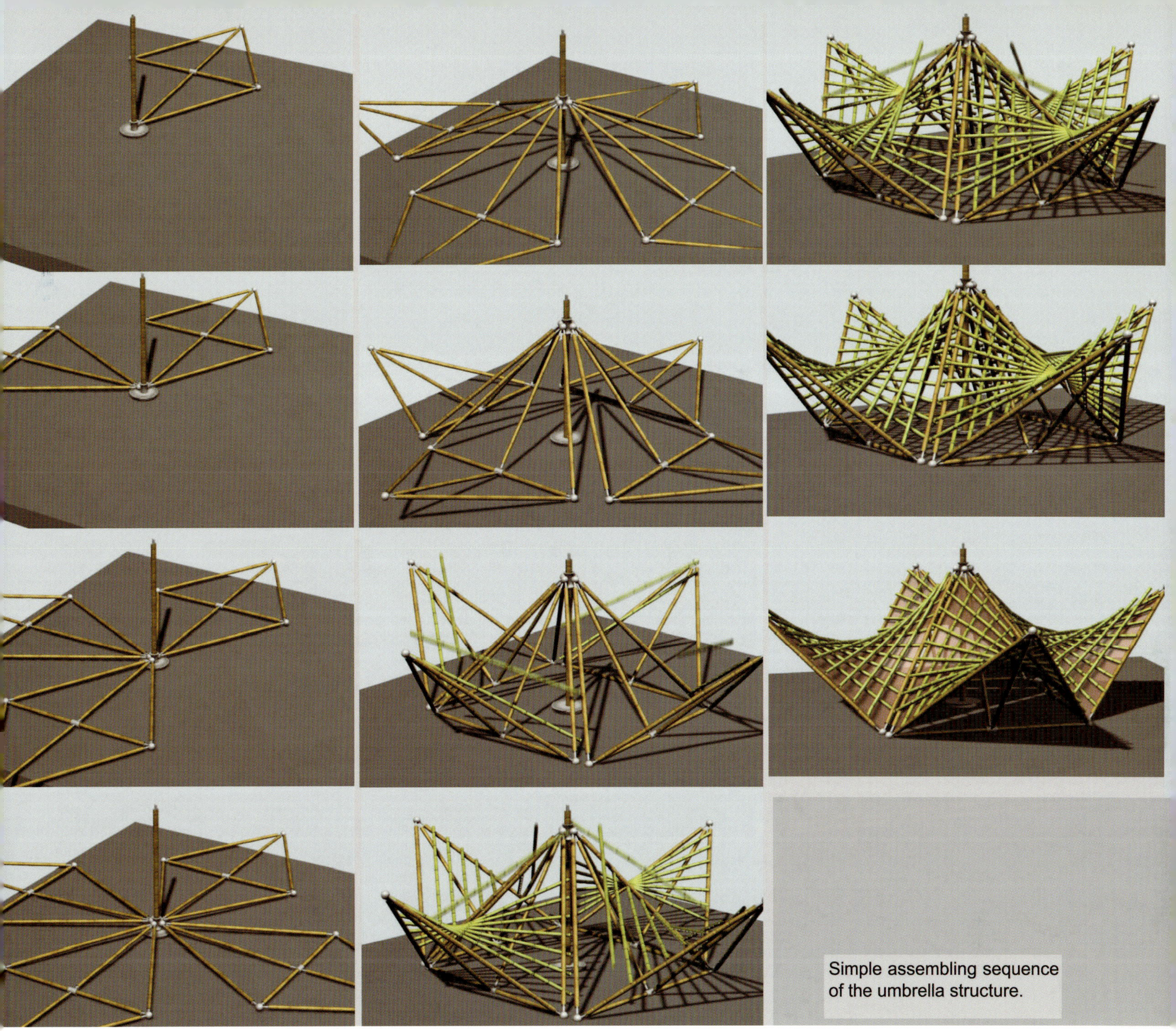

Simple assembling sequence of the umbrella structure.

the environmental impacts and energy cost of producing new construction products and material. There impacts are significant since buildings account for a large portion of our natural resources consumption, including 40% raw stone, gravel and sand, and 25% of virgin wood.

The composed of paper fibers, water, and cement can be used for a variety of construction applications. After draining the majority of the water, the slurry can be molded into tubes to be used as a construction material. The light weight paper house can be pre-assembled in the factory, folded into a small package, loaded into a truck for transportation.

By changing the angle of kinetic structure, a folded paper house can deform into unlimited forms. Thus, these process-driven methods produce the living units, community center, restaurant, and multiple use shops. Beyond these functions, spatio-temoporal organization also produces dynamic building profile, various compositions to represent diverse culture experiences and historical references.

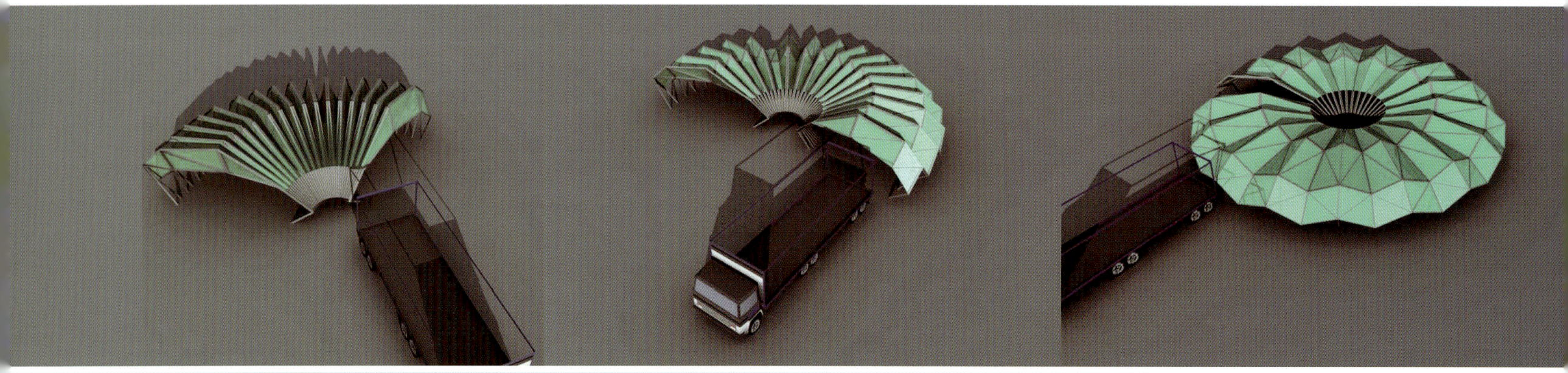

Office Name: Daniel Ventura Arquitecto

Grafishow Printing Works

Grafishow印刷厂房

Constructed in the city of Buenos Aires, on a lot measuring 8.66 metres wide and 23 metres deep, this building, a printing works covering 220 square metres, consolidates the edge of the city with a metal wall raising the height of the ground floor and first floor.
Use of resources which are very basic, in terms of both structures and construction techniques, allows the architect to employ a vocabulary that might be defined as "expressively austere".
The ground floor, a uniform, windowless white structure that seems to be intentionally limited to the function of base, barely reveals the location of the entrance door and the loading and unloading bay for printing material.
But the first floor opens up toward the sky, letting daylight in through a series of metal fins which dematerialise the building's mass to create an elaborate counterpoint of solid and empty spaces.
From the outside, the transparency of the hermetic crystal wall reveals the different degrees of depth in the building; while from the inside, a tree provides shade from the sunlight, its perennial green mass standing between the metal fins and the sky.
As the project was intended to be an elaboration on an architectural structure, space is expressed by the naked vocabulary of this structure and the elements composing it, which, playing with light, express their strong artistic value while at the same time interpreting the building's various functional requirements.
The deep volume of the printing works is illuminated by four skylights which evenly light up the whole surface, and a system of glass walls ensures the space is perceived as continuous right down to the back, where the bright light of the entrance hall dominates the void.

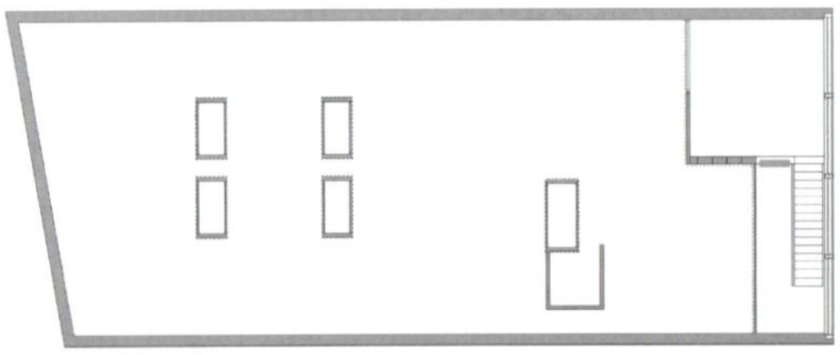

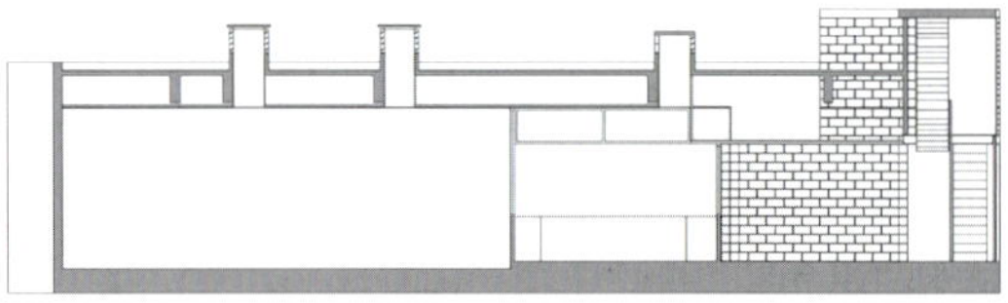

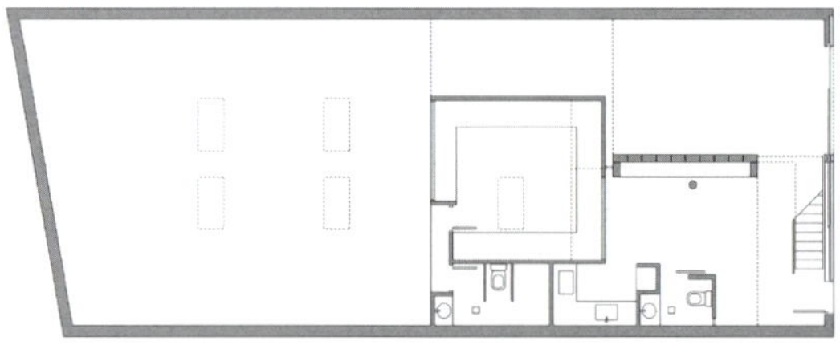

立面
Façade

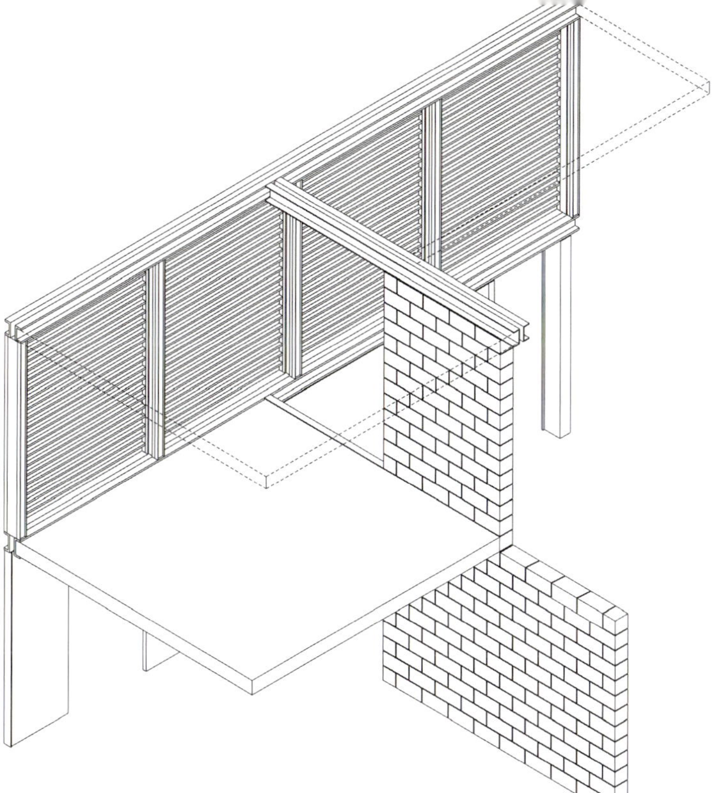

轴测图
Axonometric Drawing

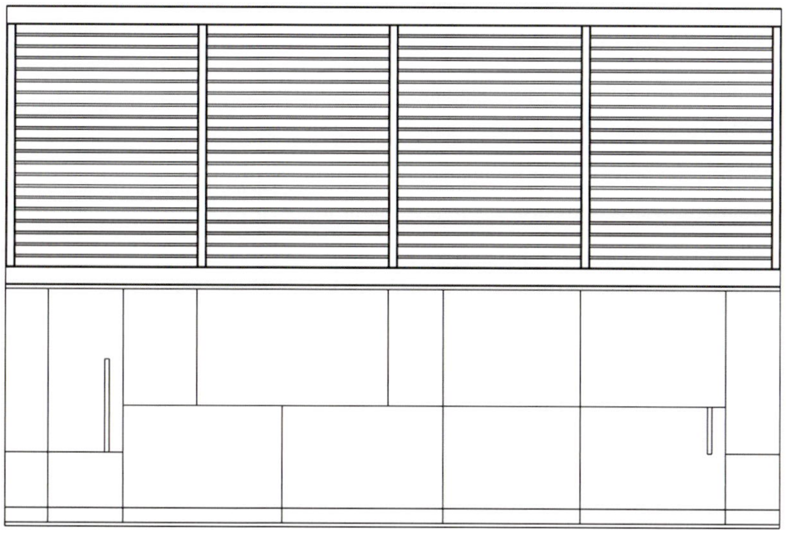

入口空间
Interior Space From Access

室内空间
Interior Space

室内空间，灯光细部
Interior Space, Light Detail

室内空间
Interior Space

Chief Architect: Kyuseon Hong

Helical Stair Gallery

螺旋楼梯美术馆

Helical stair, which have been invented by anonymous architect and used over hundreds years to create elegant spatial experiences, could be mentioned as one of the most beautiful self-supporting structure systems. This project is inspired by those structural merits of spiral stair in terms of construction efficiency by modularity with beautiful spatial quality we can obtain from. Moreover, the ultimate goal of this proposal is to extend its role from the spatial component to space itself. The structural center of traditional spiral stair is located in the center of rotational axis of structure; however, the unique characteristic of this proposal is started from the offset of 2 different center points of components' stacking to create variations and make it as one complete structure with spatial quality not as a device for vertical transportation. In terms of assembling process, more than the half of whole components are simply repeating without dimension changes as several prototypes and the others are changed from those prototypes with several dimensional changes. As a spatial experience, dominant occupiable floor keep continuing to the rooftop surface without any spatial disconnections with spiral stair steps, so occupants can have experiences inside & outside of structure simultaneously and have views over Basingstoke Canal.

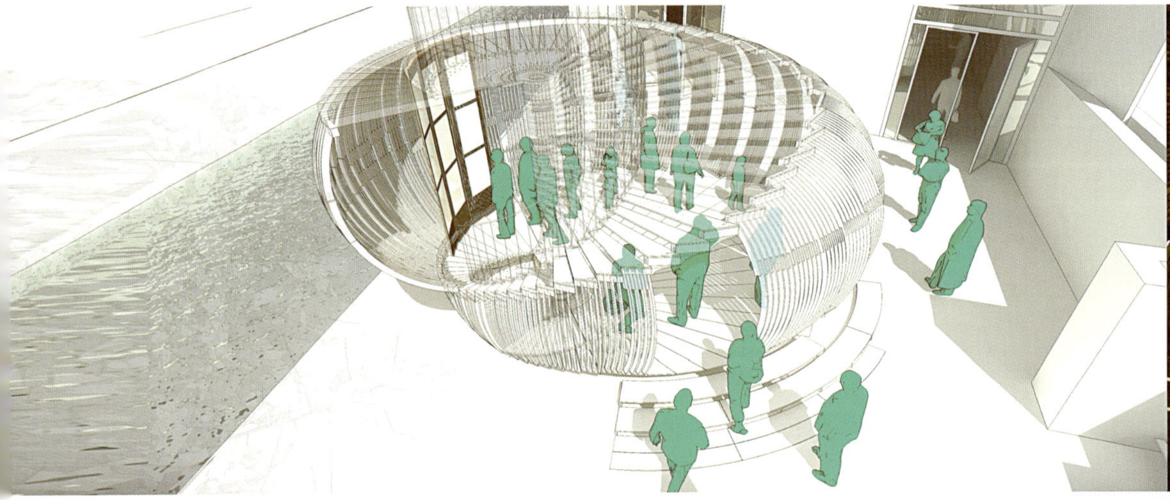

1
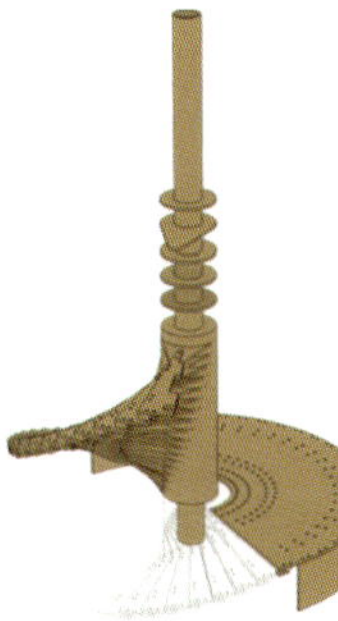
2
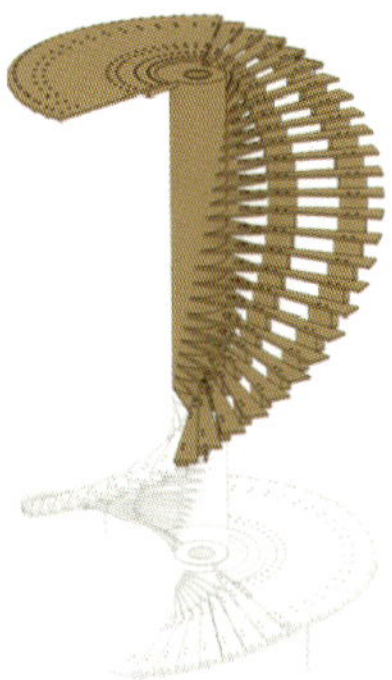
3

4
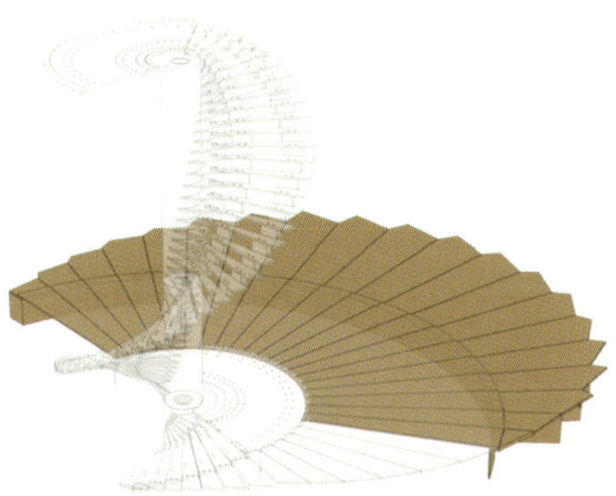
5

6

7
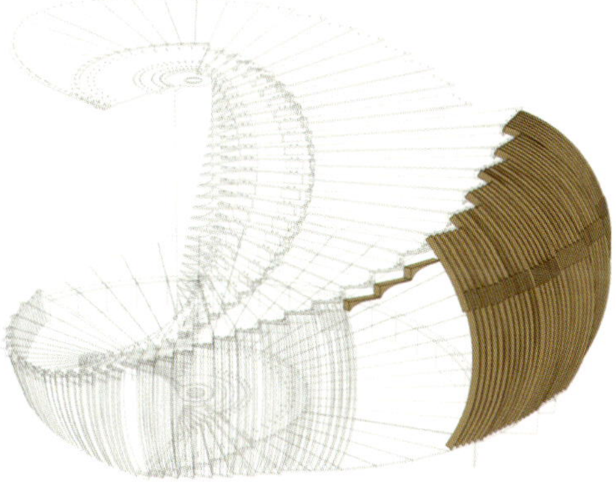
8
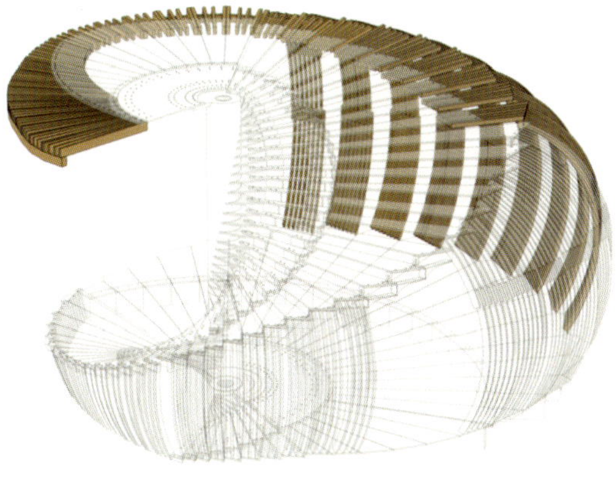
9
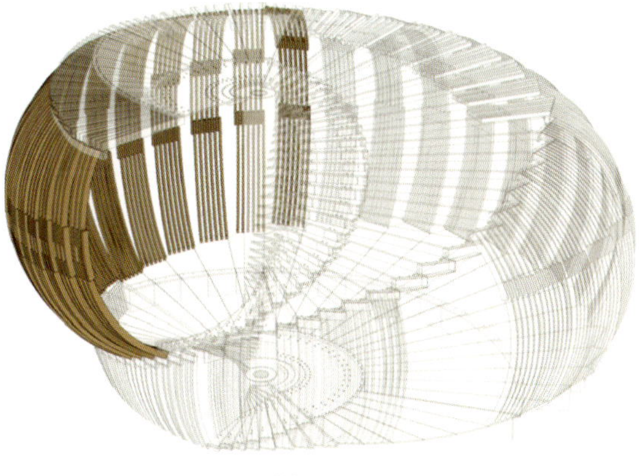
10

Plan <1:50>

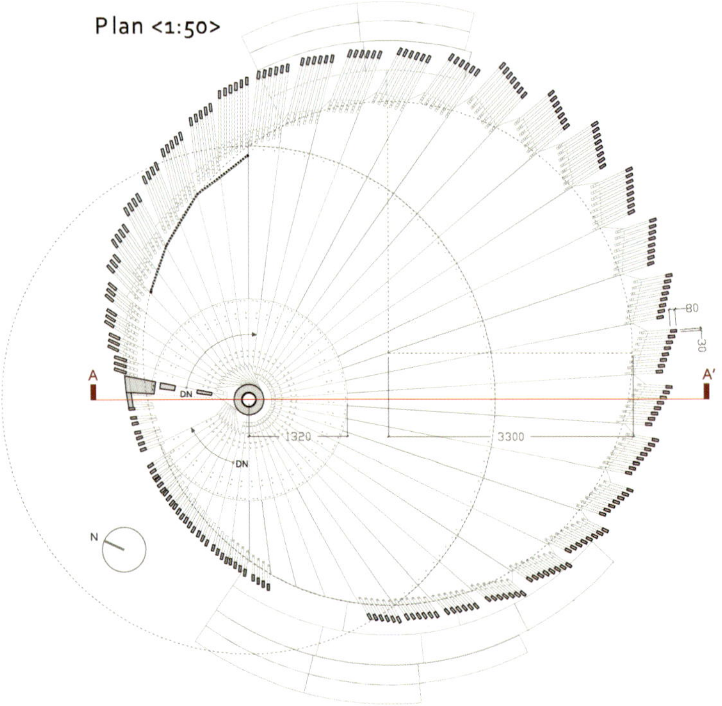

Modular components

- Flat floor
- Flat floor_center structure
- Stair steps_center structure
- Column segment 1_center structure
- Column segment 2_center structure
- Extended steps_roof structure
- Curved wood frame 1_facade

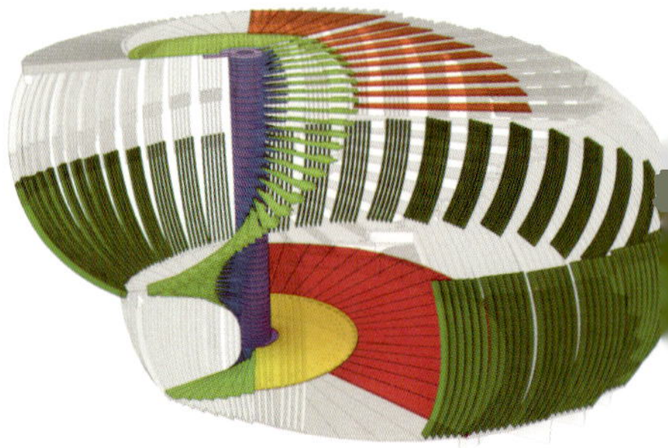

Semi-Modular components

- Stair steps_entrance
- Extended flat floor_floor + roof
- Extended flat floor 2_floor
- Extended steps_roof
- Extended steps 2_roof
- Curved wood frame 2,3_facade

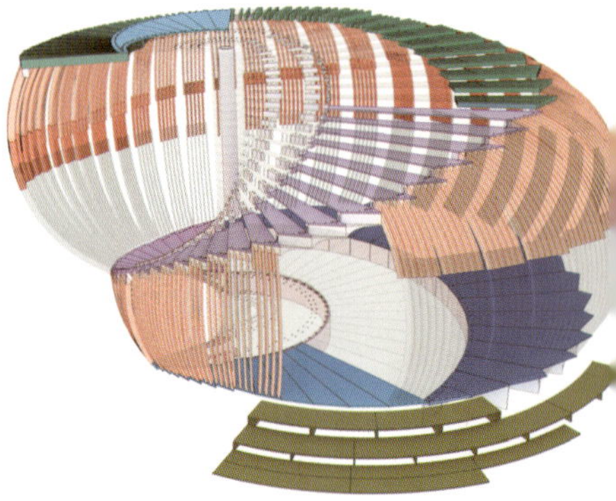

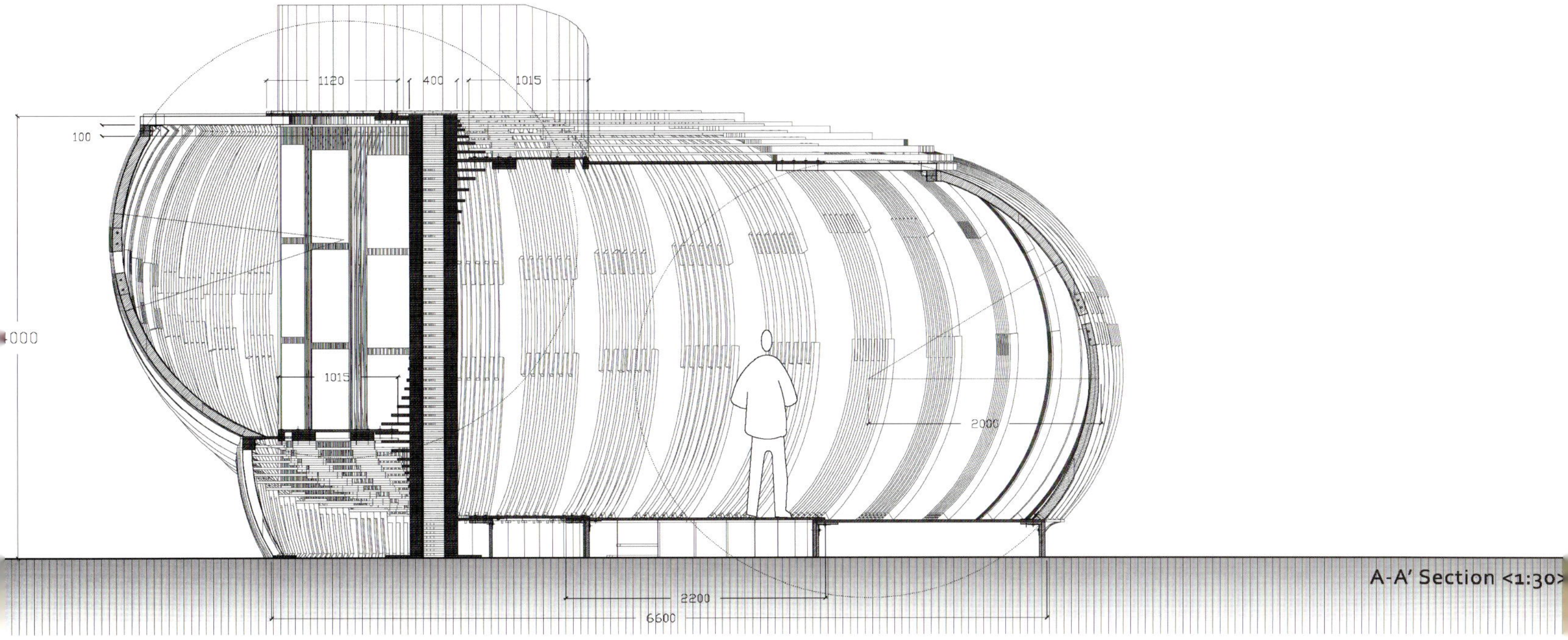
1120
400
1015
100
4000
1015
2000
2200
6600
A-A' Section <1:30>

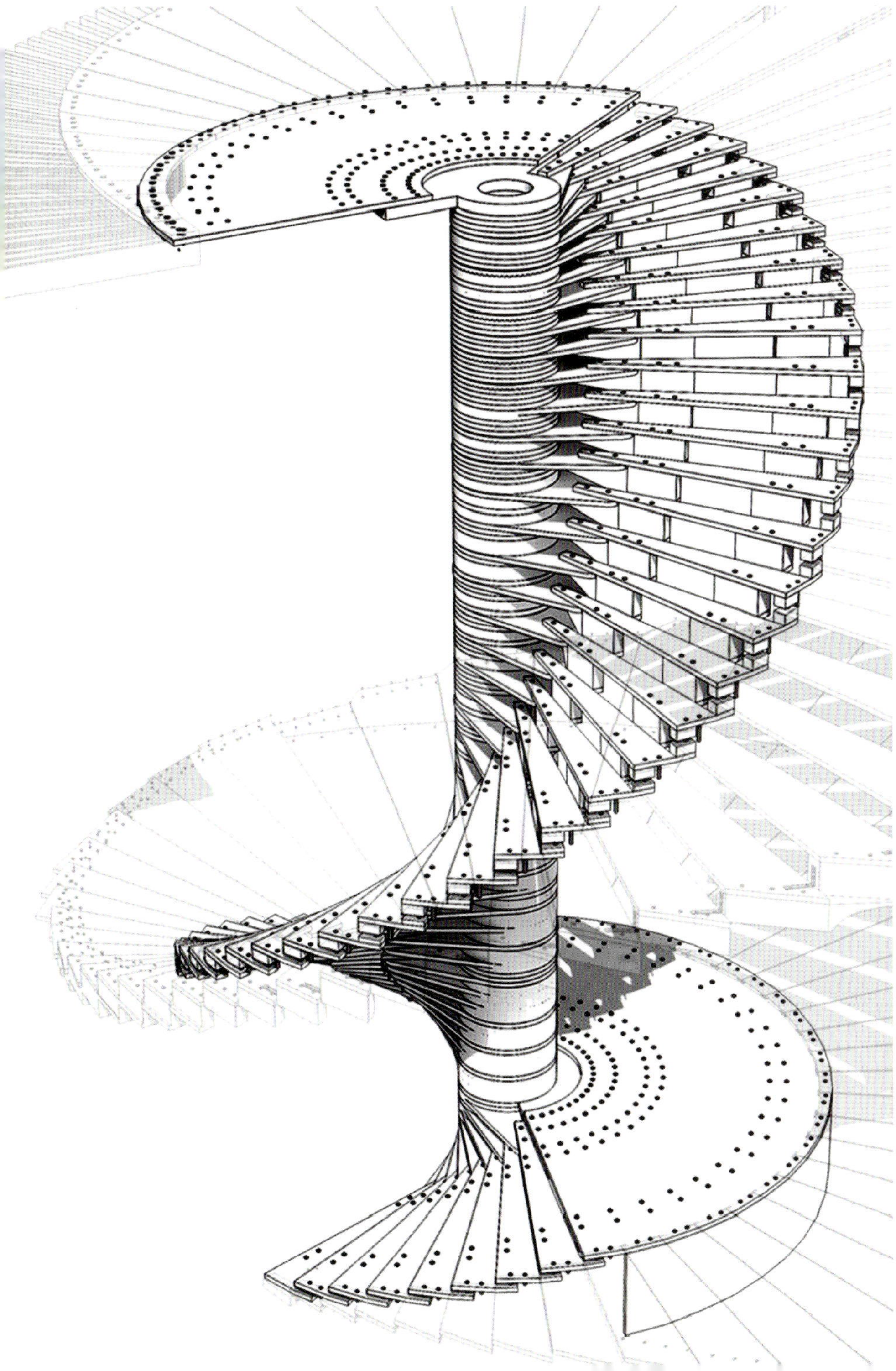

Promoter: Construction and Planning Agency

Hortus Officinalis

疾病控制中心

We link the idea of Center for Diseases Control (CDC) to the idea of Hortus Officinalis, an ancient medical research place in which the activities are strictly related to the studies and the cultivation of healing herbs.
In our interpretation, the CDC is a contemporary Hortus where the laboratories become the contemporary cultivations.
As a consequence, the general plan of the project is based on the idea of hortus conclusus in which various buildings with different role, hight and dimension are placed in the intervention area in a way to generate a central wide green court.
The court is a pedestrian area: it is separated from surrounding flows and it is placed at the lower level (0.00) than the public circulation one that coincides with the level (+ 7.00) of the scheduled central park of the Hsinchu Biomedical Science Area.
The buildings that define the space of the inner court are divided in two groups:
- The first group is composed by 3 blocks (7 meters high) spreading all over the intervention area in a way to create a wide countinous green basement.
These 3 blocks have a double role: on the one hand they define and circumscribe the area of the court (where the circulation is reserved for CDC Staff); on the other hand, they creates an artificial ground that allows a path's continuity between the level of the central park to the south and the level of the area northward;
- the second group include 6 square-shaped blocks with different hight; these blocks are rested upon the artificial ground and characterize the contour line of the court.
The allocation of functional spaces inside the two building groups is related to the feature and specific requirement of each space (Administration Center, Research&Diagnostic Center; public areas, staff areas, office areas etc.).
For this reason, the collective and multifunctional activities (exhibition areas, conference rooms, auditorium, lobbies etc.) are allocated in the basement; whereas the research and working areas (laboratories and offices) are allocated in the square-shaped blocks.
The two building groups are strictly related and create two different parts:
- the wider part located to the south of the area that hosts the Research & Diagnostic Center;
- the other one located to the north that hosts the Administration Center and creates a main front on the street.
Starting from the idea of CDC as Hortus Officinalis, the general features of the project and the materials that characterized it follow and underline environmental and landscaping principles.
As it happens in a Hortus, the natural element is predominant, but it is oriented by artificial rules.
For this reason, the 3 main elements of the project are characterized in a different way:
- The Basement (artificial ground) that defines

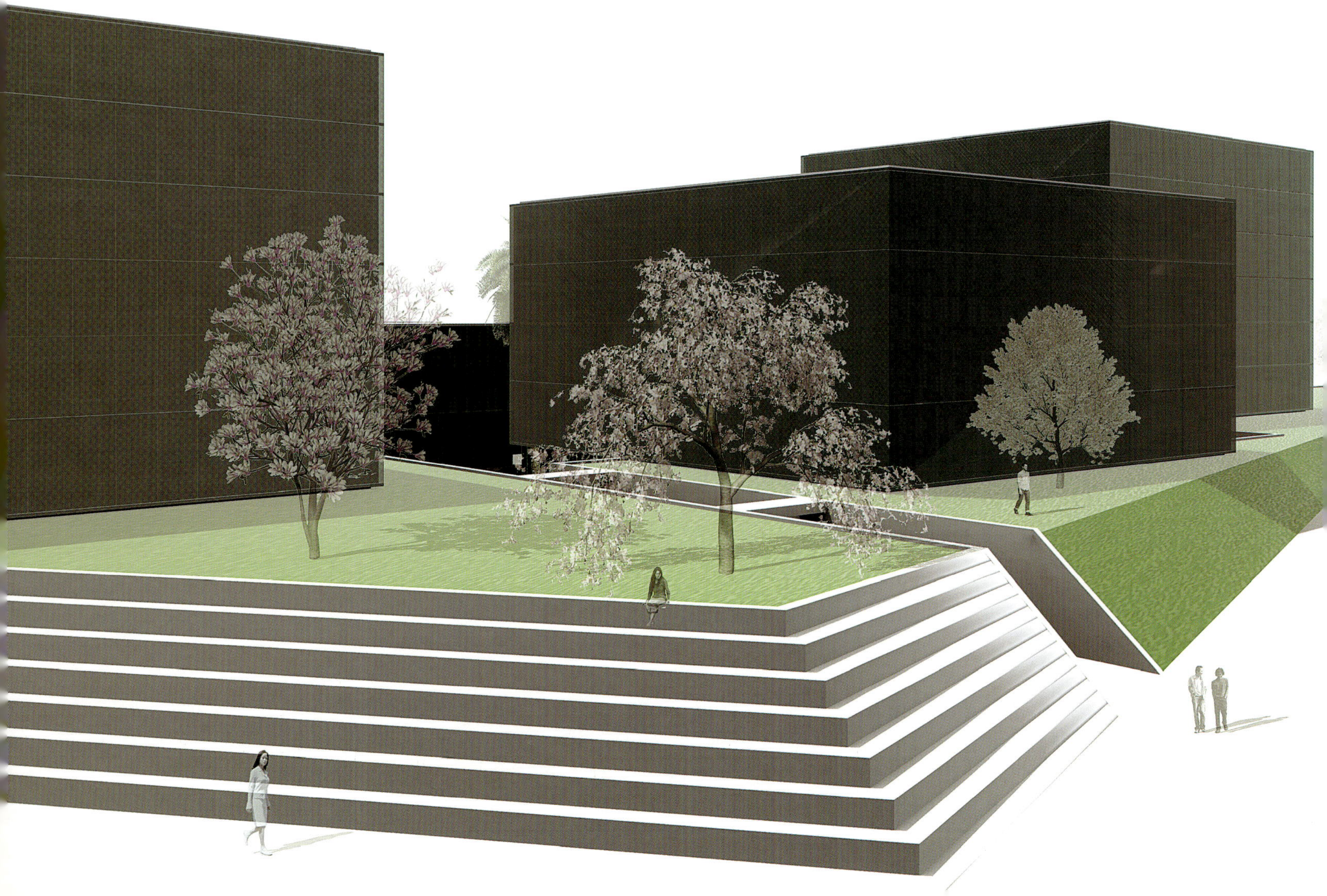

the court has a green continuous roof (with tree-lined areas). There is also a distinction between the part of basement to the south (where there is the Research & Diagnostic Center) and the other one to the north (where there is the Administration Center).

In the first, the roof of basement joins up with the streets through sloping green surfaces.

In the second, there is a sequence of tiers that connect the street level to the green elevated plaza;

- The Court is characterized by green areas and body of water areas. The other parts have a permeable paving to allow planting vegetation;
- The Blocks are double-skin covered: a metallic grid covered by climber plants is overlapped to the external wall so that from the outside the blocks seem to be made of vegetation.

Vegetation has also a bioclimatic role for controlling the quality and the level of temperature of internal spaces.

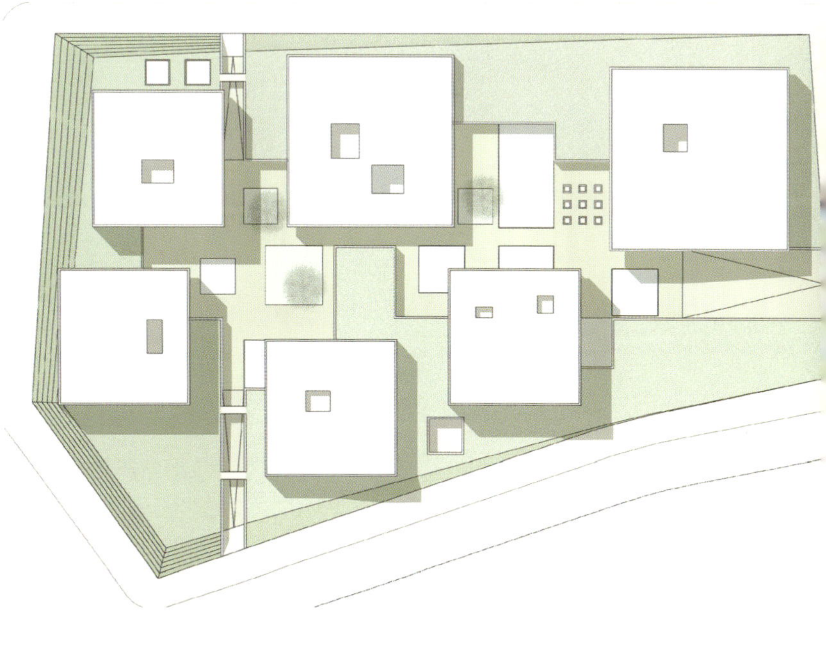

Hoshikusa
壘草

It is located at the corner of the street, where a prostitute is said to live long years ago, in an old spring hot resort "Arima". Forty five years have passed since this renovated common house was built after the Second World War. It features tiled roofs, mortar outside walls, bases of clay walls, plaster walls with cotton-wool fiber and glass sliding screens with wooden frame. Big and old maple trees in the neighbors are growing at the approach. At night stars shine over the trees.
Our client provides it as a personal guest house for his many friends in other countries. We would like the guests to relax in the atmosphere with history and nature in this place to keep a structure of the former Japanese residence. We left wooden poles and removed ceiling boards to show the existing structure such as wall bases in the ceiling without drastic change of shape and painted dirty and cracked walls as if they put on light makeup, leaving the feeling of original materials. At last we softly installed nets made of linen with light and shadow generating there on the wooden poles and at the back of paper sliding screens.

南外墙的夜景
South Outer Wall Surface (Evening)

入口通道的台阶
Approach-Terrace

通道的平台夜景
Terrace-Approach (Evening)

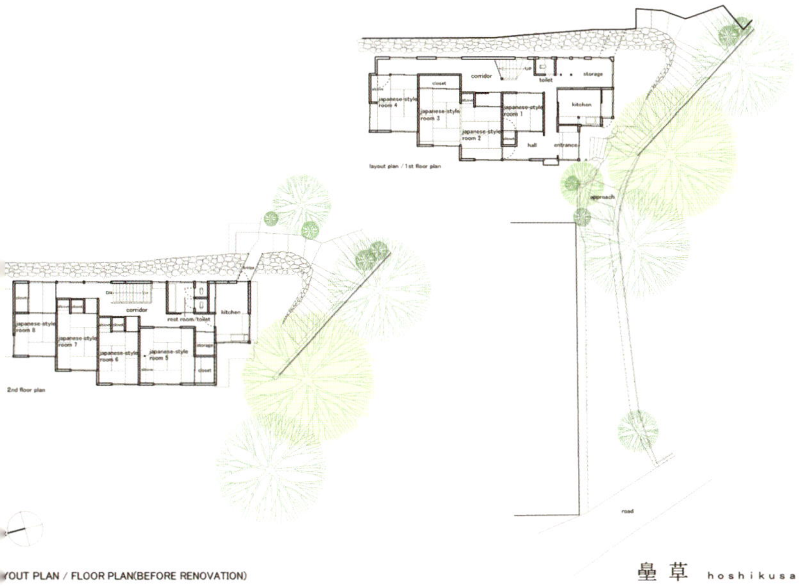

房屋翻新前的平面图
Plan-Before

房屋翻新后的平面图
Plan-After

入口通道
Approach

门廊外的平台
Entrance Hall-Terrace

通往入口通道的平台
Terrace-Approach

二层大厅的入口走道
Corridor-Entrance Hall (2F)

Authors: EQUIP; +HOSPES DESIGN

Hotel Hospes Palacio de los Patos in Granada with Luxuriousness and Beauty

豪华亮丽的格拉纳达Hospes Palacio de los Patos大酒店

The extension to this 19th century palace features 22 exclusive rooms.
The façade is covered by a spectacular alabaster curtain wall.
Barcelona, July 2007 - Barcelona-based company EQUIP Xavier Claramunt designed the recent extension to the Hotel Hospes Palacio de los Patos in Granada. Built over 18 months, the new building means the hotel now features 22 new, exclusive rooms in addition to the 20 already in existence. The recently completed, 7,100 square-metre annex is located next to the 19th-century palace and fascinating due to the impressive alabaster wall covering its façade. EQUIP Xavier Claramunt created a visual strategy for this project with the design offering a play on perspective. "Instead of merely extending the Palacio de los Patos, a series of lattice works have been erected to achieve densities designed to segregate different uses and activities. Instead of building a solid mass, a set of curtains, hangings and lattice works were raised through which the palace can be coyly observed from the new areas. On the one side there is the historic building, playing a leading role once more, and on the other a superposition of lattice works, planes and densities straining to look at the palace across a garden merged into the same lattice work which extends horizontally around the building, delineating the garden and its shadows."
Situated in the heart of Granada's commercial district, the Hotel Hospes Palacio de los Patos comprises three very different areas: the restaurant, the spa baths and the rooms. The restaurant is conceived as an extension to the adjacent garden and vice versa. The spa area consists of several spa rooms lit by natural light from above. Located on the upper floors, the rooms all overlook the palace in its capacity as the central, revered object in this project.

客房
Guest Room

客房
Guest Room

客房
Guest Room

二层走廊
Corridor (2F)

一层大厅
Hall (1F)

计算机辅助设计的鸟瞰图
CAD Model Aerial

户外休息走廊
Outdoor Sleeping Verandah

屋顶近景
Roof Detail

户外就餐场地
Outdoor Eating Space

接待室
Reception Office

挡风墙
Windbreak Walls

从大楼办公室眺望
View Across the Complex From the Office

户外住宅空间光线的过滤
Filtered Light to Outdoor Living Spaces

从户外住宅空间眺望
View Across Outdoor Living Spaces

Project Team: Brendan J Meney Architects in Association with Richard Layton & Associates

Ilpurla Substance Misuse Rehabilitation Centre

Ilpurla药物滥用康复中心

The Ilpurla Substance Misuse centre was established to provide rehabilitation through care and education to young aboriginals who are caught up in the vicious cycle of alcohol, drugs and petrol stiffing.

The building provides a safe and comfortable environment for the control, supervision, education and general health improvement of all the clients within the greater Ilpurla outstation community through dormitory style accommodation with kitchen, dining, twin share rooms, education rooms, recreation rooms and offices for the cultural focused administration of the program.

The centre at this remote aboriginal community was established over 15 years ago by Barry Abbott and his family with minimal resourcing to assist young aboriginal people to kick their debilitating and life destroying habits that had become entrenched in their everyday lives and was destroying the social fabrics of remote aboriginal communities across the region. They continue their dedication to reform through a number of successful programs that focus on removing the youth from the causes and instilling life skills and cultural self respect and self determination back into their lives.

The design of the building supports this approach through the provision of a series of semi - private living spaces for the young men and women which they become responsible for during their stay, together with the interactive communal spaces for the provision of both formal and informal education and general every day living activities.

The large roof canopy, natural ventilation system to the roof space and the thermal mass of the walls assist in stabilizing the internal temperatures which are not reliant on mechanical air conditioning. The building is powered from a remote solar hybrid power station and has efficient systems which take into account the limitations on supply. The hot water is from solar glycol heat exchange and the sewer is via septic tank black water and reticulated grey water sub surface systems.

The project was undertaken in association Richard Layton & Associates Project Management and constructed by the team from Probuild (NT).

社区里的大楼背景

Building Context Within the Community

公共空间
Common Space

儿童空间
Child Space

公共空间
Common Space

卧室与扶梯
Living Room with Staircase

透明的楼梯
Transparent Stairs

茶室周围的景致
View From Tea Room

Chief Architect/Office Name: Teruo Miyahara / Miyahara Architect Office

House TTN – Residence for an "Urban" Extended Family

TTN住宅——城市"大家庭"的新式住宅

House TTN was designed to accommodate three families – the parents and the families of their two daughters. They had decided to live together again with the birth of grandchildren. Thus, the main objective for House TTN is to provide the necessary functions for an "urban" extended family, accommodating the needs of modern nuclear families who have grown accustomed to independent life but have chosen to enjoy the benefits of being part of a large family.

The first request for this project was to have a sort of collective residence to accommodate three homes, a plan which would completely separate the families within the same building. However, after much thought on how to maximize convenience, the effective and rational use of the site, and the pleasure of each other's company, House TTN decided to take a semi-independent, sharing approach.

In order to have more than one nuclear family live together as one, it is essential to secure a comfortable distance within the design. Thus, each family has their own independent kitchen unit, bathroom, and toilet, but the homes are adjoined through the ground floor area and common deck – inside and out. The parents' living space is located on the ground floor, with a highly independent main room (that is also shared by all three families) as well as private rooms (one Japanese-style room and one bedroom) opening towards the outside. The first and second floors are divided east and west, creating living spaces for each daughter's family. Outdoor common decks in between the two sides of each floor serve as both converging points and buffer space. Transparent glass and sudare or Japanese wooden blinds are used on the common decks to separate the families but at the same time avoid complete privacy. It is possible for each family to go about their business independently, but these purposefully built common areas make it possible to achieve a higher quality of life. A comfortable distance is achieved by softly compelling the families to come together.

Another important aspect of House TTN was its structure. As the decision had been taken not to separate the homes completely, the residents wished to retain an option that would enable them to cut the building in half, left and right, in case they wished to do so in the future. In order to make this possible, the two sides of the structure including the foundation are completely independent of each other, and designed to guarantee durability after being divided. Of course, if two new separate buildings were to emerge, they would both need to pass the various building regulations. Therefore, this aspect greatly influenced the initial plan and form of House TTN. However, it may also be said that because of this requirement, it was possible to achieve a bold design, shaping the areas that would be removed if the house were to be divided into outdoor common decks. It is unclear whether this option will be taken in the future, but having an alternative will surely encourage friendly and active communication between the families.

夜景
Night View

Project Team: SuMi Jung

House on Floating Land

漂浮大地 漂浮住宅

THE HOUSE IN URBAN NATURE

This site is located in borderline between city and mountain. The view of the front side that is the east-side is clearing up, so the urban landscape of downtown Gwangju is seen. There are mountains by the rear-side as a nature and view.

DRAMATIC ACCESS WAY FROM GATE TO ENTRANCE

By opening transparent gate door, it starts to begin access into the house, along the long walkway where looks like architectural canyon. After the penetration through under space of the upper bridge-corridor, it comes to reach the inner court that is attached to entrance.

TWO BOXES ON FLOATING LAND

Two wooden boxes which are consisted of duplex room are laid on landscaped architectural mass which contain living room and dining room. The one box contains master bed and study room and another box is consisted of two bedrooms and study rooms for two daughters.

FLOATING GARDEN OVER THE LIVING ROOM - SPECIAL SPACE AND LIGHT OF "BOX INSIDE BOX"

Over the living room where has dynamic space in 9M high, double height master bed-mass with bamboo garden is floating...Through the top light of the roof, moving sunlight vary the atmosphere of the inside space of living room all day long.

FLOATING ROADS

Horizontal or vertical ways which are indoor or outdoor, all the ways inside of this house circulate itself continuously each other with the dramatic sequence of the various space.

住宅景致图
Carapicuiba House

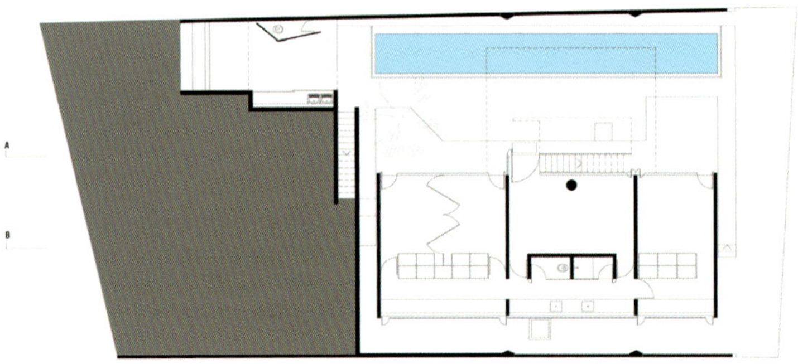

卧室地板
Carapicuiba House Bedroom Floor

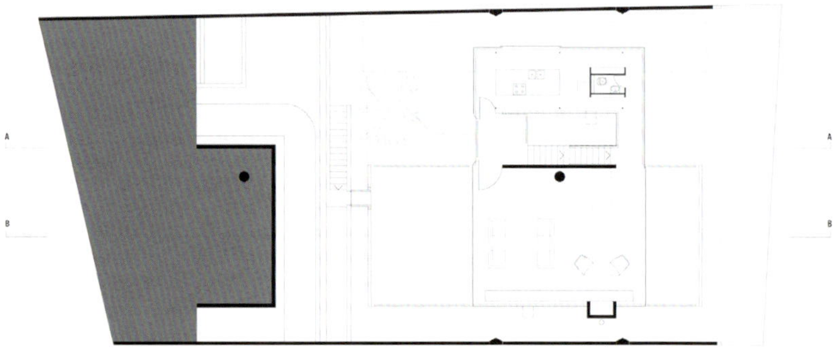

生活区地板
Carapicuiba House Living Floor

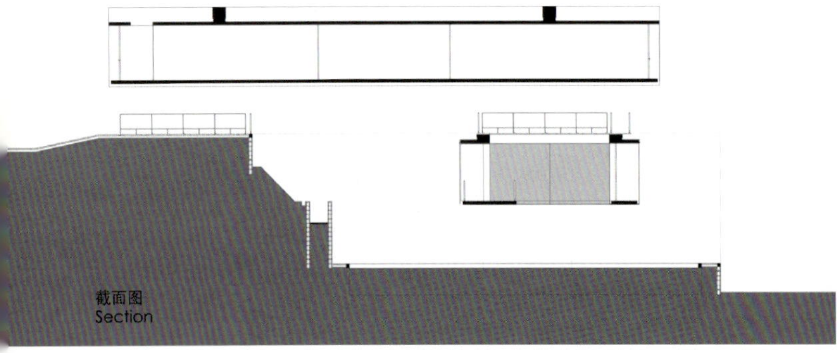

截面图
Section

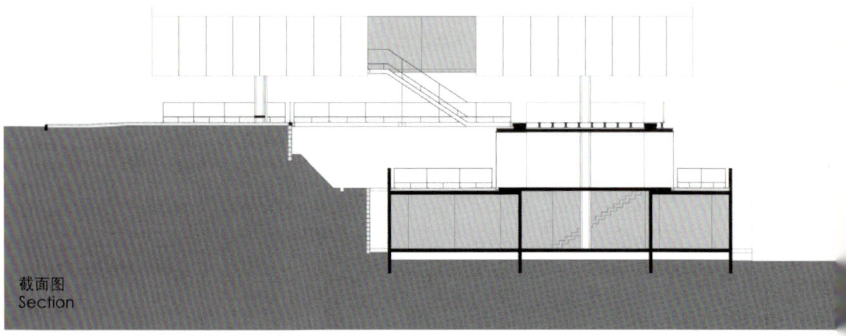

截面图
Section

Project Team: Ciro Miguel, Fernando Bizarri, Juliana Braga

House in Carapicuiba

Carapicuiba别墅

The most remarkable feature of the site where the house is located is its depression, in topographical terms. From the street, we cannot clearly see its ground level because the ground surface falls abruptly into a little valley and woods: a stage 6 m below.

The program joins two different purposes: a house and an office, it is a place to both live and work. Although these two functions share the same space, it is as separate as possible.

The different levels offered by the site were utilized to arrange the two main programs.

The street level was kept free of any enclosed space; it is a kind of "pilotis" with two different areas: the first one is on the ground, very close to the street, and the other one is aerial, as a roof terrace over the building. A bridge, made of steel, connects these two areas. The only entrance to the building is the bridge with its steel grid floor that leads over the open space: downstairs to the house or upstairs to the office.

The house is divided into two levels, both below the street level. Its spaces are integrated with the woods, valley, gardens, and pool located at the ground level. The house incorporates the outside nature indoors: a sliding glass door opens the living room into the terrace, creating one large space. The bedroom and the patios can also be integrated at the lower level.

From the street level, the office is located upstairs. Its dimensions, 3m wide and 25m long, making it look like a tube open at both extremities. Therefore, the windows offer new views: more landscape than patios, and more panoramic than an intimate space.

The "tube" only rests over two columns. Reinforced concrete supports the whole building. Besides its materials, concrete and glass, this house is essentially designed based on the site's geography and landscape. So few elements mean more concentration on the required work during its construction process. It makes it easier to control the budget and it help us focus on the necessary steps to build the house.

住宅景致图
Carapicuiba House

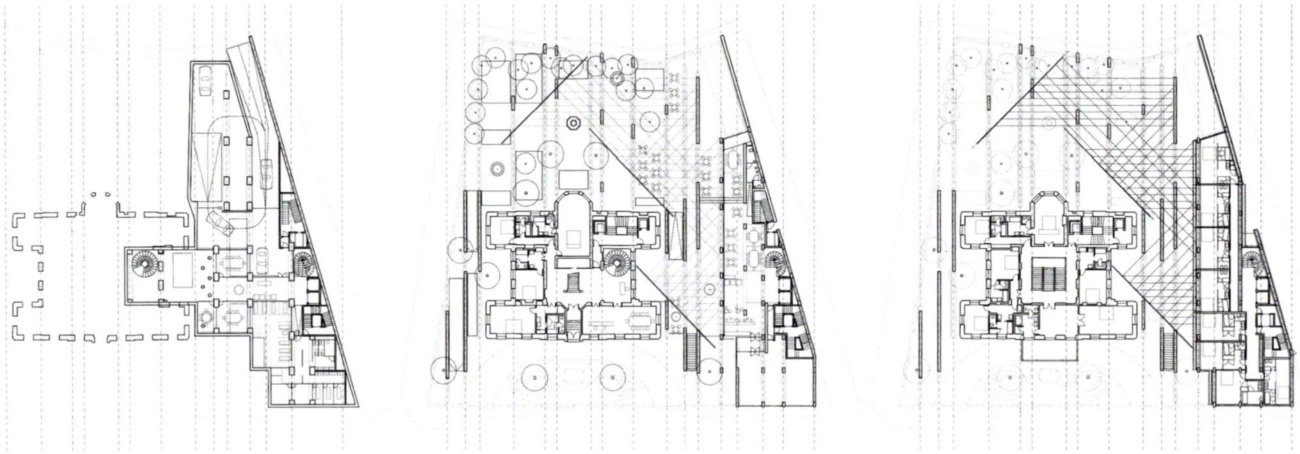

建筑规划
Plans

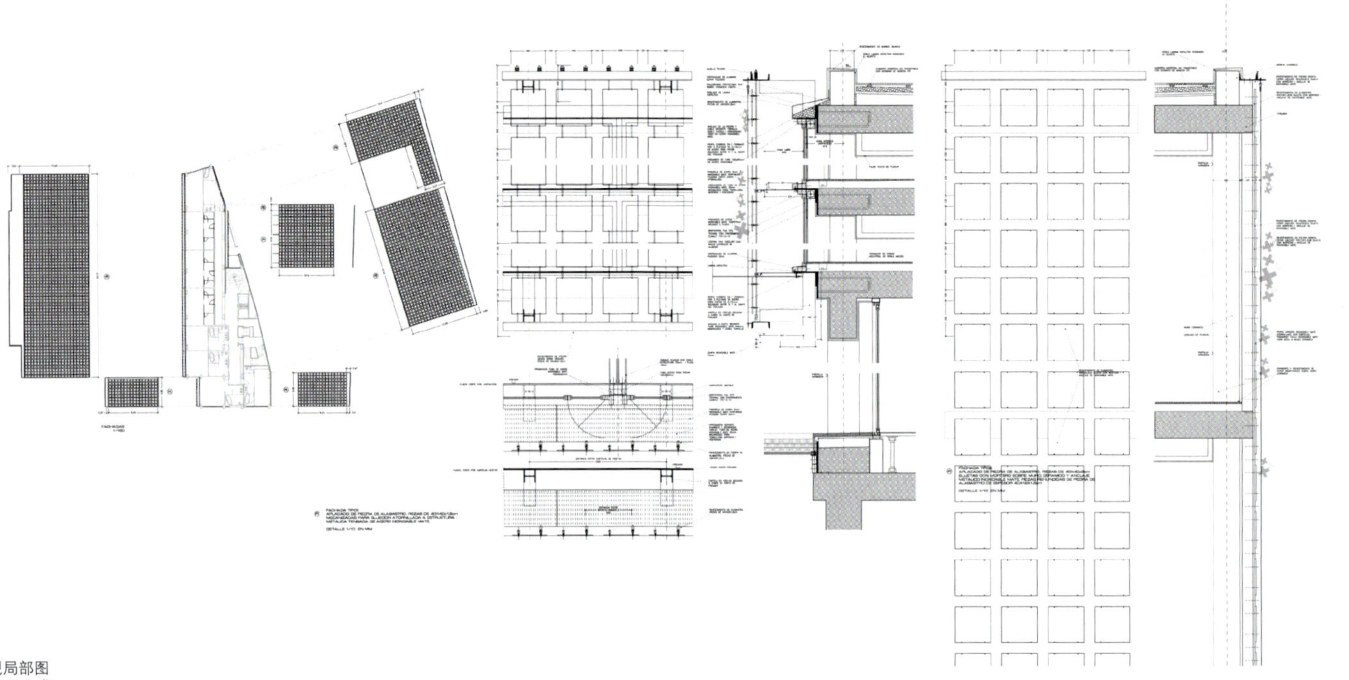

建筑外观局部图
Façade Details

户外公共空间
Communal Outdoor Space

反光的大楼中心
Reflected Light to Building Core

康乐室的壁炉
Recreation Room Fireplace

Project Team: Arch. Lorenzo Bergamini, Ing. Franco Giudici, Geom. Camillo Boccacini.

Innovation within Tradition — Cemetery of Comacchio

传统中的创新——科马乔墓地

The master plan that we are proposing for the existing cemetery aims to convey an idea of "completeness", although only one quadrant will be built presently. While the original design (XIX century) was extremely clear, recent additions made the site somehow chaotic. Our site plan aims to restore the original design clarity by reinforcing the historical south-north axis at the entrance that bisects the church.

Seven independent blocks containing burial niches are placed next to each other at a slight angle, creating an overall sense of enclosure while permitting slices of light to enter. This placement yields shadows that give rythm and measure to the massive enclosure.

Each of the 7 blocks of burial niches is made out of concrete, coloured slightly different matching the colour of the bricks. Our goal was not to replicate the colors and proportions of the existing brick wall, but to create chromatic and proportional harmony with the entire environment.

We've used tall, elegant cypresses - specifically cupressus pyramidalis - so beloved and familiar to our people, to provide both a natural element and to give dimension to the monumentality of the traditional, walled Italian cemetery. From a distance, these sculptural trees define the landscape, form the skyline, and beckon to the living, "here is the city of the dead, here are your ancestors."

A small wood of these cypresses is the heart of our plan. At the center of the wood and the terminus of a long axis is a large cross, "a cross of air" created by the negative space carved from two adjacent walls.

We encourage the individuality of the design of each mausoleum - even if this seems to break the homogeneity of the architecture we propose - and we provide for the privacy of the visitor. Each individual family mausoleum uses diverse materials, colors, shapes and architectural styles. Traditionally, burial niches are positioned perpendicular to a lengthy corridor, providing one very long and public place. Instead, we've broken that line and have created privacy by positioning the chambers around smaller courtyards with partial walls where one can mourn privately. Here, the widow dressed in black can bring her folding chair, sit quietly with her beloved in a serene, small court to talk about the present and the past, to visit and cry in peace.

轴线
Axonometry

"空气的交错"景观
Cross of Air

柏树与门廊
Cypress and Porch

壁龛墙面
Wall Burial Niches

用于回收雨水的石景
Gargoyle Recycling Rain Water

轴线北侧
Axonometric North

十字形景观
Crossing the Cross

墙面上光的投影
Blades of Light in the Wall

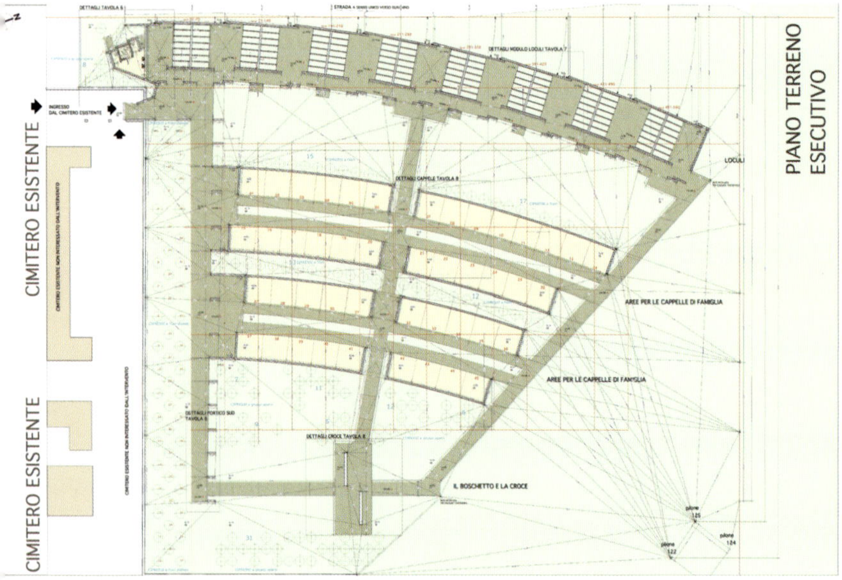

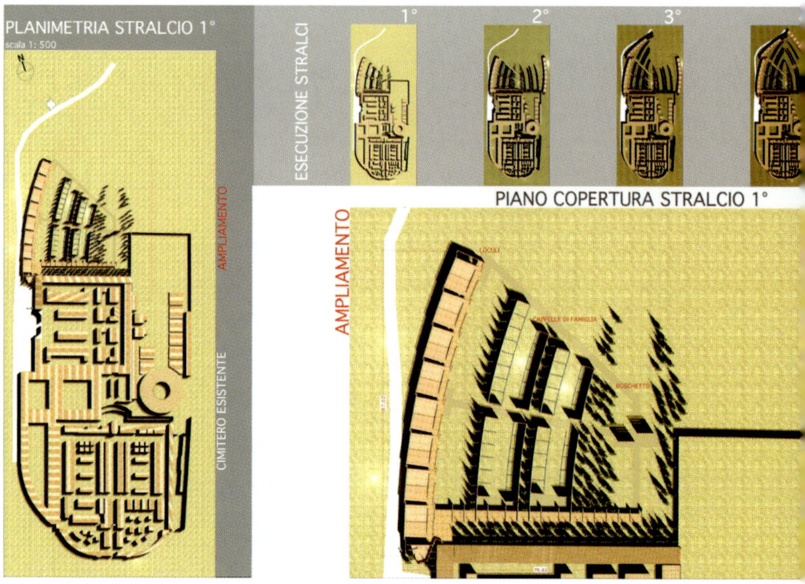

SEZIONI

SEZIONE A-A — SEZIONE A-A

VEDI dettagli TAV. 7 — VEDI dettagli TAV. 8

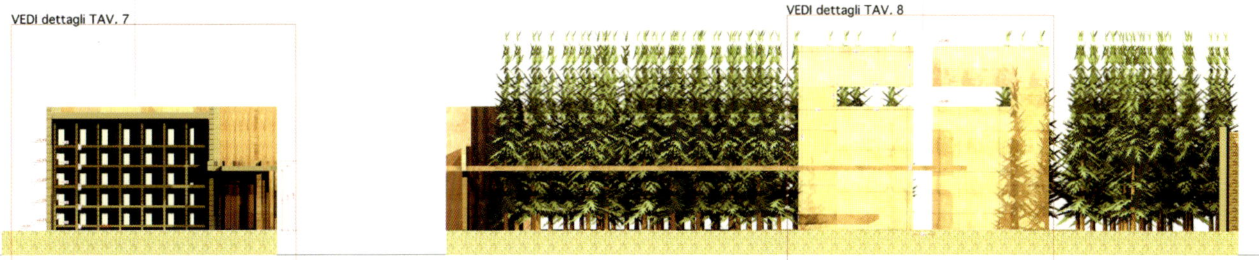

DETTAGLI — DETTAGLIO LOCULI : vedere tavola 7 — DETTAGLIO CROCE. vedere tavola 8 — DETTAGLI

SEZIONE B-B — SEZIONE B-B

VEDI dettagli TAV. 6 — VEDI dettagli TAV. 7

DETTAGLI — DETTAGLIO LOCULI: vedere tavola 7 — DETTAGLIO LOCULI E LAPIDI: vedere tavola 7 — DETTAGLI

colorazioni

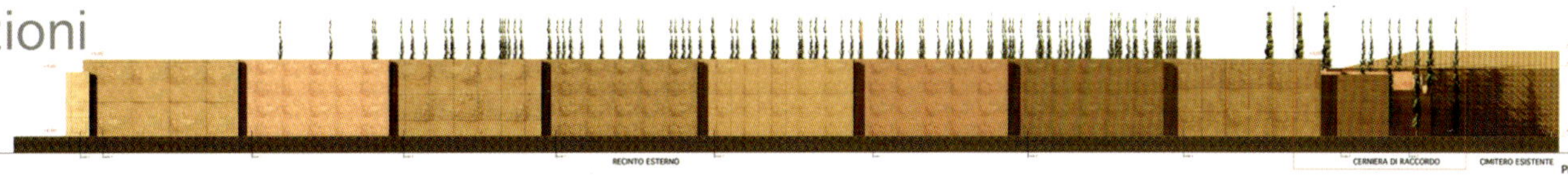

PROSPETTO NORD-OVEST — PROSPETTO NORD-OVEST

PROSPETTO NORD-EST — PROSPETTO NORD-EST

PROSPETTO INTERNO SUD-EST
FRONTE LOCULI — PROSPETTO INTERNO SUD-EST
FRONTE LOCULI

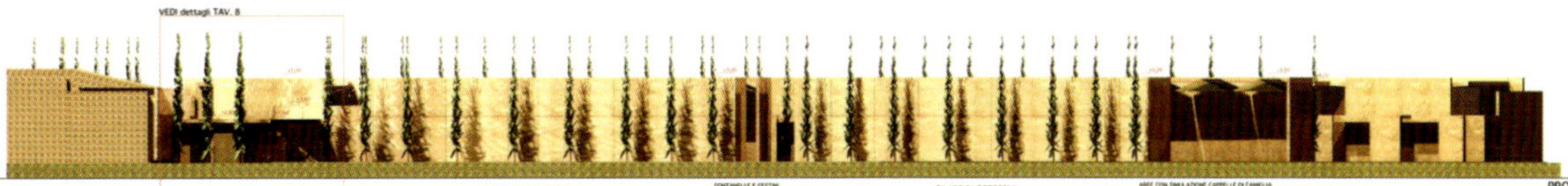

PROSPETTO SUD-OVEST
FRONTE FILARE CIPRESSINI — PROSPETTO SUD-OVEST
FRONTE FILARE CIPRESSINI

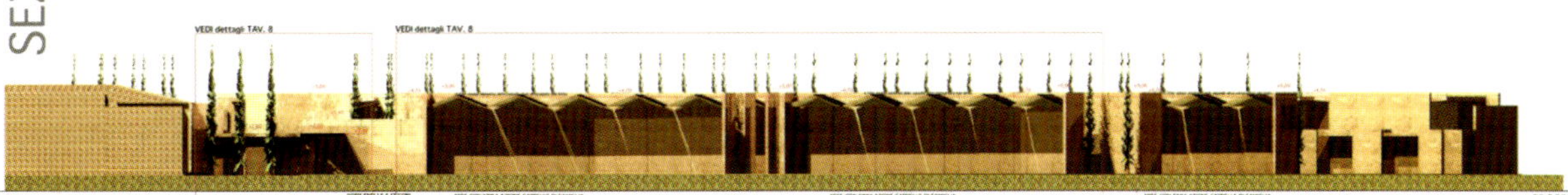

PROSPETTO INTERNO SUD-EST
FRONTE CAPPELLE DI FAMIGLIA — PROSPETTO INTERNO SUD-EST
FRONTE CAPPELLE DI FAMIGLIA

Project Team: Gianluca Aresta (architect); Gianpiero Conserva (Architect); Paolo Tarì (Designer)

Instant House

创新性临时住宅

The famous analyst James Hillman (1926 - 1989) in his "The politics of beauty" wondered how a neglected design, cheap colors, senseless sounds, structures and spaces could cost in terms of physical comfort and psychological balance. According to such reflection we wanted investigate how to find the right balance between the need to provide an accommodation, which is temporary—"instant"—and therefore answering to requirements of low spatial impact, and the need for settlement characterized by biunique capacity, both suitable for the nature of man that in the sense of optimal enjoyment linked to all the sensory aspects.

Through the location of the project within the park we want create the potential relationship between man and nature within the city. In this case we chose the urban Parco delle Cave, strategically favorable not only for its landscape quality, but also because it is easily reachable thanks to the networks of connection of the city, among which some are included in the city planning program of Milan (for example, the Via dell' Acqua).

One of the most influential appliances deriving from the relationship with the nature within the project is that of the modular aggregation. Specifically it has only one living module, containing two beds and a dining area, that can join with other units in order to reach the capacity of eight people, a toilet and kitchen module, which at the beginning and at the end of the aggregation changes into bathroom space with a technical room, assuming in the composition the function of opening and closing the aggregative system.

So the system adapts itself to the natural shape of the place, of which we try to respect the environment through the mimesis deriving from the use of wood and green roof. The aspects of sustainability are closely linked to these two materials: it is enough to think of the low visual and natural impact of the use of wood, especially if it comes from reforested reserves, according to FSC certifications, it is enough to think of the aspect of low energy consumption deriving from the insulation offered by the green roof, on the one hand, and the strategic section of the longitudinally inclined roof, which hinders the summer rays inclined at about 60° and admits the solar rays inclined at about 30°, which heat the environment through the solar greenhouse created.

But the planning result meets also the needs linked to the human sensory perception. Indeed we can remake how the longitudinal direction of the staves damps the perception of reduced spaces and how the gradual chromatic lightness relieves the sense of oppression caused by restricted spaces and infuses a pleasant sense of warmth. Finally, the attention paid to the availability by the disabled within each module is no less important: it is possible thanks to apposite external ramps, in case of need.

地景模拟
Landscape Simulation

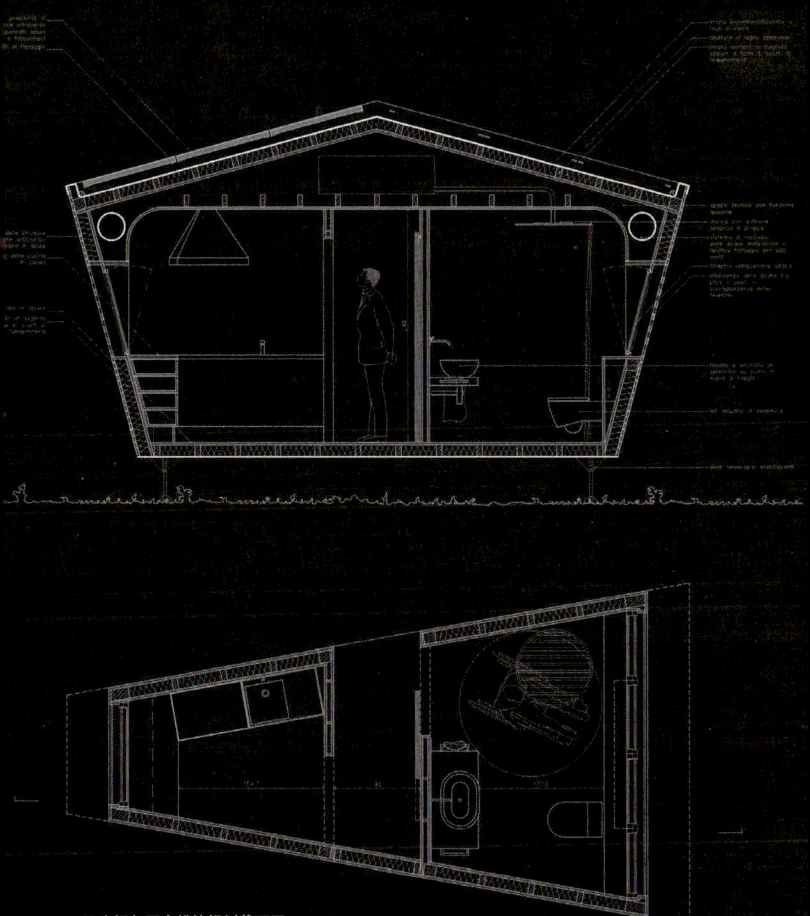

卫生间与厨房模块规划截面图
Plan and Section of Toilet and Kitchen Module

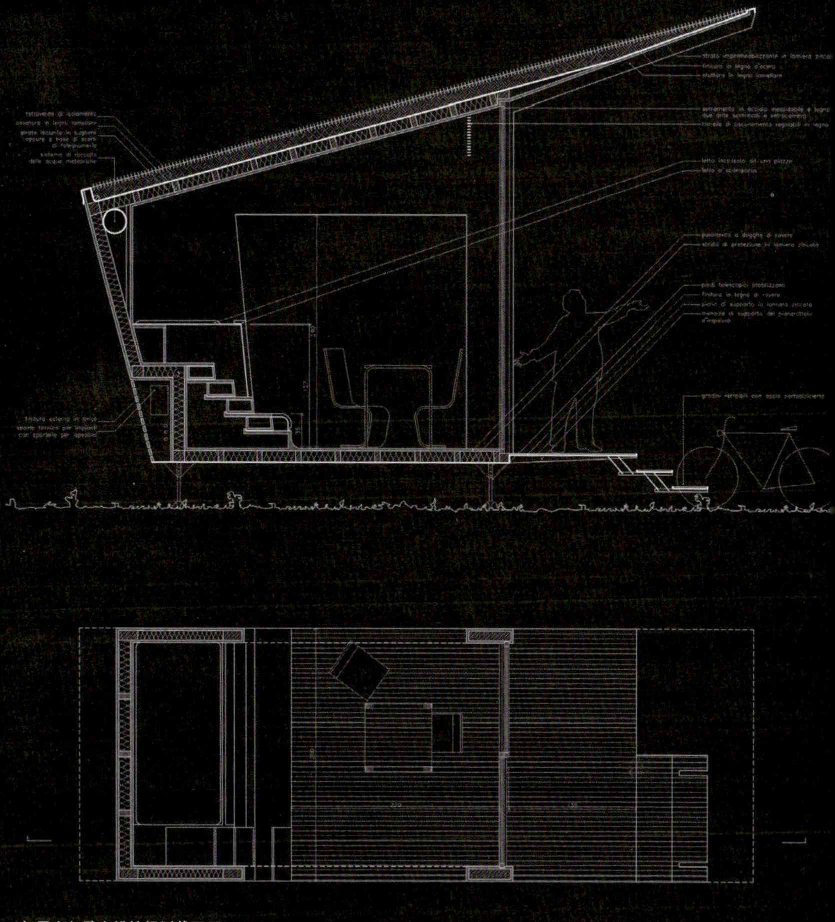

起居室与卧室模块规划截面图
Plan and Section of the Living/Sleeping Module

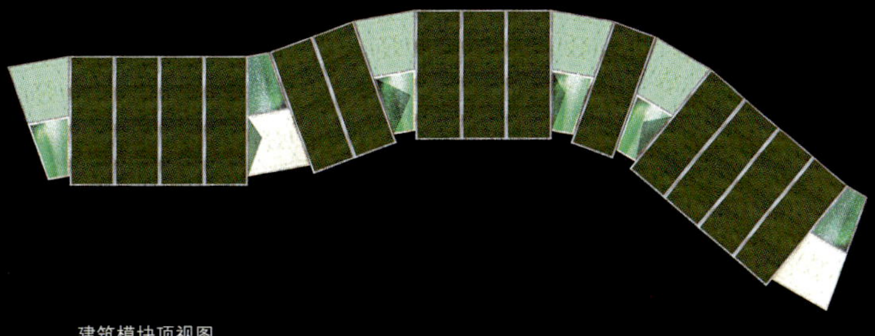

建筑模块顶视图
Top View of Aggregation

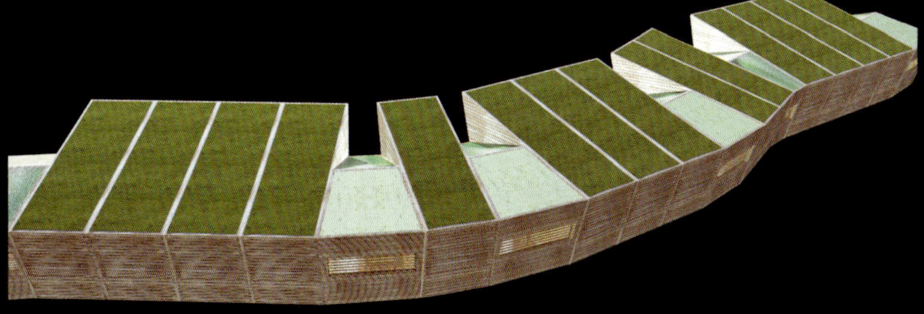

建筑模块透视图
Perspective View of Aggregation

Integrated Parking Construction in the Historic Centre of Thessaloniki

塞萨洛尼基历史文化中心停车场综合设施建设

The PARKinPARK is conceived as a response to the ongoing increase of car traffic in the historic centre of Thessaloniki and the future plans of the city council to address the problem by constructing a number of underground car park stations around it. The main challenge of the thesis is to prove that architecture can have a positive effect when applied to the rigor of transport pragmatism. The project reconsiders the conventional parking garage typology as a potential urban apparatus. By exploring the latent flexibility of its function and intensifying its station-character, the parking garage acquires the qualities of a hybrid multi-layered urban plaza ready to accommodate a variety of events together with the streams of cars. The scheme integrates an underground car park station with a digital film school, an urban plaza and a bus station, all of them seamlessly connected with each other in terms of form and structure, and with the surrounding urban tissue in terms of circulation and views, resulting into a catalyst of unpredicted urban events.

My thesis is set at the edge of the city centre of Thessaloniki Greece. The site is currently used as an open air car park although it is called a plaza by the people and it has a bus terminal. My project for the site consists of an underground car park with a capacity for 500 cars and a digital film school combined and set under an urban plaza with a bus terminal. My intension was to create a transport hub for people arriving by car and continuing by public transport in the city centre and a place of interest for the local community that could function as something more than just a plaza on top of a car park. For that reason I started rethinking the conventional car park typology and how it could finally acquire a more station-like character and not just a storage area of cars. I tried to achieve a programmatic intensification in the building organised in a way that changes the process of leaving and taking a car or even delays or postpones it.

The form of the car park achieves to embody a diversity of programs into the parking experience in two ways. First by the way that the parking ramps are interwoven with the film school and second by the flexibility of the space on the ramps themselves.

The zigzag form of the ramps with the connections between them allows for multiple possibilities of how they can be used. Every ramp can be separated from the rest of the network of ramps and function autonomously inviting different activities such as open markets, exhibitions, installations or any other event.

建筑模型
Model

建筑模型
Model

建筑模型
Model

建筑模型
Model

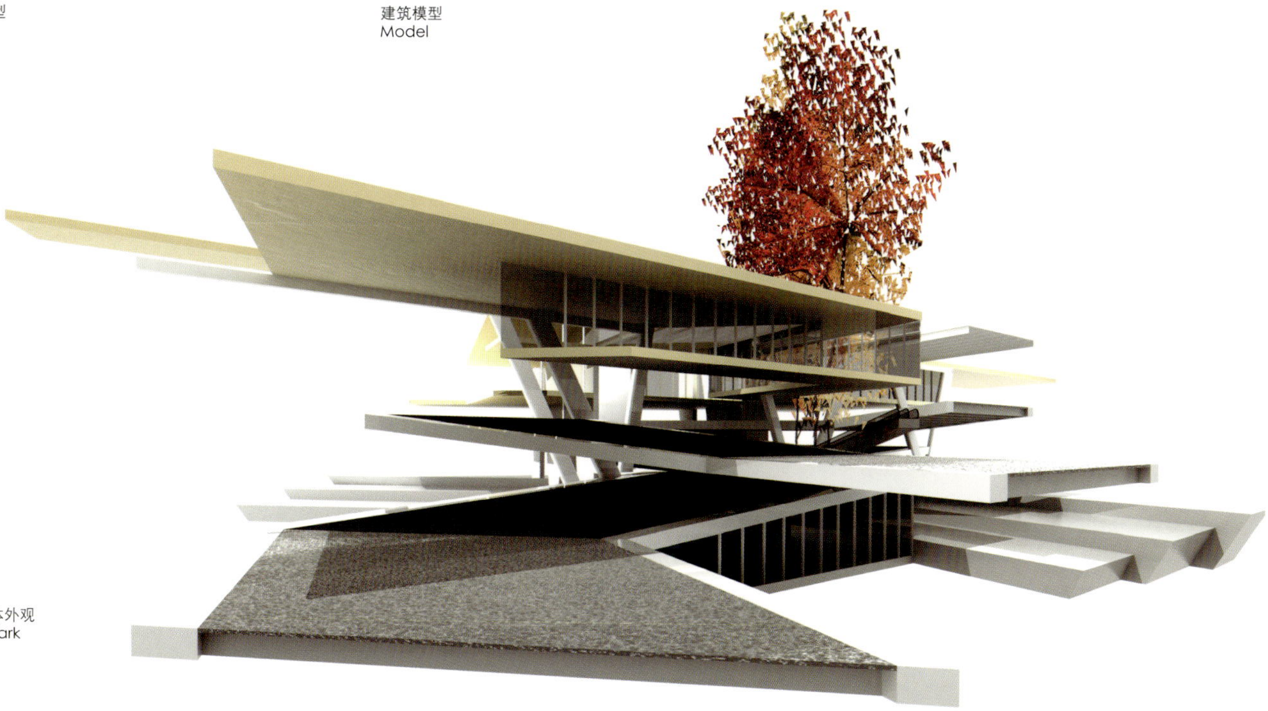

建筑整体外观
Parkinpark

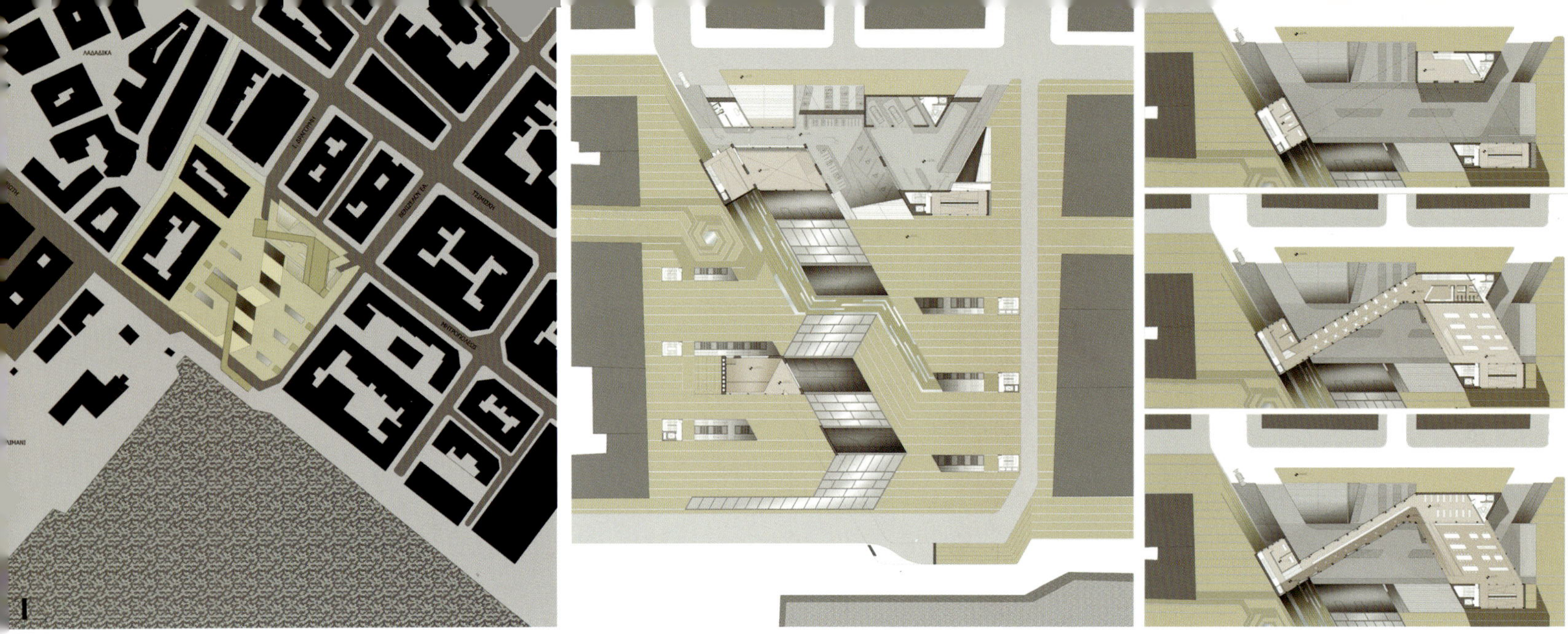

地面连接规划图
Overground-Plans

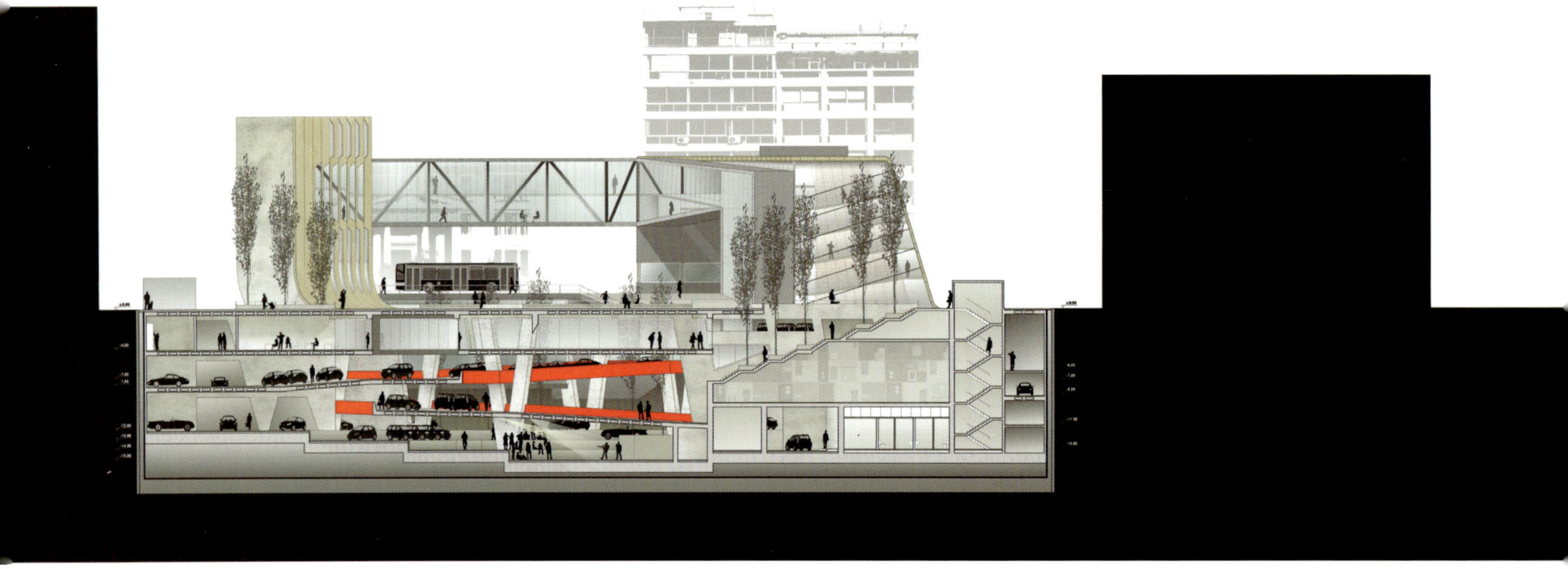

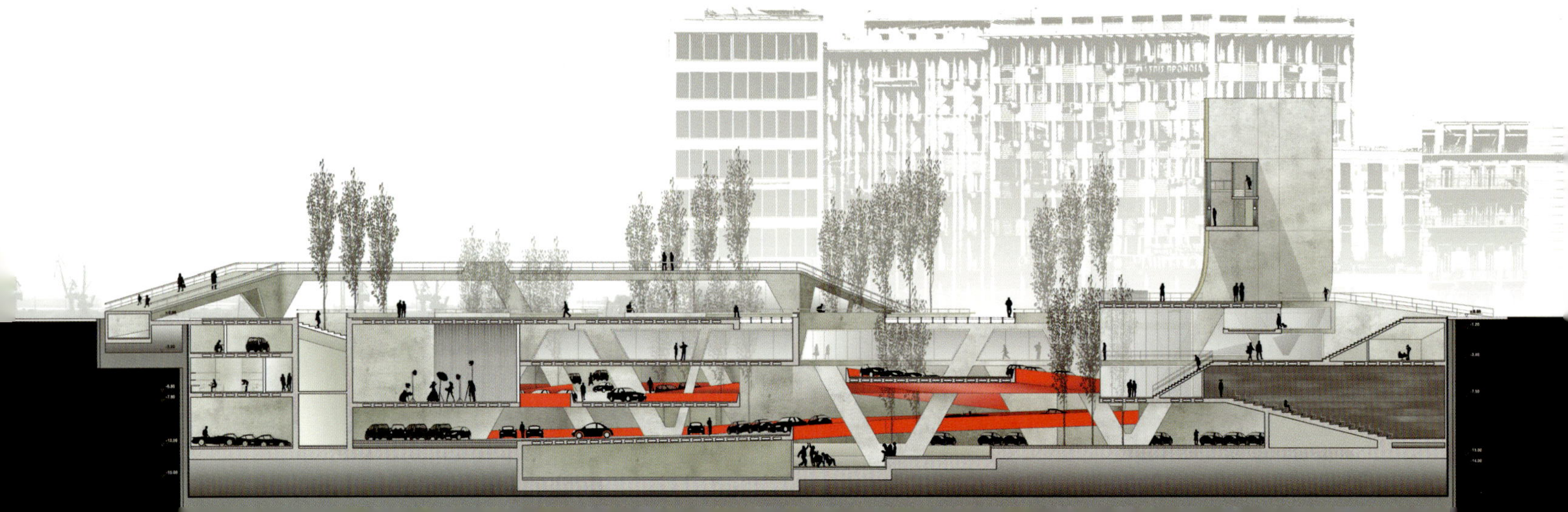

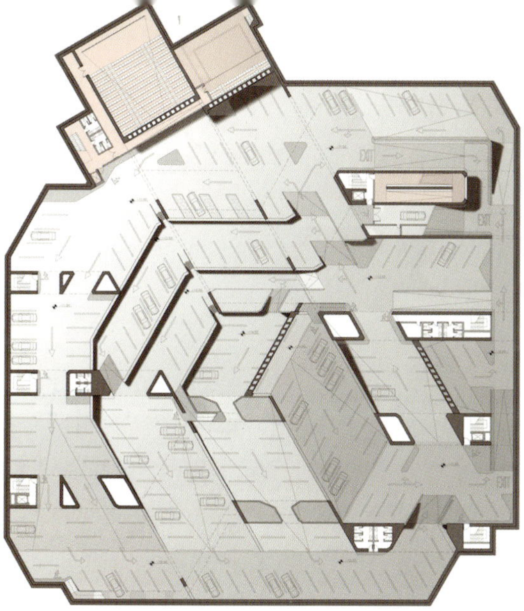

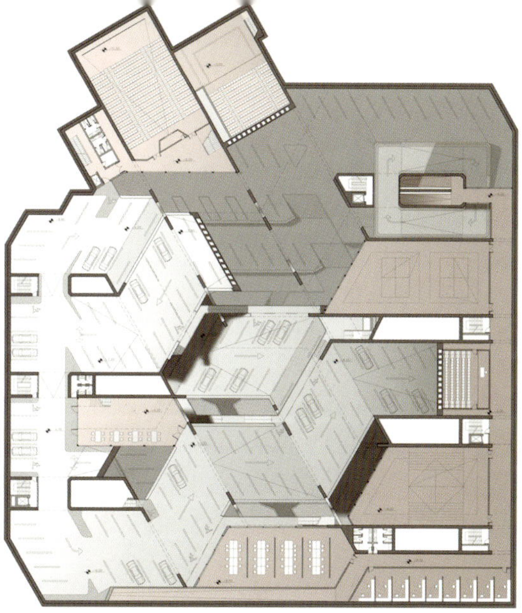

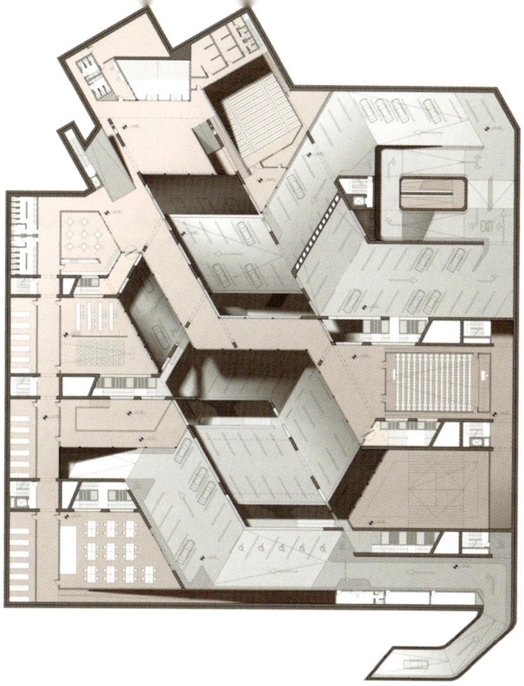

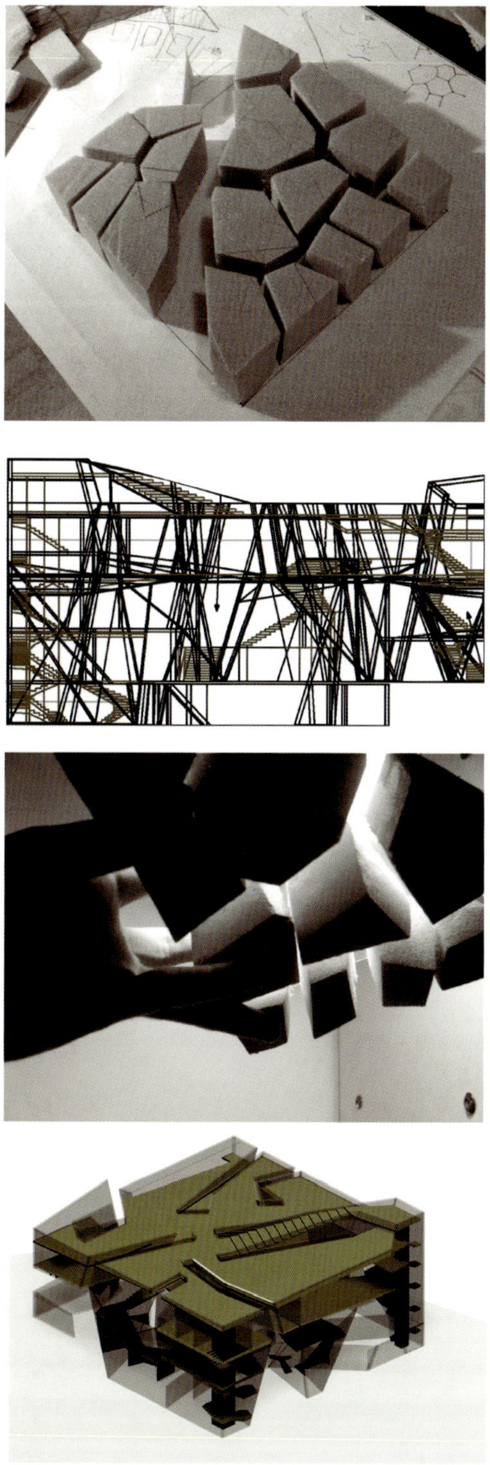

设计步骤
Design Process

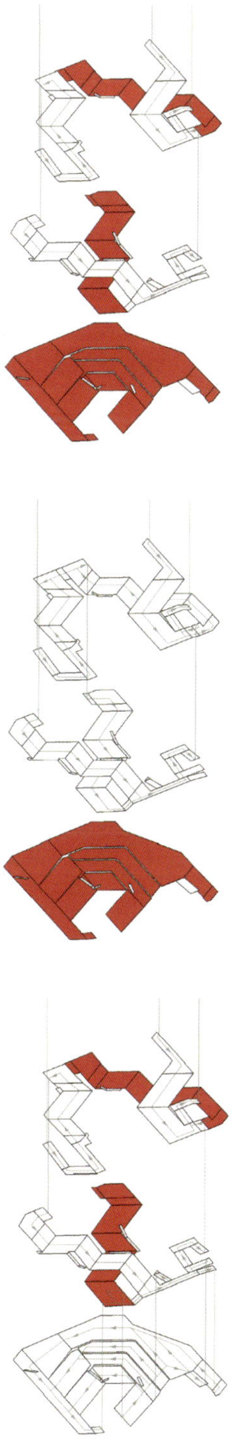

规划性场地灵活度设计
Programmatic Flexibility

音乐厅
Music-museum

设计步骤
Design Process

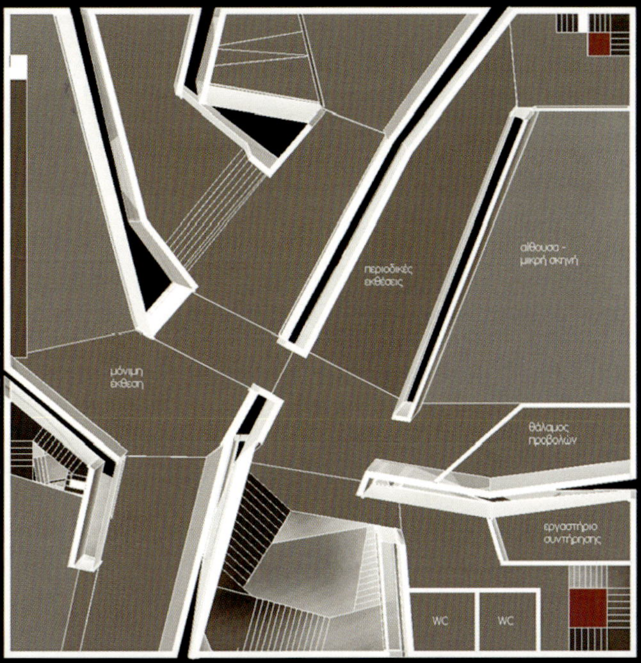

περιοδικές εκθέσεις
αίθουσα - μικρή σκηνή
μόνιμη έκθεση
θάλαμος προβολών
εργαστήριο συντήρησης
WC
WC

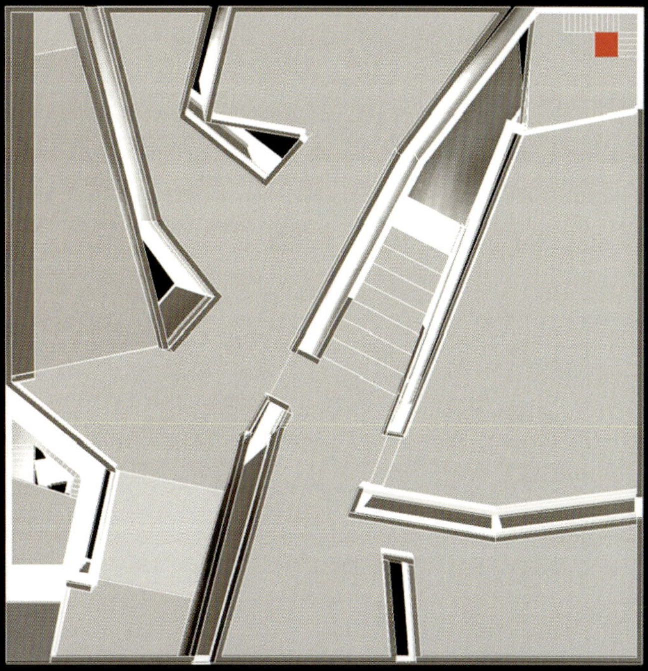

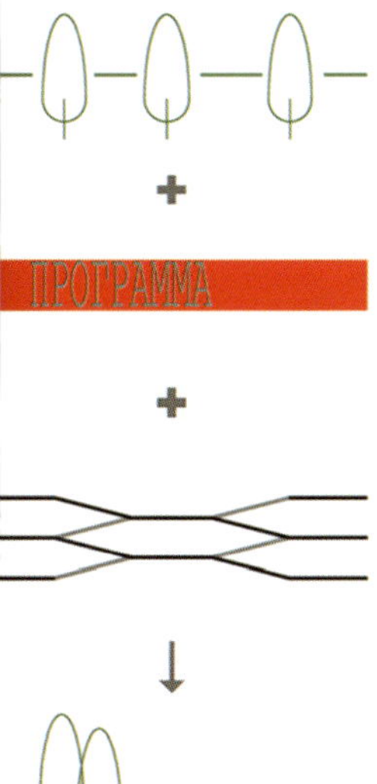

ΠΡΟΓΡΑΜΜΑ

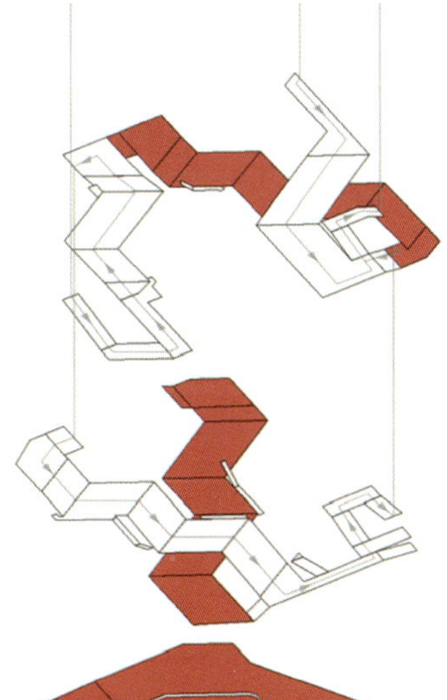

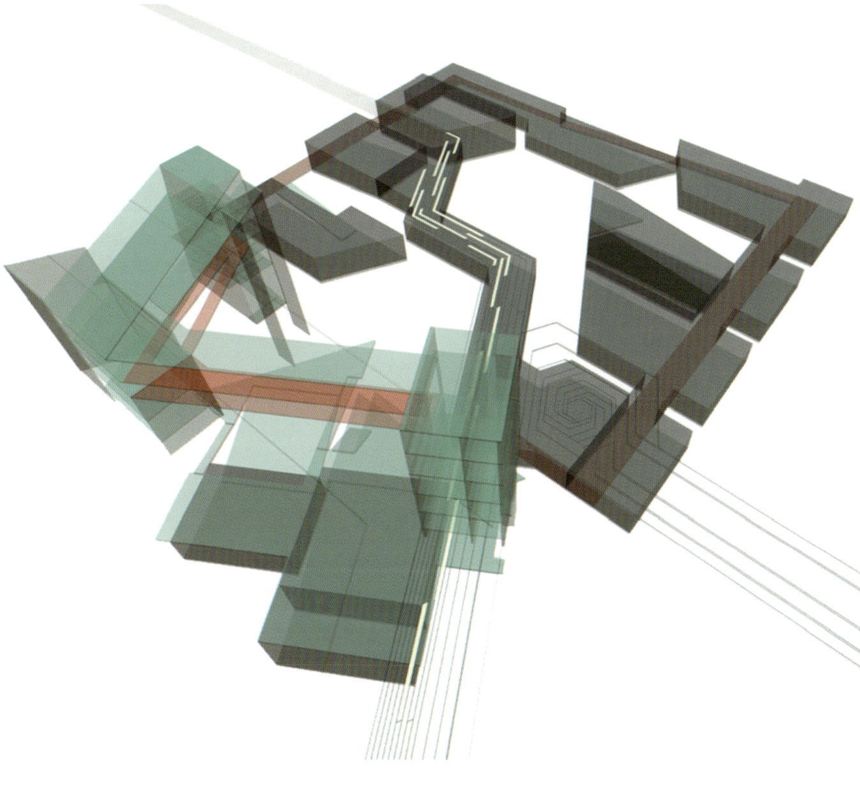

Project Team: Piotr Urbanowicz, Yoko Sano

Krakow Conference & Concert Centre Krakow, Poland

克拉科夫会议中心&音乐厅

The building is located in Krakow near the Japanese Manggha Technology and Arts Museum on the Vistula riverside. Its form is an inherent part of the corner of the building plot at Grunwaldzkie Roundabout. Its modern image is determined by the organically shaped building with a large glazed foyer, which opens the perspective to the river, to the old city centre and the Wawel Castel. The building gets lower towards the river to reduce its scale in the embankments panorama.

It will allow congress and concert participants to see the city from the multi-functional foyer and be very well seen at the same time. Foyer will be visible together with the "fifth elevation" of the building from the Wawel terrace. The view will cover the building together with the nearby building of The Manggha Museum with which architectural form corresponds well.

The elevations will be mostly finished with glass and titanium-zinc, sheeting, complemented with individually designed ceramic and stone cladding (granite, limestone and sandstone).

These elements refer to typical materials creating the history of architecture of the Wawel Hill. Combination of these various materials with the glazing creates a mosaic on the building elevations and determines its image.

The building with 18,000 square meters of space (+10,000 m^2 of underground parking), will have halls with capacity of 1,800,600 and 300 seats and a number of conference rooms sized about 500 square meters, stock rooms and technical back rooms and an underground parking garage with about 320 spots.

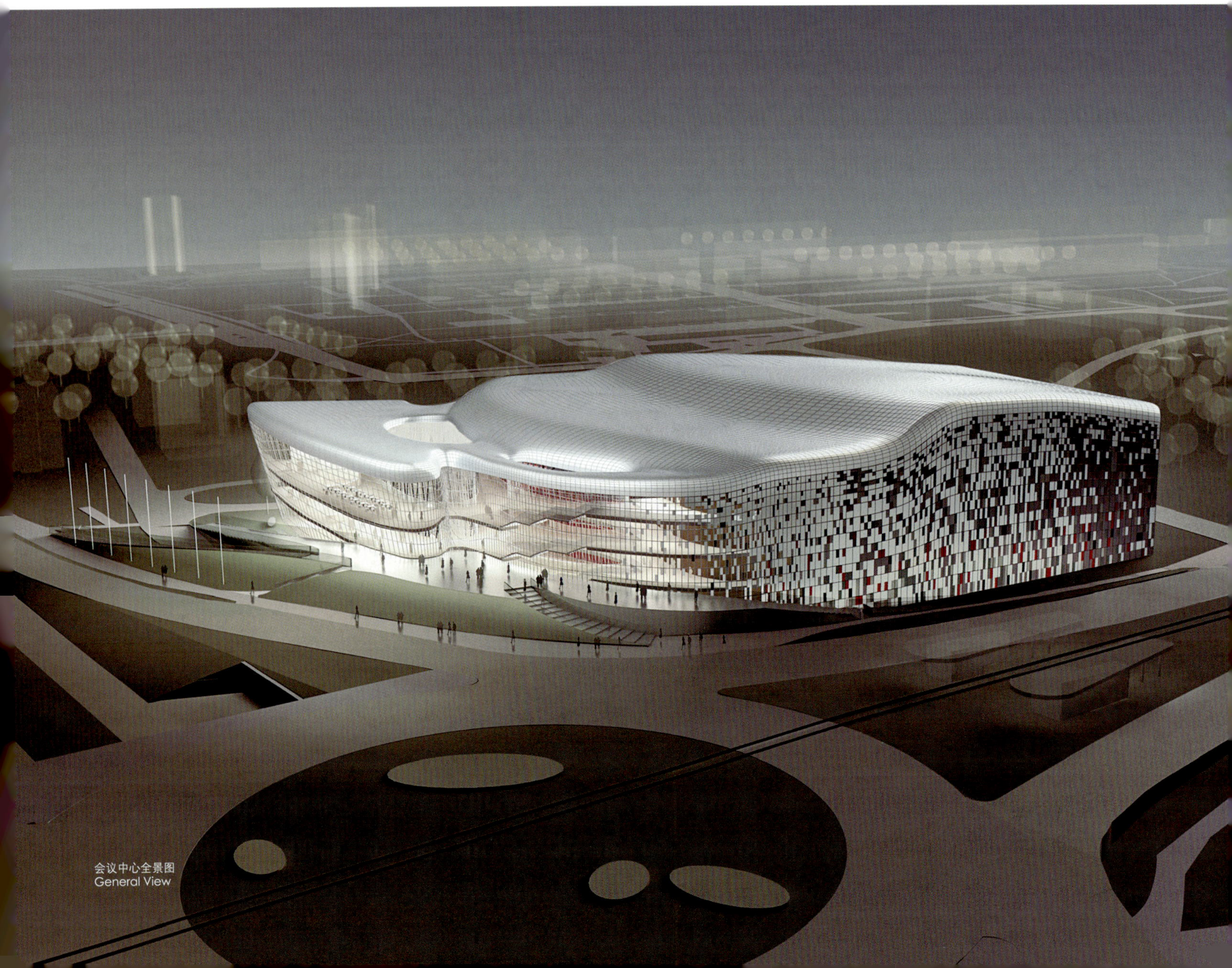

会议中心全景图
General View

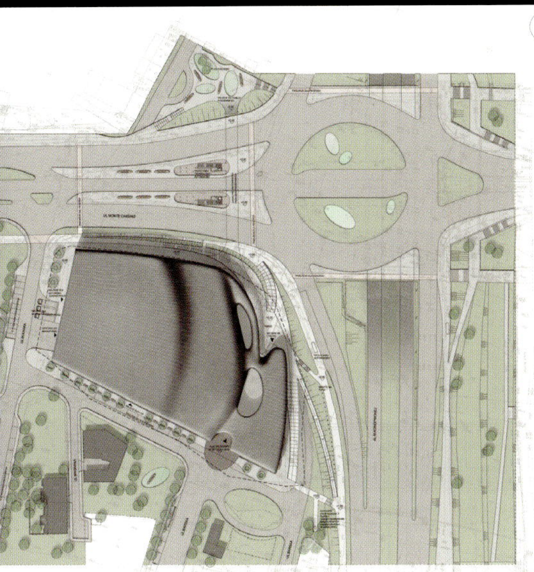

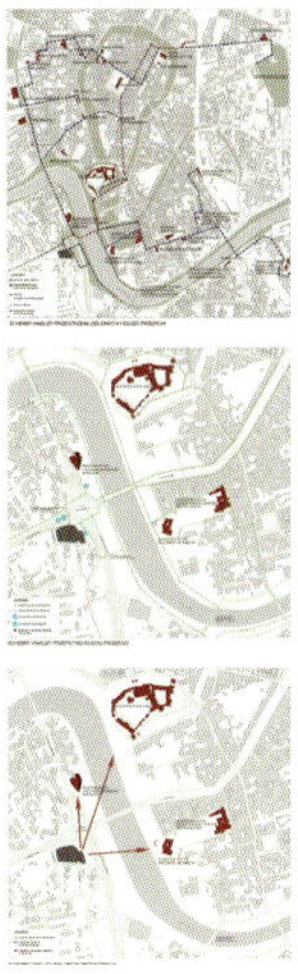

场地
Site

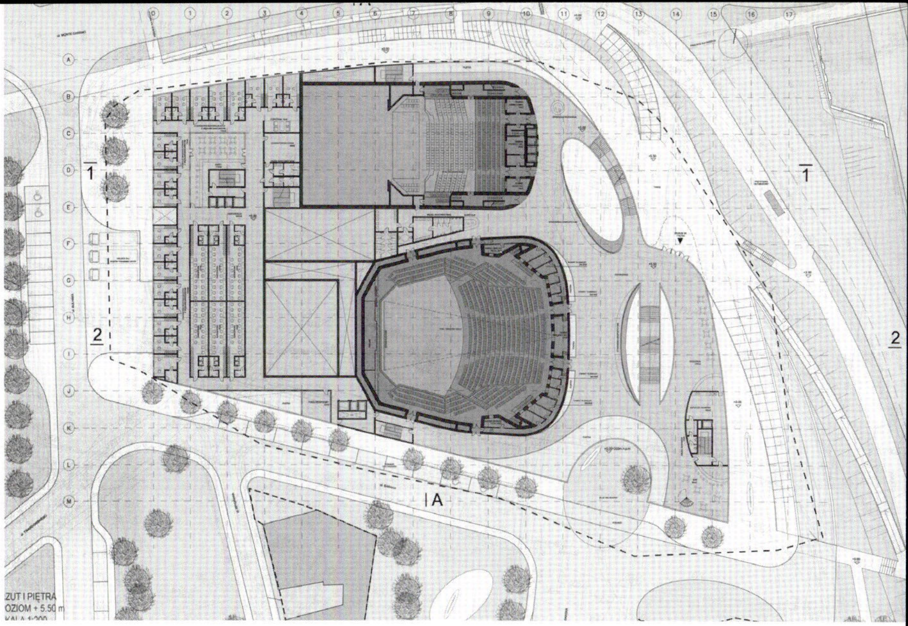

F1平面图
F1 Plan

立面图
Elevations

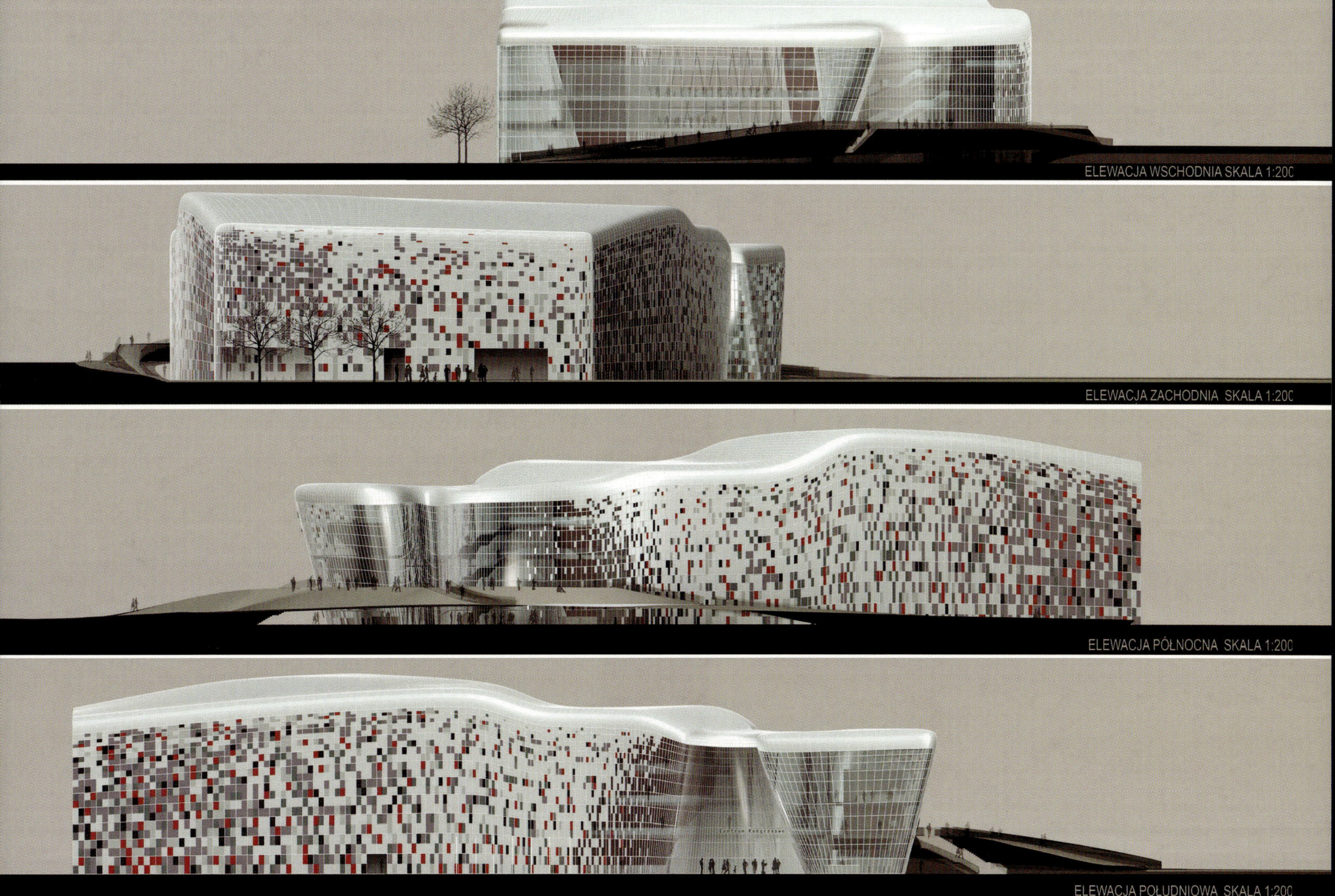

入口景致
Entrance View

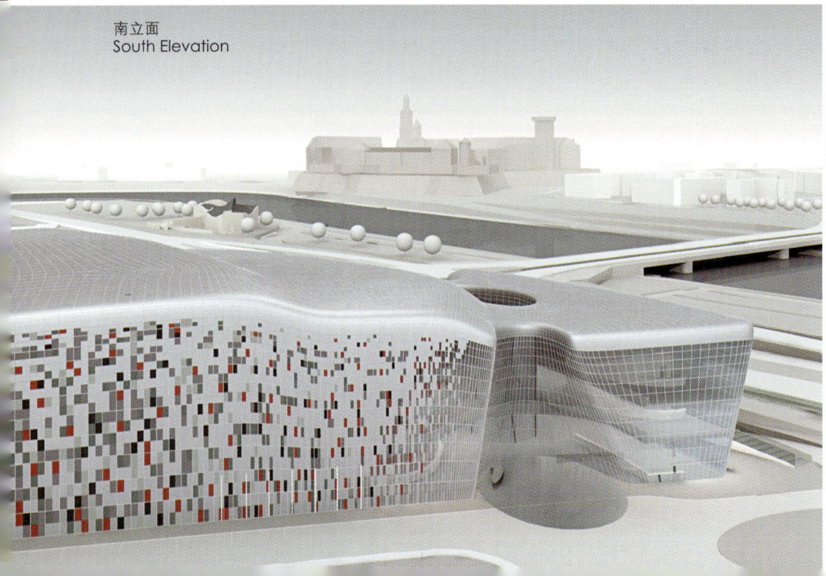

南立面
South Elevation

北部景致
North View

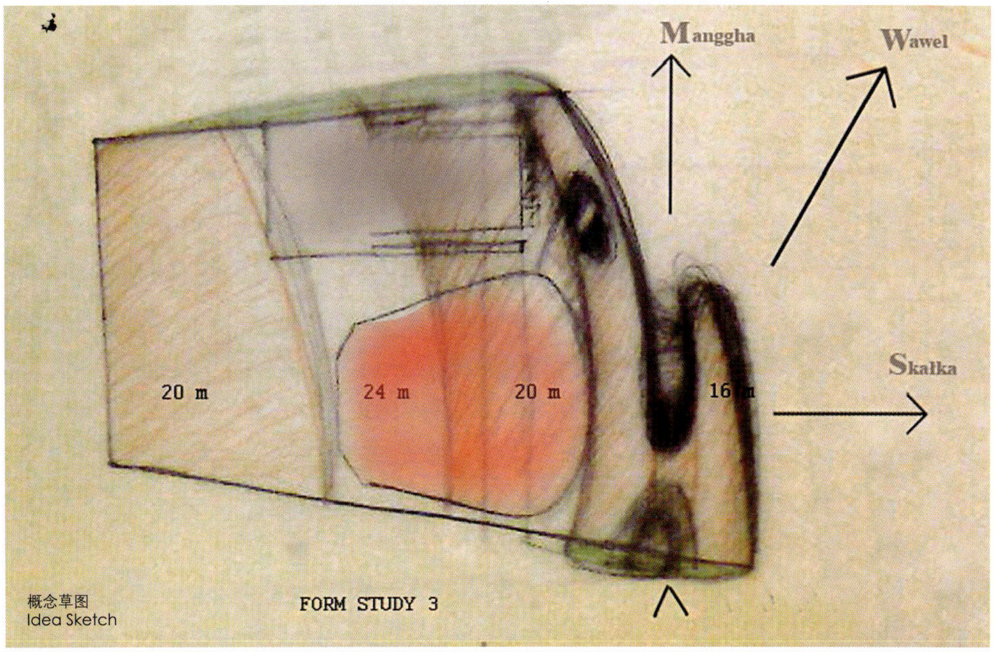

概念草图
Idea Sketch

一楼休息大厅
Foyer 1

二楼休息大厅
Foyer 2

Project Team: Dimitris Gourdoukis

KU School of Architecture

美国堪萨斯大学建筑学院教学大楼

The design/research project was for a new school of architecture for the KU. The beginning of the research was as an exploration of the properties of lattice structures with rigid joins and the ways that they can be applied to architecture. Starting point was Frei Otto's research on the same subject and it was evolved into a study of how a structural system can "overcome" its main property (that of being structurally efficient) and became a system that is at the same time an "ornamental" element and the element that defines spaces (and of course the structural system as well).

The results of the research were applied on a specific design problem: A new school of architecture located in the Kansas University Campus. Kansas University is planning to build a new building to house the department of Architecture and was looking for innovative ideas in order to see what the needs of the project are and what directions it could follow.

The solution proposes an organic structural system that creates clusters of space to host the program. Even though the nature of the project is highly "irregular", it is a the same time framed inside the volume of a long box, that is essentially floating above the site leaving this way the ground floor open in order for various "open air" activities to take place there. Fiber carbon is used as the main material for the structural system, which combined with the extraordinary structural properties of lattice structures which can allow large spans that minimize the places where the building touches the ground.

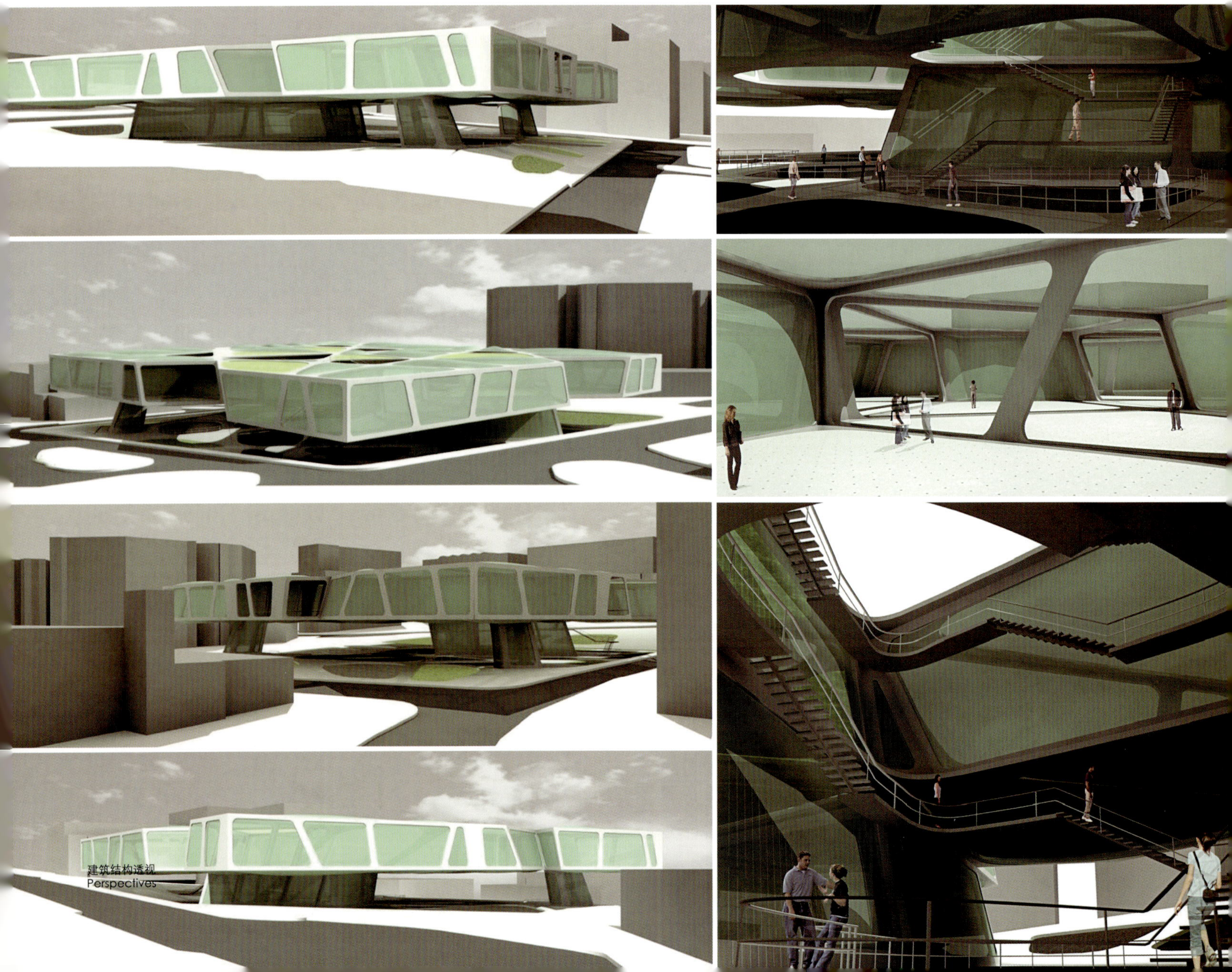

建筑结构透视
Perspectives

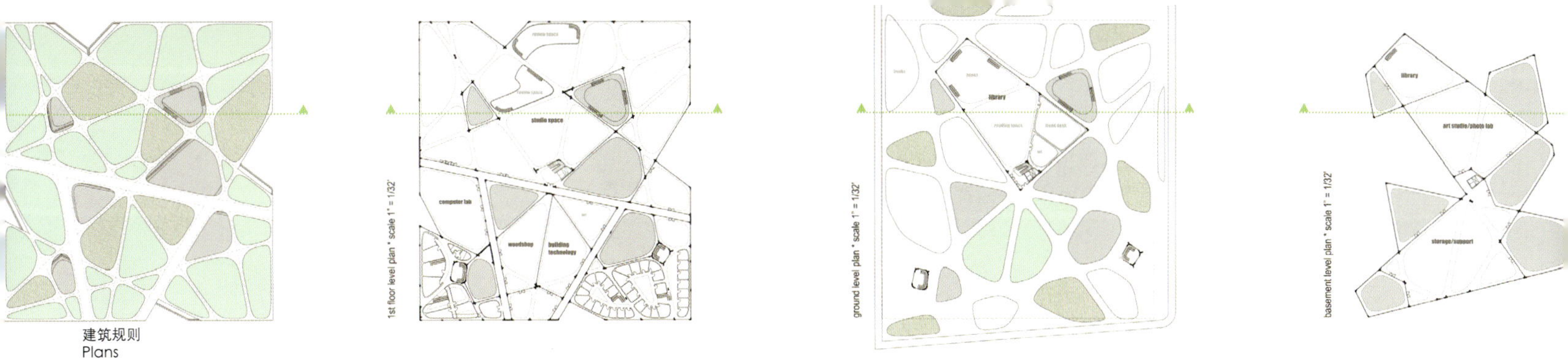

建筑规则
Plans

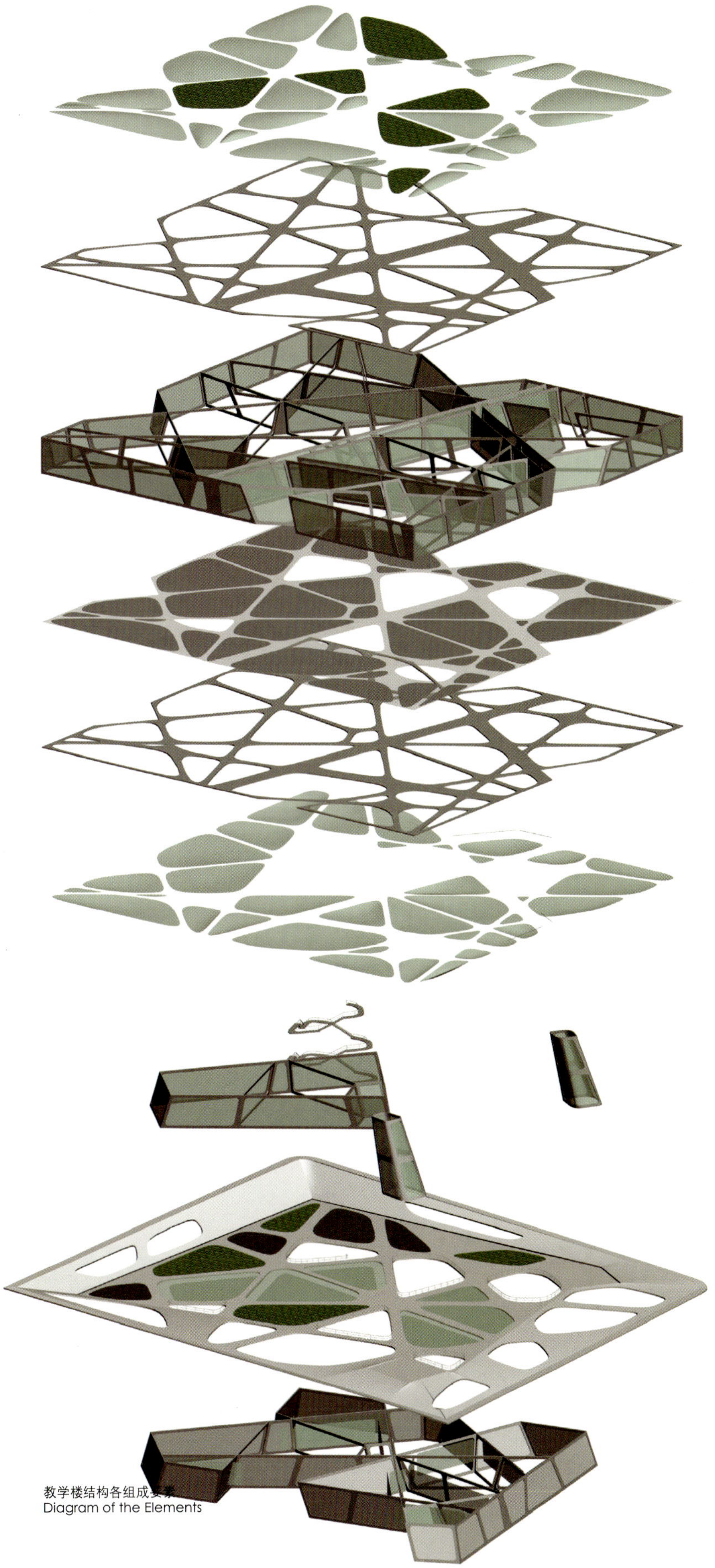

教学楼结构各组成要素
Diagram of the Elements

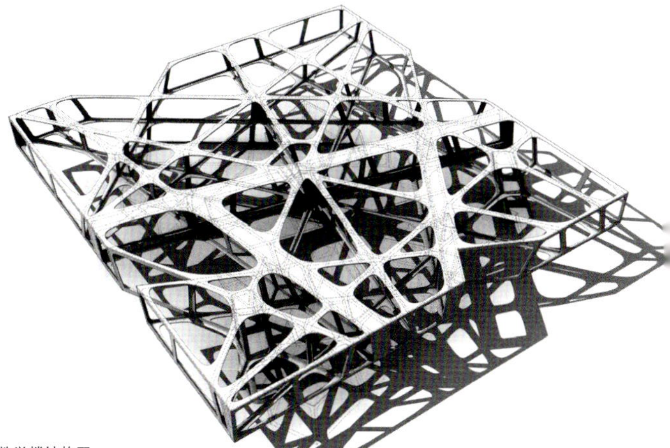

教学楼结构图
Structure Diagram

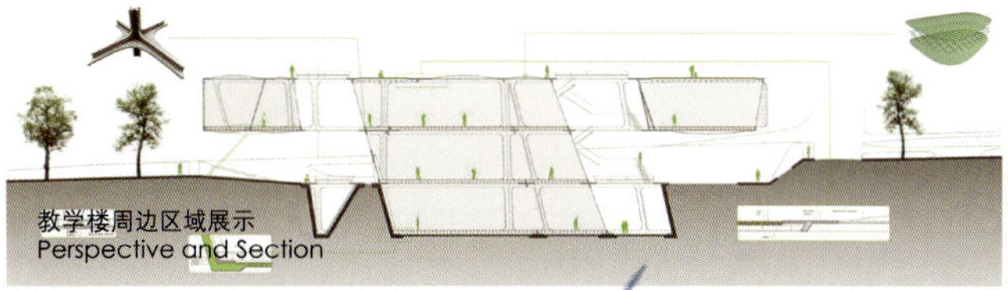

教学楼周边区域展示
Perspective and Section

教学楼结构各组成要素
Diagram of the Elements

Project Team: Johan Roman, Miikka Hirsimäki, Pia Sabelström, Marcus Upmeier, Elina Juureva, Mika Väisänen, Sauli Luttinen

KUMU – The Main Building of the Art Museum of Estonia

KUMU——爱沙尼亚艺术博物馆主楼

The project is based on the winning proposal of an international architectural competition held in 1994. Further design work began immediately after the competition, and revised preliminary designs were submitted to the museum in January 1995. During next few years there was a long pause until funding was secured in 1999 and the actual design work could begin. After various delays, a building permit was issued in summer 2002. Almost 216,000 cubic metres of limestone was excavated and removed from the site, allowing construction work to begin in autumn 2003. The construction work was completed at the end of 2005 and the museum was opened to the public on 18 February 2006.

The impressive site with a 20-metre-high lime stone slope is located at the south end of the Kadriorg Park, some 3 km from Tallinn city centre. In order to leave the park as intact as possible and not to threaten the dominant position of the adjacent Kadriorg Palace, the large museum building was placed in the slope, partly underground. The round shape came about as if by itself, as a result of the ramp solution that connects the different levels of the museum. A curved wall encloses the courtyard and an outdoor sculpture exhibition. The ramp divides the museum building into two different parts. Outside the curve are, for instance, the administrative personnel's rooms and the conservation facilities, and inside are the lobby

公园夜景图
Night View from Park

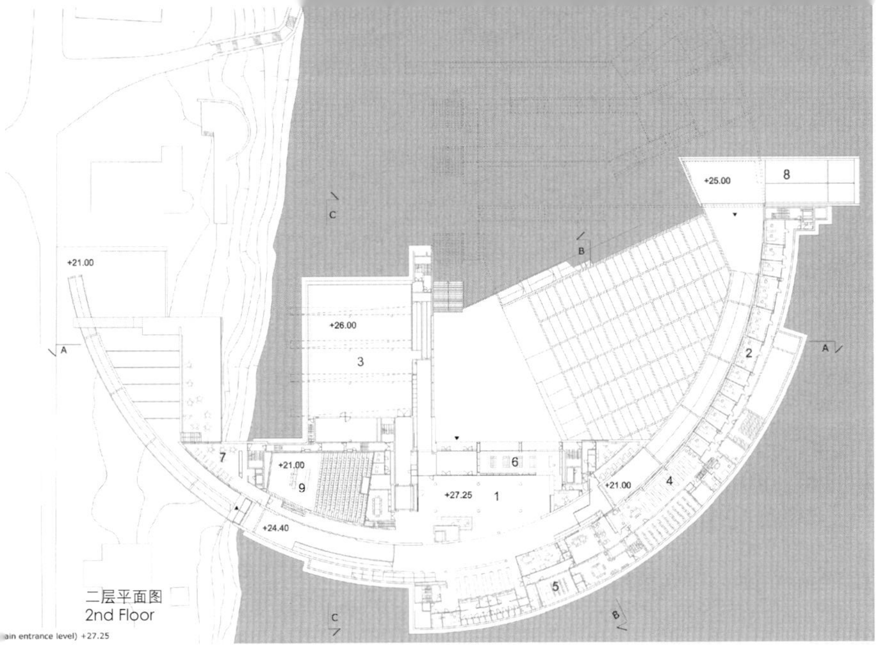

二层平面图
2nd Floor

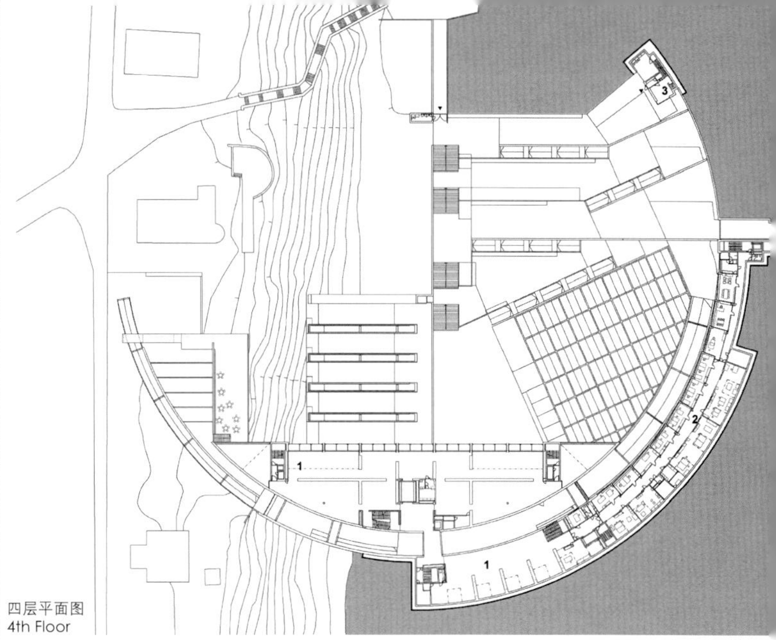
四层平面图
4th Floor

and exhibition halls.
The parking area and bus stops are on the uppermost level of the site. Visitors walk down the stairs of a pedestrian tunnel and through the outdoor sculpture exhibition area to the main entrance. On the entrance level are ticket booths, museum shop and entrance to temporary exhibitions. The other entrance to the museum is on the ground level from the Kadriorg park side. The ramp takes visitors up to the foyer of the auditorium and further to the main entrance lobby. The high lobby area that divides the museum into two parts is dominated by the connecting bridges.
The design aims at simplicity and clarity. The exhibition halls are simple and unassuming, placing the artwork at centre stage. The ascetism of the interior continues in the exterior, which relies on the power of plain geometric forms. The main facade materials are limestone, green-patinated copper and glass.

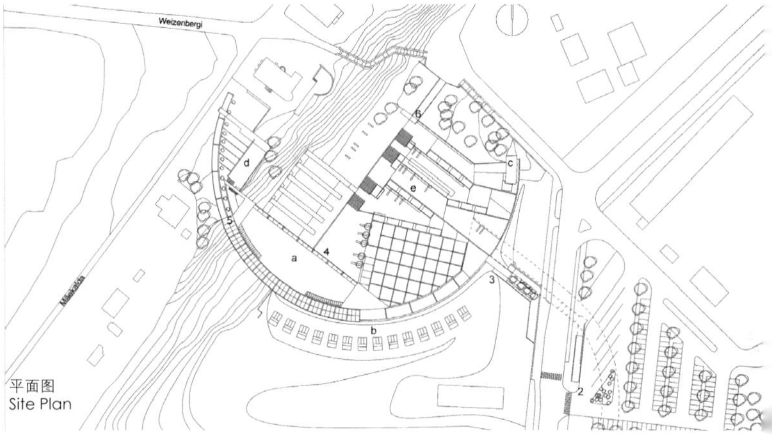

平面图
Site Plan

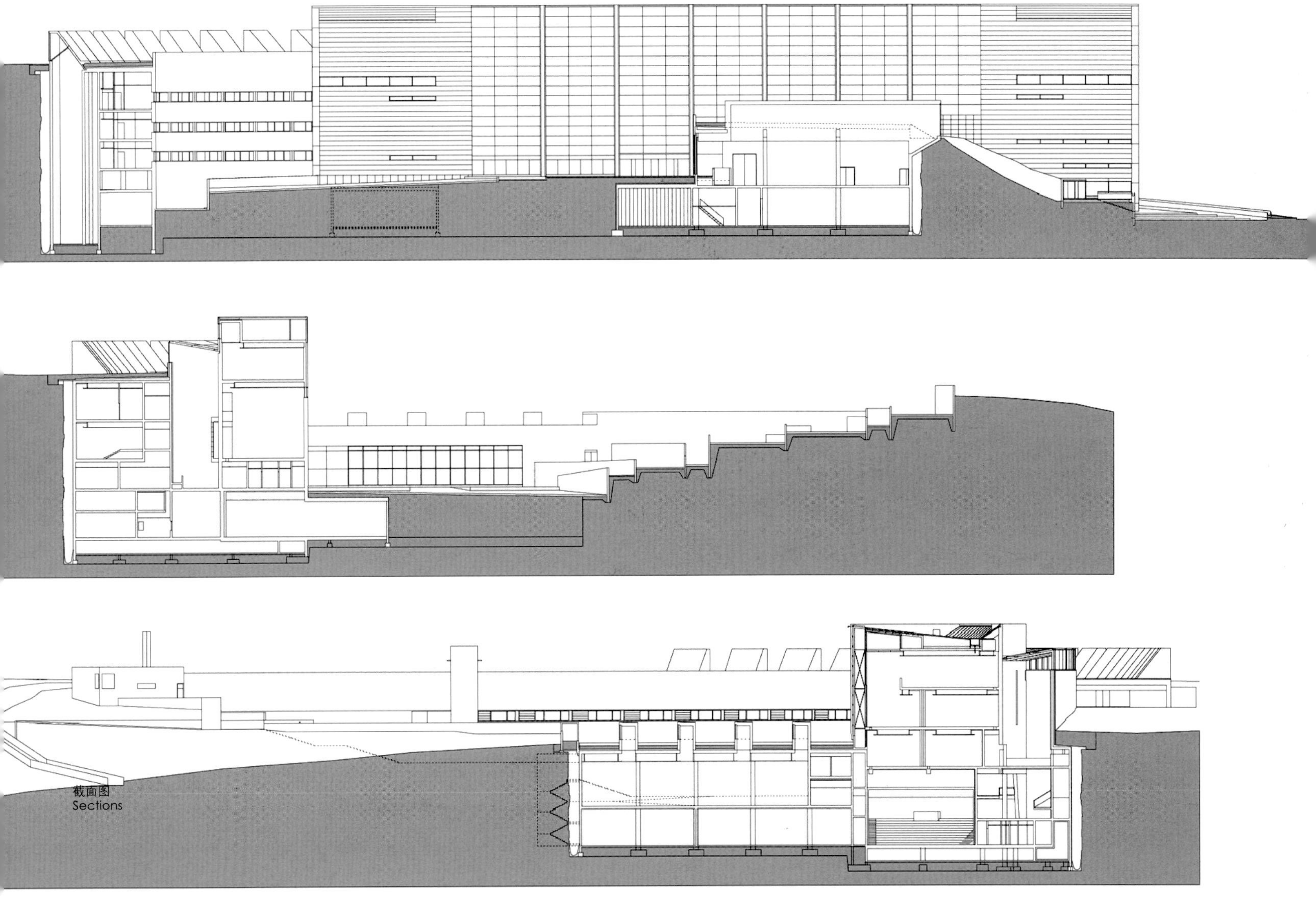
截面图
Sections

公园中欣赏到的外部景观
View from Park

建筑二楼
2nd Floor

仓库入口
Storage Entrance

建筑三层
3rd Floor

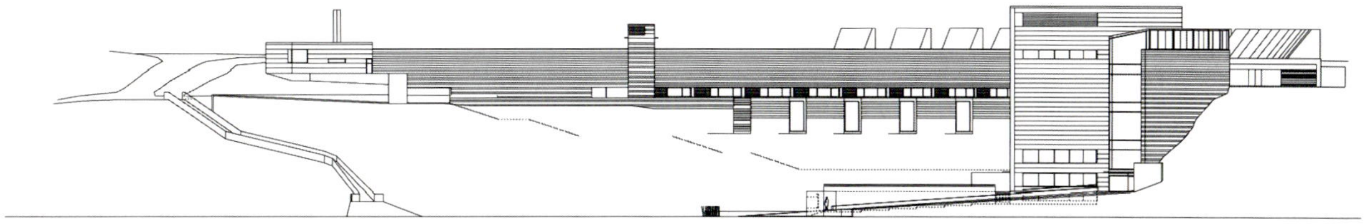

northwest elevation

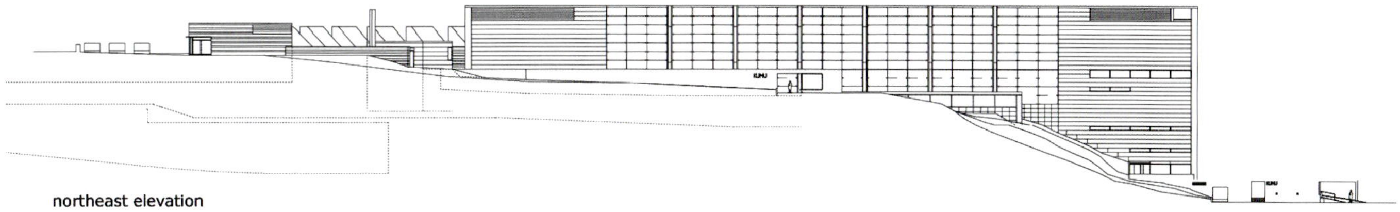

northeast elevation

southeast elevation

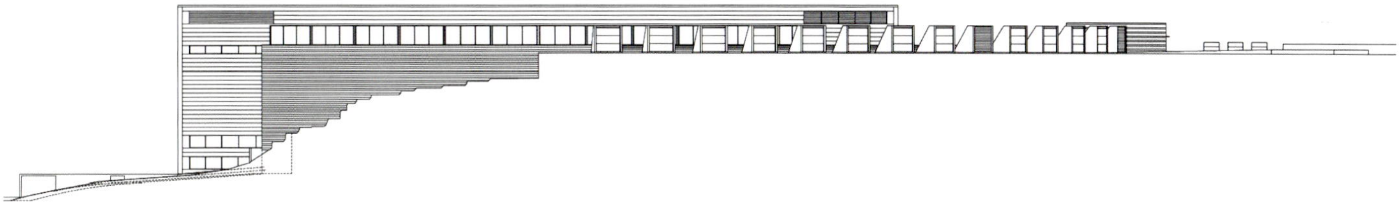

建筑四层
4th Floor

不断变化的展览区
Changing Exhibition

不断变化的展览区
Changing Exhibition

Project: Residences & Mixed Use

LEGO

乐高木建筑

During the Marshall years when post war Denmark was being rebuilt, the state chose to favour prefabricated concrete over all other forms of construction. As a result the in-situ cast concrete industry has almost vanished, leaving the entire building industry based on prefabricated modules and concrete elements. Contemporary Denmark has become a country entirely made from LEGOs.
When asked to do a dense mixed use development comprising parking, retail, offices and housing in to a single super dense development of a rather small but central site, we knew that the program wouldn't fit in the generic perimeter block.
We knew that we once again had to step in to the minefield of public tower phobia. Once again comparing the popular spires to the modern monoliths, we attempted to diagnose the demise of the Copenhagen high-rise. It appeared that until the early 1900s architects had been capable of designing imaginative towers in the honour of the church or the king. But upon the arrival of functionalism, with all its good concerns for daylight, views, functionality and production technique, our imagination seemed to have been reduced to the repetition of identical rectangular slabs. Could we harness all the good intentions of functionalism and turn it into a platform of the creative process, rather than its straightjacket. Rather than starting by designing crazy shapes and then figuring out how to squeeze it in to the mold of prefabrication, we decided to turn the project in to homage of the modular building industry.
We decided to cover the entire site with the favourite module of the building industry: 3.60m x 3.60m. Rather than designing with a composition of built volumes, we considered the site an elastic moldable substance, defined by zones of varying density. And like saving for the web, we then adjusted the soft topography of the urban massing, to a rough pixelated landscape of peaks and valleys. Like a Maya Lin sculpture the towers rise from a stepped planes into cliffs and ledges of human occupation. A modular wet dream.
Seen from above the entire site is accessible, like a 3 dimensional field of a thousand plateaus of varying degrees of privacy. At the foot of the

建筑模型
Model

规划图
Plans

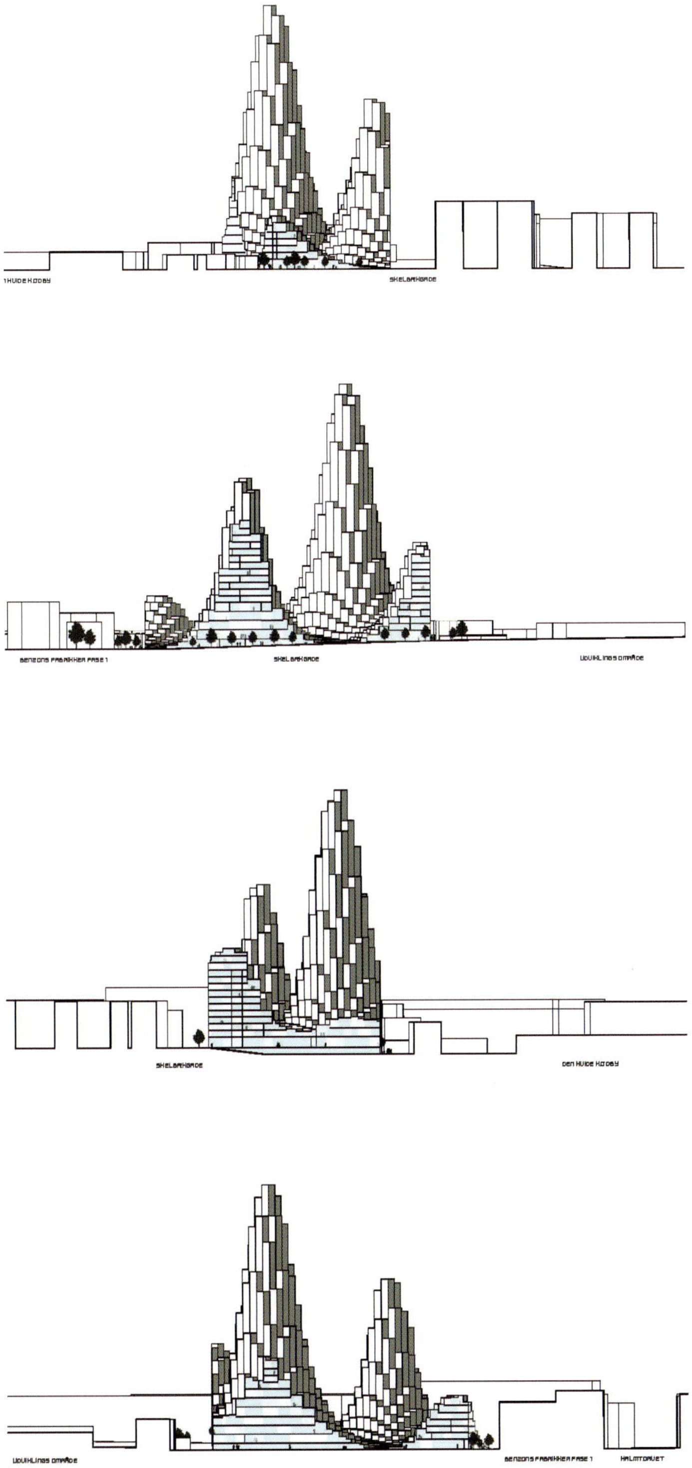

建筑外观
Facades

规划图
Plan

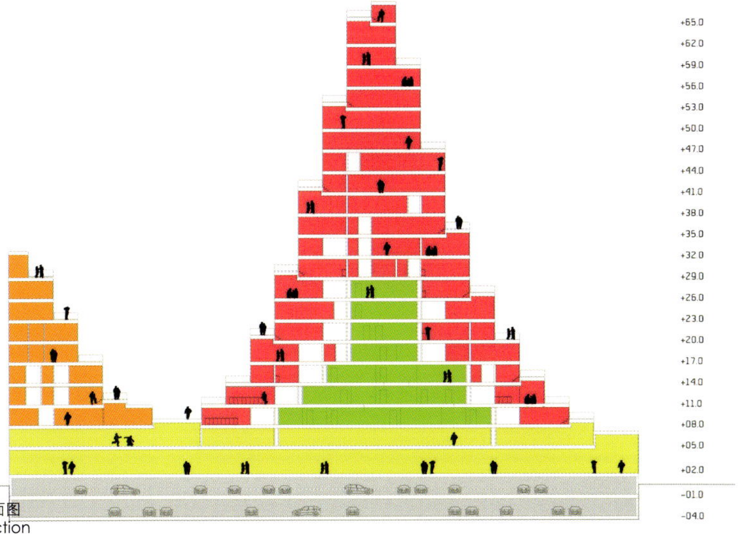

断面图
Section

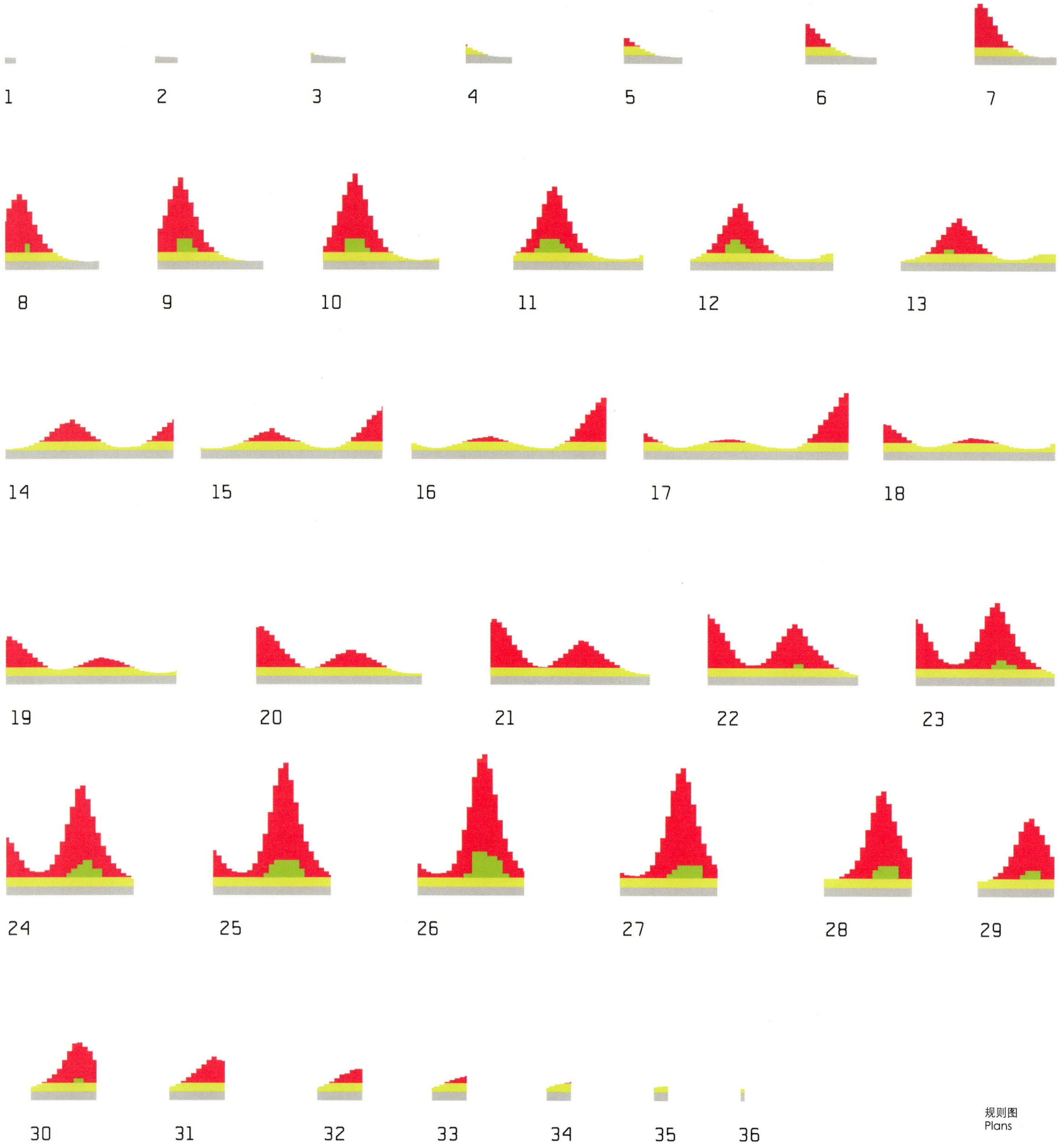

规则图
Plans

towers the plateaus merge forming flat plazas or gently stepping amphitheatres. Light and air seeps between the steps in to the parking and retail below. At higher levels the terraces become inaccessible to the general public and provide a wide range of balconys for the working population. Finally they turn in to private terraces for the residences, some of which have up to 3 or 4 in different directions.

A frequent criticism of high-rises is that they fail to relate to the human scale. In this case the consistent module makes human proportions (and accommodation) omnipresent throughout the building envelope.

When you ask an experienced functional architect to design apartments he will introduce a multitude of protrusions and recesses, to create and maximize daylight penetration. We get all that for free. When

建筑模型
Model

you talk to a real-estate agent he will tell you that the most attractive apartments are the corner apartments. In this house we only have corner apartments. Essentially designing homes, becomes the addition of modules, to reach the desired flat size and configuration. It's a bit like buying land on an urban hillside. Why even bother to design the flats? Why not leave it up to the users to configure their own domestic setup?
Since Danish salarys are high, the amount of time you save will mean massive savings or increases in quality. As a consequence bathrooms come as preassembled sealed pods, complete with tiles and fittings. We decided to take that principle one step further and consolidate all utility: bathroom, toilet, kitchen, bedroom wall and closets in to one condensed super pixel. Once you have chosen your territory all you need to do is place a super pixel or two and you house is done.
As a last act of persuasion we knew that even though private developers might be excitable by clear thinking and good looks, they are more occupied by the bottom line than the skyline. So to prove that this was really doable with standard techniques we decided that nothing would be more disarming than if you could build it in LEGO. It happened that on the 1:500 scale model of the building, the size of a single pixel was identical to the size of the smallest one-dot LEGO brick. It even turned out that Lego had a website where you can download free software and built virtually with all the LEGO pieces available. When you are done, you click submit, get a price quote and 3 days later you receive a box from LEGO factory with an image of your design on the cover and all the bricks to build it inside. So, one of our architects spent 2 days of his life building it on-line, and another 2 days to repeat the trick real-life. So at the end of the presentation we gave the client his own LEGO-project (preassembled). He passed it on to his son, and gave us the commission to do it. So with a little luck (and considerable speed) in three years the Copenhagen skyline will be populated by a new silhouette that combines the stepped figure of the traditional spire with the rationality of a functional structure. Gaudi meets Herzberger - or blob meets LEGO.

整体表现图
Rendering

1800 1900 2000

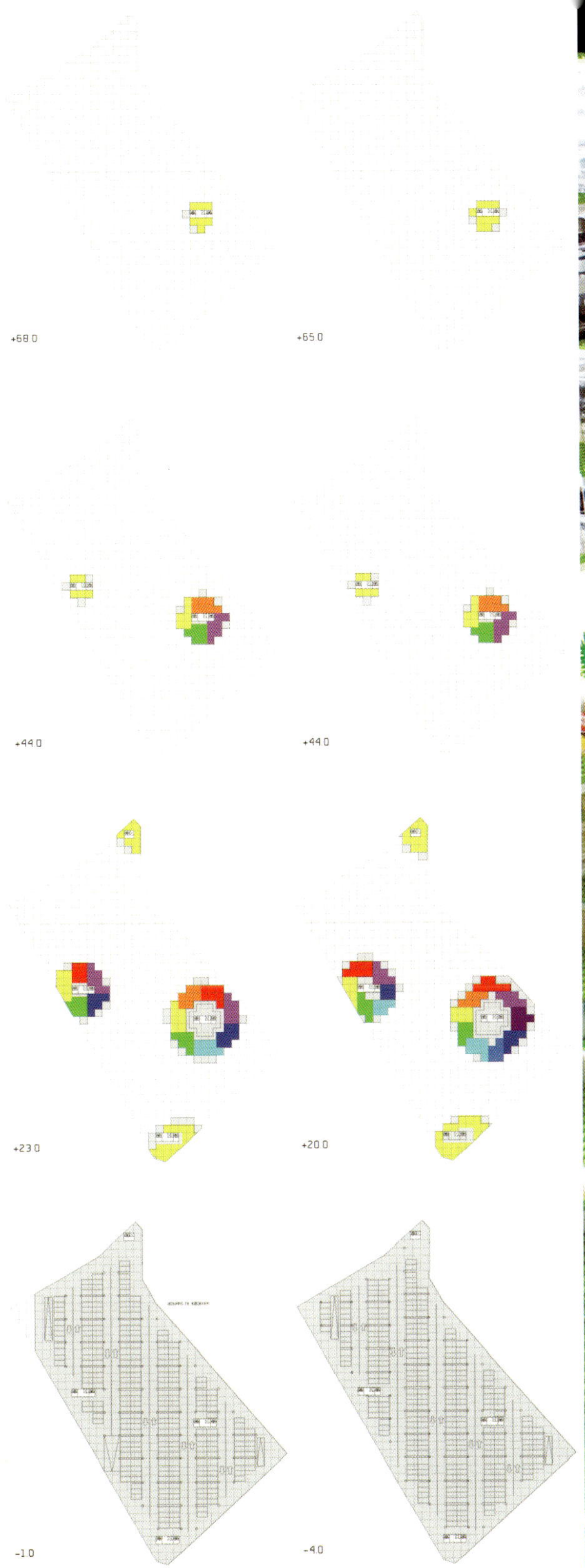
+68.0
+65.0
+44.0
+44.0
+23.0
+20.0
-1.0
-4.0

Chief Architect / Office Name: HyoMan Kim, IROJE KHM Architects

Lim Geo Dang

森林生活

TRADITIONAL BUT MODERN

"Traditional but modern!" That the requirement of clients, in their thirties, who worked as authors and publishers. For a long time, successful translation of tradition of tradition expressed in a modern vocabulary has been a goal of design for Korean architects presently. That has been a very complex puzzle and a difficult problem to solve. However, it's essential to study the heritage of tradition, since tradition could be the motive of creation.

DRAMATIC SEQUENCE OF SPACE & PICTURESQUE FRAME

The spatial design concept set for this small house was based on is dramatic circulation, which was characteristic of traditional architecture in Korea. The circulation of ancient Korea continues infinitely in space, overlaps and penetrates each other asymmetrically, and finally, produces several picturesque frames by surrounding nature and building.

EIGHT PLACES OF "MADANG" (TRADITIONAL INNER COURT)

At this time, the house has to be a shelter from the outer world. It is necessary to contact with nature which brings the urban house rich changes and variation during its life time (i.e., witnessing the blossoming and decaying of flowers). "Madang" (open court yard) was the very "nature" in ancient Korea.

SPATIAL KOREAN TRAITS

Eight places of "madang" located in different levels were planned to produce attractive and interesting drama of spaces. All the major spaces of this house were designed to have a direct relationship to "madang". Entrance court, basement courts & , jungja (pavilion), service court, inner court, and upper court & were all the "madangs" at different levels that linked each other with both horizontal and vertical circulation. There are interesting changes, variety, through the circulation between the "madang".

It is a very small site to have exterior spaces of desirable size by common layout of rooms at ground level. To have as wide an area of "madang" as possible, only a living room and a dining room are placed on the ground level; the others are either below ground or above ground. Moreover, the bed-zone-mass of the upper floor was built with a piloti structure; therefore, the space below can be used as "madang", too.

This so called "RU" type program brings the house several merits of architectural condition. First of all, it gives the inner court a comfortable atmosphere because of the surrounding "madangs" which play important roles. Secondly, it creates another "madang", which was a part of the inner court, covered with a roof. Last, without any fence (a solid fence cannot be built in this area according to the regional building code), it separates the house from public road, and, at the same time, it solves the problem of relationship between the "privacy" inside and the neighborhood outside. "Madang" is a buffer zone between the house and the public road. It is also the place to contact with the neighborhood.

Project Team: N / A

Living on Display — Harmonious Combination of Private Housing and Public Sales Site

私人住宅与公共销售场所的和谐共存

"Living on Display" is a "store and storage" prototype that incorporates housing units and a refrigerated warehouse, which also houses retail and wholesale programs. Its ambition is to learn from, reinvent and thus preserve the industrial. It also strives to raise awareness about refrigeration and its related energy consumption.

By having the kitchens of housing units and the warehouse share a refrigerated space, this prototype radically exposes the private side of people's daily lives to the public. The retail and wholesale program adopts the 'cutting out the middleman' business model which shortens the food supply chain. The instal¬lation of a conveyor belt not only minimizes waste by serving in small portions but also examines the notion of movement, display and spectacle.

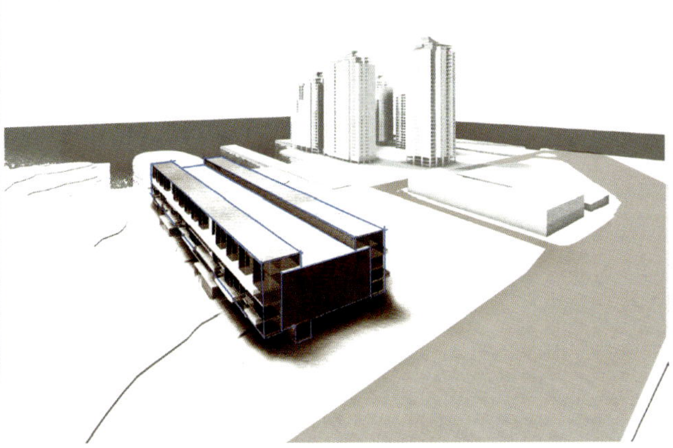

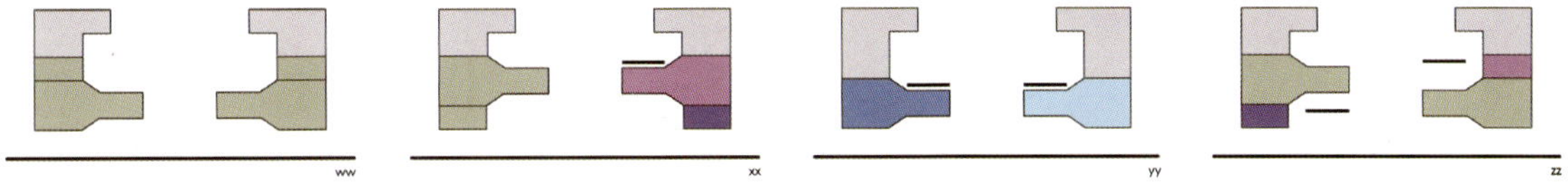

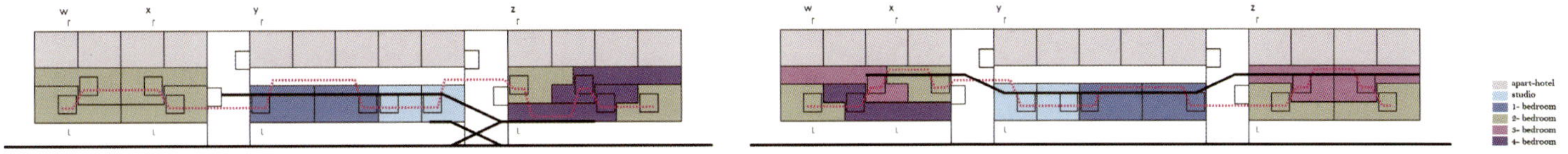

Distribution
Scale 1/32"=1'

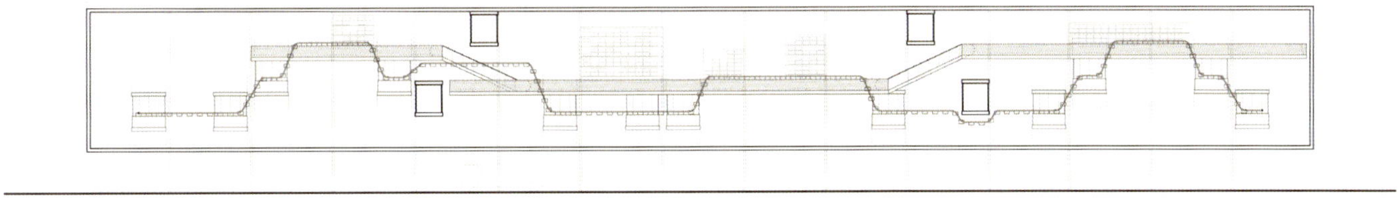

Section AA
Scale 1/16"=1'

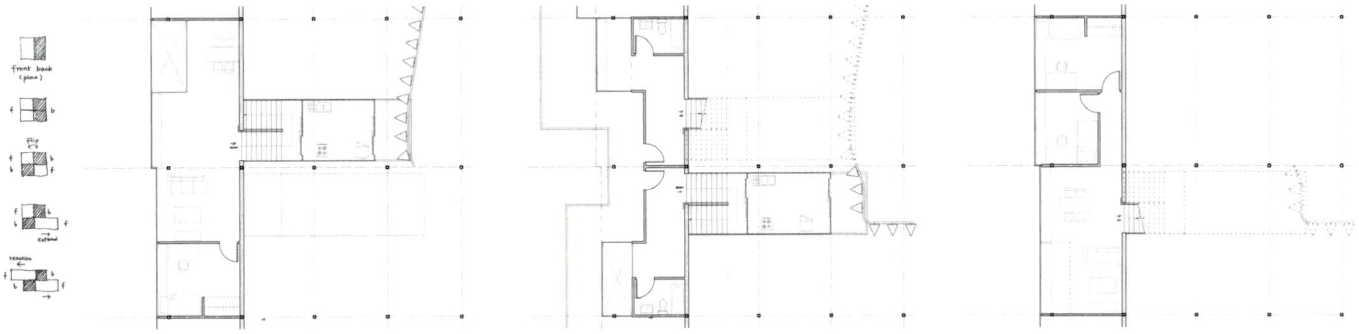

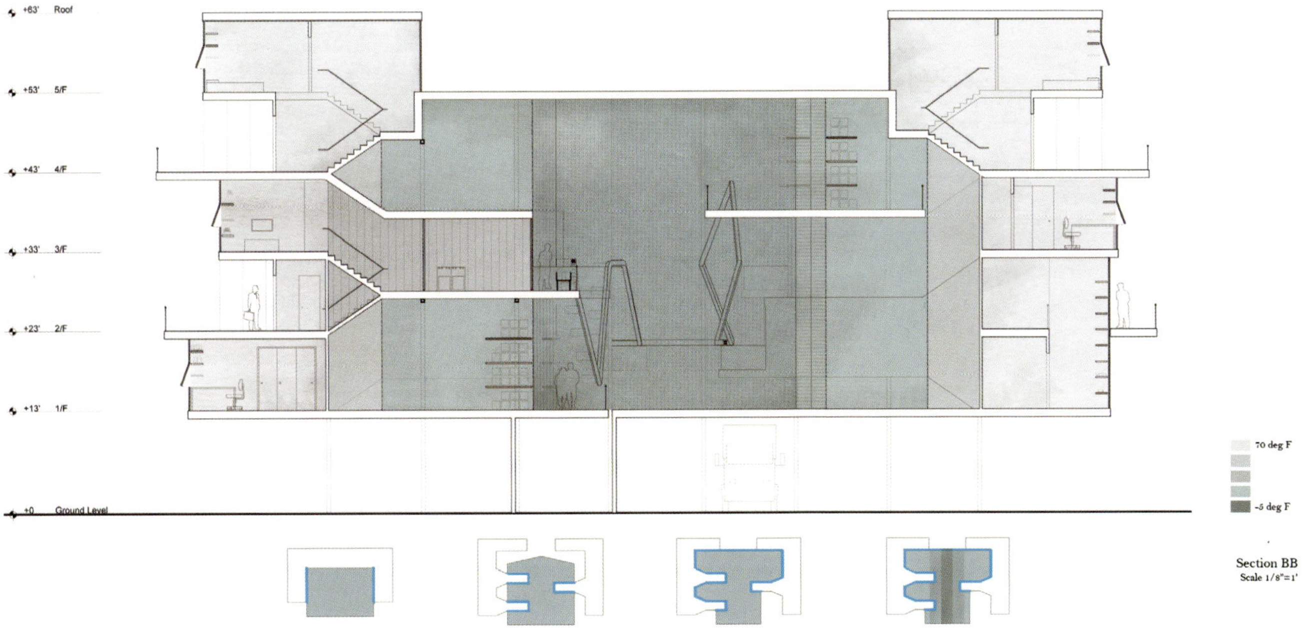

Section BB
Scale 1/8"=1'

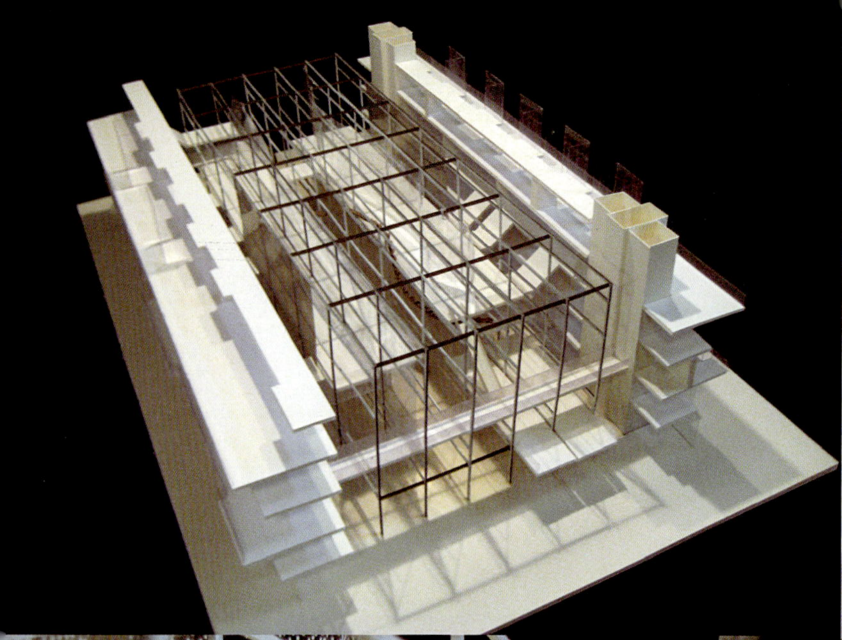

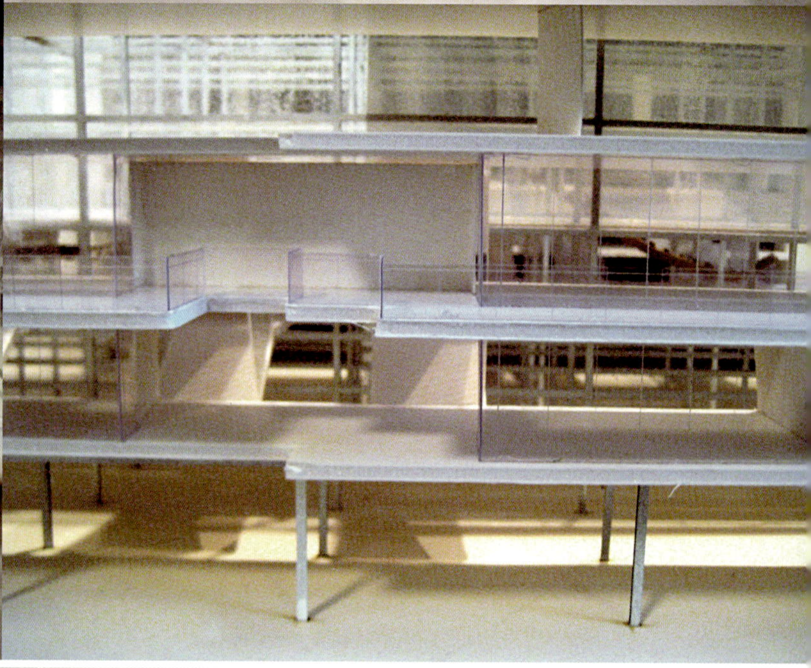

Crayons

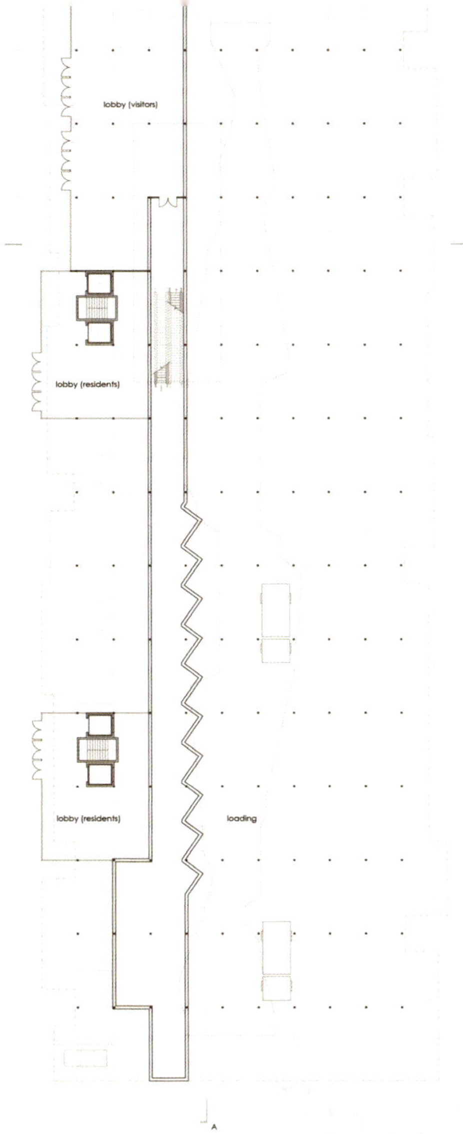

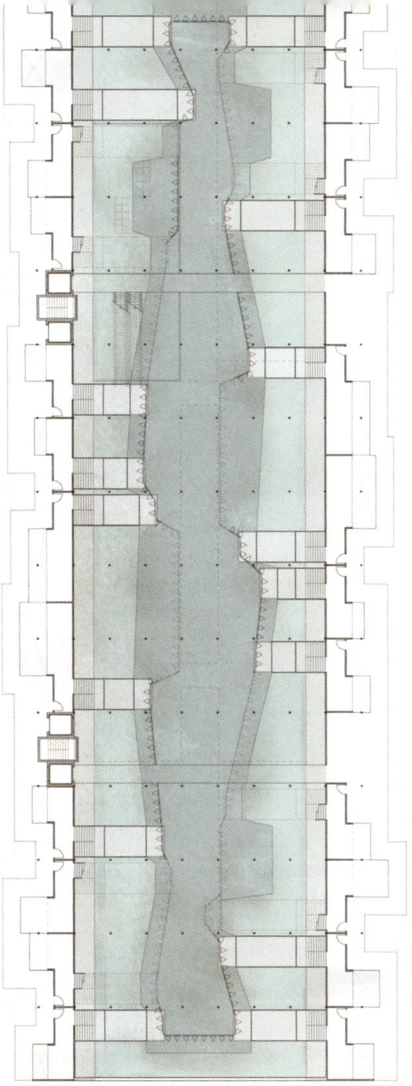

Ground Floor Plan
Scale 1/16"=1'

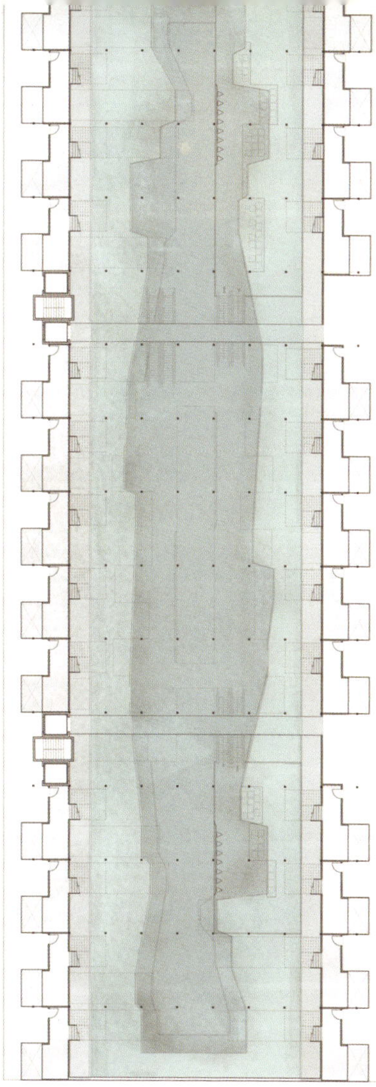

3/F Plan
Scale 1/16"=1'

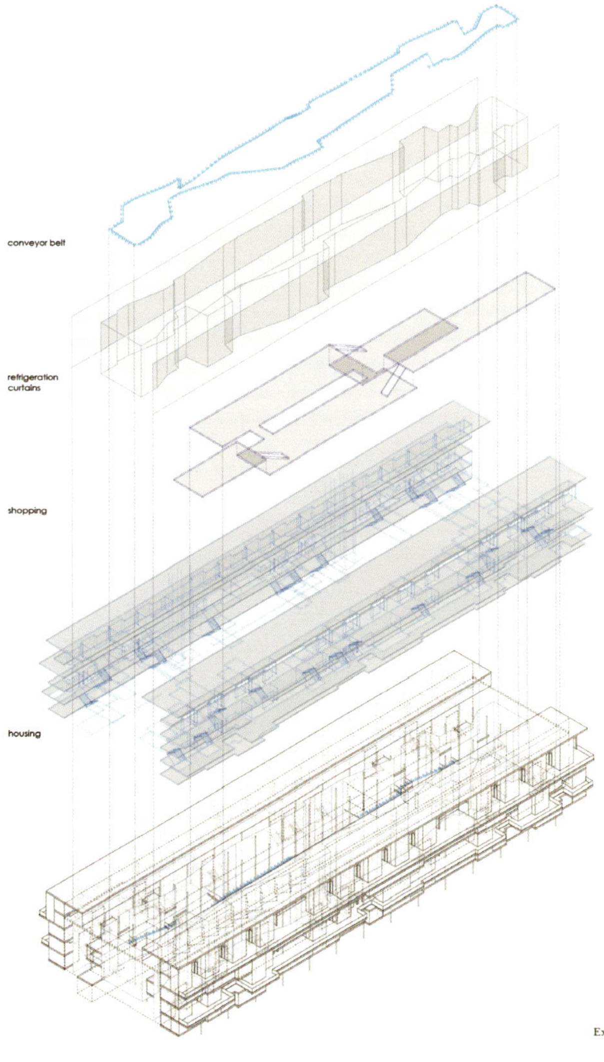

Exploded Axonometric
Scale 1/32"=1'

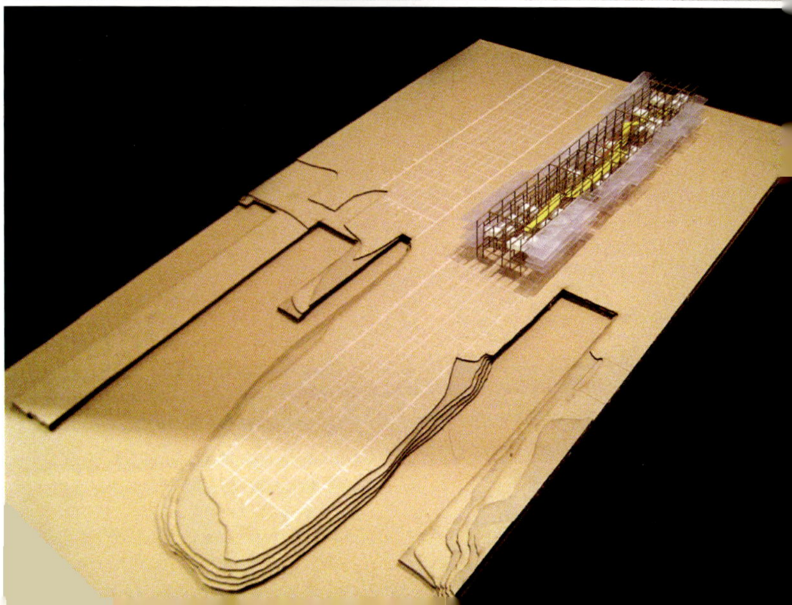

Low-cost Warm Homestead – School and House for HIV Orphans in Uganda

低造价的温馨家园——乌干达艾滋病遗孤校舍

The client is a non-profit organization.
Though the contributed budget is so limited ($30,000) that I aim at a space where the orphans can create their community like a family, by using domestic low-techniques (brick wall and wooden roof structure). The orphans also join the construction to learn the way to build it.
To the great deal of land of 9 ha without any infrastructure in the project, I found myself planning the required rooms, 2 big bedrooms, 3 lecture rooms, 2 offices and a restroom by avoiding such a way as modern zoning which works well in an urban situation. I rather consider that the architecture should be divided into 8 primitive huts and mixed up to a village-like shape as much as possible while they are considered to be functionally related with each other.
I respect the fact that the children gather and have classes under a big tree, so the tree should be situated in the center of the new architecture. Then the 8 huts surround the tree and link with each other. The in-between spaces have canopies which connect the all huts together and create triangular outdoor spaces, so the court is pleat-like open space with the triangular spaces in the plan. The canopies also can make shadow on the court and trigger the activity of the children. They can run, play and work there comfortably even under the strong sunlight in Uganda.
The center court has a well and the roof has solar panels producing electricity, so that the children will be able to stand their own feet here with their farming.
In the future, as the children grow, they extend this orphanage with a lot of huts and it will become a real village, which lasts changing. I expect that this architecture should be a focal point of the village. This concept can be transferred to the architecture of any developing countries as no special technology is necessary.

建筑面积扩长草图
Sketch: Expanding

建筑模型：地面视角
Model: From Ground

村[illegible]不断向外扩[illegible]
Expanding Village

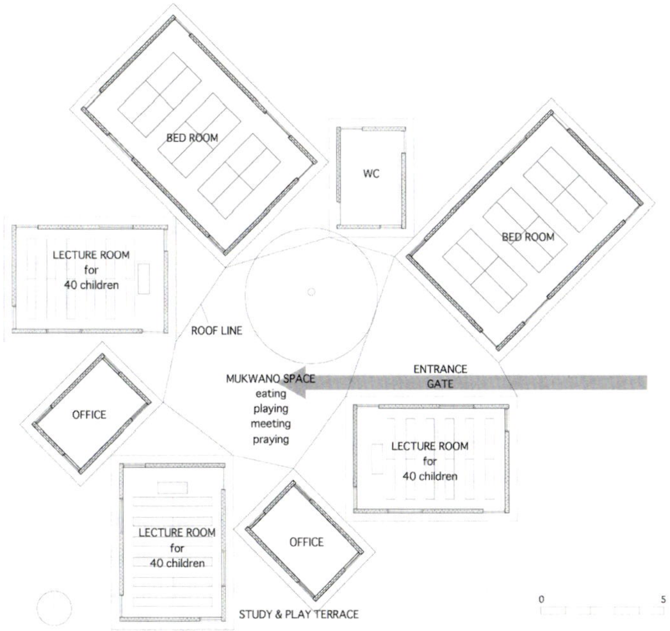

总体规划
Plan

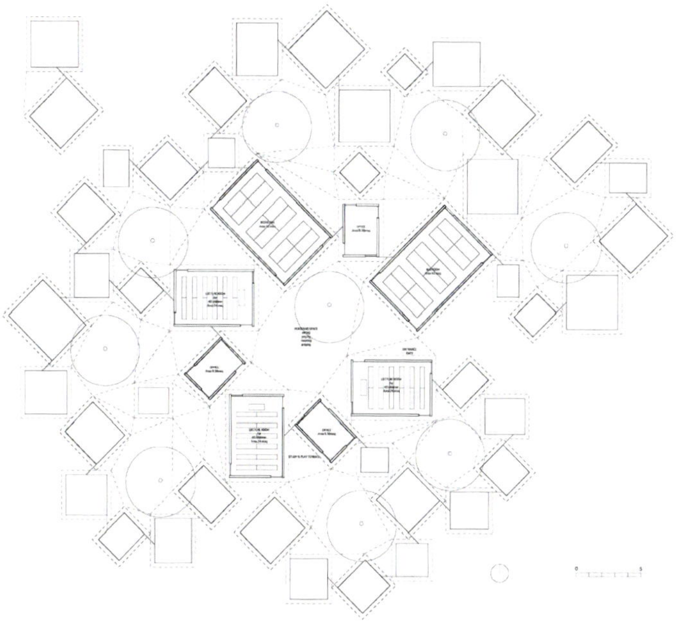

总体规则：扩大建筑面积
Plan: Expanding

简陋的小屋群成为他们的家园
Primitive Huts Gather and Become Their Home

建筑模型：砖墙与木质屋顶结构
Model: Brick Wall and Wooden Roof Structure

Project Team: Jedrzej Lewandowski / Lukasz Skirzynski

Luxurious House on the Water

水上豪华住宅

Luxury House on the Water located in the Zante, Zakynthos, Greece Island.

The site of the House on the Water gives a great opportunity to challenge the theme of the water house by making it an icon of endurance and solitude.

The proposed building is designed as a single family house.

Form of the House on the Water is loosely based on the form of a yacht. It is accessible only from the water through the boat landing-stage which is movable in order to respect the water level differences.

Facing the water The H.O.T.W. provides an outstanding sea view from every room space. Admire the beauty of the surrounding water and cliffs while having a rest on the breeze in the inner garden or the top deck.

The architecture of exterior and interior should explain everything.

House On The Water is a fully sustainable design. Research on water desalination, energy accumulation, ventilation methods, water recycling, heat and energy consumption, tidal and solar energy was made during the design process.

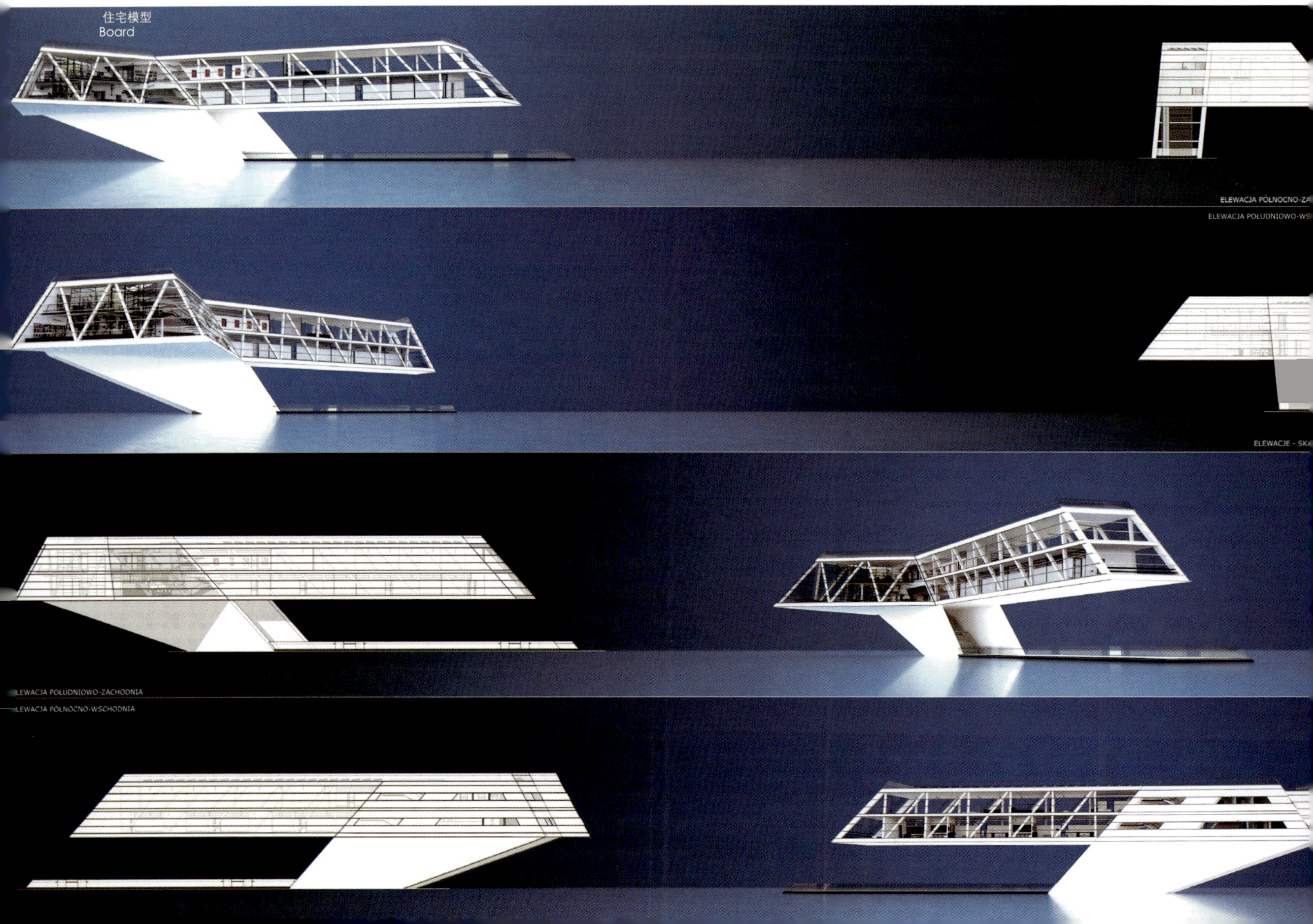

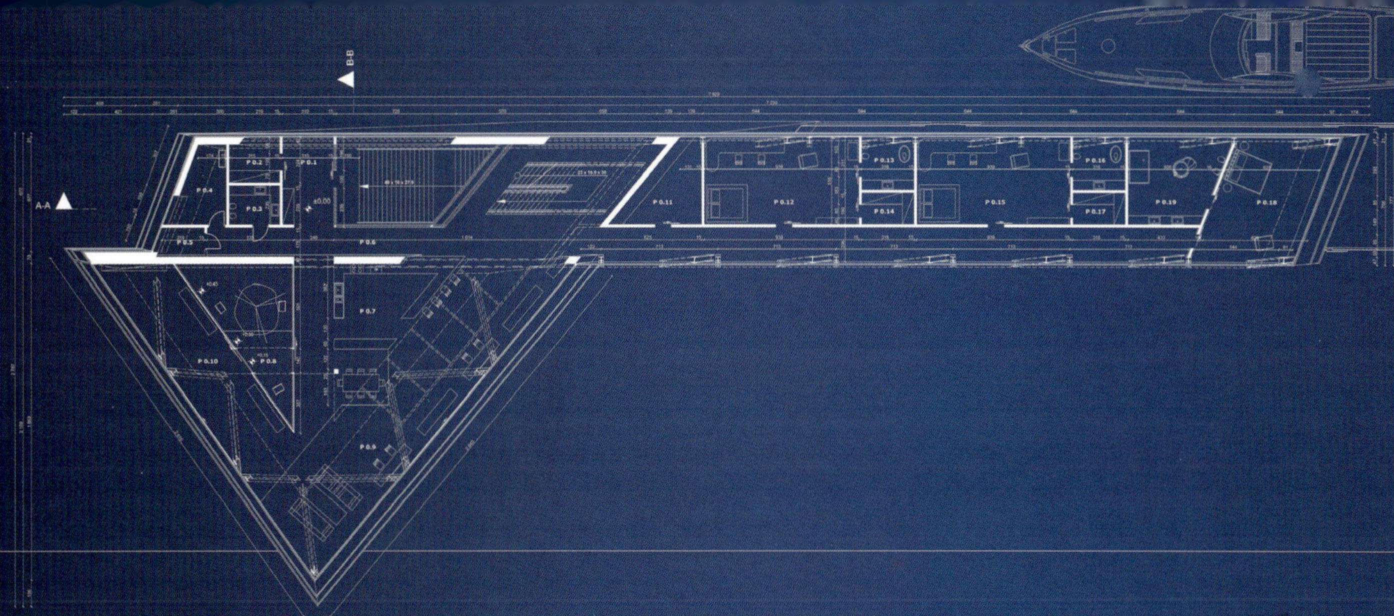

OZN.	NAZWA	POW. [m2]
P 0.1	HALL WEJŚCIOWY	15.04
P 0.2	GARDEROBA	7.05
P 0.3	TOALETA	7.39
P 0.4	CHŁODNIA	12.56
P 0.5	POM. NA ODPADKI	4.24
P 0.6	KOMUNIKACJA	184.37
P 0.7	KUCHNIA	39.58
P 0.8	OGRÓD	36.68
P 0.9	SALON	114.18
P 0.10	CZĘŚĆ BIBLIOTECZNA	62.27
P 0.11	POM. GOSPODARCZE	13.44
P 0.12	SYPIALNIA 1	48.38
P 0.13	ŁAZIENKA 1	9.85
P 0.14	GARDEROBA 1	6.06
P 0.15	SYPIALNIA 2	48.38
P 0.16	ŁAZIENKA 2	9.85
P 0.17	GARDEROBA 2	6.06
P 0.18	SYPIALNIA 3	45.92
P 0.19	ŁAZIENKA 3	27.42
K 0.1	KEJA	357.74

RZUT PARTERU - SKALA 1:100

WIDOK DACHU - SKALA 1:200

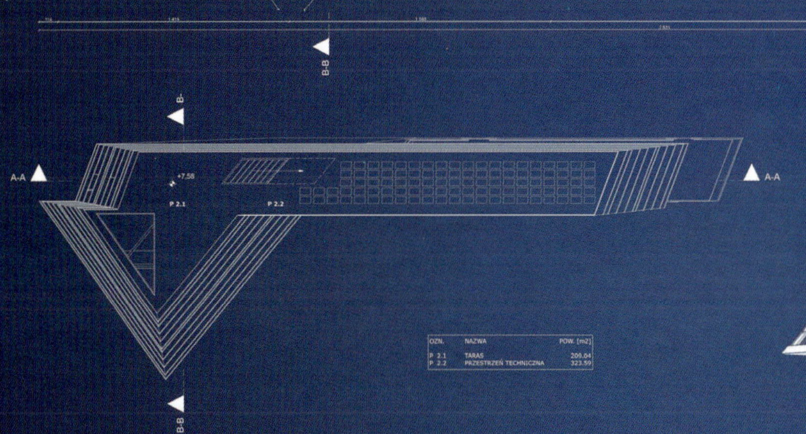

OZN.	NAZWA	POW. [m2]
P 2.1	TARAS	209.04
P 2.2	PRZESTRZEŃ TECHNICZNA	323.59

住宅模型
Board

DOM NA WODZIE

PRZEKRÓJ A-A - SKALA 1:100

PRZEKRÓJ B-B - SKALA 1:100

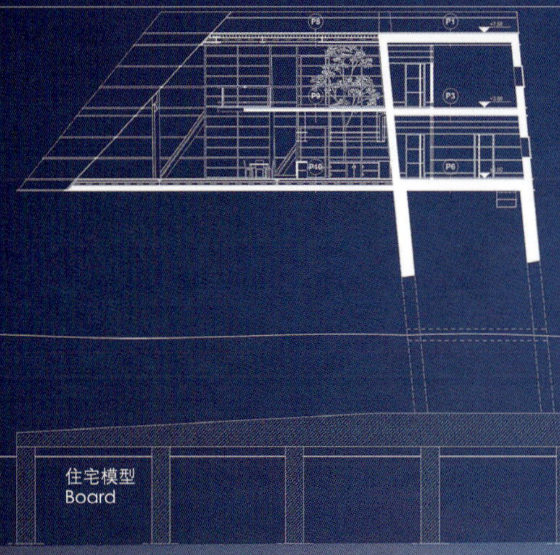

住宅模型
Board

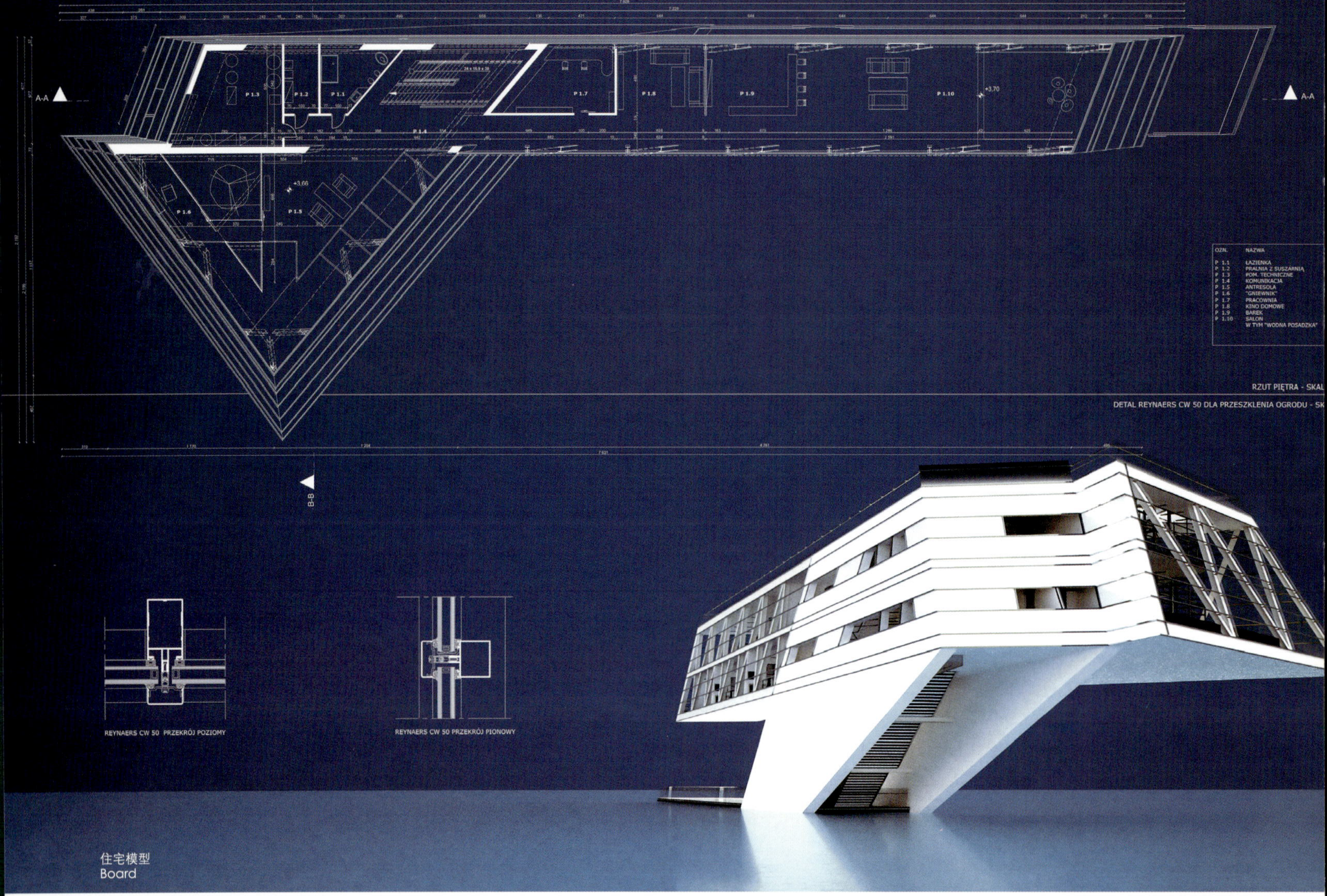

住宅模型
Board

住宅室内布局
Interior

住宅室内布局
Interior

住宅室内布局
Interior

住宅室内布局
Interior

Team: Christoph Jessnitz, Libby Jessnitz

M House (Module House)
模块住宅

The idea to invent a concept like the m_house occurred by looking at trailer parks, which are used in the U.S. as an alternative form of residence for low income families.
The problem was how to provide a dwelling that is affordable, flexible in size, and movable if necessary while being attractive enough to be interesting to people of all income levels. By doing so, this would weaken the ability to judge the financial situation of a family by the appearance of their home.
The concept for the m_house is based on the functional requirements of temporary exhibition pavilions. The structure is independent from the ground condition, consists of uniformed members, and provides all necessary connections like electricity and water. The wall modules are attached only to the structure and therefore easily exchangeable. They are made of various materials with an endless variety of designs and functions. Inspired by the concept of Open Source software, the question evolved that if it would be possible to develop "open source architecture" and have it available to companies that usually have nothing or little to do with architecture such as companies that produce electronics, appliances or furniture.
This expanded involvement creates a new way of how to design, provide, use and recycle residential buildings.
After deciding on the function and overall form, the users have the ability to choose their modules from a vast number of recourses to meet their needs and to give the rooms their desired function. Depending on how much the users plan to invest, they can purchase modules produced for retailers like home improvement stores, order modules online which allows a more personalized selection or order individually designed and produced modules for a high-end customization.
The erection of the building is simple enough to be done by the user, some friends and one trained m_house coordinator within a very short amount of time. The only requirements are the preparation of the ground wherever the "feeds" of the structure come into contact with it, a small scale crane and the connection to electricity, water and plumbing.
A major feature of the m_house is the possibility to add, reduce or simply to exchange wall modules at any given time. If the number of family members live in one unit changes, it might be necessary to enlarge or reduce the available space. The change of income could allow for an upgrade or downgrade to save money. Furthermore, it makes it possible to keep your building up to date in regards to energy standards and personal comfort. A worn out module can easily be replaced or repaired.
What makes the m_house different from other module buildings is the idea of involving multiple companies and organizations of various interests.
It is imaginable, that modules can be donated by companies and sponsors and that used modules are resold by services like Good Will.

模块娱乐区
Module Entertainment

单户家庭住宅
Single Family House

屋顶住宅
Roof House

浴室
Bathroom

模块娱乐区
Module Entertainment

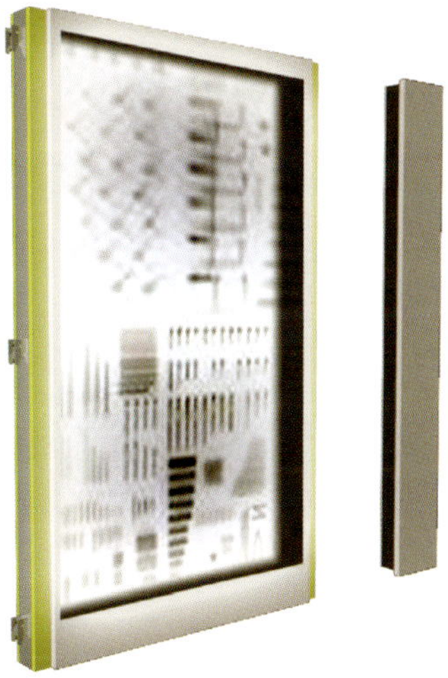

车库的设计
Module Garage

图片墙
Module Photo Wall

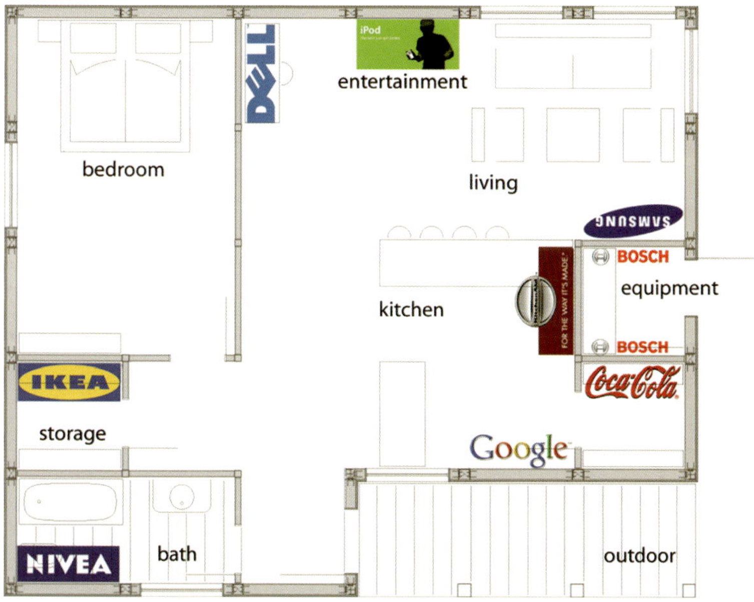

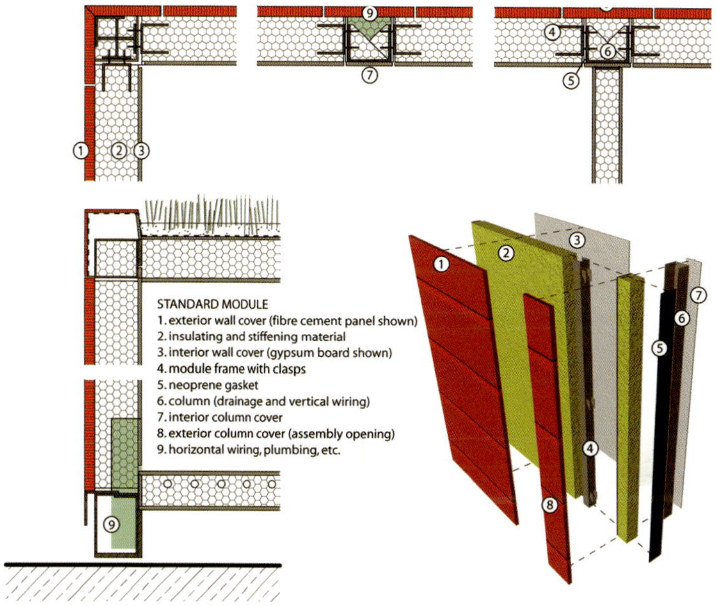

细部图
Details

截面图
Section

MACRO Museum of Contemporary Art in Rome

MACRO罗马当代艺术博物馆

Inserted into the complexity of an ancient industrial estate and confronted to a historical patrimony, the new Museum is resonating to the existing static condition of the site with dynamism, movement and evidence.
In the architectural work the section reveals the non-visible. In the Museum the section is shown through the translation from horizontal to vertical, from inside to outside, from the foyer to the roof-landscape-garden.
The old and the new, the exhibition spaces and the other activities connected with it are at the same time articulated and specifically distinct. In the system of transition in which the new contaminates all around, every part wins a complex clarity and becomes attractive.
The roof surface is a landscape. It is a projection surface of the art trail below. It creates an abstract continuity.
It's an abstract art garden. This landscape terrace is offered to the visitors of the Museum and to the city. It offers multiples textures under the feet of the visitors: roughness, deepness, smooth, softness, density, brightness...The central volume of running water is there for refreshing the air. Sand, gravels, smooth limestone, bright surface of water, gramineous... the landscape terrace becomes a place of calm, freshness and sensual textures.
The exhibition room, by its nature and the courses and flux introduced inside is a landscape. The all Museum is as a running space, from the public space to the private space to the public space in a continuous fluidity.
In opposition with the dynamism of the spaces around, the exhibition spaces are perceived as more neutral. They are developed in a non regular, but simple, space given to the artists, offered to multiple ways to exhibit: hanging, suspending, floating, on/in the floor...

Architectural Design: Yatin Pandya (Principal Architectand project incharge-Vastu Shilpa Foundation)

Manav Sadhna Activity Centre — Ahmedabad

艾哈迈达巴德市的Manav Sadhna活动中心

CONTEXT

Nearly 27.4 million tonnes of waste is produced daily in the urban centers of India. Cities like Ahmedabad alone produce 2,750 metric tonnes. Unfortunately nothing really gets processed of the same. This waste is simply dumped openly in the landfill sites, which uses enormous volumes of fossil fuel, creating an altered, polluted, unsafe and unhealthy landscape. Thankfully India has a well-established tradition of waste recycling which is clearly demonstrated in daily practices and lifestyle. By giving waste or surplus food to beggars and animals, the leftover food waste goes beyond its primary life cycle. Food along with many other objects, are given added value for their multiple uses and diverse applications. Can the building industry not learn from these applications? An activity centre at Rama Pir Tekra, Wadaj in Ahmedabad has been one small attempt in the direction of recycling municipal / domestic waste into building materials.

The activity centre is located amidst the largest squatter settlement of Ahmedabad, and was created under the initiative of the social NGO, Manav Sadhna. The multi-purpose activity centre serves as an informal school for young children, provides evening education for adults and serves as a training centre and activity workshop for the manufacturing of craft based products by women and elderly. The campus also includes a dormitory, an administrative unit and an all-religion meditation unit. The campus is built using components prepared through recycling municipal / domestic waste. This process simultaneously addresses environmental concerns, economic issues and affordable housing; As municipal waste from the domestic sector is used for producing building components, it helps to reduce waste as pollution. Through value addition processes of recycling the waste, it provides a means of economic activity for the poor as well as a sense of empowerment. Finally as the recycled building components are cheaper and of higher quality than the conventional materials, they provide affordable and superior quality building alternatives for the urban poor.

CONCERNS

Non-polluting environment, economic empowerment and affordable built forms are the three key dimensions of this initiative. The project is an outcome of over three years of empirical research at the Vastu Shilpa Foundation for Studies and Research in Environmental Design, with the goal of converting municipal waste from the domestic sector into building components. First hand experiments and onsite explorations have led to the development of innovative building components that use waste, simple hand-operated tools and local resources and know-how. The project also demonstrates that building can become an economic activity, empowering the poor. It shows potential of becoming a cottage industry for economic self-reliance and possibilities to improve the quality of their homes using the affordable alternative building components.

CONSTRUCTION

The campus is built as a live demonstration for the application of recycled waste as affordable,

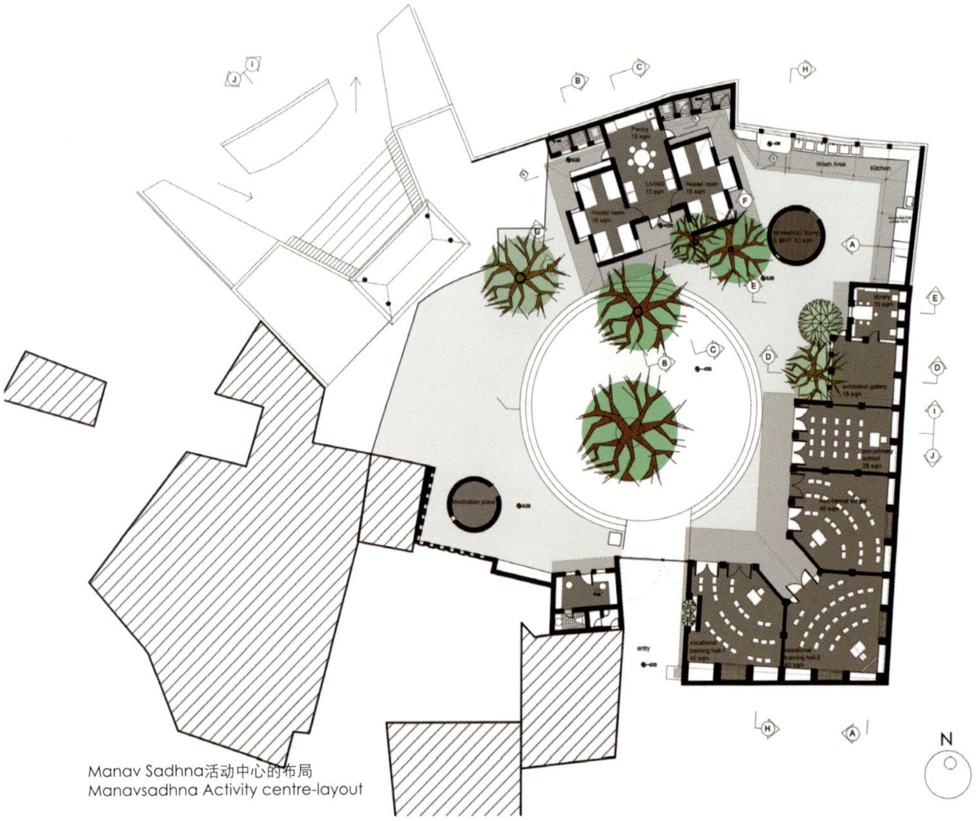

Manav Sadhna活动中心的布局
Manavsadhna Activity centre-layout

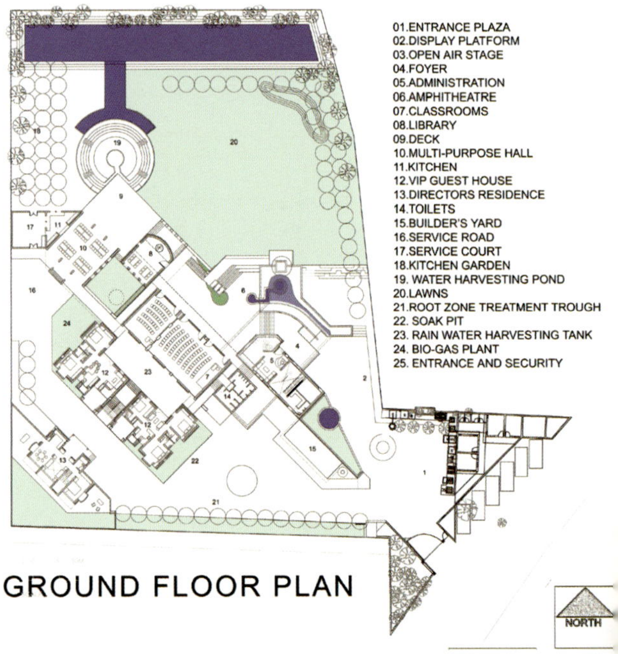

aesthetically pleasing and efficient building components. The products developed for this project, which incorporate municipal / domestic waste and are prepared with simple hand operated tools, are demonstrated in the walls, roofs / slabs, doors and windows. There are six types of materials and techniques used in the making of the walls. These include: cement bonded flyash bricks, mould-compressed bricks made from landfill site waste residue, stabilized soil blocks, recycled glass bottles, recycled plastic bottles filled with ash and waste residue, and vegetable crate wood paneling in the inner partition walls.

Similarly the floor and roof slabs as applied in the activity centre include: filler slab with glass bottles, plastic bottles and bricks, stone slab, cement bonded particle board with clay tile cover, as well as light conduit pipe truss with G.I. sheet with clay tile roof. The door paneling uses shredded packaging wrapper and coated paper waste as reinforcement substitute for fiber reinforced plastic (FRP). Vegetable crate wood as a frame and oil tin container as blades make the ventilation louvers in the toilets. A paneled door using vegetable crate wood and oil tin containers for the frame and cladding respectively is also provided in the administrative block office toilet. Flyash and waste residue moulded tiles with inlaid ceramic industry waste as china mosaic (applied during tile moulding itself) is also applied in patches for their demonstration.

All of theseW products are developed and produced first hand. The products thus produced have been lab tested for their engineered performance and they prove to be economical, environmentally friendly, participatory and aesthetically pleasing solutions and express alternatives to contemporary practices.

利用诸如分层充填开采法、互为遮阴、空心墙、根区废弃物处理、屋顶蓄水、钢丝网水泥屋顶、通风的空心墙的施工技术来适宜地装点校园的环境
Cut and Fill, Mutual Shading, Cavity Wall, Root Zone Waste Treatment, Roof Top Water Harvesting, Ferrocement Roofing and Ventilated Cavity Wall Construction are Some of the Strategies Adopted to Render Campus Environment Friendly

Architectural Project: Pascal Arquitectos, Carlos Pascal & Gerard Pascal

Meditation House — A Tranquil and Classic Spiritual Homestead

冥想住宅——恬静而典雅的心灵家园

This is a project with very strong emotional implications. We had to understand the mood of the user, who at such this moment would not care very much for an aesthetical analysis of any place, but at the same time we wanted to create a space that can create a spiritual mood, for this we referred to ancient buildings that were designed for this purpose, as the Egyptian "Mastabas" or some Mayan buildings in Palenque. A project of this nature must encourage introspection and peaceful visual harmony through a discrete use of materials and lighting. The construction is located in a residential area and being planned as an isolated construction from the surroundings, an indoor - illuminated yard was built. The building façade is completely covered by Grissal flamed granite.

The building welcomes de visitor with a 6´4" wide and 30´ high triangular shaped wooden door which leads to an access tunnel in the same shape, creating a solemn atmosphere as you enter. This darkening experience at the entrance ends when the hallway opens to a large, double height granite hall illuminated by the northern light coming from the indoor courtyard with a tall Dracaena at the center and a symbolical abstract sculpture by artist Saul Kaminer as the only artwork piece for decoration.

No furniture was used inside this place, only a floating bench surrounding the room that is made out of the same wood used in all the building. It also serves to hide all of the air conditioning, speakers and recessed lighting which add a dramatically touch. This way the rooms show no added ornament but the light and shades playing on the granite volumes. The ceiling is made of dark kumara wood and floats without touching any wall and a skylight puts a highlight on this detail.

建筑外观
Main Facade View

入口通道
Access Tunnel

Sual Kaminer雕塑
Sual Kaminer Sculpture

室内天井侧面
Indoor Yard Side View

建筑北面光线取自室内庭院
Northern Light Comes From the Indoor Courtyard

Associate Architects: Md. Roushn Ul Islam, Khandaker Zakaria (Jewel)

Meghna Residence — Living in Delta

位于三角洲的梅克纳住宅

Due to lack of a proper master plan, construction rules and awareness, Dhaka, one of the most densely populated cities in the world has become a city of urban mayhem. Dhaka, having less than five percent green area for a city of more than 400 square miles with a growing population of disparity and persistent difficulties with other utilities, is fast losing its living ambiance.

Dhaka is a densely populated city where most of the single storied residential buildings with open green spaces are fast disappearing and multistoried (six storied) apartment buildings are replacing them.

The Meghna Residence is a single extended family house located at Dhanmondi - a planned residential area of the early sixties. This house has been conceived in three layers-- Its surrounding context, the climate, the traditional typology.

THE SURROUNDING CONTEXT

This is a corner plot connecting roads at the northeast and northwest. The northwest road has median with large mahoganies acting as a screen providing privacy for the house. The house is facing a large open play field and a swimming pool belonging to the women sports federation across the northeast road. On the southeast is a six-storied apartment with a facade having 49 windows peeping over the plot and a two-storied residence on the southwest with potentials of high-rise construction in future.

CLIMATE CONSIDERATIONS

Dhaka is on the 24.5-degree latitude near the tropics of cancer. Having summer winds from the east, southeast and the south shifted towards the southwest carrying monsoon rain approximately 60 inches yearly. The summer solstice does not seem to be critical that starts from the northeast to the vertex at midday to the northwest in the afternoon. But in winter when the sun inclines further towards the south needs careful intervention.

TRADITIONAL TYPOLOGY

In Bangladesh the courtyard is a traditional typology creating space for living in privacy along with cross ventilation, ample daylight and vegetation.

Design considerations and building configuration

To create a screen against 49 windows on the southeast particularly for guarding the swimming pool on the 5th level of the house with much privacy and to ensure winter solstice that is critical

住宅实体
Building With Context

达卡城——城市中被破坏的一角
Context Dhaka - An Urban Mayhem

and summer cross ventilation for the living spaces that is mostly concentrated in the northwest and southwest side of the land was quite a challenge. The effort was further to achieve a traditional living ambience with layers of greeneries and water to achieve a deltaic condition. For this, a "U" shape configuration has been adopted.
Also a "stacking" design in different floors has been adopted to provide much space on the ground and other levels. The part-semi basement along with most of the ground level has been allocated for service provisions i.e. water treatment plant, pressure vessel system, power generation, parking and lift lounge etc. The second level for the formal spaces such as formal living and dining, guest lodge, client's home office and the main kitchen and the third level is especially for private family spaces i.e. all bedrooms, lounges, dining, library, water bodies and the gardens. The forth level used for the swimming pond tray and other electro mechanical support facilities. The fifth level contains sauna, gymnasium, partly covered swimming pond, deck along with traditional "GHAT" and an adjacent large garden. In addition the swimming pond provides a beautiful view of the green field with large trees caught by the sight afar.

位于五层的游泳池
Swimmimg Pool at Level 5

三楼水管
Water Body at Level 3

主卧室阳台一角
The Living Screen from Master Bed Veranda

混合材料
Mixing Materials

热带植物美景
Poetics of Tropics

宁静中的声音
Sound of Silence

通往书房的空间和楼梯
Void and Stairs to Library

让空气自由地流通
Let the Air Flow

家庭生活视角
View From Family Living

Project Team: Vahid Pariz & Afsoon jaberi & Sina Abdollah poor

Memorial of Flightno.655

655次航班纪念馆

Iran Air Flight 655, also known as IR655, was a civilian airliner shot down by US missiles on Sunday 3 July 1988, over the Strait of Hormuz, toward the end of the Iran-Iraq War.

The aircraft, an Airbus A300B2 operated by Iran Air as IR655, was flying from Bandar Abbas, Iran, to Dubai, UAE, when it was destroyed by the U.S. Navy's guided missile cruiser USS Vincennes, killing all 290 passengers and crew aboard, including 66 children, ranking it seventh among the deadliest airliner fatalities. Vincennes was traversing the Straits of Hormuz, inside Iranian territorial waters, at the time of the attack and IR655 was within Iranian airspace.

Freezing history "Memorial of the Martyrs of Iran's passenger of flight No.655"

Memorial the architecture design in more than anything else the sequence of space phenomenon related to happen to the picture, this catastrophe that continuity to the sea that can be objective meaning and place in the search phenomenon. In this thought that even though his concepts (poverty, singular and resistance). All the life's sea to have such a period of history, frozen and seized Most of life and the lost in screen and beautiful ultimate something to build a strange, that into to passing through the ultimate evil. Human society that its continuity to the existence of the debtor afloat in an element quadruplet (Sea), to prove it debt that the life's sea in his time and on the other hand with his glory endorsed the width sea. Kind of space sequence that observers during the access to the catastrophe, sea with two pages in a frame that join mean it has happened in the passage of the two pages ends. The entrance route and that the lines between life and death are drawn to silence in the end, they width sea in a frame between the two pages will be watching a point that the direct and infinite width of the martyrs of this disaster is that lack of it. Tanis in their rotation to sunset during the year also supervisor for the picture, and it muddied lasting pages and worn out by that name engraved martyrs. Pale elevations are element in the design of a massive bizarreries sense and oppression in mind the observer that this feeling has been emanates from the massive bizarreries and oppression of the martyrs of this disaster. Space sequence this path is in the event that as an observer in a trip symbolic disappeared and unknown, the voice of death from a trip that humans, which without any of the destroyed.

This way the thin on the screen of water will be the climax of recklessly before entering, while movement and to see the water hole in the heart of that are shining light. In this sequence supervisor, noticed the bed and reliable tight formation and a soft bed, where the meaning of movement and the body of the other is clear and the presence of all the ideas in space, empty of crystalline light that is an indication of lack of divinity in this sequence.

建筑东北面
North-eastern View

建筑东南面
South-eastern View

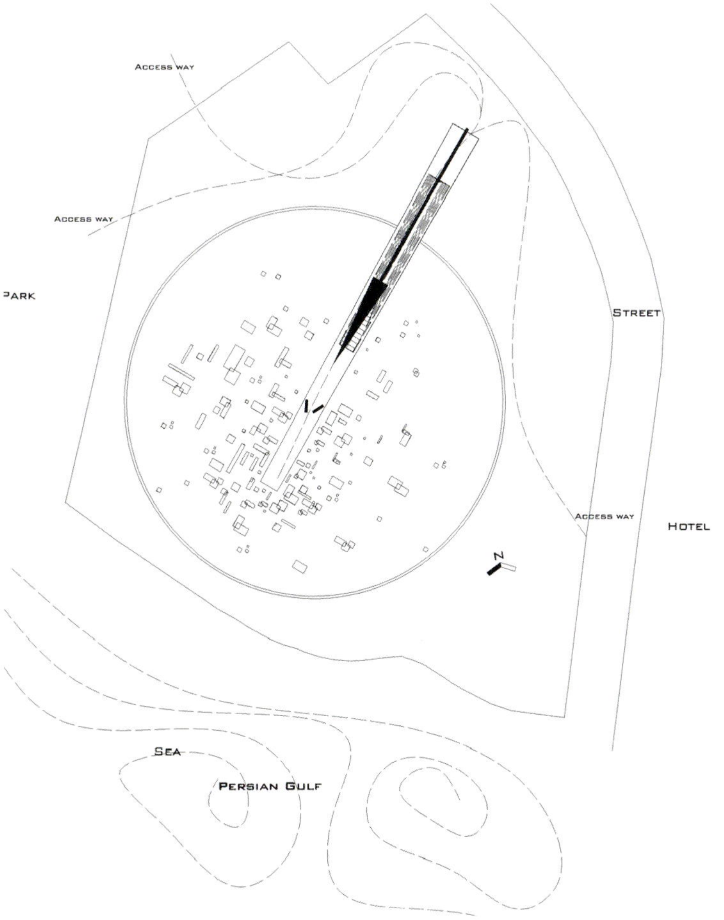

建筑主体
View from the Main Way

建筑东南面
South-eastern Elevation

大海景观
View from the Sea

建筑规划透视图
Plan Perspective

Milhundos House
Milhundos住宅

The ground places in the outskirts of the city of Penafiel, in the valley of the river Cavalum and is inserted in an allotment near the park of Quinta das Lages.
The proposal modifies the implementation anticipated in the original allotment and resumes the duplication rule, not observed in the buildings of the lots L8 and L9, seeking a more articulated solution with the nearest volumes and simultaneously taking the vast ambiguities and disarrangements between the allotment and the lot drawings into account.
As for the house itself, it is organised in two functional, though complementary, autonomous groups, to which correspond two distinct volumes. Each one of these volumes has different relations towards the exterior space: one white and plastered seeks the most distant landscape; the other one, made with dark stone has a more direct contact with its interior ground.

花园及南面处观看到的景致
View From the Garden, South Façade Detail

天井及东面观看到的景致
View from the Patio, East Façade

花园及南面处观看到的景致
View from the Garden, South Façade

入口及西面观看到的景致
View from the Entrance, West Façade

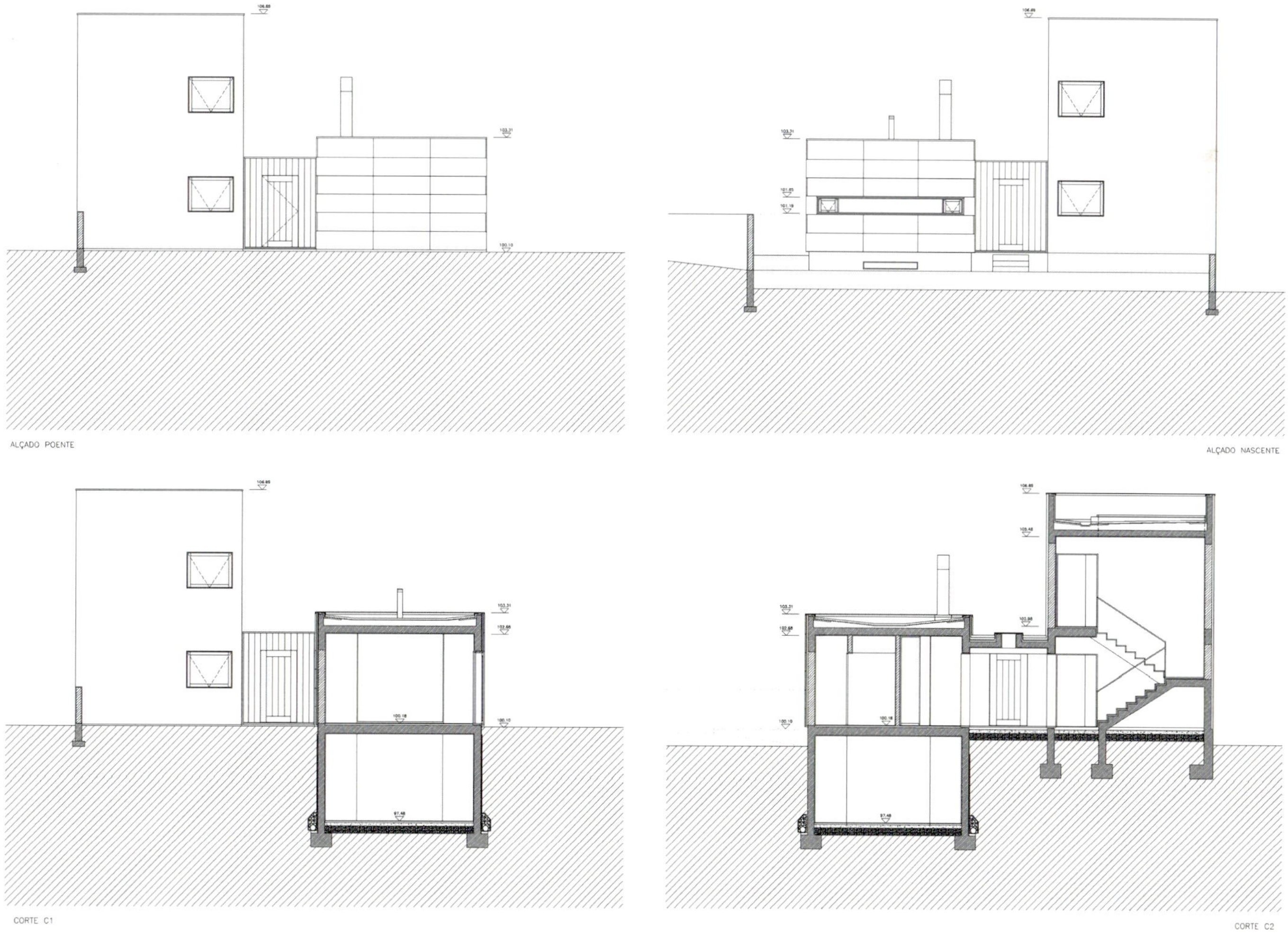

C1、C2及东西面截面图
Sections C1, C2, West and East Façades

西面、厨房及卧室
East façade, Kitchen and Living Room Nucleus

东面及卧室
East Facade, Rooms Nucleus

南面、起居室窗户细部
South Façade, Living Room Window Detail

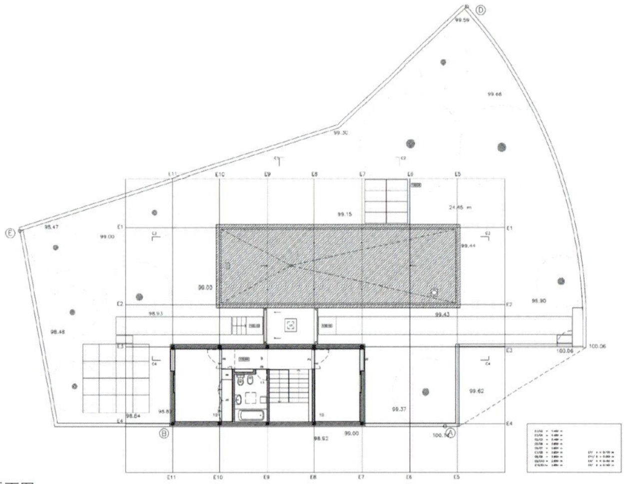

二楼平面图
Second Level Plan

底层平面图
Ground Level Plan

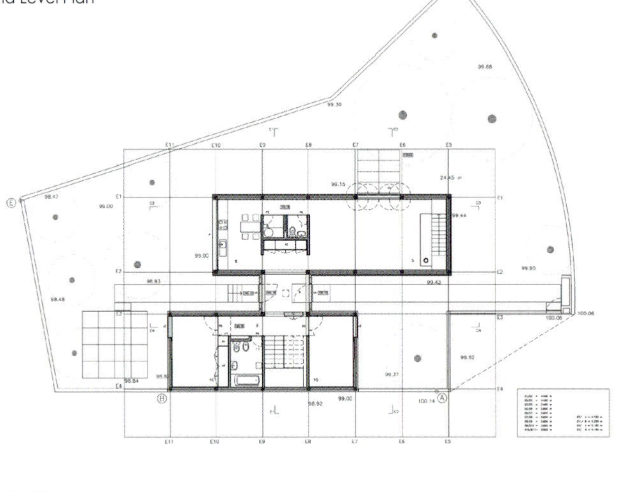

扶梯及走廊观看到的景致
View from the Stairs and Corridor

入口大厅到起居室的景致
View from the Entry Hall to the Living Room

室内景致
View from the Interior

入口大厅到扶梯的景致
View from the Entry Hall to the Stairs

Project Team: Suleiman Alhadidi

Moon Project – Villa

月球度假区

The Moon Villa is an imaginary project which was done in 2007. After an intensive study of the possibilities of making these kinds of projects on the Moon land due to the environmental constrains and moon's gravity we end up with a theme "Shallow".
The site of the moon is shallow with no signs of life on it. The organic shape we designed is to break down the silence of the context and to respond to the environmental forces.

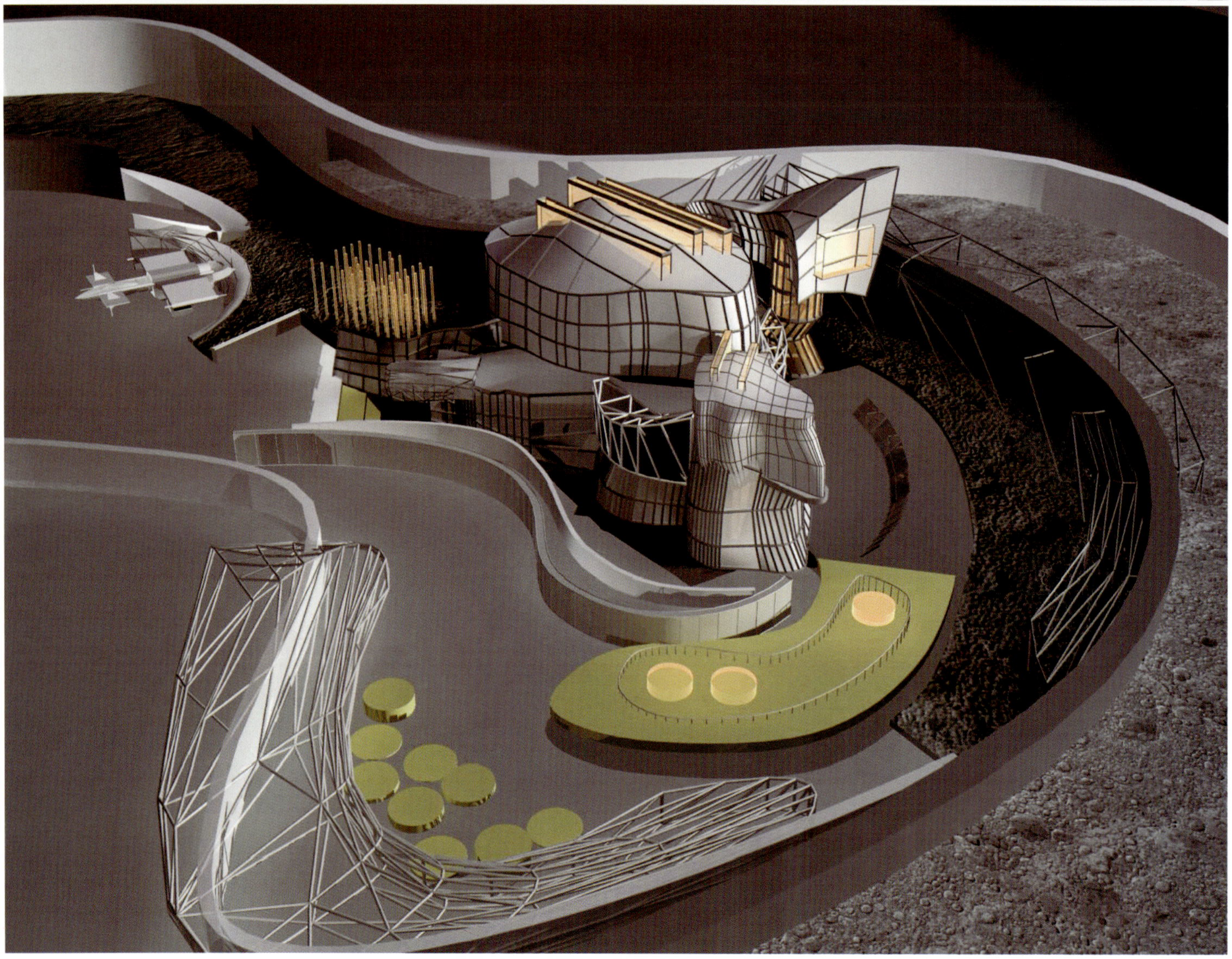

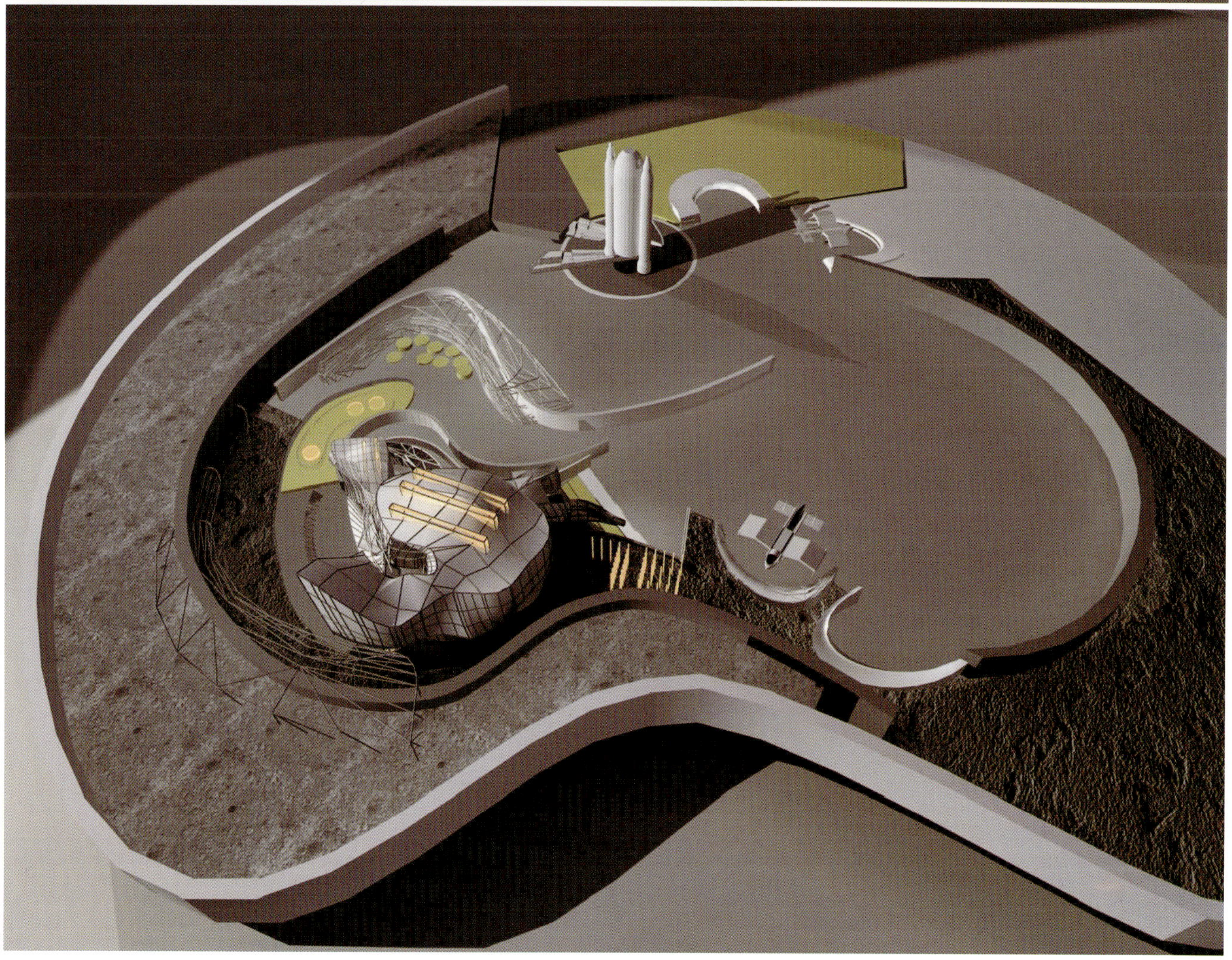

Project Architect: Markus Prossnigg

Musee Des Conflluences, Lyon, France (2001 - 2010)

法国里昂博物馆——现在与未来的完美结合

CRYSTAL CLOUD OF KNOWLEDGE

The future society will be a society of knowledge. However this knowledge can hardly be split into clearly defined fields.

Innovations develop within interspaces, within indistinctness, within the overlapping and hybridising. Questions regarding the future will be decided within transitional fields situated between technology, biology and ethics that are the central themes of the Musée des Confluences. Mutations of form, penetrations, deformations, simultaneities, breakdowns and variabilities have an effect on architecture. The resulting architecture is characterised by the interactions, the fusion and mutation of different entities constituting a new shape.

The Musée des Confluences does not consider itself as an exclusive temple for the intellectual bourgeoisie but as a public place providing access to the knowledge of our age.

Stimulating a direct and active use, it is not only a museum site but also a venue in town. The architecture hybridises the typology of a museum with the typology of an urban leisure space.

HARD SPACE —SOFT SPACE

The concept of two complexly connected architectural units is a result of the striking interface-like situation of the building site.

The crystal rising towards the side of the town is conceived as an urban forum and entrance hall for visitors. Its shape that can be read clearly stands for the everyday world. In contrast to this the cloud hides the knowledge about the future; it is a soft space of hidden streams and countless transitions.

Within the Musée des Confluences the present and the future, the known and the still unknown are conceived as a spatial arrangement trying to "spur public curiosity". As an extension of the park located on the Southern top of the island a new urban space formulates itself; a landscape consisting of ramps and surfaces merging the inside and the outside and resulting in a dynamic sequence of spatial events.

This movement is also followed by the alternating spatial structure of the exhibition halls. Closed Black Boxes and free exhibition areas are alternated by exploiting the double room height of two levels. The architecture is as changeable as the content entrusted.

The idea of the permanent reinvention of an urban event enables Lyon to perfectly position itself within a regional as well as within a global context.

Project Team: Viktor Zotov, Mariya Andonieva, Aleksandr Zhidkov, Sergii Gorlov, Sergey Ferley, Liudmila Bielashova,

Museum of Ivan Marchuk
伊万马尔丘克博物馆

The main urban task is to connect two central public zones: Andreevski slope and Mikhailovska square – this pedestrian way lays on the building body. Main part of the building cover becomes public; this is amphitheatre and outdoor museum at the same time (prolongation of internal exposition).

Very precious city territory defines "hidden" architecture: All vertical surfaces made from glass, all horizontal ones have natural grass cover. Sharp angle of visual perception reflects the park. Part of the existing trees is saved inside the building.

Premises that are adjoined to the cover (with natural light) are offices; museum and picture galleries situated "inside" the volume.

Decorative surfaces, interiors are made according to Ivan Marchuk paintings.

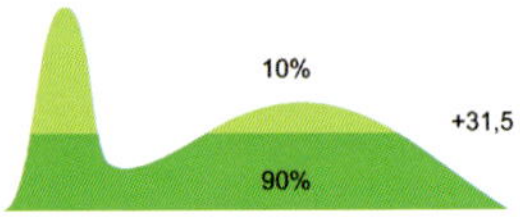

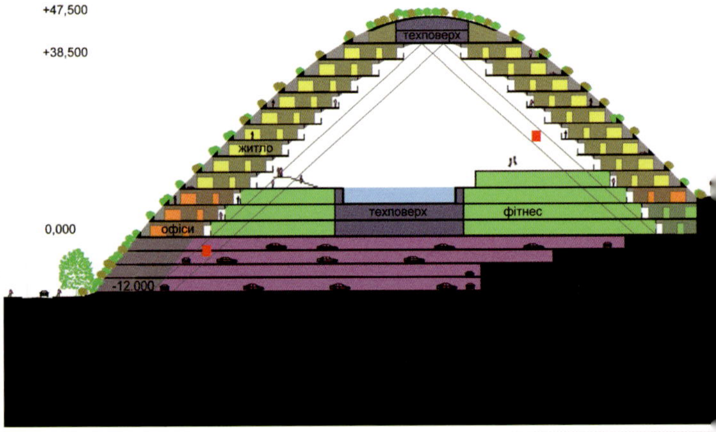

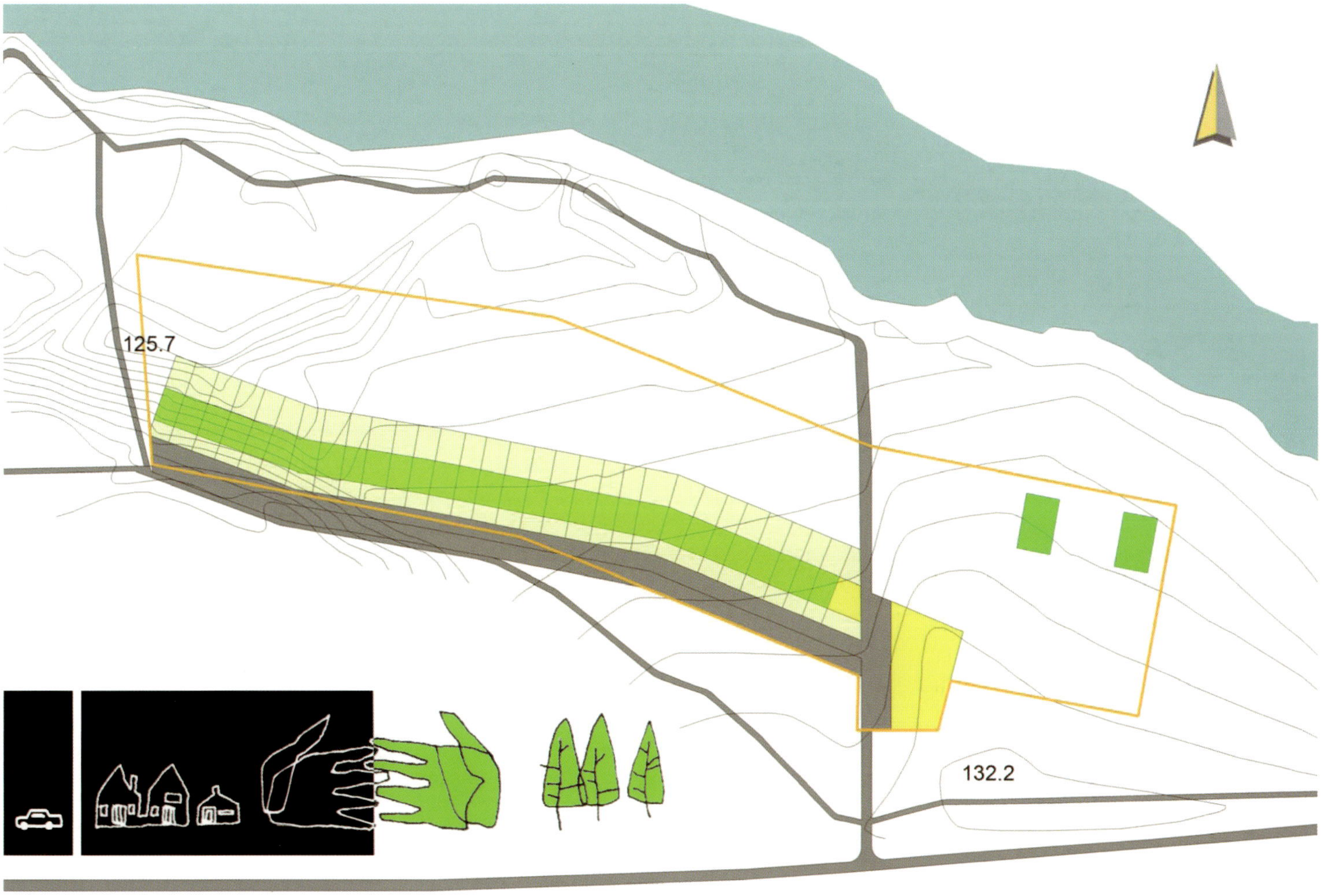
125.7
132.2

ЗЕЙ ІВАНА МАРЧУКА
IVAN MARCHUK MUSEUM

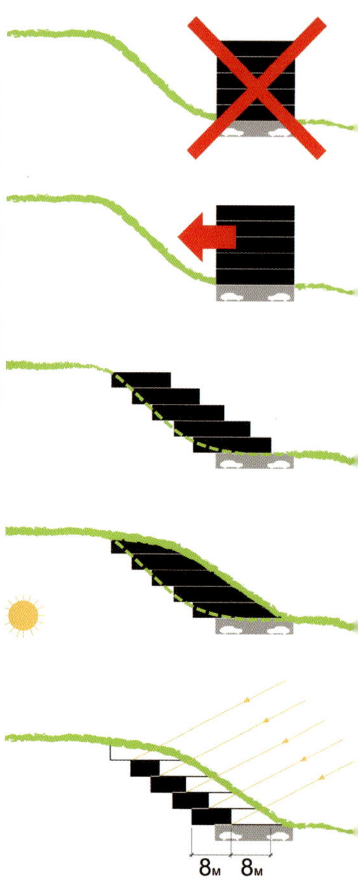
8м 8м

MUSEUM

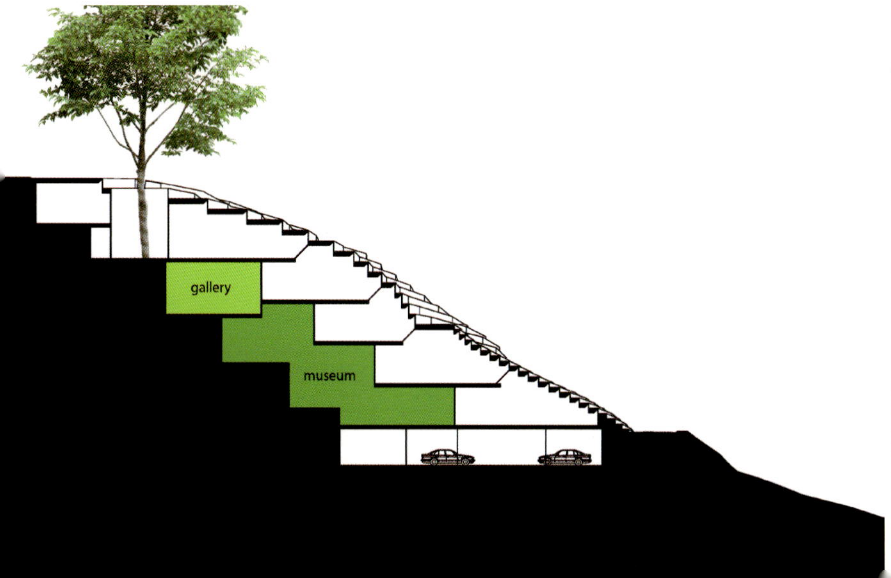
gallery
museum

Project Team: Ali Ghaffari

My House

我的小屋

My house is minimal space for seeing nature, without any traverse between my eyes and my around...
This house set above hills to verge a beautiful mountain who calls avrin in Azerbaijan, in this house i see avrin every day...
It is a minimal idea and minimal design and minimal space.
Less is more...

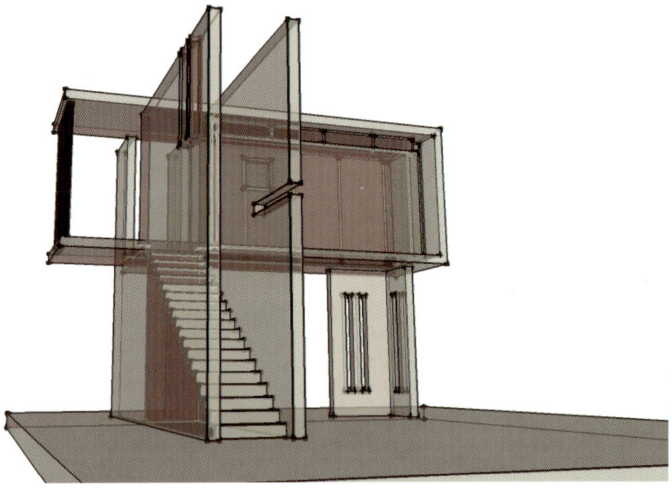

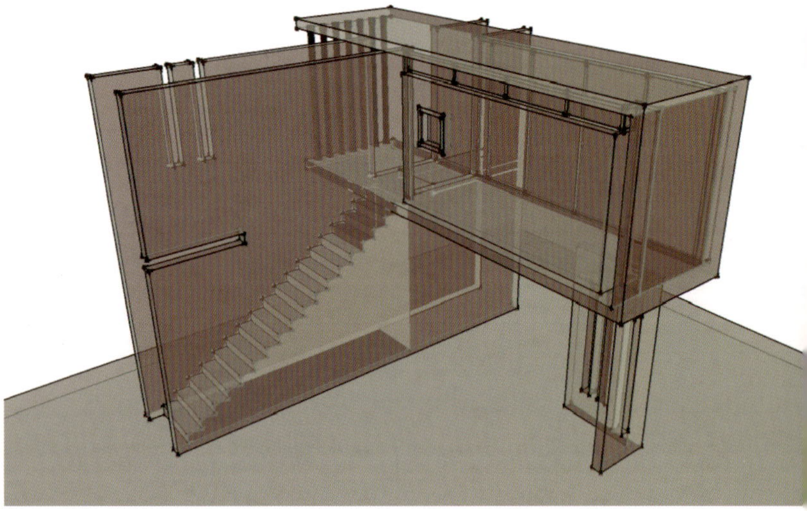

外部景致
Avrin View

透视图
Perspective

透视图
Perspective

全景图
General View

全景图
General View

室内景致
Interior View

Chief Designer: Reza Pourvaziry

Nashr Yadavaran Administrative Building

Nashr Yadavaran行政大楼

The main concept behind the design of this project can be expressed through several layers. The science of "space" is concerned with the different dimensions of both the society and the single individual.
The outer surface of the building creates a transparent effect by allowing the entrance way to be visible in great depth. This is a combination of the various elements of both social and bureaucratic space, which is capable of promoting space quality. The large 2x2 partially see-through gate opens by rotating on one leg, and connects the outside public space with the private inside building space creating a unique type of "shared space".
Likewise, the wooden-roofed platform, on top of the ground floor stairways, creates an ideal setting for holding events such as book presentations or group readings.
The use of one particular material, the red travertine rock, was accentuated. Its use in the overall structure gave tone to the local community and can possibly create a "remembrance space".
Limitations met were the regulations of urban development in Iran which divide the land into two proportions of 4/10 and 6/10. The regulation states that 6/10 of the land can be allocated for the structure, while at least 4/10 must be allocated for the yard. These regulations do not favor architects who are trying to create the highest quality space possible.
In the interior, spaces are blended together and the borders of the walls are vague, nevertheless when the entirety of the building is viewed the divisions are clearly evident.
The extent of sunlight that enters the building varies according the level of necessity of each floor. The workroom receives a great deal of sunlight; the meeting room receives a bit less, while the upstairs attic receives the smallest amount of sunlight.

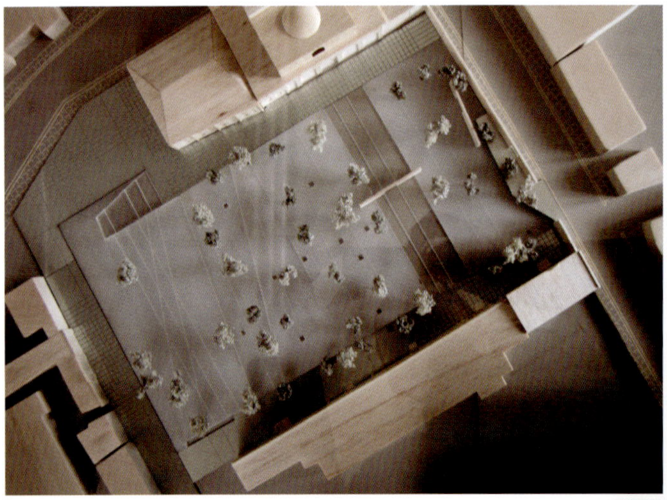

白天街景
Street View at Day

室内布局
Interior

室内布局
Interior

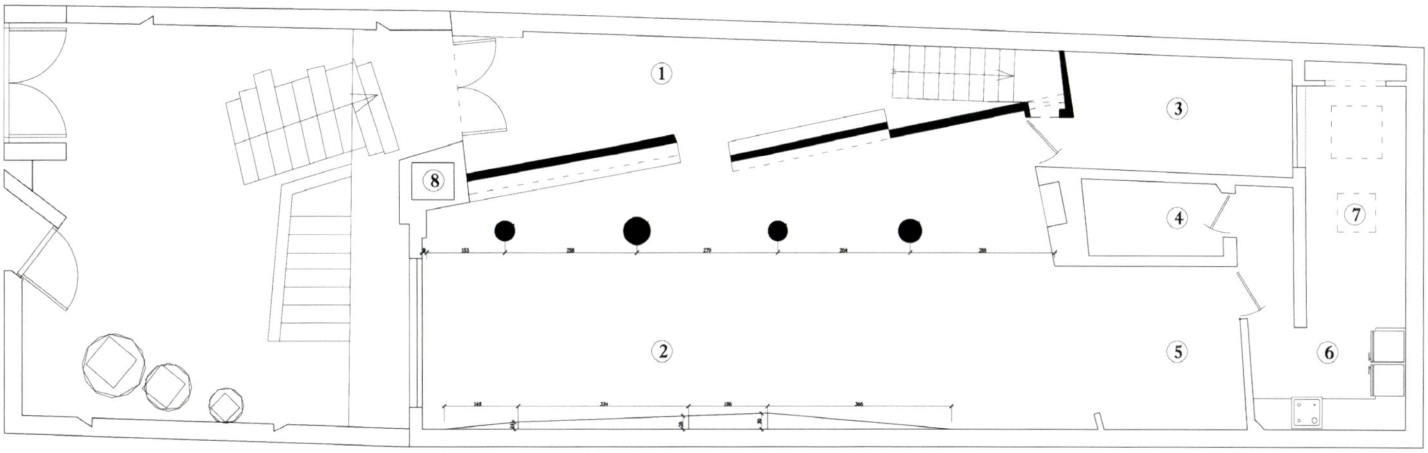

一楼规划图
Ground Floor Plan

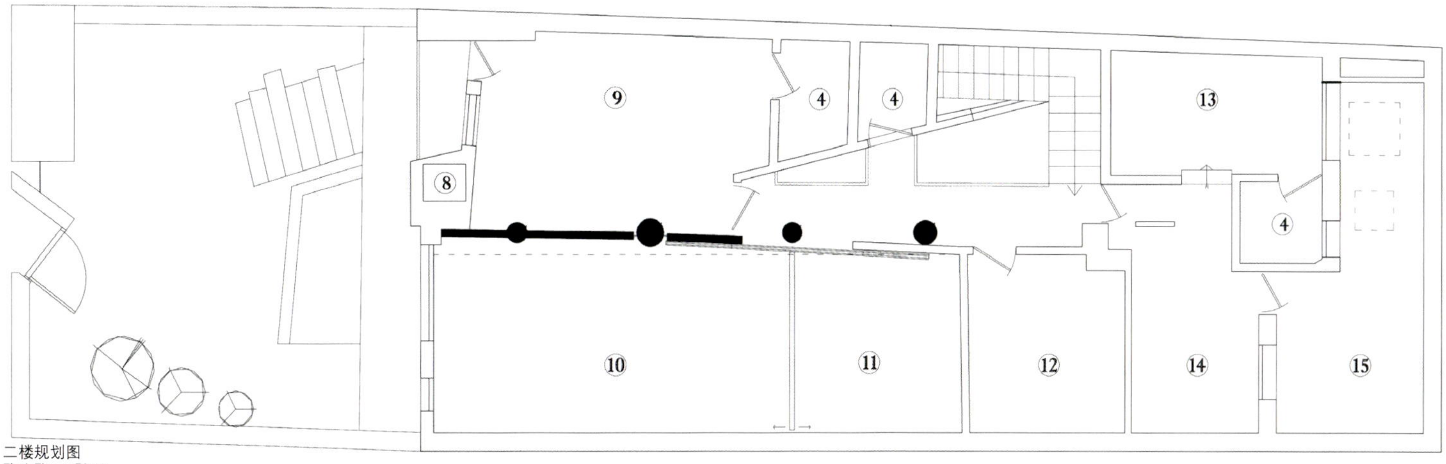

二楼规划图
First Floor Plan

Project Team: Saqib Ejaz, Muzaffar Shah, Farhan Ahmed, Imran Ahmed.

National Museum of Pakistan

巴基斯坦国家博物馆

CONCEPT

A history museum is the ultimate showcase of a nations' life through centuries. Museums can build morale and shape opinions. They also often create a lasting input on the future.

Following basic thoughts have gone into coming up with the design:

a. It is of utmost importance for the building to be user friendly.

b. The building should be cheerful and not drab. The form should be such that it is ageless. At the same time it is conscious of the contours and its context and stands as an un-intrusive structure in the woods of Shakarparian.

c. The flow inside the galleries is regulated yet informal and non-compartmental. It provides flexibility in display spaces that merge into each other. The inspiration for the flowing plan comes from the Rivers that flow through this land.

d. Interior space will be kept neat and uncluttered with straight lines and simple finishes. From inside the emphasis is on the display and not the architecture.

e. The vast cross-section of history that it proposes to display makes it impractical for the building to reflect one era, even symbolically. So the reference for the form is borrowed from the land itself. The jagged Potohar Plataue is what inspires the angular, tapering walls. The use of various colored earthly toned bricks as exterior finish reflecting the soil strata that is a witness to thousands of years of history steeped in this part of the world.

THE PLAN

The building has been divided into two distinct zones joined together by a large pedestrian space. Towards the North are the galleries, allied activities and admin offices, while towards the South we have the library, auditorium and the public restaurant.

The hall itself is divided into levels and has internal light wells and courtyards. There is also a deep verandah in the North-East for artifacts that can be placed in the semi open.

To the East of the galleries are located the workshops. From the office floor there is a mezzanine level that hangs into the gallery and looks over the entire hall. It also provides a connection to come down into the gallery.

The auditorium and library on a lower level can be accessed through a set of stairs and ramps. The restaurant has been elevated so visitors can enjoy the enchanting views of the woods. The area beneath the restaurant provides a covered and

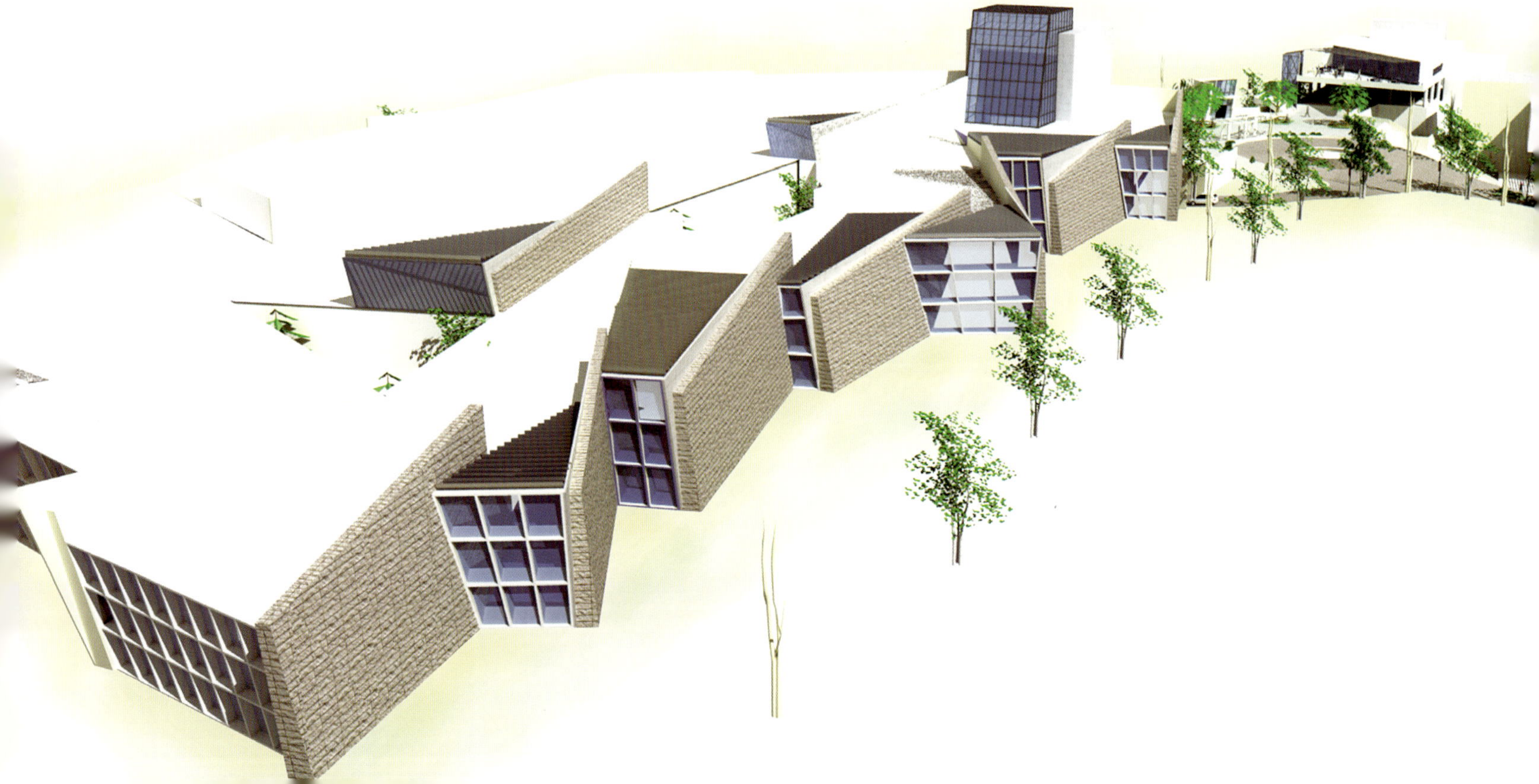

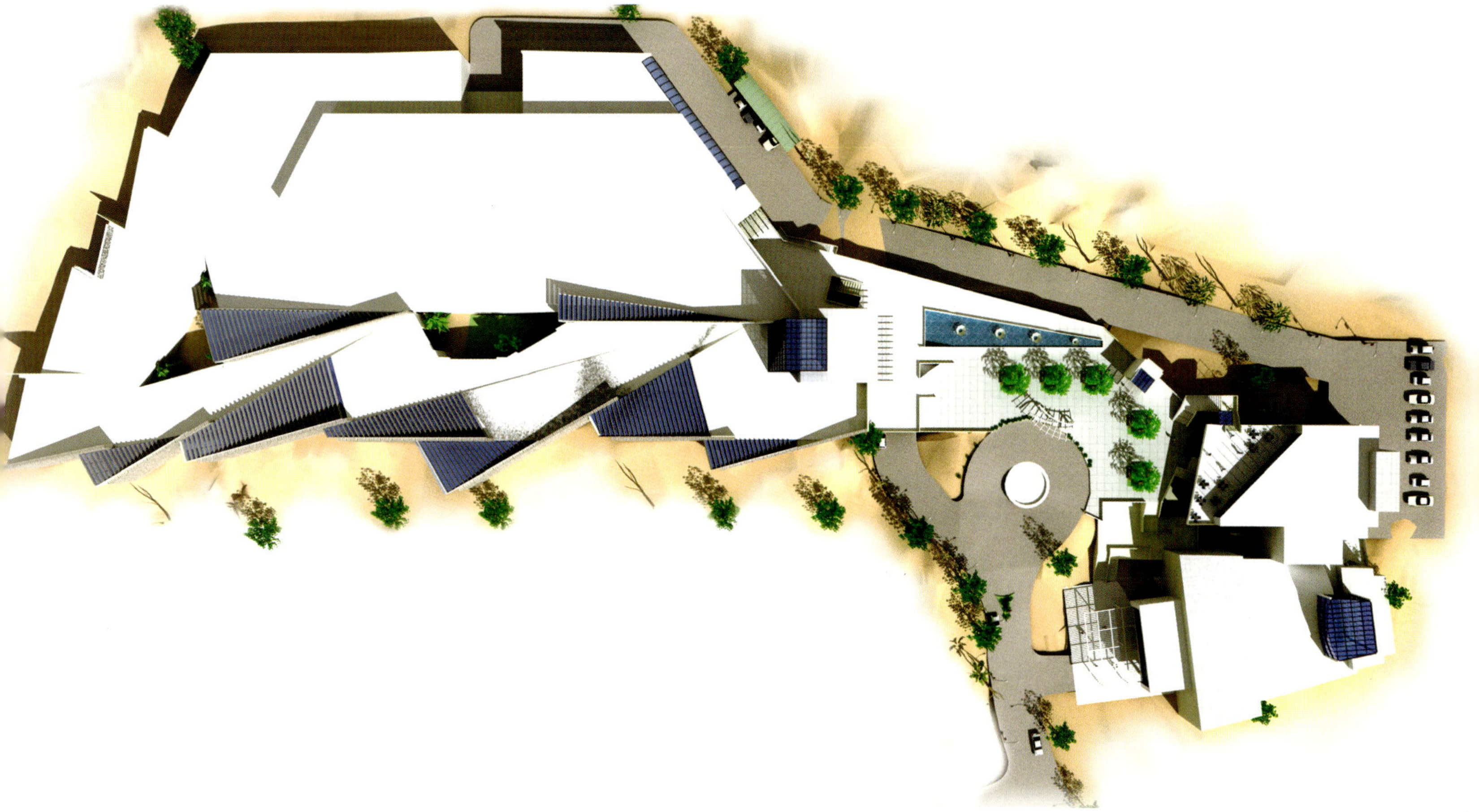

partly protected extension to the Pedestrian Plaza. There is a large (appox 3 acres) tract of land that is left to the extreme South side along the main road. The idea is to use this space for any temporary exhibits that can either be displayed in the open or under temporary structures.

ACCESS AND CIRCULATION

Access to the building and its circulation is what will determine its functional efficiency and user friendliness.

1. For visitors: The Pedestrian Plaza leads to all the facilities. The plaza has an information booth and access to the museum, library, cafeteria & auditorium. The underground parking has an independent stairway to the Pedestrian Plaza. The bus and coach parking is provided near the main entrance to the left. There is also a wide side walk for pedestrians coming by public transport.

2. For management: They can get to their offices through the drop-off at Pedestrian Plaza or through the staircase connecting their specified basement parking to the office area.

3. For services: There are three service areas:

a. The electrical and mechanical spaces located at the lower ground (semi basement) level opposite the public parking.

b. The deliveries to the museum through the registration area located at the rear North Side. Workshops are also located here.

c. The deliveries to the restaurant and solid waste disposal located at the rear South side. Auditorium back of the house is also served from here.

All these areas are accessible through an underpass that goes under the Pedestrian Plaza so there is no conflict between vehicular and pedestrian traffic.

ENVIRONMENTAL CONSIDERATIONS

The museum is conceived as a green building. The building design and its orientation inherently protect it from the strong summer sun. The thick cavity walls and the position of the main service areas towards the West and South-West sides make it energy efficient. The maximum light is let through the North Side and through the roof with North facing louvers. The office windows on the East side are deep set and protected. There is practically no glazing on the South and South West side.

CONCLUSION

It is obvious that a number of finer details and minor changes will have to be incorporated at a later stage. These drawings show the design intent and should be treated as such. But most of all it's a modern building responsive to elements of nature and its user demands.

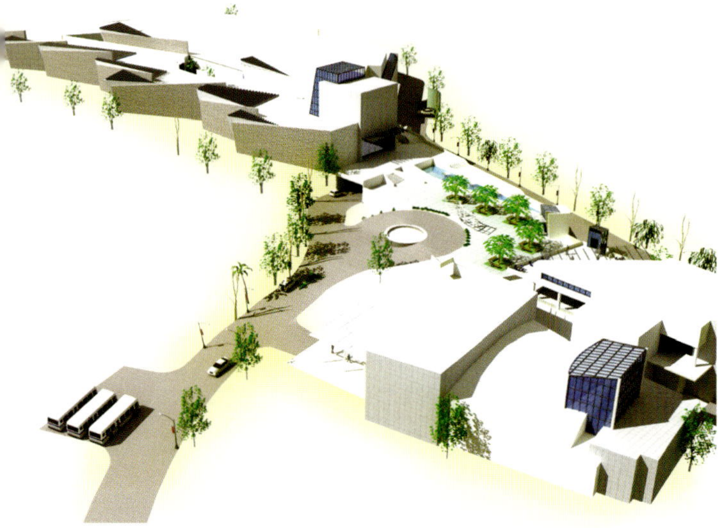

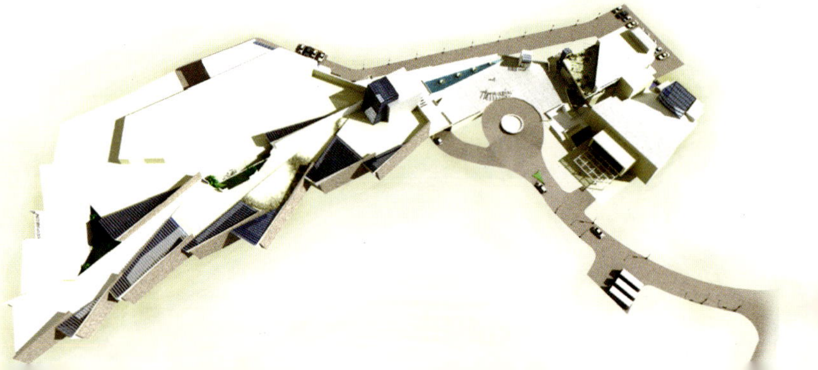

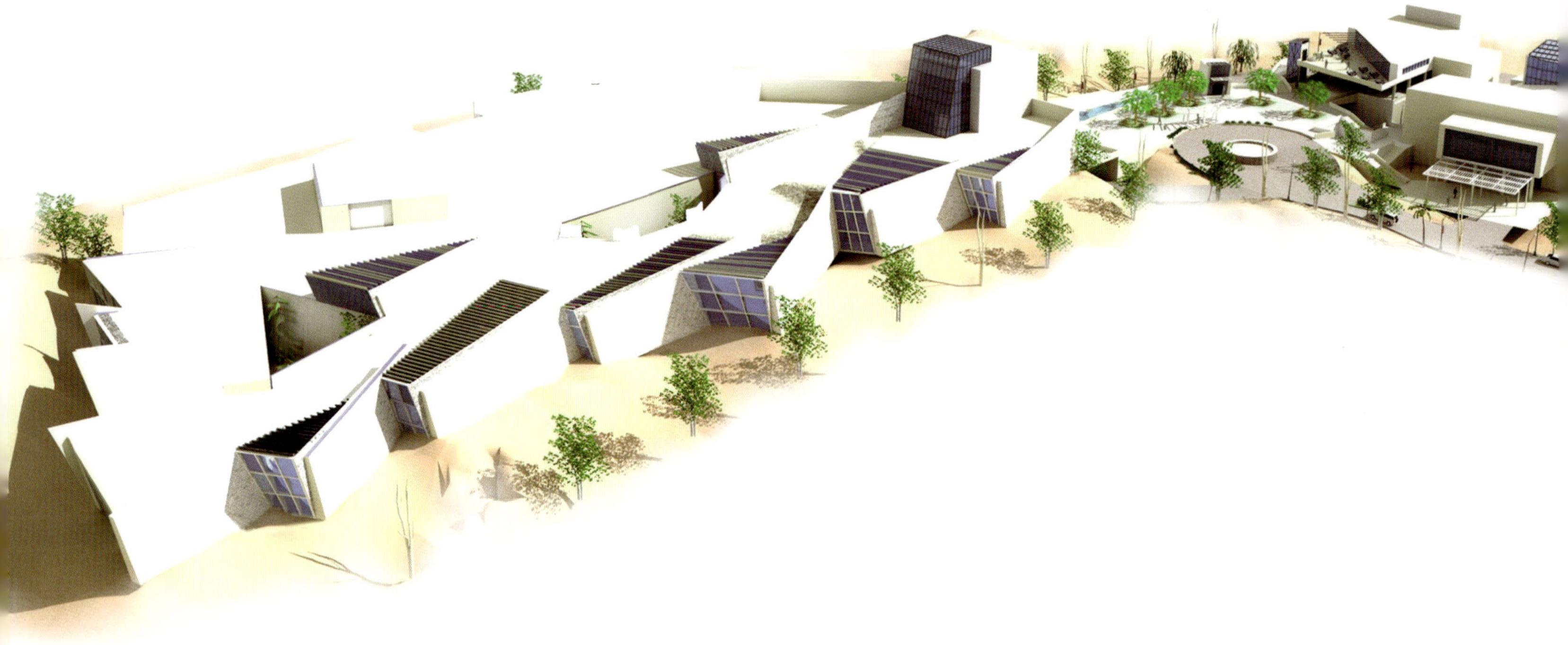

RESTAURANT TERRACE

CONNECTION OF PEDESTRIAN DECK TO RESTAURANT

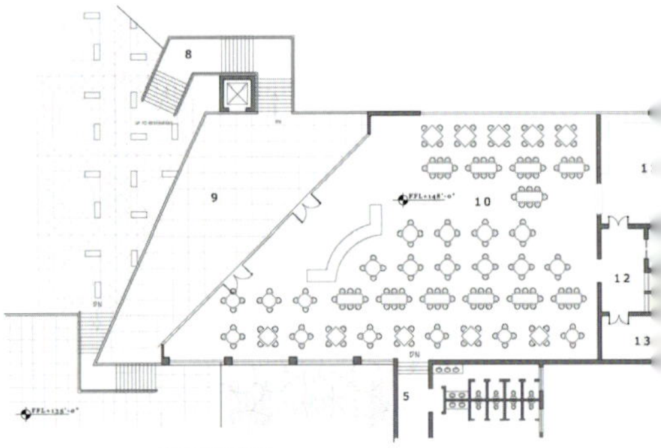

RESTAURANT
SCALE : 1'-0" = 1/16"

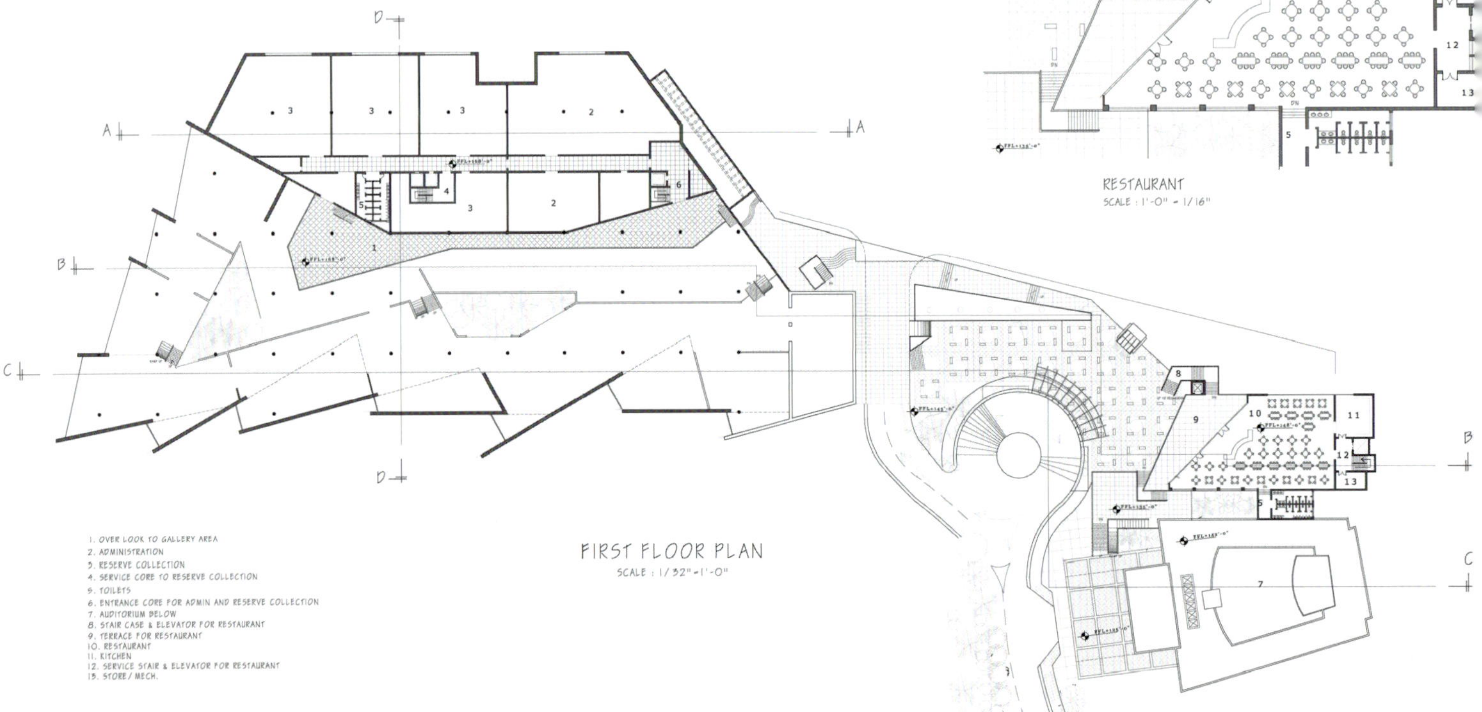

FIRST FLOOR PLAN
SCALE : 1/32"=1'-0"

1. OVER LOOK TO GALLERY AREA
2. ADMINISTRATION
3. RESERVE COLLECTION
4. SERVICE CORE TO RESERVE COLLECTION
5. TOILETS
6. ENTRANCE CORE FOR ADMIN AND RESERVE COLLECTION
7. AUDITORIUM BELOW
8. STAIR CASE & ELEVATOR FOR RESTAURANT
9. TERRACE FOR RESTAURANT
10. RESTAURANT
11. KITCHEN
12. SERVICE STAIR & ELEVATOR FOR RESTAURANT
13. STORE / MECH.

ACCESS TO LOADING/ UNLOADING DECK

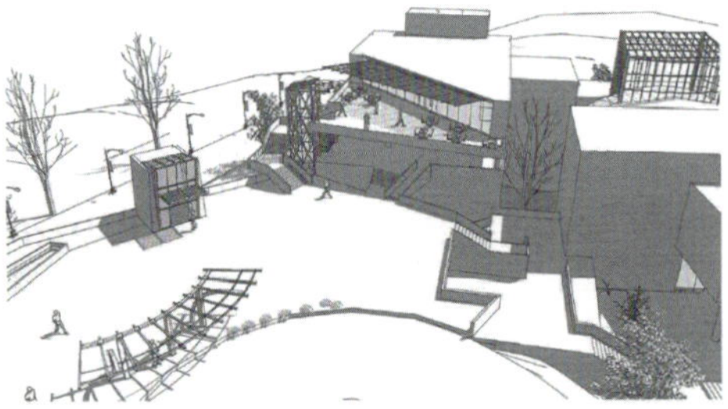

AERIAL VIEW LOOKING TOWARD RESTAURANT

ENTRANCE TO AUDITORIUM AND EXHIBITION SPACE

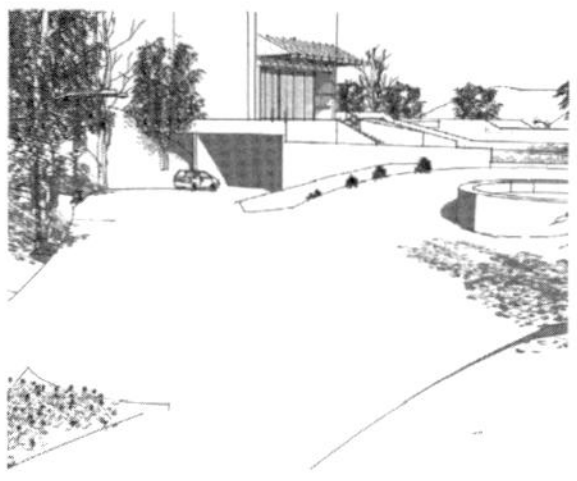

ENTRANCE TO BASEMENT PARKING

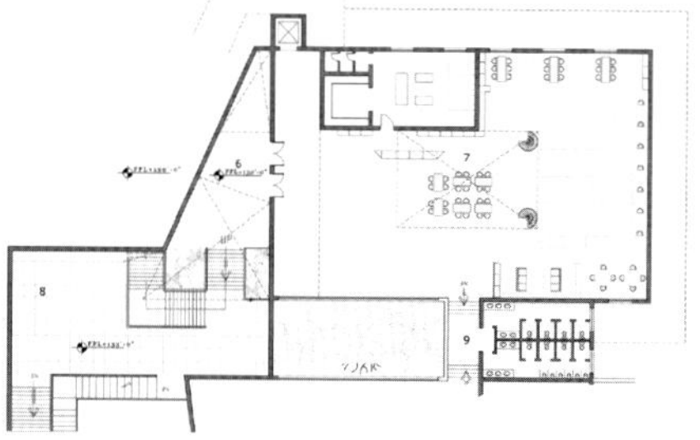

BLOW UP PLAN OF LIBRARY
SCALE : 1'-0" = 1/16"

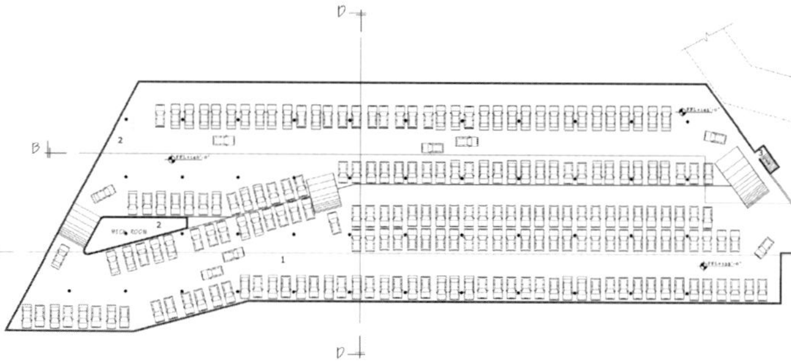

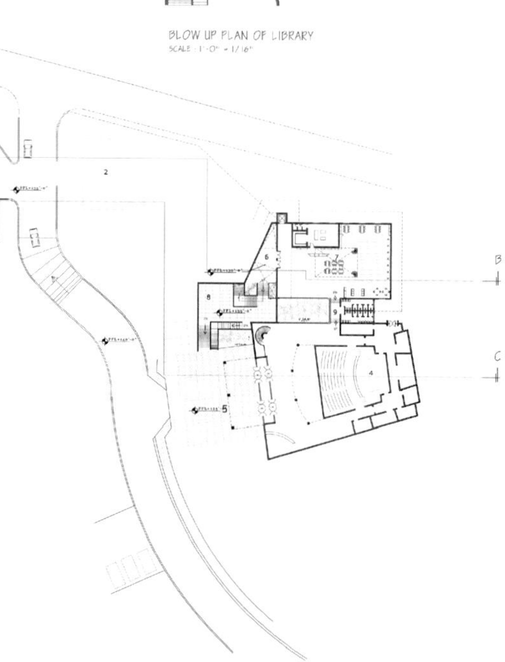

1. PARKING
2. ELECTRICAL AND MECHANICAL ROOM
3. STAIRS TO PEDESTRAIN DECK
4. AUDITORIUM
5. DECK ENTRANCE TO AUDITORIUM
6. ENTRANCE TO LIBRARY
7. LIBRARY
8. MID LEVEL DECK
9. TOILETS

LOWER GROUND FLOOR PLAN
SCALE : 1/32" = 1'-0"

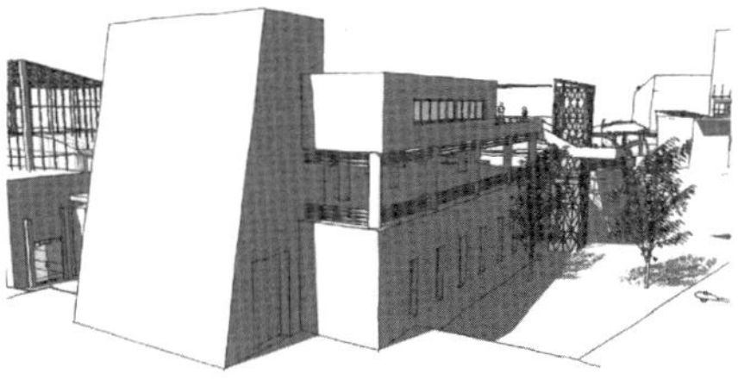

SERVICE ENTRANCE FOR RESTAURANT AND DECK

SHEET 03 OF 12

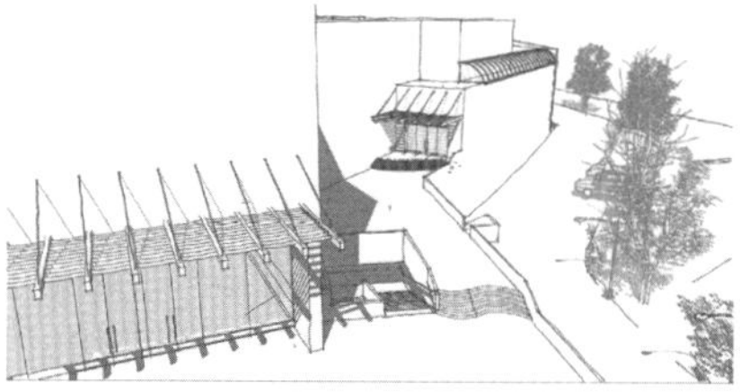

ENTRANCE TO OFFICE AREA FROM PEDESTRIAN DECK

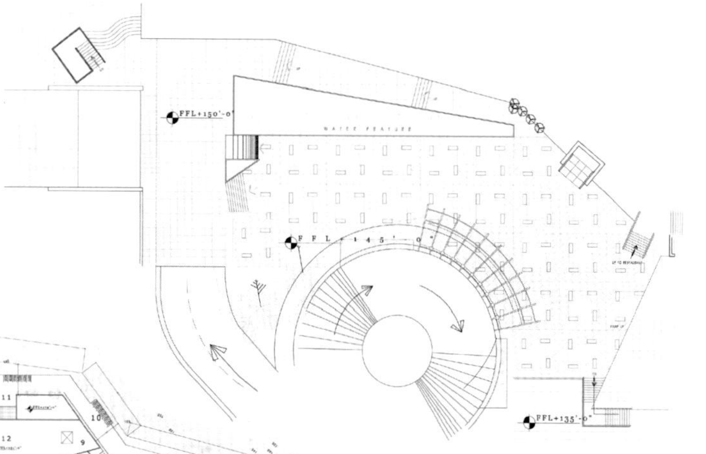

BLOW UP PLAN OF PEDESTRIAN DECK
SCALE : 1'-0" = 1/16"

VIEW OF INFORMATION BOOTH FROM DROP OFF POINT

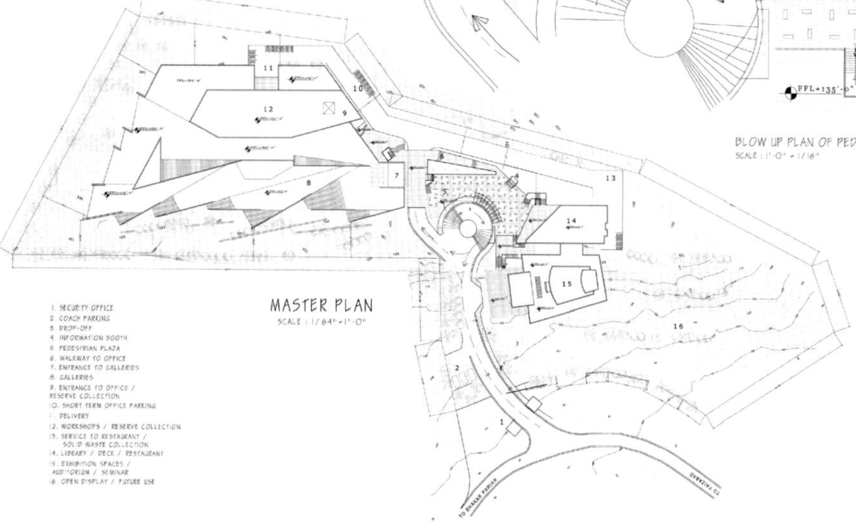

1. SECURITY OFFICE
2. COACH PARKING
3. DROP-OFF
4. INFORMATION BOOTH
5. PEDESTRIAN PLAZA
6. WALKWAY TO OFFICE
7. ENTRANCE TO GALLERIES
8. GALLERIES
9. ENTRANCE TO OFFICE / RESERVE COLLECTION
10. SHORT TERM OFFICE PARKING
11. DELIVERY
12. WORKSHOPS / RESERVE COLLECTION
13. SERVICE TO RESTAURANT / SOLID WASTE COLLECTION
14. LIBRARY / DECK / RESTAURANT
15. EXHIBITION SPACES / AUDITORIUM / SEMINAR
16. OPEN DISPLAY / FUTURE USE

MASTER PLAN
SCALE : 1/64" = 1'-0"

VIEW OF COMPLEX FROM RESTAURANT

WEST SIDE ELEVATIONAL VIEW

SHEET 02 OF 12

Author: Wellington Costa Lima

Co-authors: Larissa Leite Tredezini and Rodrigo Pessoa de Mendonça

Nautilus

鹦鹉螺大厦

Building designed for residential and commercial use. "Nautilus is a concept of Ecological and Sustainable Contemporary Architecture." The Nautilus building as well as its environment will be like a second skin for those who live there. A project fully dedicated to environmental concerns, offering appropriate and feasible solutions. The new consciousness of the modern man that will be living in the NAUTILUS BUILDING will be: "By caring for the environment, we are taking care of ourselves." And what can we do so the contemporary world can be aware that we have these concerns and try to deal with them in our project? - Capture, filtering and use of rainwater; - Green roof top for leisure, landscaping and permeability; - Solar energy; - Wind energy with vertical axis; - Ecological bricks; - Double glass for better comfort and acoustic insulation; - Mini station for sewage treatment; - Organic waste composting; - Selective waste; - Small manufacture of soap, detergent and disinfectant with recycled cooking oil collected by the residents of the building, for their own condominium use; - Domes for better use of sunlight in the garages and in the basement of shops. 1. Description Inserted in a block defined by four roadways, the Nautilus building has its main volume defined by a large nautiluses section with four axles, located on a platform. To streamline the circulation, the main volume was divided into three contiguous blocks of flats with independent vertical circulations and three water tanks. These towers are grouped in a single volume. Up to the 5th floor, the apartments' floor plan are the same, from the 5th to the 27th floors the apartments staggered to meet some goals of conception: permeability on the terraces; the final shape allowing free flow of dominant east-west wind barrier or without offering resistance, resulting in a great movement of cross air, all apartments filled with views of the two facades, front and rear. This staggering also broke up the noise that could generate turbulence in the area of coexistence on the foundation pavement, and excellent light was given in all apartments. Platform: - 1st subsoil: Stores' garages. - Ground floor: Shops, some geared to the level of the tracks, up the front facade of the building in the lower level area. - Higher floor: Covered garages for residential units. Residential Tower: - Foundation, with direct access to the roadway, in the region of highest level. - 1st to the 5th floor – Standard type apartments. - 6th to the 27th floor - Duplex apartments with terraces. Number of apartments = 59 units of one to six bedrooms. Number of residential garages = 120. Number of stores = 16 units. Number of commercial garages = 1,202. Concept and Geometric Design: The concept of the project and its geometric design was inspired by the geometric shape and structure of the Nautilus,

mollusks which begin their lives in microscopic sizes and are wrapped in a calcified shell. As they continue to develop, they leave the old shell and calcify a new shell on its back and so on up to major proportions. The Nautilus always occupies the last chamber so the others may be filled of air, making it floats. It is said that, by presenting the targeted body in a spirally shape, the nautilus is one of the living kinds that remarkably shows the golden ratio in their development, so called Spiral of Gold. Those successive shells that will be calcified around the Nautilus to their new home, establish between themselves this gold relation or gold ratio that is 1,618. The same relation will be adopted between the apartments of the building. It is frequently used in Renaissances paintings compositions, such as the master's Giotto Leonardo da Vinci and others. This gold number is involved with the nature of growth. It can be found in the proportion of shells (the nautilus, for example), humans (the size of the phalanges and finger bones, for example), even in the relation of males and females of any beehive in the world. Because it's so involved in the growth, this number becomes so frequent. And because this frequency exists, the gold number won a status of "almost magical" and is subject to researchers, artists, antiquity architects and writers. Despite this status, the gold number is only what is due to the contexts in which it is inserted. The fact that it's found through mathematical development is that makes it fascinating. That is why we adopted as the tonic resonance in the "NAUTILUS" project.

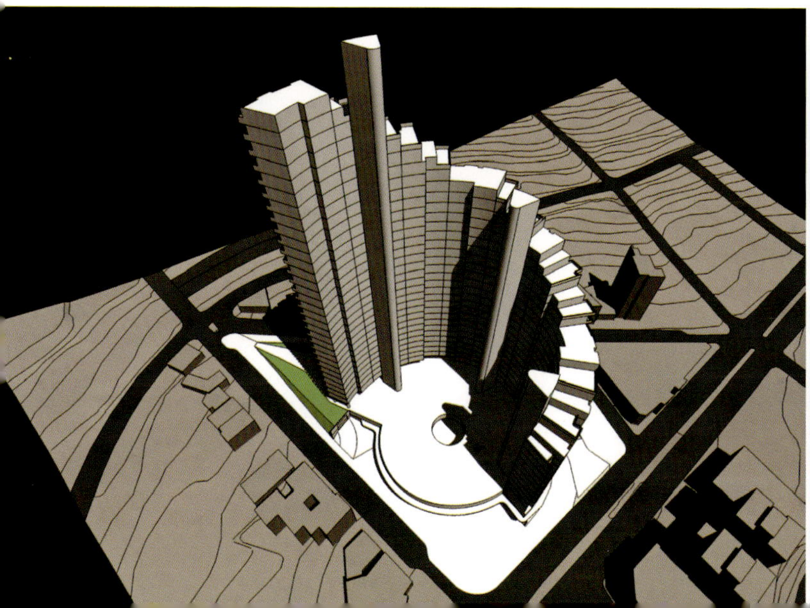
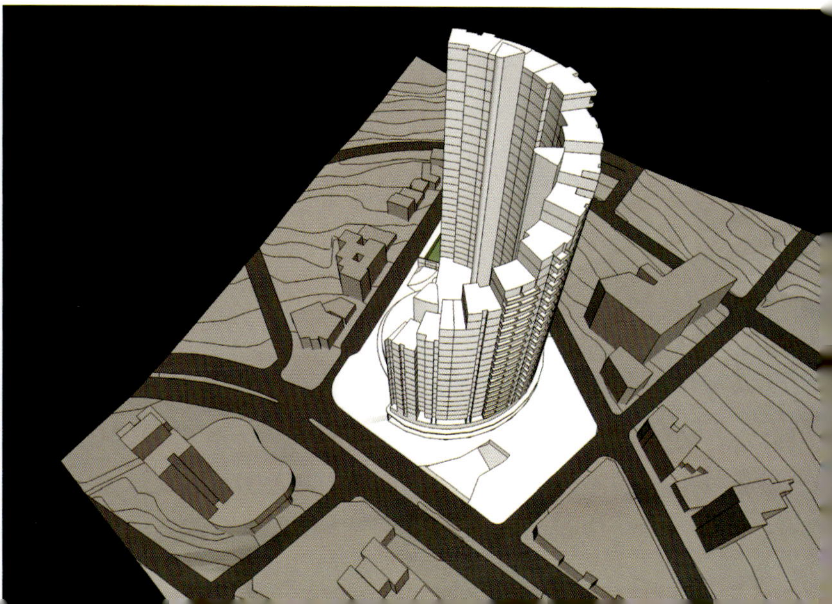

27º
26º
PORCH
BEDROOM
BATHROOM
HALLWAY
VERTICAL CIRCULATION - ELEVATORS
KITCHEN
SERVICE ROOM
TERRACE
LIVING ROOM

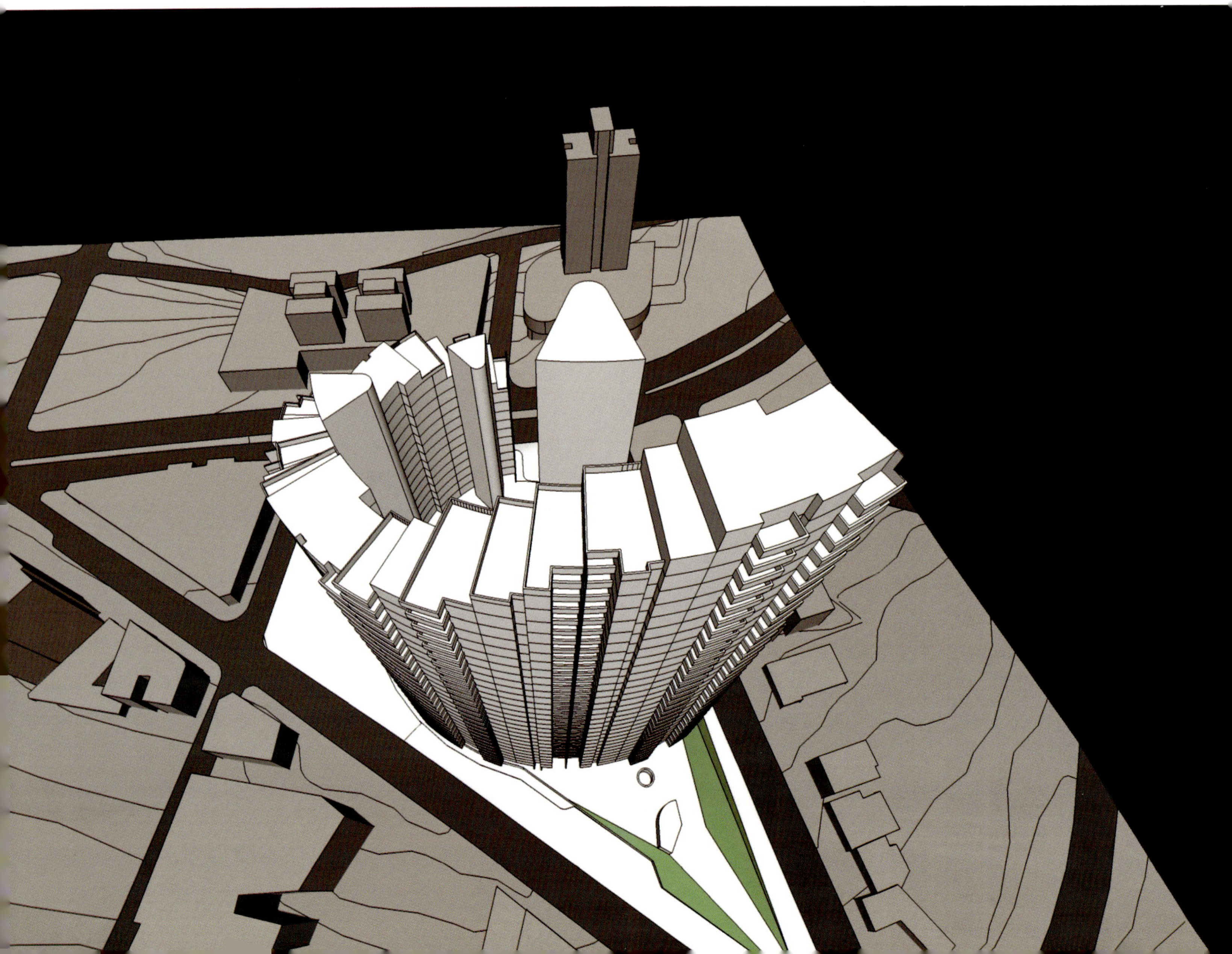

Sophisticated project design where form, function and concept are in harmony without compromising the accuracy of the geometry."

GOLDEN RATIO Phi a/b=1,618 b/c=1,618

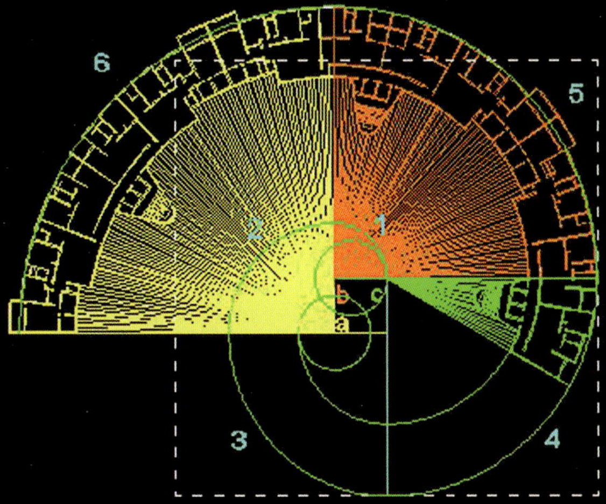

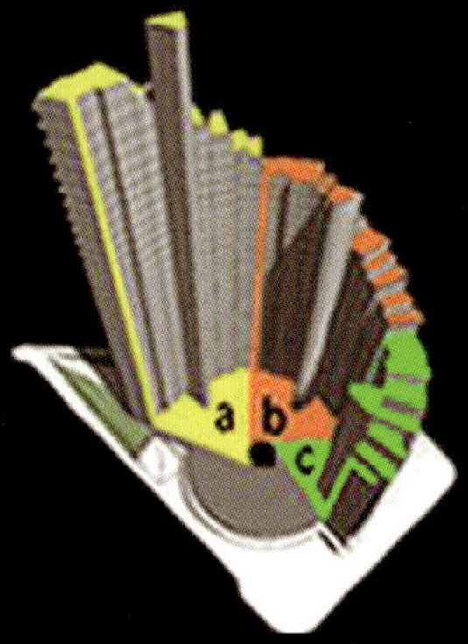

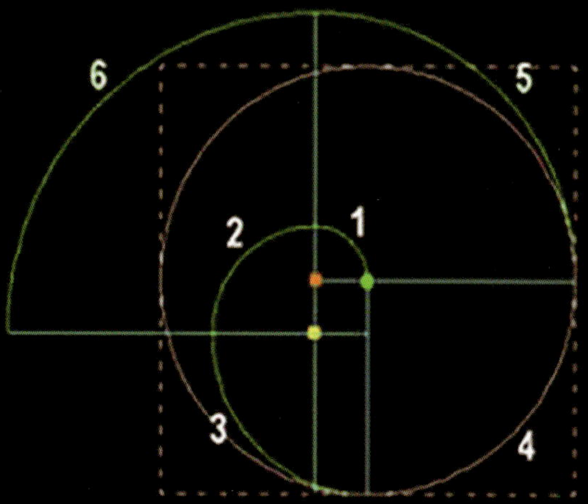

THE NAUTILUS AND THE MAN ON ITS GOLDEN RATIO OF 1,618 = Phi

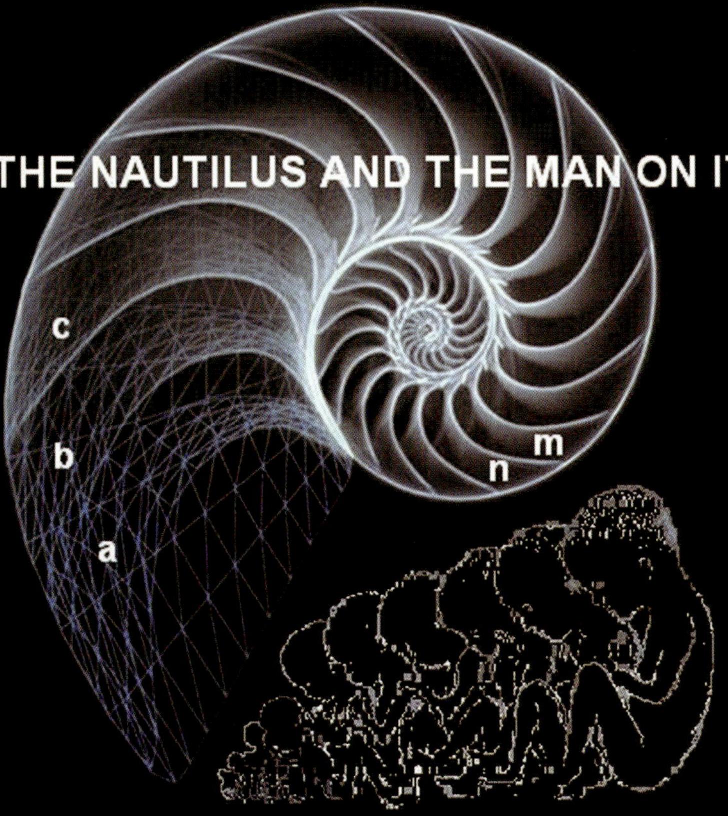

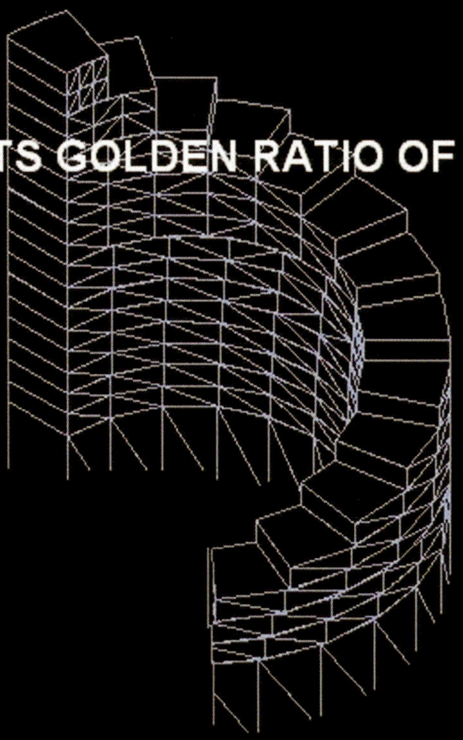

a/b = 1, 618
b/c = 1, 618
m/n= 1, 618

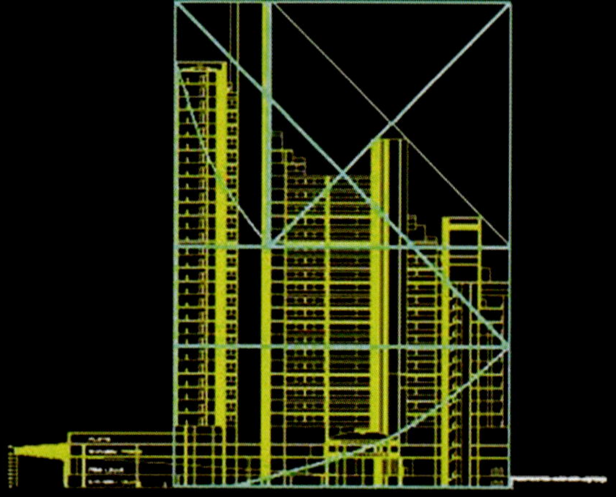

REGULATOR TRACK

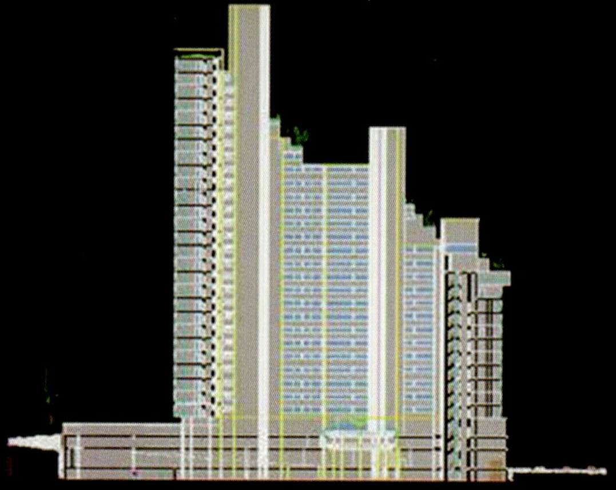

SECTION

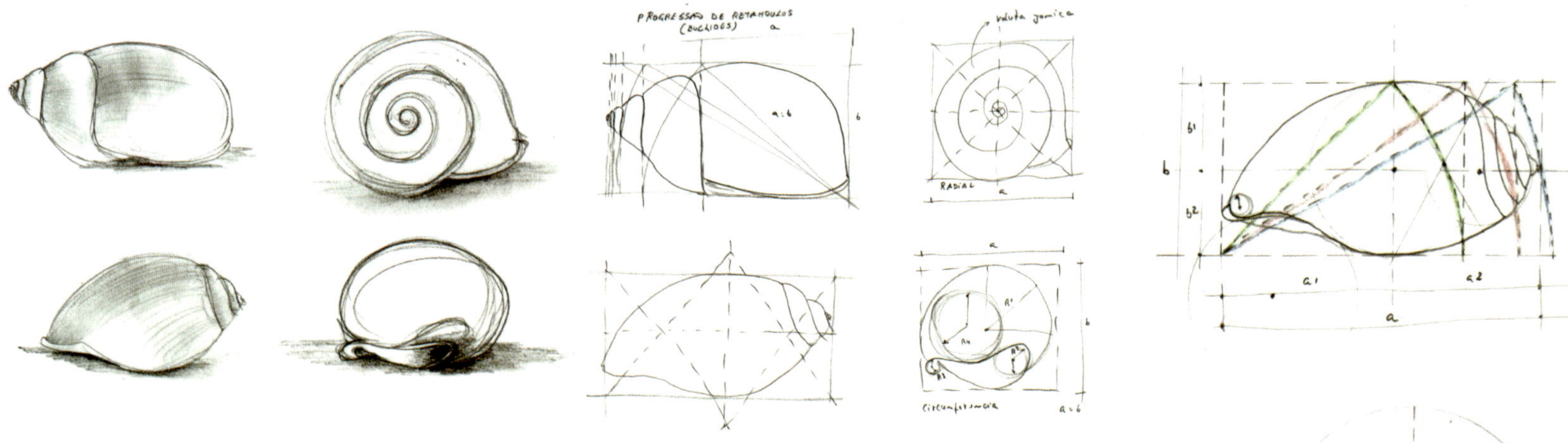

FRAMING STUDY IN THE NAUTILUS
ORGANIC GEOMETRY

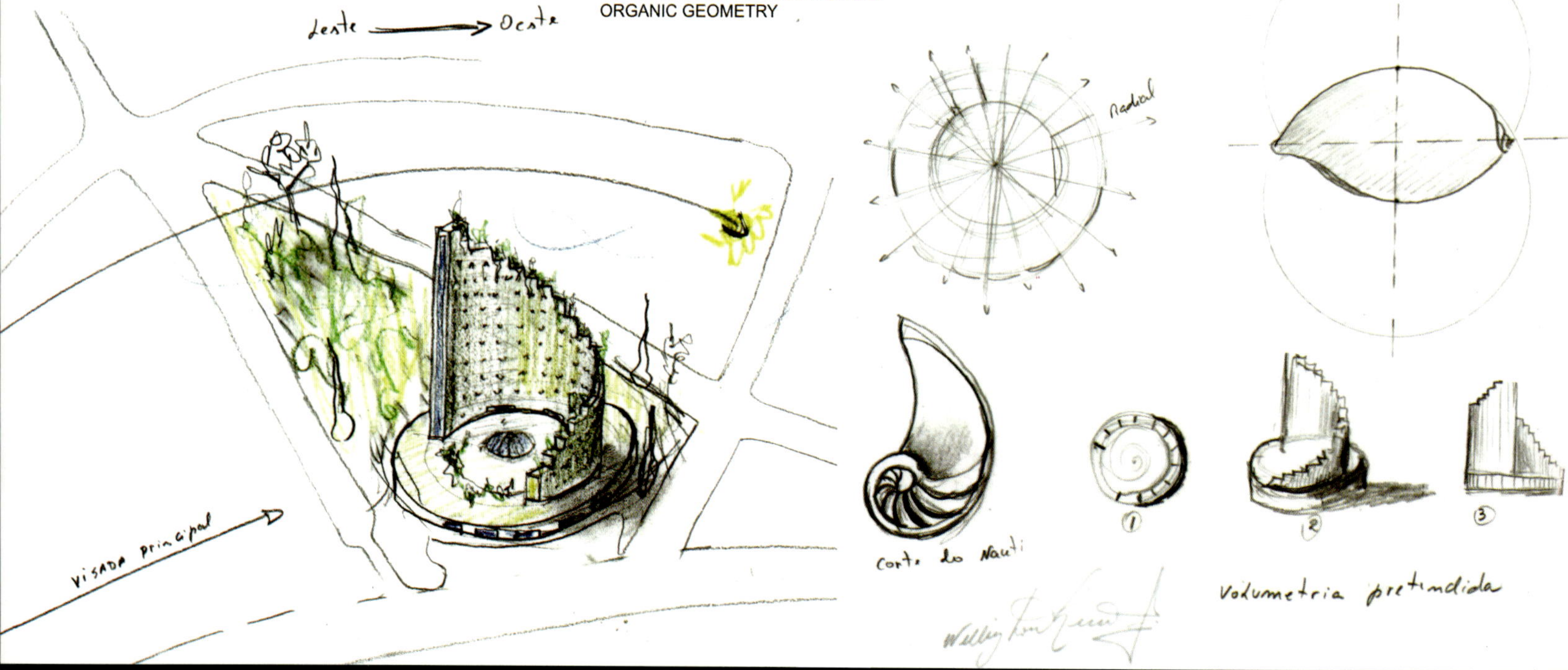

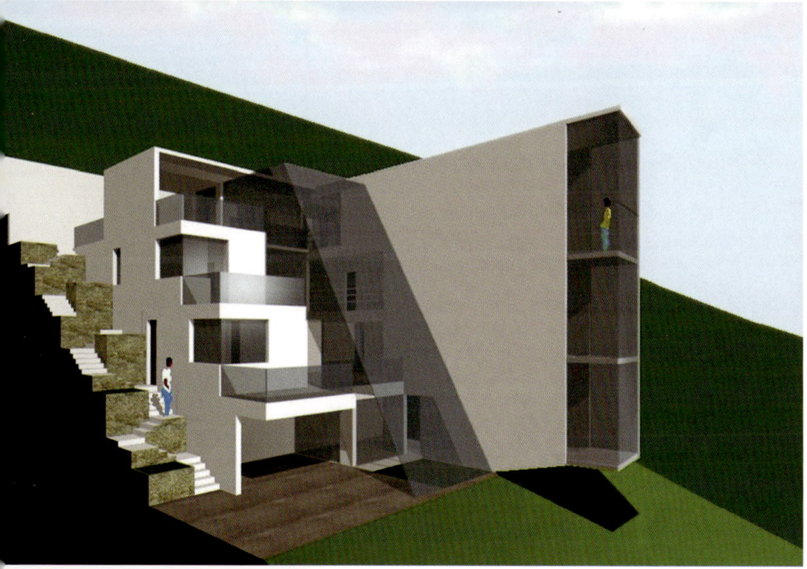

Chief Architect: Elliot Justin Giovacchini & Stephen Foulkes

New Sicilian Deli Corner — Food Experience

城市饮食广场中的建筑哲学

The urban interior becomes space. The city, under the influence of light, creates moments of life, fluid and dynamic, and architectural elements individual buildings. The philosophy of the project and its concepts express the sense of living, the tradition in its most classic conception of the Italian Piazza, source of history and passion as Renzo Piano says in an interview with Charlie Rose in the U.S:

(Renzo Piano) "Transparency and light a small miracle that creates a public space, a small square inside a closed space"

(Charlie Rose) "Is this your Italian soul?"

(Renzo Piano) "The piazza is an icon; it is the city of my dreams: made of light and transparence, full of water, greenery and trees together with bridges and streets. Piazza is the place where people meet, disappear and move.. So the idea that you can make a little piazza and you can create that sense of participation between the street and the building is part of the story ... this is not really what you should call style it is more a kind of desire to create a space more open where you can start an adventure..an adventure means ... to understand and listen the site ... and every site has a story to tell..."

From this the cue to tell a story about the piazza's life and its flavour, odour and materials atmosphere Characterizing the Sicilian Italian culture a food-experience, in a conceptual space not a mere reconstruction of a cross-false and unreal cutaway. Try to create the genius loci is not reconstruction but rather create a feeling of knowing what only a small part you can make it live and experience. Here the use of panels and animated images of a cross section of a reality, distant in time and space, smell, taste and materials that create memories that bring to mind the historical civilization and its customs, and finally the use of panels white and neutral without lifeless, in contrast with real life represented by visitors-customers who come together as people of the same plastic-model of Sicilian food experience.

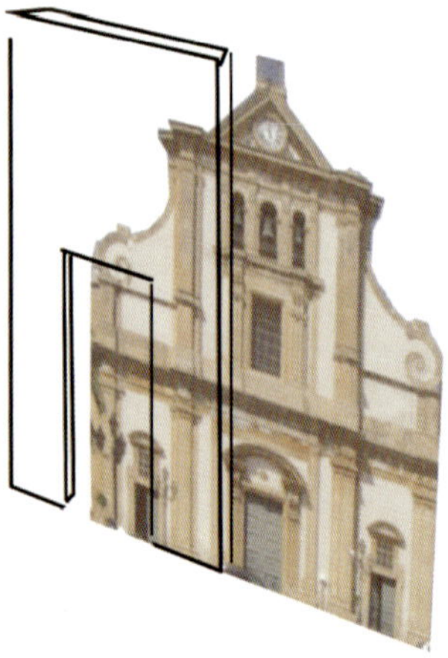

设计理念
Concept

coffee bar
ice-cream
TRATTORIA RES

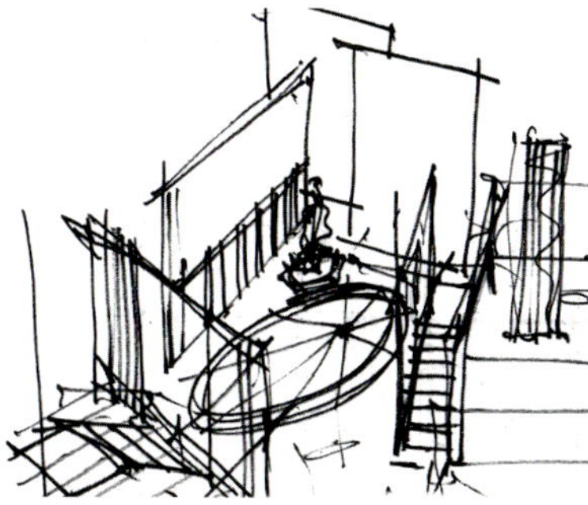

TRATTORIA RESTAURANT

market

Project Team: Eleftheria Serghidou, Vasilis Pashiourtides

Olympic House and Park, Nicosia, Cyprus

塞浦路斯奥林匹克之家及公园

The design for the Olympic House and Park sought to achieve an architectural composition that would give the fullest possible expression to the Olympic ideal. The following three fundamental considerations were addressed and in due course served to frame the proposal:

•The global dimension inherent in the encouragement and inculcation of the Olympic spirit imbued as it is with ideals such as brotherhood, noble emulation and world peace.

•The historical dimension of preserving and reviving a venerable institution of great antiquity.

•The athletic dimension that pursues training of "body and spirit" in equal measure as an essential human activity.

It was felt that the shifts from "individual to collective" and from "ancient to modern" could be achieved by means of architectural transitions deployed in space, and thus be translated, through mediating voids, into the layout of the building in plan, and its dialogue with the ground in section respectively. On master plan the concord of body and spirit was reflected en the concord of built and non-built, of edifice and park. With respect to the plan the quest for a layout that would express the global parameter led to the image of an ancient stadium. The building is developed perimetrically around a central void, which opens out to the town, calling upon the "external" to enter and allowing the "internal" to be viewed.

In section, the transition from "ancient to modern" is realised through raising the building high and creating a ground-floor void, where the promenade takes on the significance of museum space for showcasing the ancient and modern history of the institution.

At the master plan level, the building is not juxtaposed to the park, but neither is the reverse true. Initially the built penetrates the un-built part of the plot centrally letting the un-built surround it, while subsequently the un-built penetrates the build centrally and by piercing it, this is incorporated into it. This interpenetration and co-existence of built and un-built, of edifice and park, on an equal footing, expresses the need for material and spiritual to coexist: training for "body and spirit".

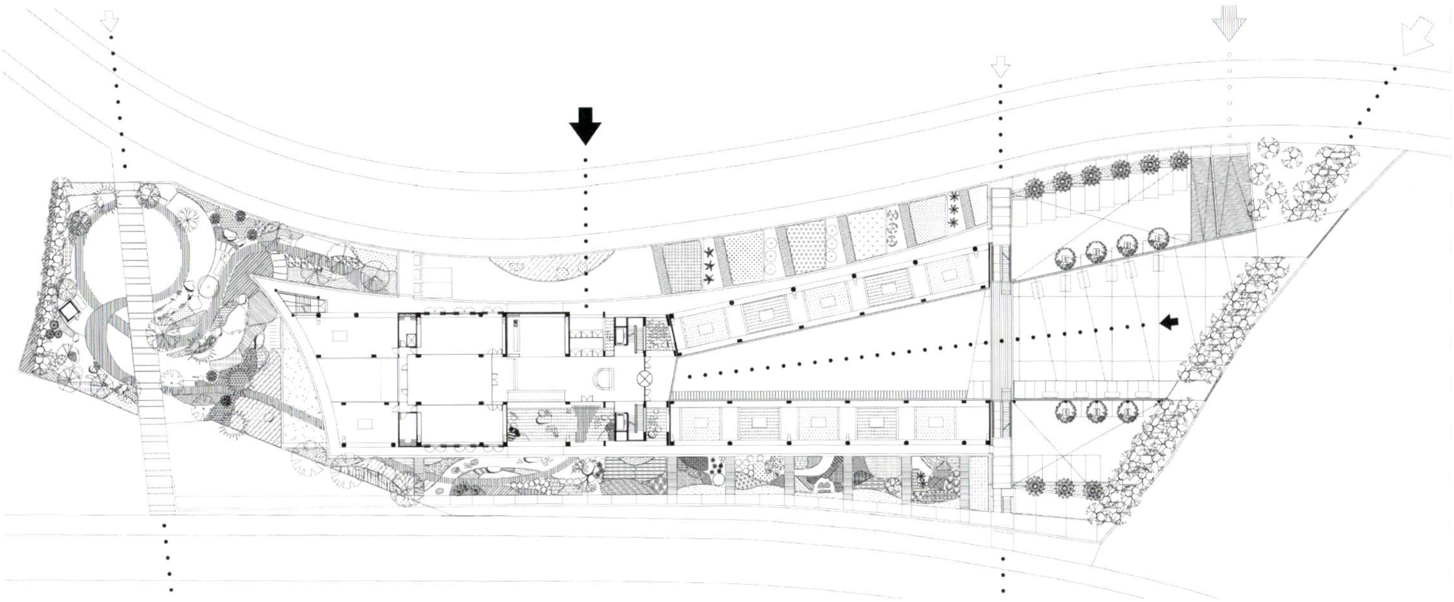

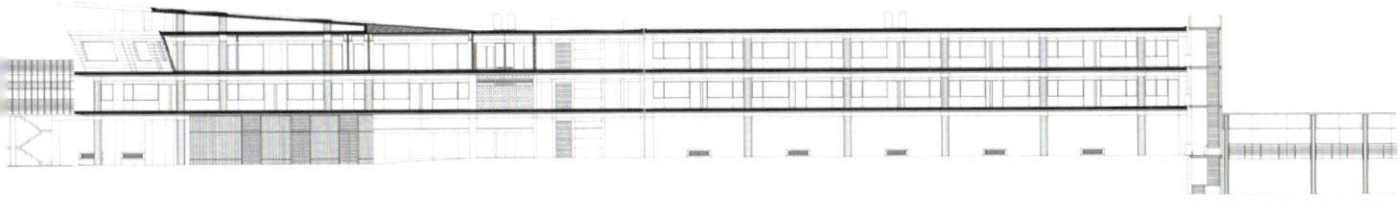

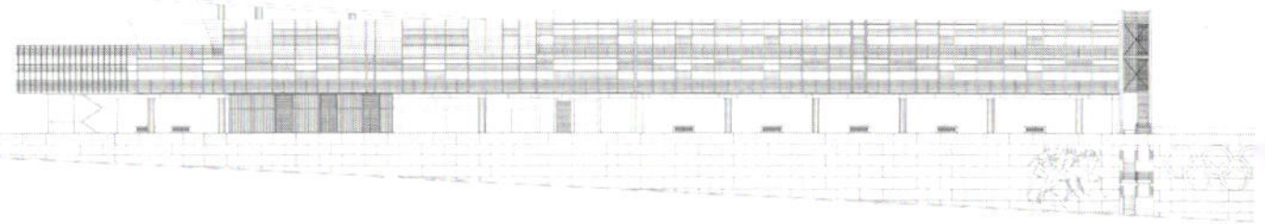

SOUTHEAST ELEVATION

The central void, a large space for gatherings, is developed sequentially from the park through the shaded atrium to the entrance lobby and then through the multi-purpose hall once again towards the park. Circulation in the offices is carried out externally so that the spaces be aligned on either side of the atria, just like spectators and, also, of the events taking place within them.

Project Team: Saimum Kabir, Md. Rabiul Islam, Md. Mizanur Rahman, M Azizul Mohit

Orphanage at Chandpur, Bangladesh
孟加拉孤儿院

The Orphanage is situated in a rural context, surrounded by low land and water body. The project is an extension of exiting complex to accommodate rooms for 24 orphans, a computer class and indoor and outdoor game facilities. Rural architecture of Bangladesh has its distinct characteristics as regards planning, use of materials and locations. Like urban architecture, rural architecture is also subject to change, but in Bangladesh rural architecture has clung to tradition. For centuries, rural architecture has been constructed based on experience by using local materials and technology. The complex is designed with sympathy to the culture and tradition of rural area. The main challenge of the project was to adopt a design solution which will not only be technically superior but socially acceptable and properly suited to the genius of rural people. Besides, environmental performance of the building was critically judged in respect to natural light and ventilation. The site is east west elongated and surrounded by agricultural land on its two sides. On the west side there is a canal used for irrigation purpose. The building is basically L-shaped in plan oriented towards an open space. Dormitory blocks and class room are located in two different arms of the L-shape. In order to avoid the excessive heat and to get the maximum benefit of cross ventilation, dormitory block is slightly tilted along south east direction. A screen using local bamboo is designed to cut the direct sunlight. At the same time window behind the screen was designed to act as a light-shelves to ensure diffused distribution of daylight. The class room block is located along the north south direction and is elevated on stilts. The objective was to link the inner open space with the surrounding serene natural landscape which at the same time creates room for indoor games.

草图
Sketch

Orphanage

草图
Sketch

景致
View

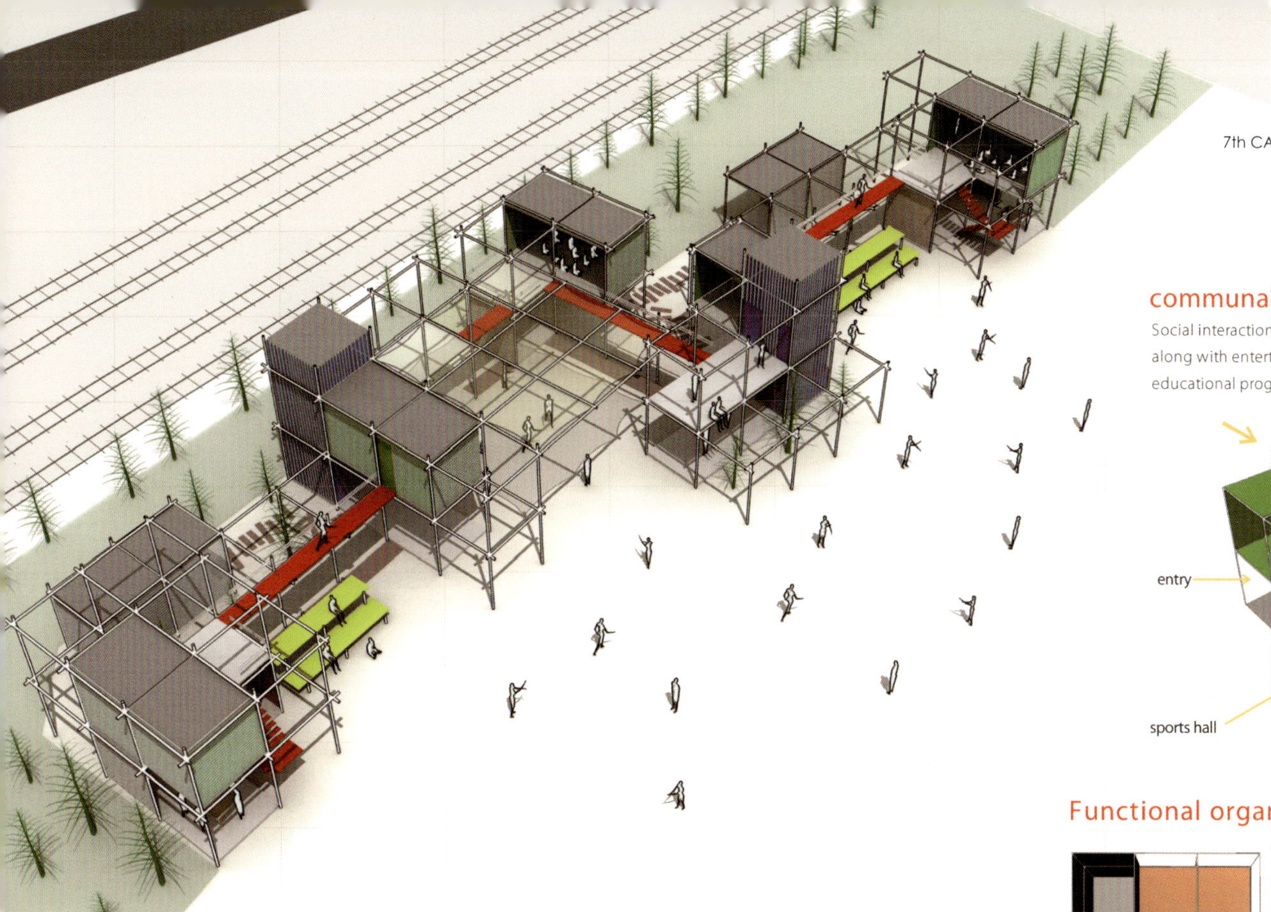

communal space

Social interaction ,counselling & motivation, along with entertainment facilities -like showing educational programmes & documentaries

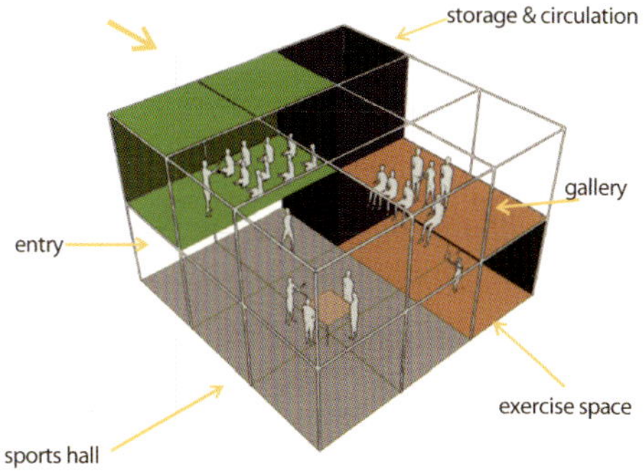

Functional organization & basic module:

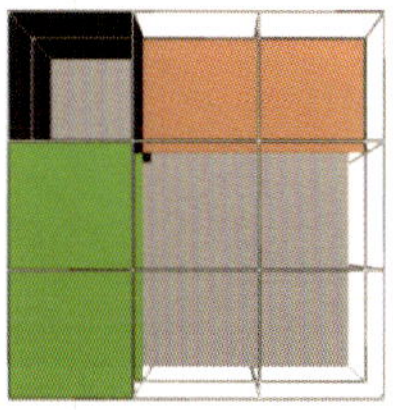

The basic unit

Square in plan having 2 levels containg-

1. A sports hall -50% of total volume with gallery , exercise room & storage facility.

2. A commnal space & a stair way connecting these function

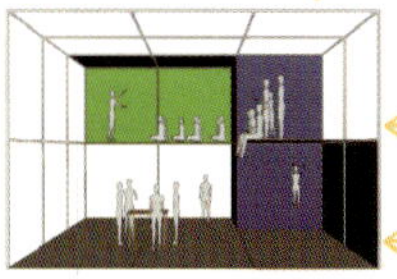

...tability and flexibility

...erent sites the permutation and combination of the basic unit can generate larger sports complex according to the need.

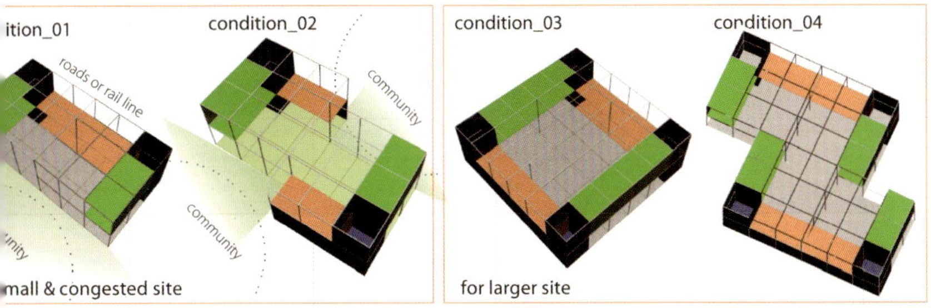

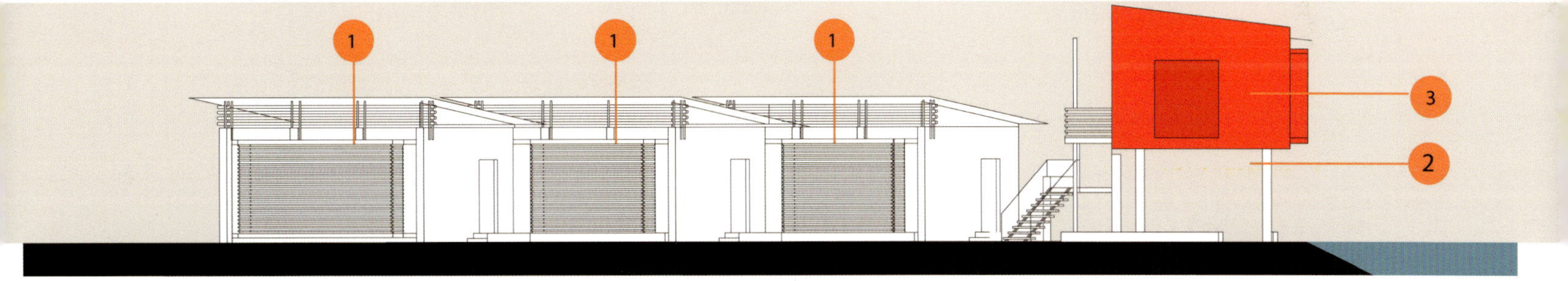

西立面
West Elevation

1 DORMITORY

2 MULTY-PURPOSE SPACE

3 COMPUTER CLASS

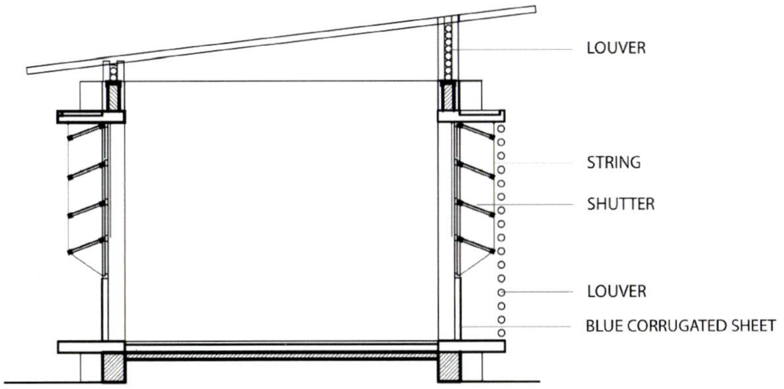

平面图
PLAN (Photographer: KHROMA)

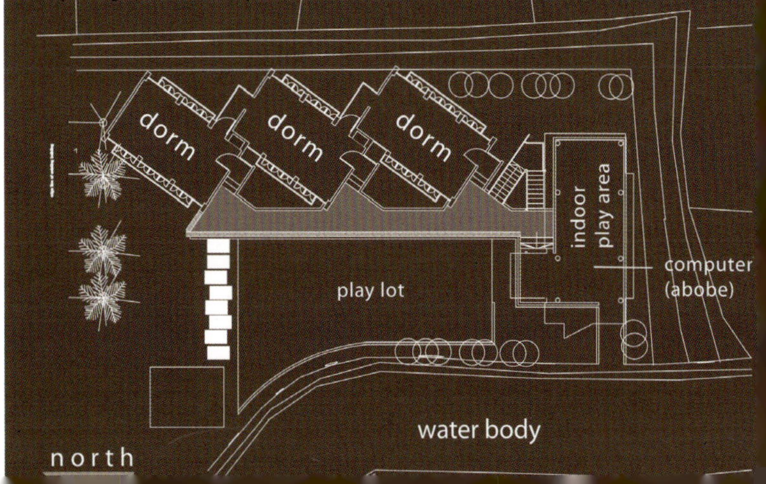

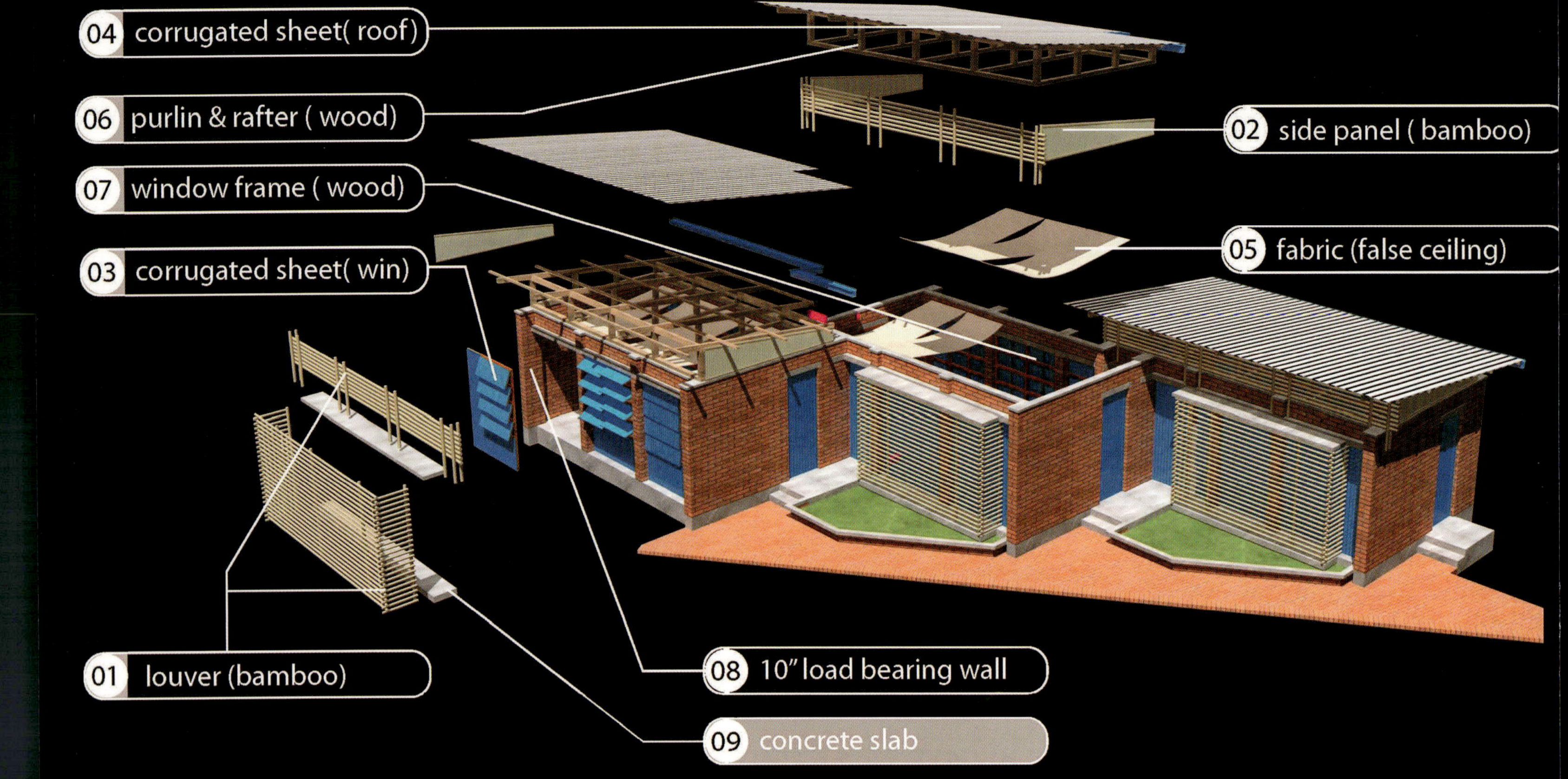

竹节细部图
Bamboo Joint Detail

景致
View

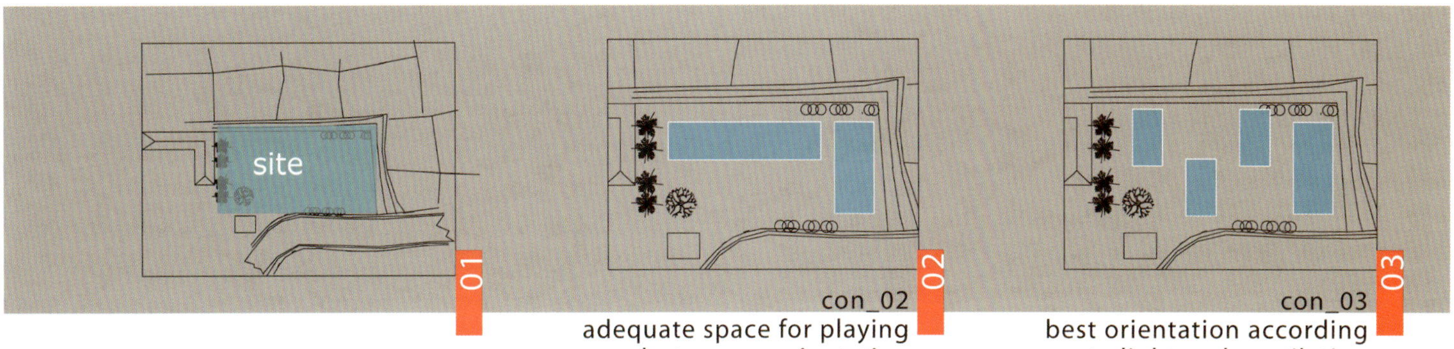

adequate space for playing
but wrong orientation

best orientation according
to light and ventilation
but no space for playing

north

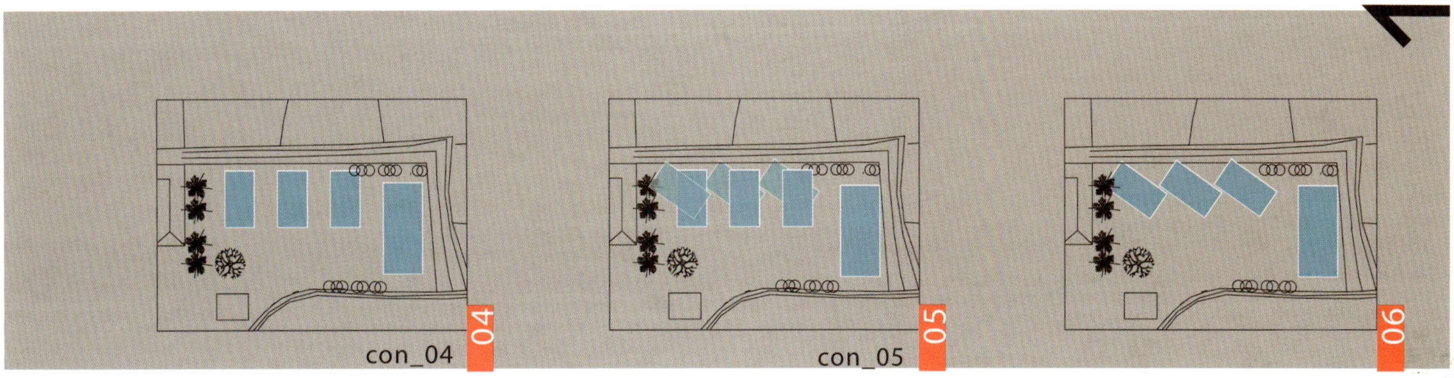

第七届CAA学生设计竞赛
7th Caa Student Design Competition

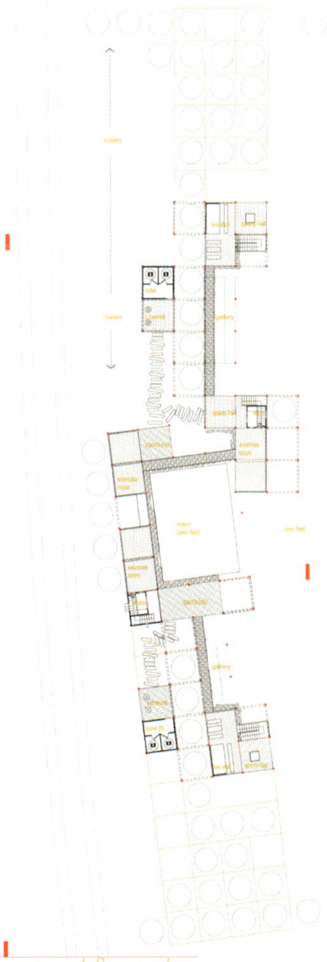

ground floor plan

2nd floor plan

Project Team: Carla Tamariz, Veronica Schereibeis, Christian Bottger

Pachacamac House

秘鲁巴恰卡马隐藏别墅

A hill in Pachacamac, located 40km south of Lima at Peru's coast, is the site for the retirement home for a couple both philosophers. The response to the site's conditions was to bury the house inside the hill, trying to create a balanced dialogue between architecture and landscape, where inside/outside becomes a constant interpretation of materiality with a strong sense of protection and appreciation of the dark and the light. A glass box sticks out of the hill symbolizing architectural intervention on untouched nature.

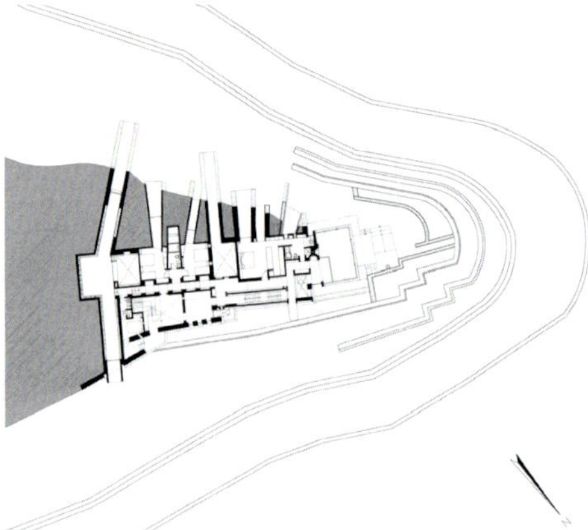

地面平面图
Floor Plan

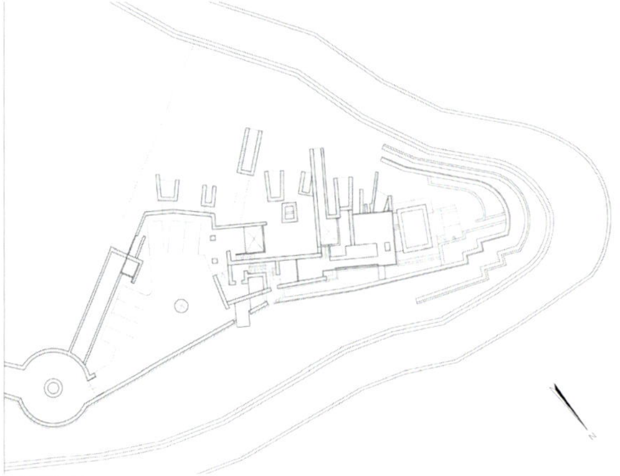

屋顶平面图
Roof Plan

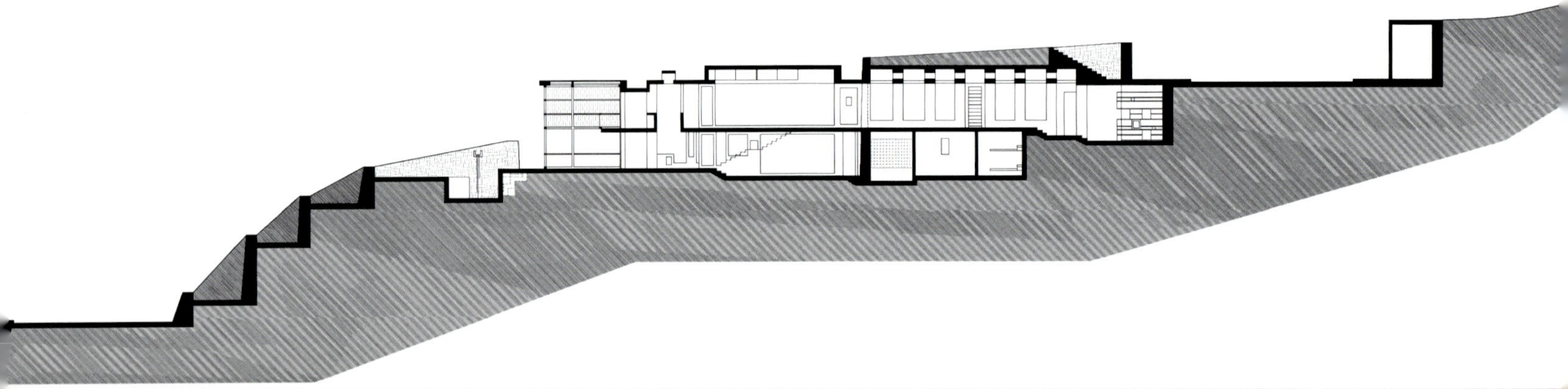

西南方鸟瞰图
Eye Bird View from the South-West

进入西面的通道，卧室及溪谷
Entry to west tunnel, bedroom 1, 2 and Service Canyon

男性入口
Male entrance P

主卧窗户
Window at Master Bedroom

东面隧道
East Tunnel Ready for Shelves and Books at Underground Studio

Project Team: Aaron Beyers

Passive Solar Habitat

被动式太阳能住宅

In an effort to better educate those less familiar with concepts of sustainable living and being able to co-exist with our natural environment, this project expands upon techniques of passive solar housing by utilizing local resources and investigating methods of operability.

Inspired by its' natural habitat while simultaneously consistent with techniques in passive solar design, the project takes advantage of immediate local resources to sustain itself by using the sun's energy for heating and cooling living spaces and collecting abundant, discarded materials as a way of re-thinking how the building is constructed. The design is perceived as a composition of recyclable materials such as salvaged steel from local industrial zones, timbers, aggregates from quarries and slag. The housing structure is intended to be assembled as a series of prototypical, pre-fabricated components minimizing cost, labor and site impact. In highlighting the use of materials, the design becomes a representation of sustainable re-use principles.

Equally relevant to the project's conceptualization of meshing with its' natural environment are the dynamics of operable building components. Although many passive systems often do not implement moving parts, the design seeks to investigate means of operability and layers allowing it to open up to natural breezes and sunlight or close itself off from harsh weather,

CLOSED POSITION

OPEN POSITION

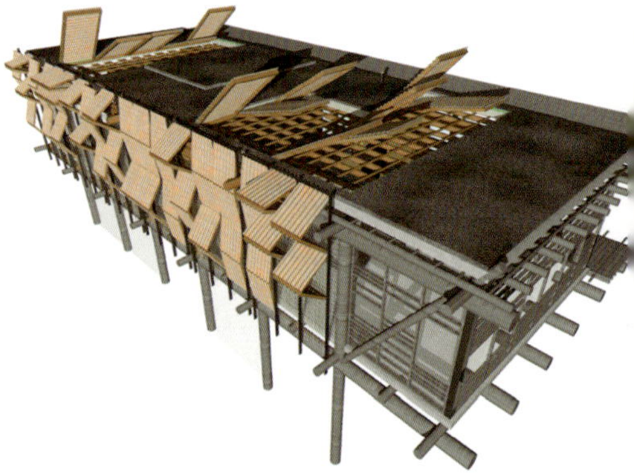

CUSTOMIZED BY OCCUPANT

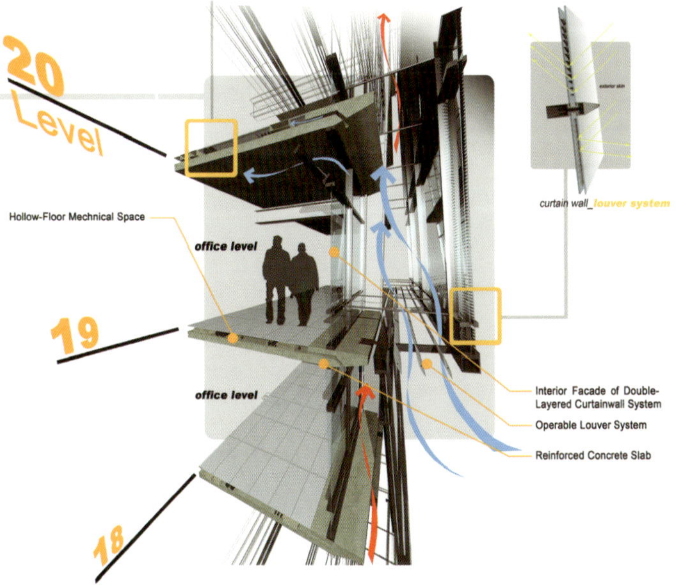

20
Level
19
18
office level
office level
Hollow-Floor Mechnical Space
curtain wall_louver system
Interior Facade of Double-Layered Curtainwall System
Operable Louver System
Reinforced Concrete Slab

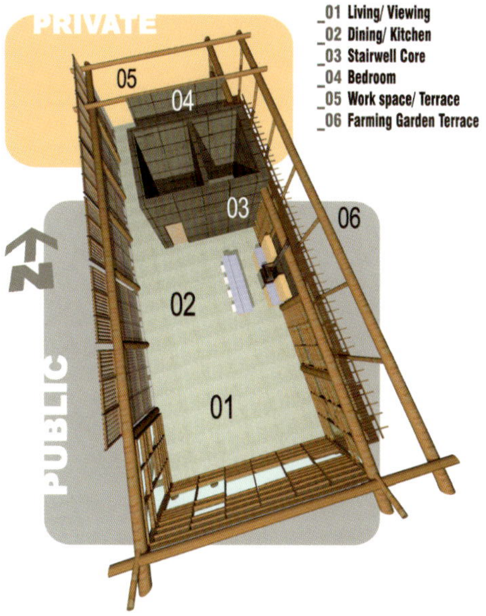

PRIVATE
PUBLIC
_01 Living/ Viewing
_02 Dining/ Kitchen
_03 Stairwell Core
_04 Bedroom
_05 Work space/ Terrace
_06 Farming Garden Terrace
01
02
03
04
05
06

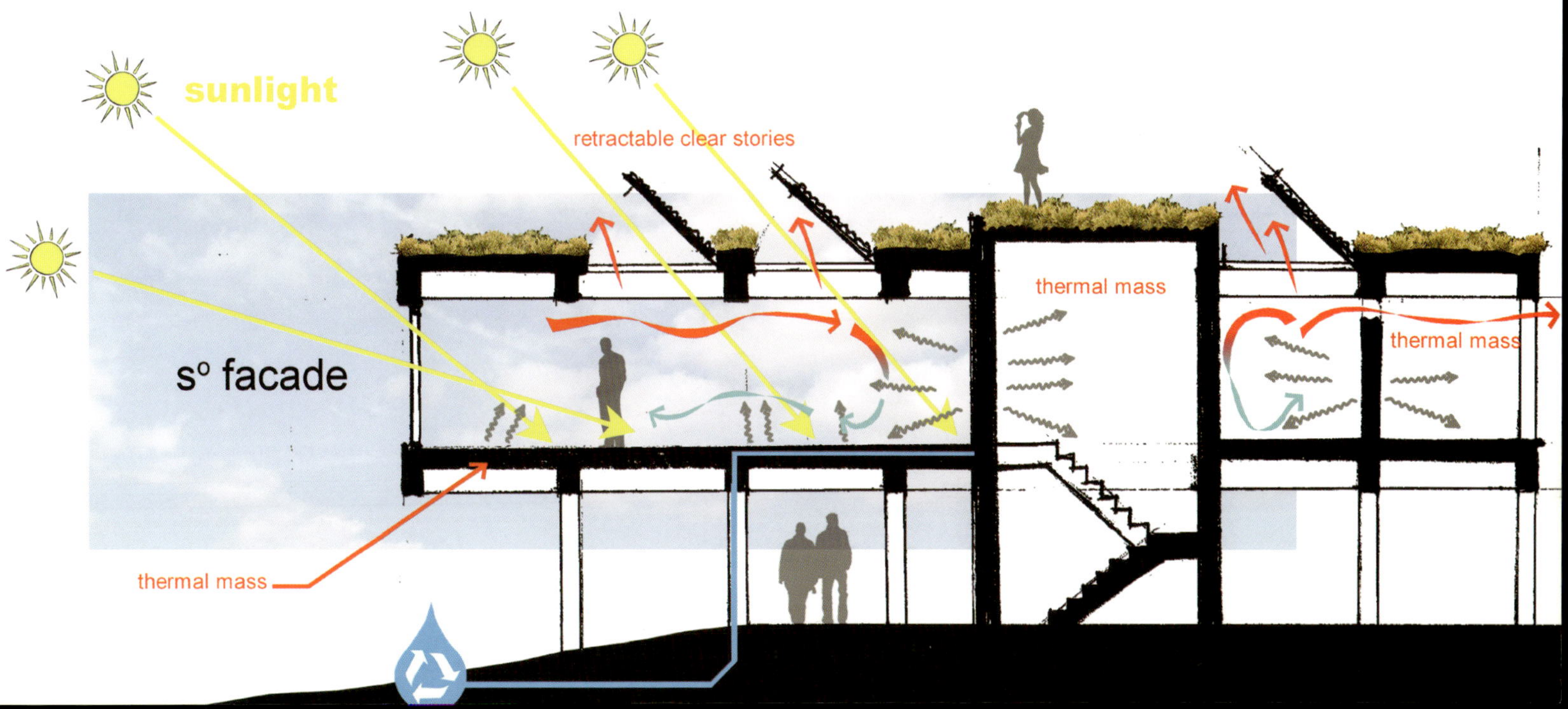

sunlight
retractable clear stories
thermal mass
thermal mass
thermal mass
s° facade

winds or western sun. The building tectonics is seen as an exploratio of possibilities in ventilation and natural day-lighting that come with breezeways, operable walls and clerestories. A series of pre-fabricated façade and roof panels are used and are independentl operable to maximize interior comfort. The majority of window glazi utilizes south and east facing sun to collect and distribute heat by means of the design's thermal masses, which are centralized as the floor slabs and circulation core.

The ground level serves as the main entry plaza below the interior living spaces. By elevating the primary living/dining/sleeping zones, the functional spaces are doubled without doubling the resources, site impact is minimized and contextual views are enhanced. The ro incorporates an accessible green roof system used for both cooling purposes within the house and outdoor recreation.

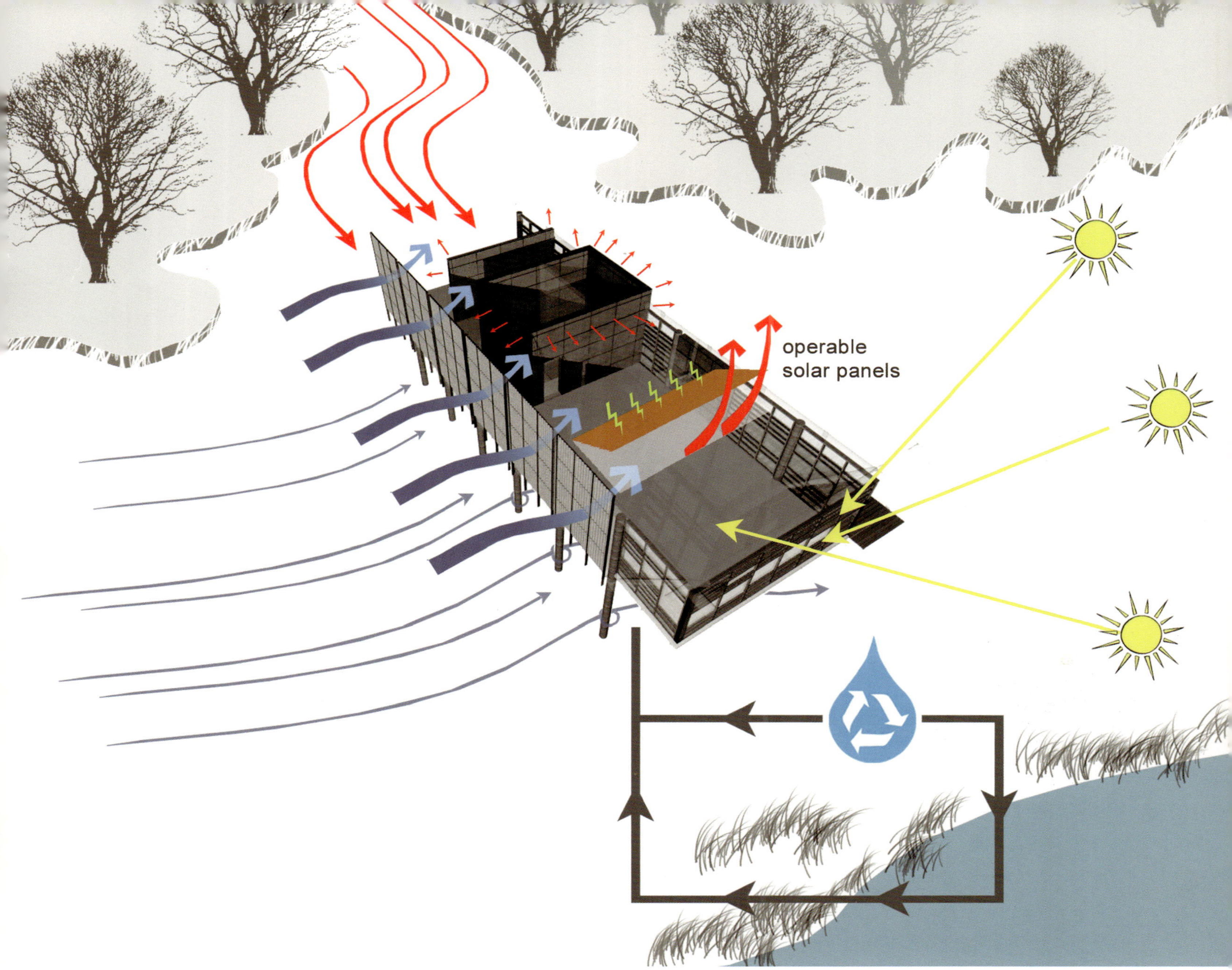
operable
solar panels

Project Team: Dadatsi dominiki, Fountoulaki Rena, Pavlidou Eleni

Plan[T]is

未来城市住宅新模型之典范

The project team proposes a housing project as an example of adaptive architectural environment. Based on ecology systems, the team investigates how an urban development can be explored as a simulation model of natural growth, negotiating and adapting to the existing urban fabric. Borrowing rules and functions of the natural world, such as growth, and phyllotaxis, the project is investigated as a proposal of parametric phasing development according to different urban needs.

According to the phasing logic of the project and the housing capacity of the site, the proposed housing units range between2.500-4.500. The plan{t}is project aims to the creation of an open urban parametric system capable of adjusting to different urban needs and conditions. It explores a way of producing architectural interventions based on understanding the nomadic way of living in London taking five different communities of people and creating an interactive housing environment through a nomadic habitation. Through reflexive systems the urban design is developed through generative computational processes as time-based phasing logic by collaborating the generative tool of agent based system with the distribution tool of phyllotaxis natural model. The project strategically located between the London City Airport, the Thames River, the Docklands and the Millennium Dome negotiates the different territorial boundaries and

transportation networks by interweaving all synergic sources of the site. As London is one of the most Nomadic cities of the world this projects aims to fulfil the need of the temporary leaving of our days. As today's rhythms of life have created the need for a new model of city, more flexible, ephemeral and more prepared for population's changes and needs this project aims to fulfil this demands, seeking to create a new way of living , creating different lifestyles of habitation for different urban needs. Through this project, and taking the tower as a case study, RED.pdf will try to defy the logic of the present, formulate new questions, and perhaps will give a new type of building. A building for the next generation, for the new type of living that allows new potentials and perceptions.

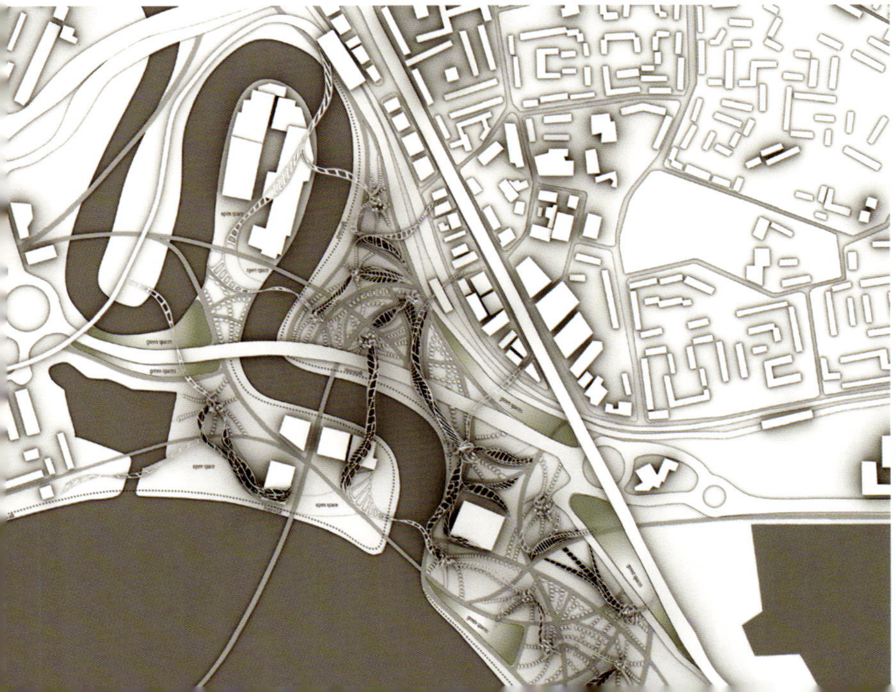

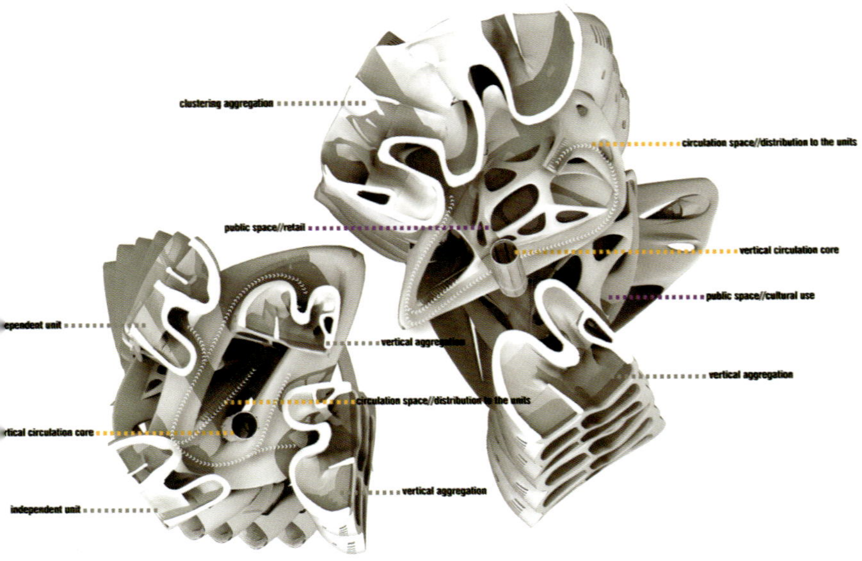

basic component
type of users: bussiness people
number of users: 1-3

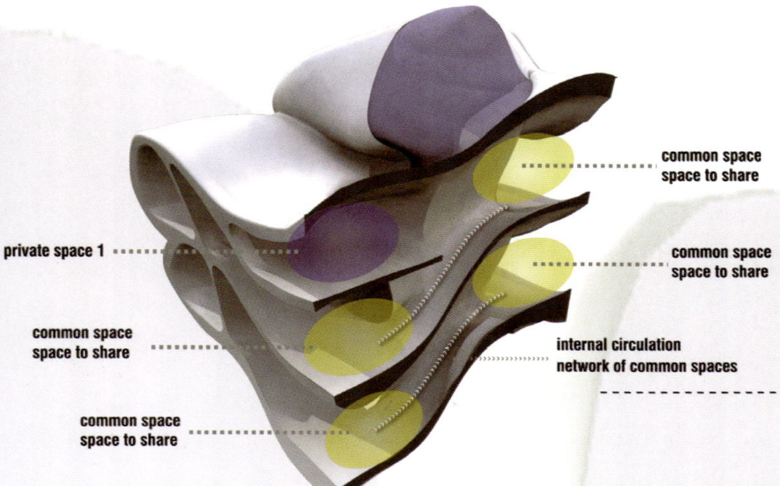

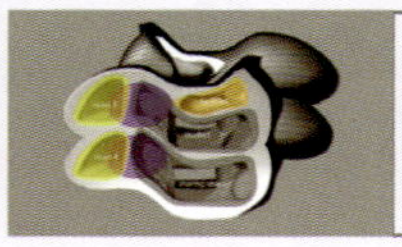

type of aggregation: back to back
number of components: 4
type of users: artists
number of users: 6

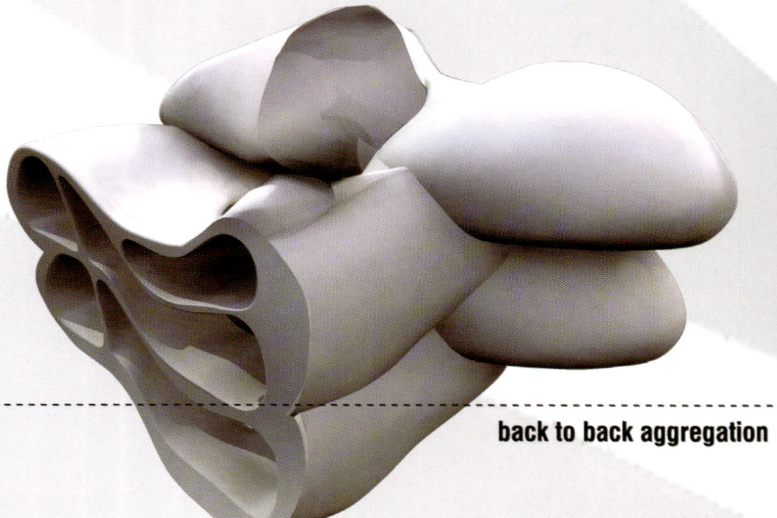

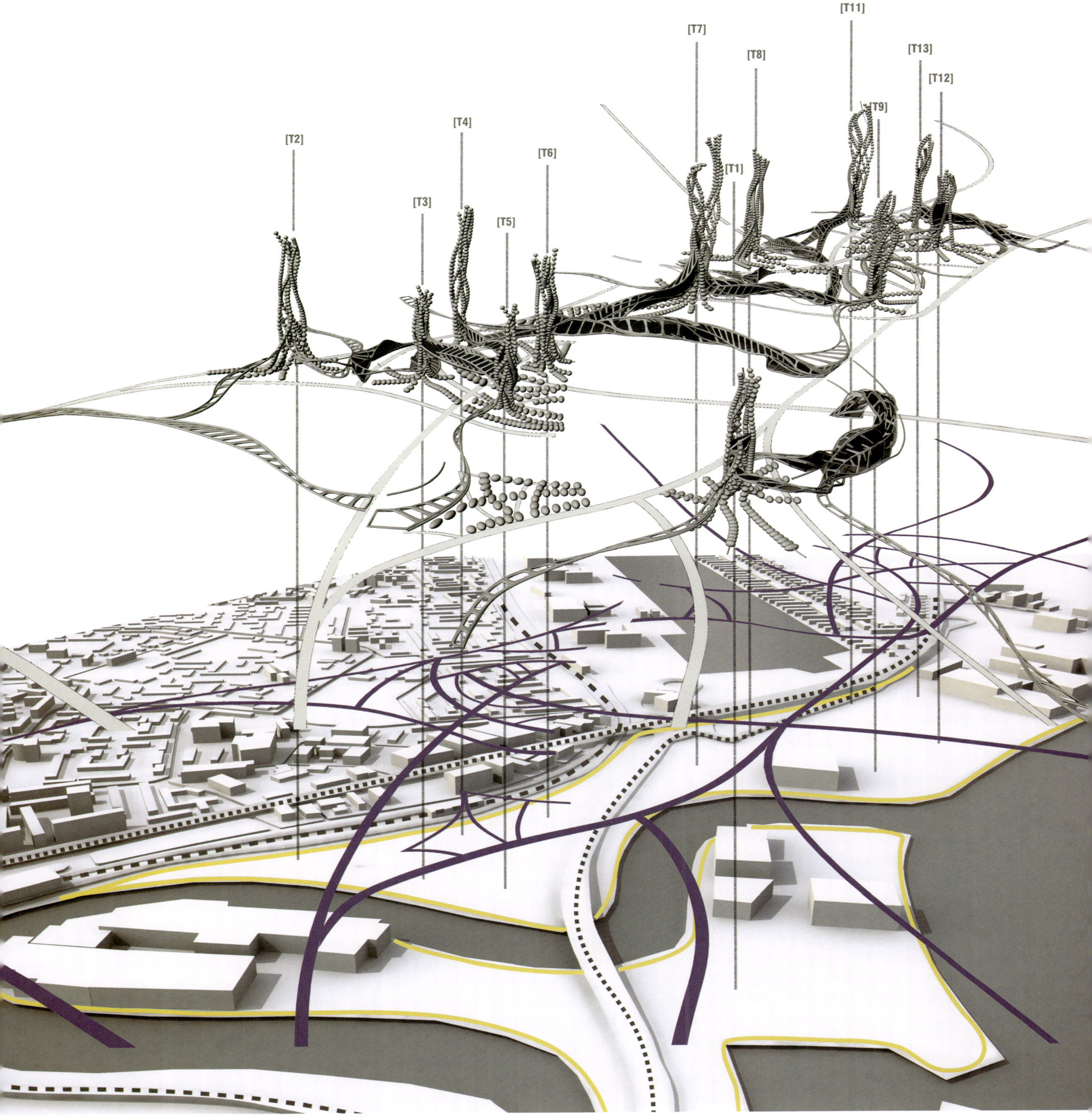
[T2]
[T3]
[T4]
[T5]
[T6]
[T7]
[T1]
[T8]
[T11]
[T9]
[T13]
[T12]

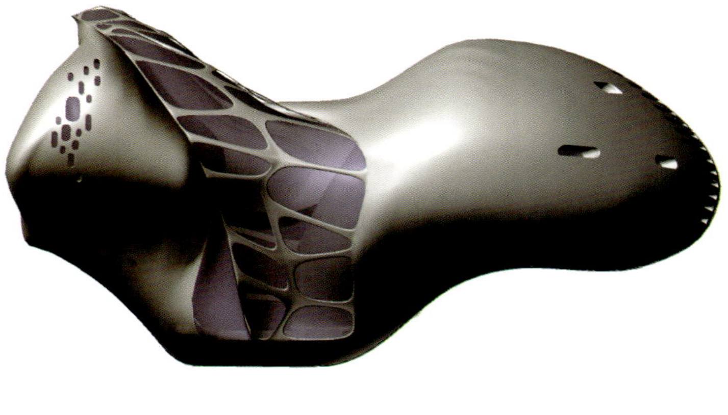

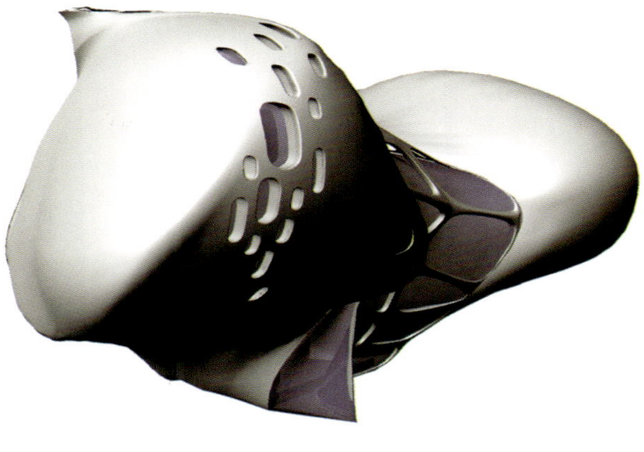

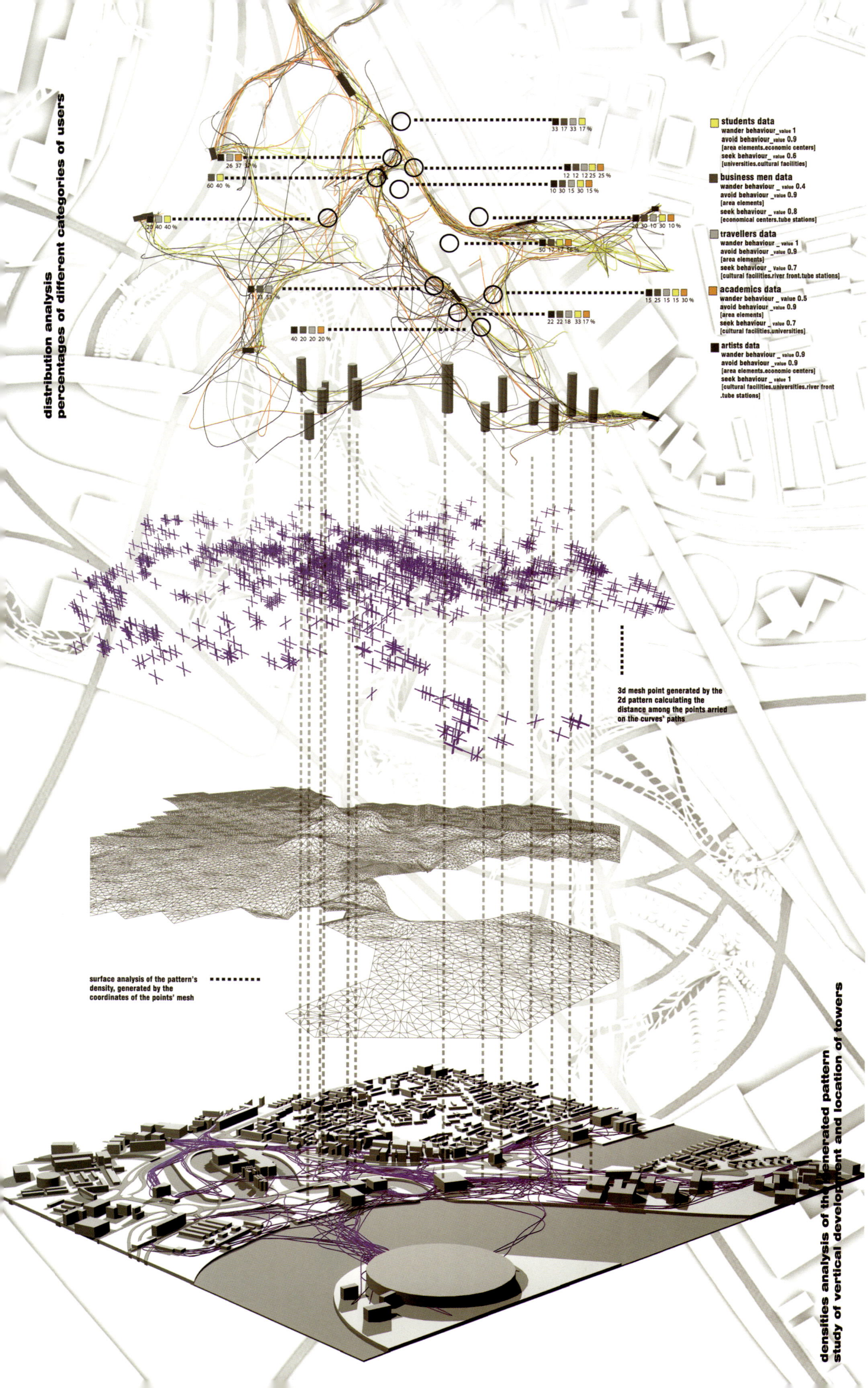
distribution analysis
percentages of different categories of users
33 17 33 17 %
26 37 37 %
60 40 %
12 12 12 25 25 %
10 30 15 30 15 %
20 40 40 %
20 30 10 30 10 %
50 17 17 16 %
33 33 33 %
15 25 15 15 30 %
22 22 18 33 17 %
40 20 20 20 %
students data
wander behaviour_value 1
avoid behaviour_value 0.9
[area elements.economic centers]
seek behaviour_ value 0.6
[universities.cultural facilities]
business men data
wander behaviour _ value 0.4
avoid behaviour _value 0.9
[area elements]
seek behaviour _ value 0.8
[economical centers.tube stations]
travellers data
wander behaviour _ value 1
avoid behaviour _value 0.9
[area elements]
seek behaviour _ Value 0.7
[cultural facilities.river front.tube stations]
academics data
wander behaviour _ value 0.5
avoid behaviour _value 0.9
[area elements]
seek behaviour _ value 0.7
[cultural facilities.universities]
artists data
wander behaviour _ value 0.9
avoid behaviour _value 0.9
[area elements.economic centers]
seek behaviour _ value 1
[cultural facilities.universities.river front
.tube stations]
3d mesh point generated by the
2d pattern calculating the
distance among the points arried
on the curves' paths
surface analysis of the pattern's
density, generated by the
coordinates of the points' mesh
densities analysis of the generated pattern
study of vertical development and location of towers

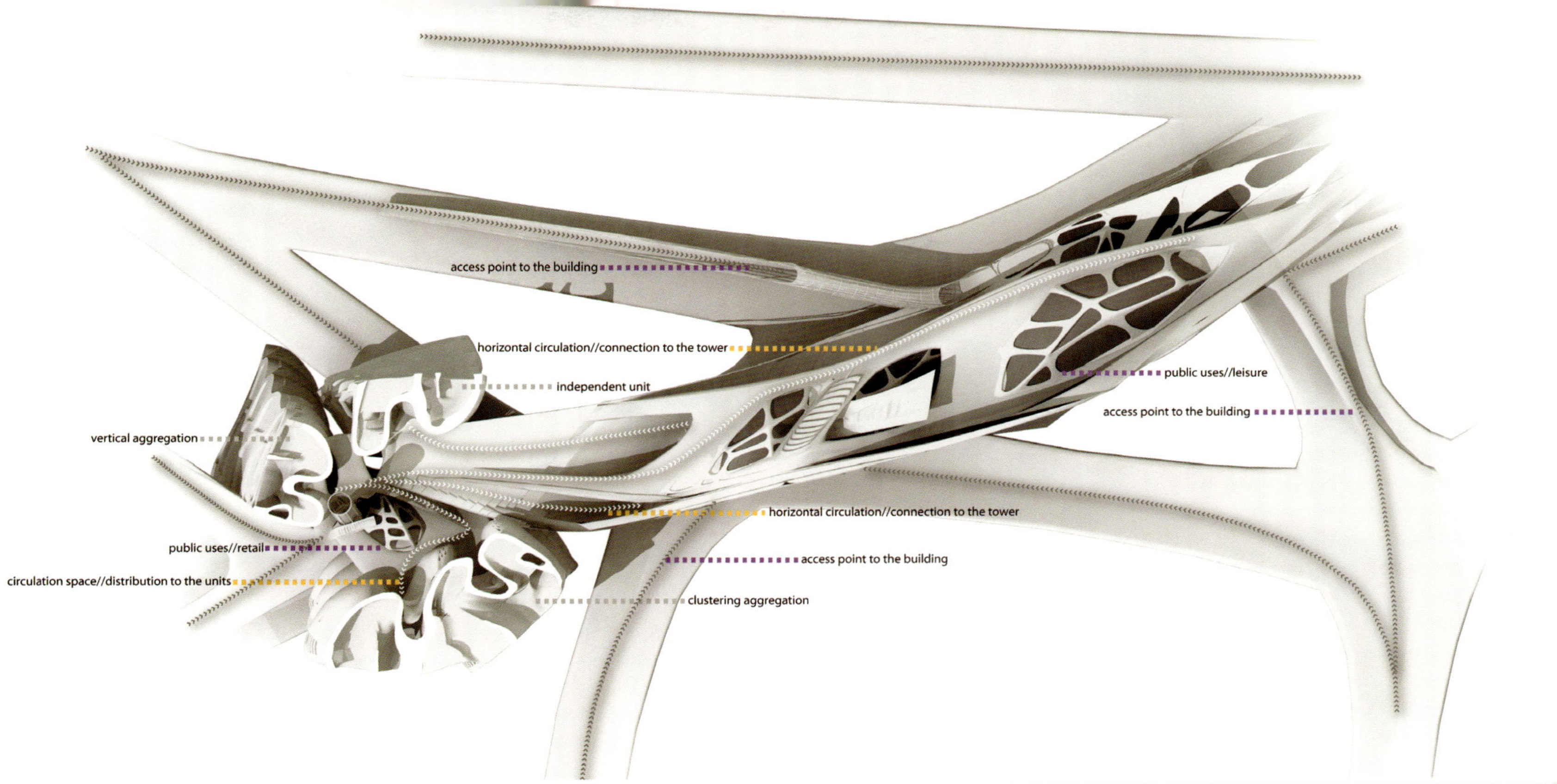
access point to the building
horizontal circulation//connection to the tower
independent unit
public uses//leisure
access point to the building
vertical aggregation
horizontal circulation//connection to the tower
public uses//retail
access point to the building
circulation space//distribution to the units
clustering aggregation

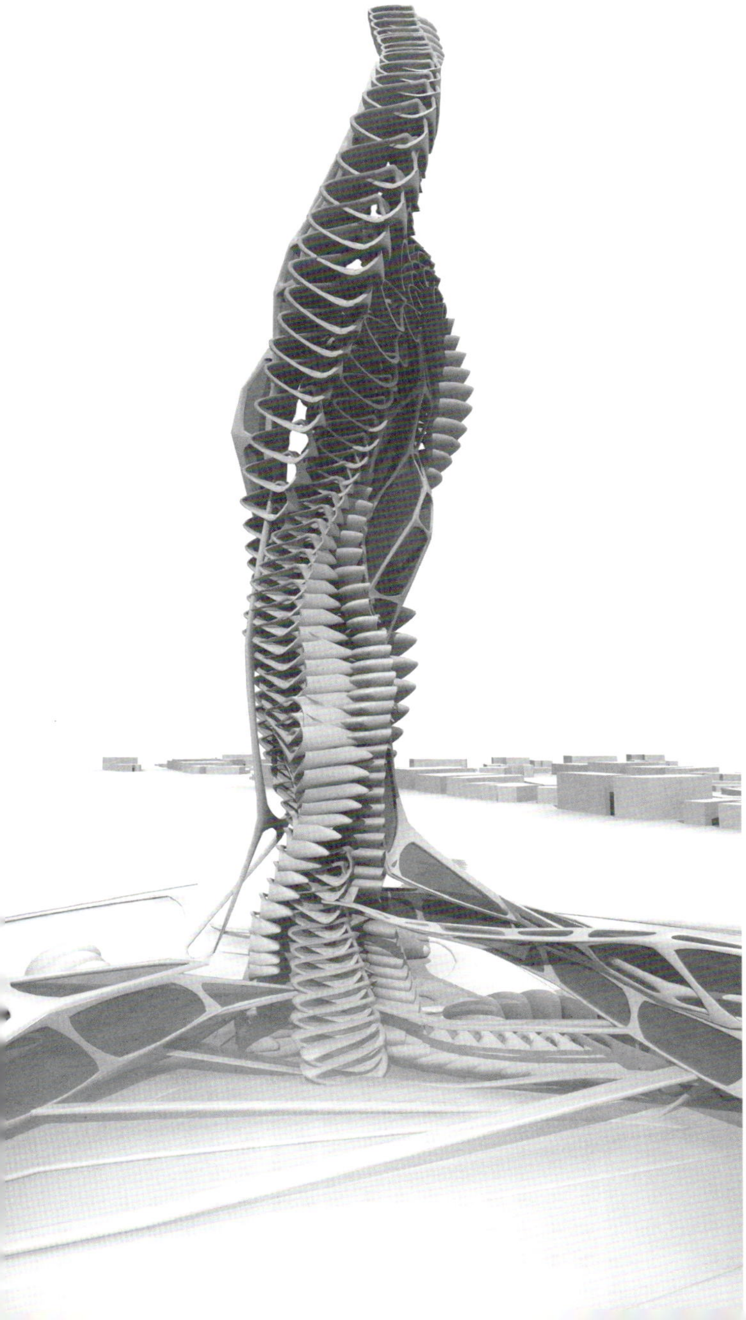

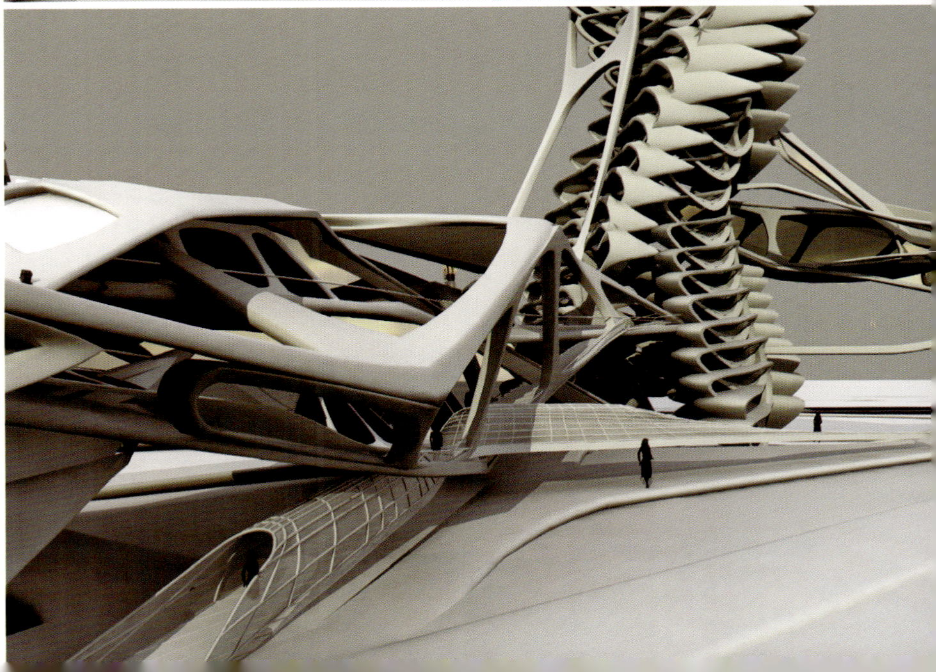

Design: LOT-EK, Ada Tolla + Giuseppe Lignano, Principals; Keisuke Nibe, Koki Hashimoto, Project Architects

PUMA CITY – A Nimble Dismountable Container Building with Convenience

PUMA CITY——便利灵巧的可拆卸式集装箱建筑

Twenty-four shipping containers are retrofitted and transformed into PUMA CITY, a transportable retail and event building that is traveling around the world along with the 70-foot long Puma sailing boat — il Mostro — during the one-year long 2008 Volvo Ocean Race, just started in Alicante, Spain in early October.

The building is fully dismountable and travels on a cargo ship along with the sail boats; it will be assembled and disassembled a number of times once it reaches the different international harbors.

The building is conceived as a three levels stack of containers, shifted to create internal outdoor spaces, large overhangs and terraces. The stack is branded with the super-graphic logo of the company — fragmented as a result of the stack shift.

Puma City is comprised of two full retail spaces on the lower levels, both with large double heights as well as with 4-container-wide open spaces to challenge the modular box-quality of the container inner space; offices, press area and storage occupy the second level and a bar, lounge and event space with a large open terrace is at the top.

The building uses 40-foot long shipping containers as well as a number of the existing container connectors to join and secure containers both horizontally and vertically. Each module is designed to ship as conventional cargo container through a system of structural covering panels that fully seals all of its large openings to be removed on site to re-connect the large, open interior spaces.

Puma City is a truly experimental building that takes full advantage of the global shipping network already in place. At 11,000 square feet of space, it is the first container building of its scale to be truly mobile, designed to respond to all of the architectural challenges of a building of its kind, including international building code, dramatic climate changes, plug-in electrical and HVAC systems and ease of assembly and operations.

Design & Research: Suprio Bhattacharjee

Reclaiming the City — Urban Nexus 2020

改造城市——城市枢纽2020

BRIEF

Sustainable Infrastructure and Growth Model for dense cities such as Mumbai, station redeveloped with speculative development for retail, a hotel, cultural facilities [two 500-seater auditoria and a 1000-seat concert hall] along with a multiplex cinema with 5 screens. Project involved the use of green roofs over concrete slabs acting as heat sinks cooling the station below, self-shading strategy in the orientation of towers, bio-climatic design principles for the office towers such as east-west orientation, narrow floor plates [average width 12-14m] to maximize daylight penetration and natural ventilation, use of terracotta and GRC jaalis on the south façade of office towers and planted terraces [sky gardens] on sunward sides.

ISSUES

This prospective global city faces crises at multitude levels. Placeless objects in the name of new age architecture are slowly obliterating a sense of cohesive social structure and unique scale specific to each place and community – does the city loose its layered identity in favour of a place-less plastic imagery? Besides, the city needs to address other issues important for its survival – such as sustainable models of growth, the issue of sprawl; and the indiscriminate destruction of verdant land and natural eco-systems for development purposes when brown-field and grey-field sites are more than available.

THE CITY AS A HYPER-NETWORKED ENTITY

The premise of this project works on the reality that the city functions as a series of node-places

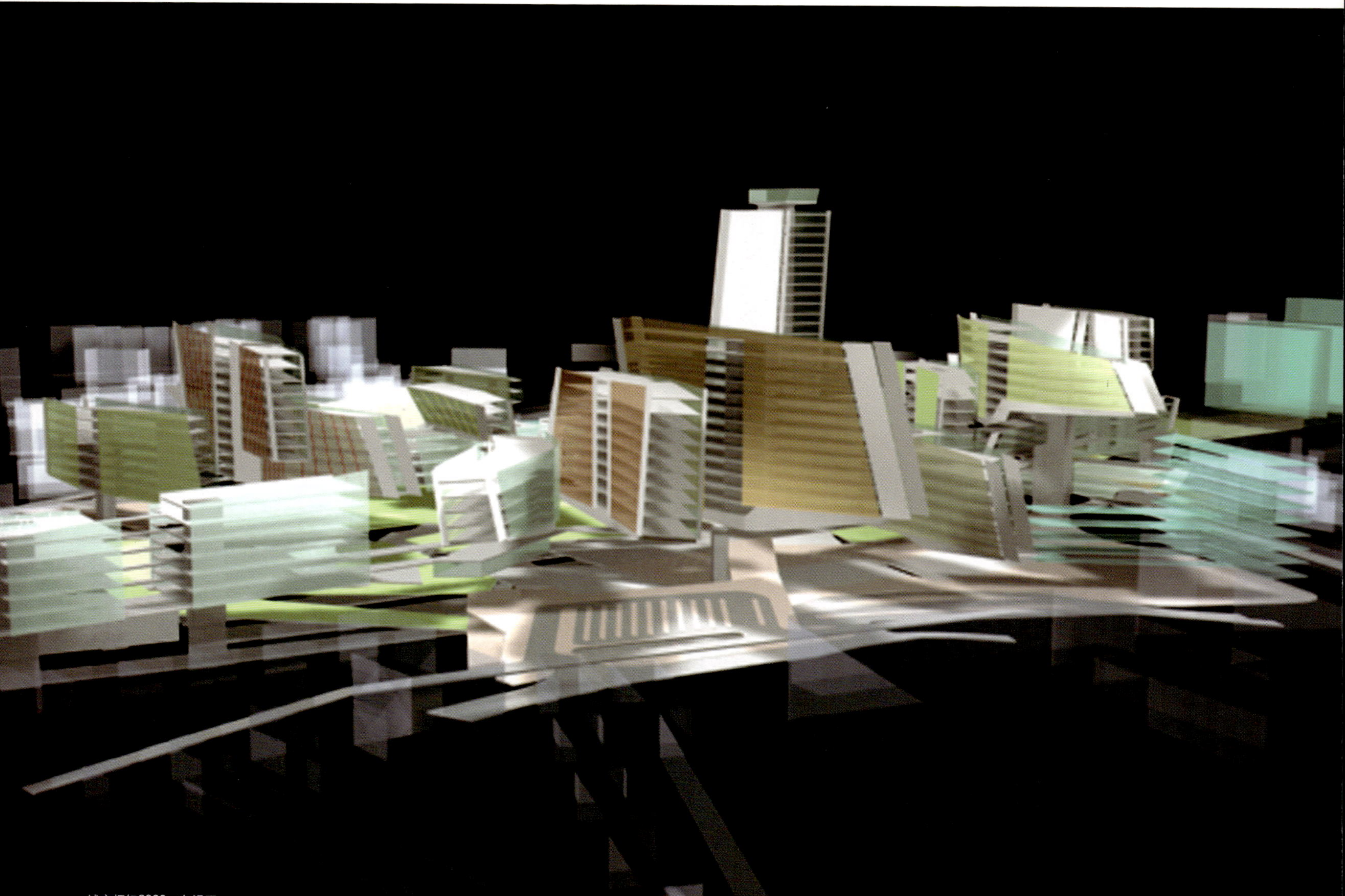

城市枢纽2020—东视图
Urban Nexus 2020 - View from East

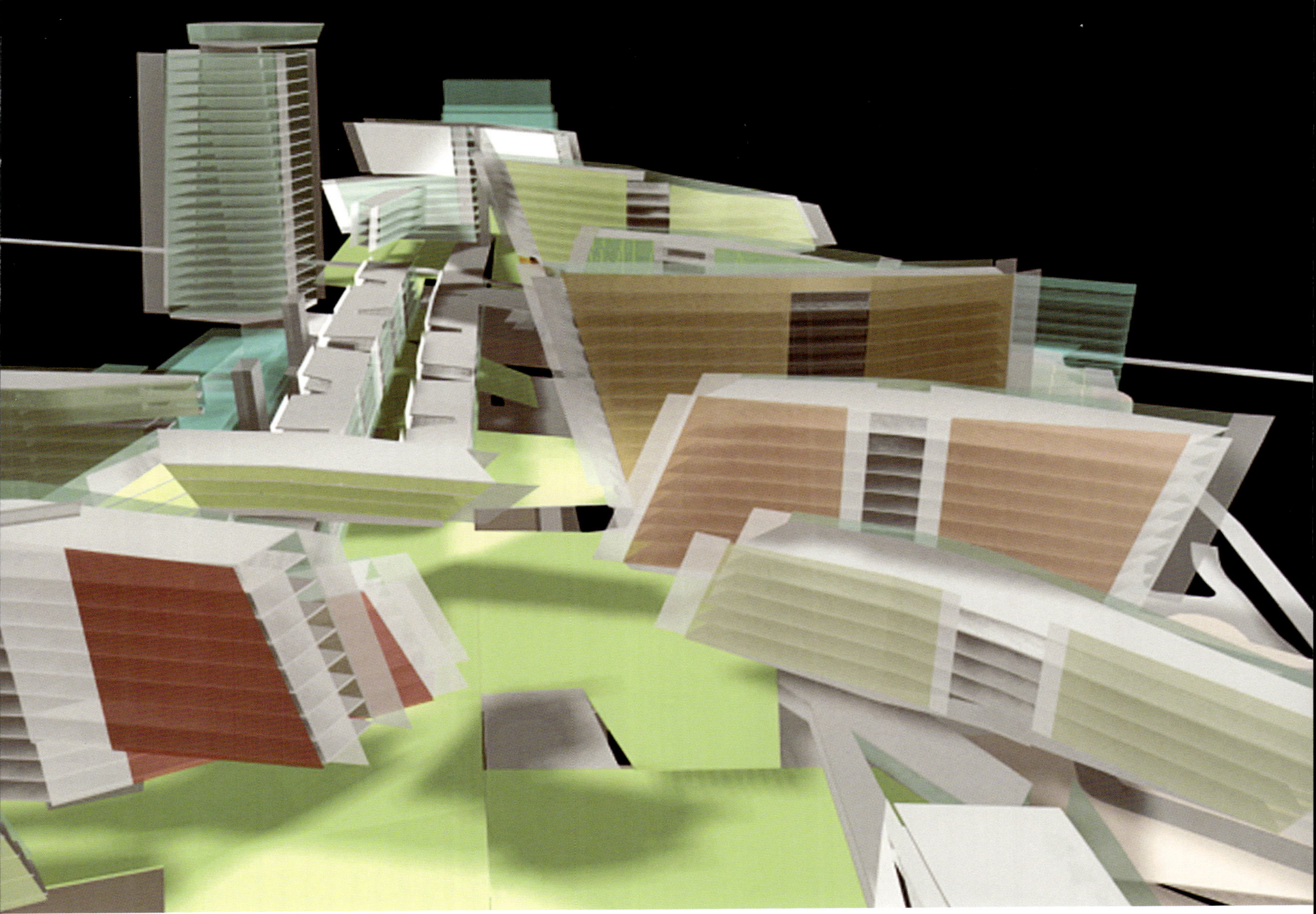

城市枢纽2020—纵向截面图—手绘—方案2001
Urban Nexus 2020 - Longitudinal Section - Hand Drawing - 2001 Scheme

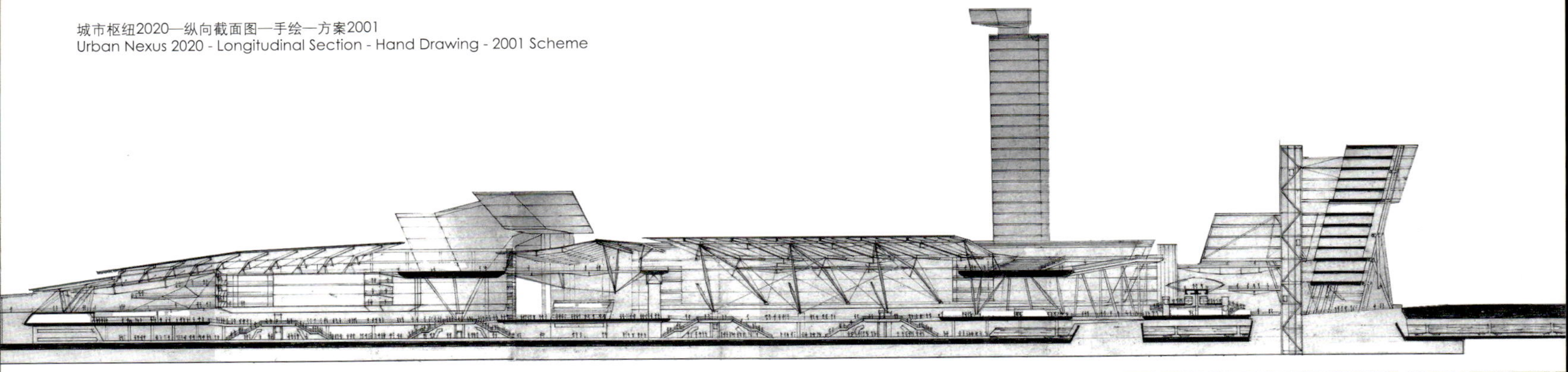

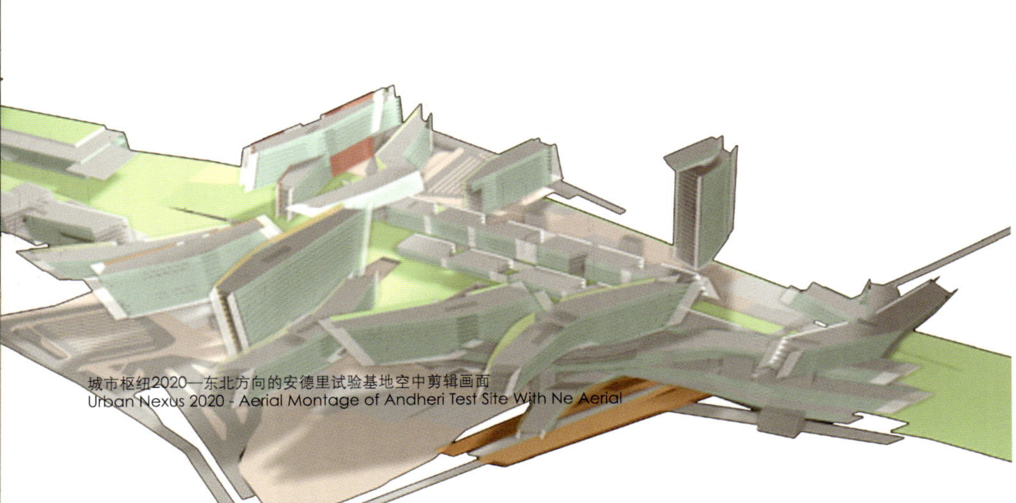

城市枢纽2020—东北方向的安德里试验基地空中剪辑画面
Urban Nexus 2020 - Aerial Montage of Andheri Test Site With Ne Aerial

where each railway station becomes the hub of each suburb – since it is most accessible. Going on the dictum of maximum density at the place of maximum connectivity, a city structure for Bombay can be envisaged wherein each suburb shall act as a self-sustaining fragment of a larger urban system – connected by a web of non-linear mass movement system, creating a hyper-networked entity. This would encourage and initiate balanced growth within the city and would ensure a sustainable vision for the city into the future.

ARCHITECTURE AS BEING EMBEDDED

Urban Nexus – Reclaiming the City – Project for the Andheri Intermodal

Large-scale architectural interventions can evolve from the forces that inhabit the site to become extensions of the city, embedding themselves in diverse immediate conditions while being differentially scaled so that each part responds to a specific and changing urban situation – pieces of cityscape rather than an imposed monolith.

About a million and a half people use the Andheri Station Precinct every day. In addition to being a center of small-scale industry and a prominent dormitory suburb, Andheri has become the unofficial hub of the service industry in the city, a fact aided by its strategic geographic centrality, its proximity to the international airport and a growing hotel and business district.

This project was conceived in anticipation of such a growth. Contradicting the official stand of the city government, it proposed an intensification of the urban experience at the station through a densely layered spatial organisation – all integrated into one seamless whole.

The station site was imagined as a crossing point or a "nexus" in a city traditionally divided into inaccessible easts and wests by broad swathes of railway tracks cutting across it. The project would need to suture these deep gashes - healing the cityscape and reclaiming precious land lost.

Conceptually, it begins with stretching across a membrane of connectivity forming a fluid public space allowing for visual clarity and easy orientation. This membrane becomes the crossing or nexus accommodating a number of super-imposed built forms hovering above.

These would be of two kinds – low-scaled linear multi-use buildings oriented N-S along courtyard spines to maximize daylight and natural ventilation – and narrow-plate office towers oriented E-W, raised on massive concrete undercrofts cantilevering over the activities below.

The two orientations are meant to express the two important flow patterns inherent on the site. The relentless N-S motion of the railways intersecting the much-desired attempt to bind the E-W divisions created.

The taller buildings along the E-W axis huddle and congregate, becoming a visual barometer to indicate the main crossing points on the site, while the linear buildings at their base act as their entry points. This orientation helps in reducing solar heat gain as well as maximizing natural ventilation and daylight, while adequate plantations on open deck levels at each floor moderate internal temperatures. These towers impart the project with strong colours when seen from the south through the use of an outer scrim of terracotta or GRC panels in vivid earth tones that act as mega-scaled jaalis modulating the light quality inside.

The idea of the nexus evolves into an undulating landscape - a shallow 'mound' formed by surfaces that gently heave up to form the roof/deck, marked by vertical shafts of towers, that would bring back memories of rock formations and lost geological treasures - unique basalt cliffs with hexagonal rock formations soaring over 300 feet high that once marked the landscape of Andheri before they were all mindlessly quarried away in the early 1900's.

Spatially, it generates an indeterminate urban topography of generous ramps and raked surfaces - which blurs boundaries and the distinction between levels, and accommodates the irregular nature of the site.

From this roof deck, the presence of the station below is discovered through incidental gaps between the raked surfaces. The drama of the space unfolds as one descends into the bowels of the station (the platform levels), from where the complexity of the layered space above is revealed through unexpected glimpses and shafts of light and vision.

The individual becomes immersed in an intense urban experience that is always incomplete – and in the process of unfolding through repeated uses.

A lasting comment that this project also attempts to convey is that Bombay/Mumbai is not flat . . . as most people and planners have considered it to be (the July 2005 floods awoke everyone from their slumber to realise this) and giving back a sense of its lost topography in an urban condition that has forgotten that the ground in Bombay was never meant to be flat, is a dire need.

城市枢纽2020—大楼理念原型—手绘—2001
Urban Nexus 2020 - Prototypical Tower Concept - Hand Drawings - 2001

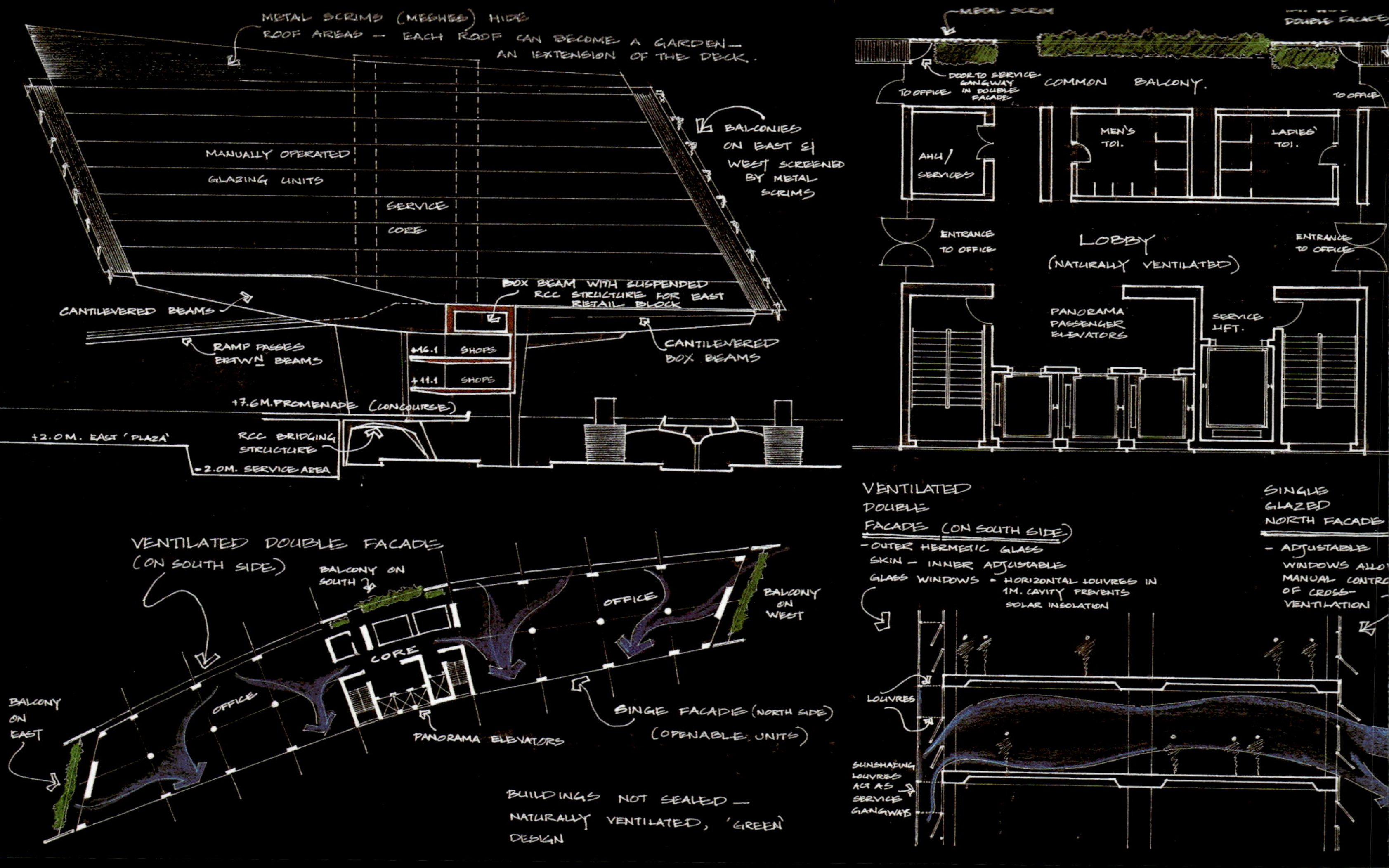

城市枢纽2020—北广场上升的火车站视图
Urban Nexus 2020 - View From Elevated Train Station Over North Plaza

城市枢纽2020—东广场视图
Urban Nexus 2020 - View From East Plaza

城市枢纽2020—西北方向视图
Urban Nexus 2020 - View From North-west

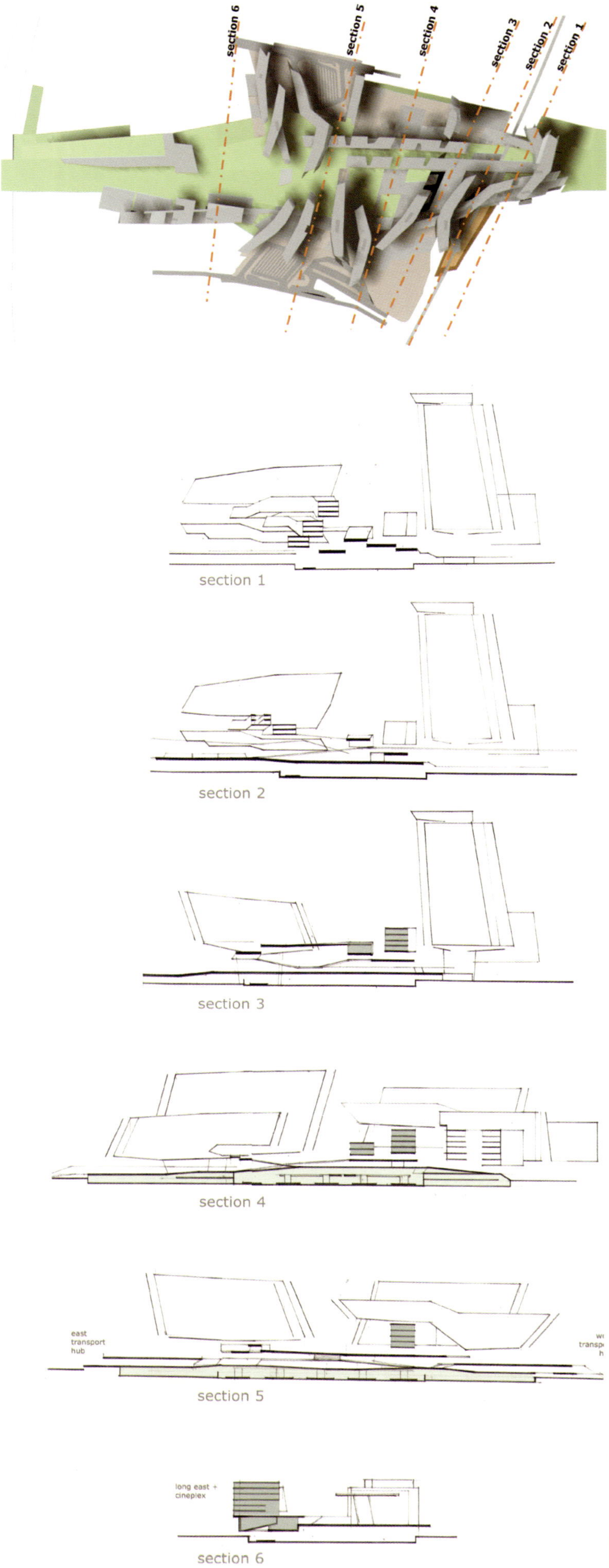

城市枢纽2020—安德里试验基地的地形纲要截图
Urban Nexus 2020 - Schematic Topographical Slices Through Andheri Test Site

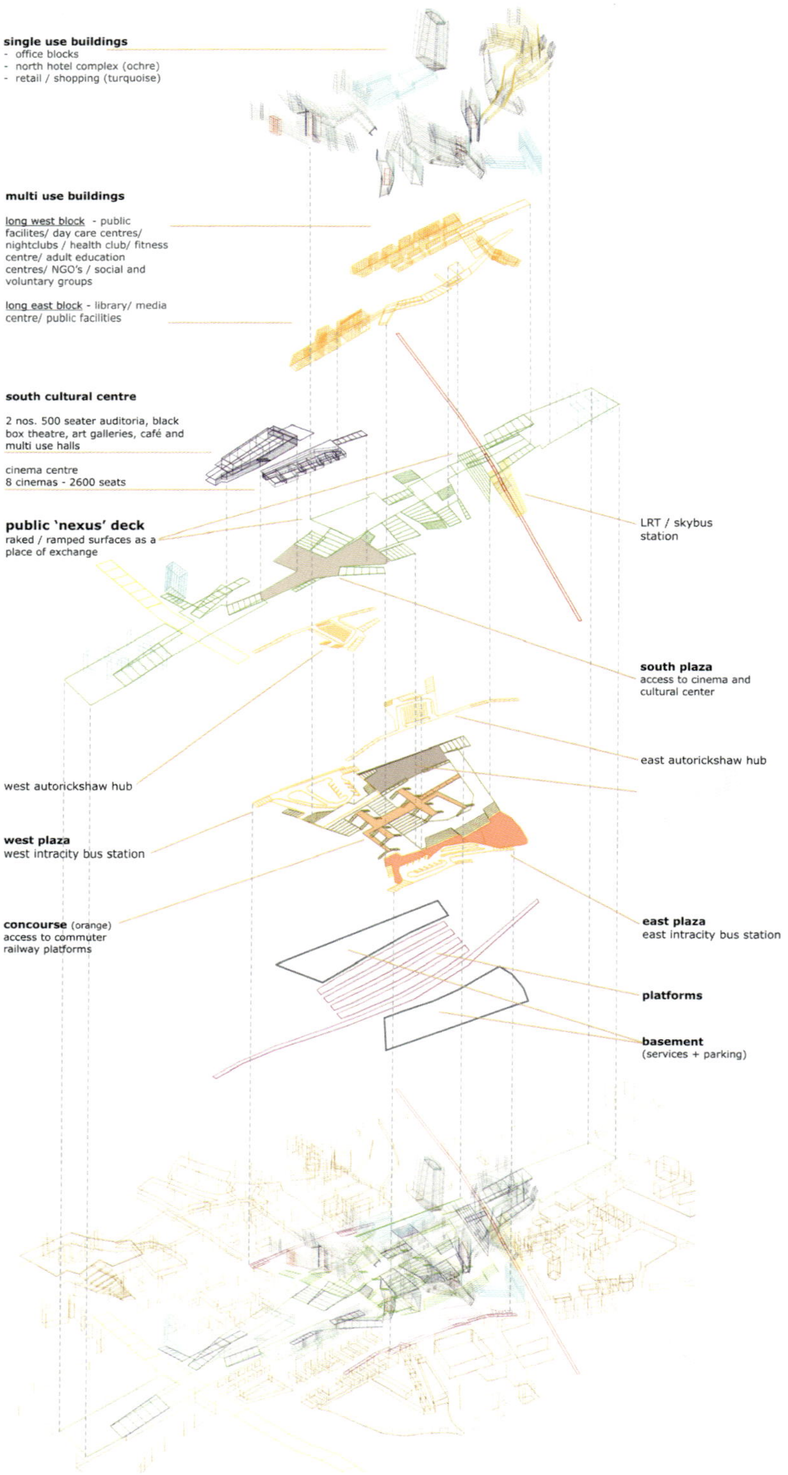

城市枢纽2020—后阳台向北视图，透过后阳台缝隙可以看到西站入口
Urban Nexus 2020 - View At Deck Level Towards North With West Station Entrance Through Deck Fissure

城市枢纽2020—东北方向视图
Urban Nexus 2020 - View From North-east

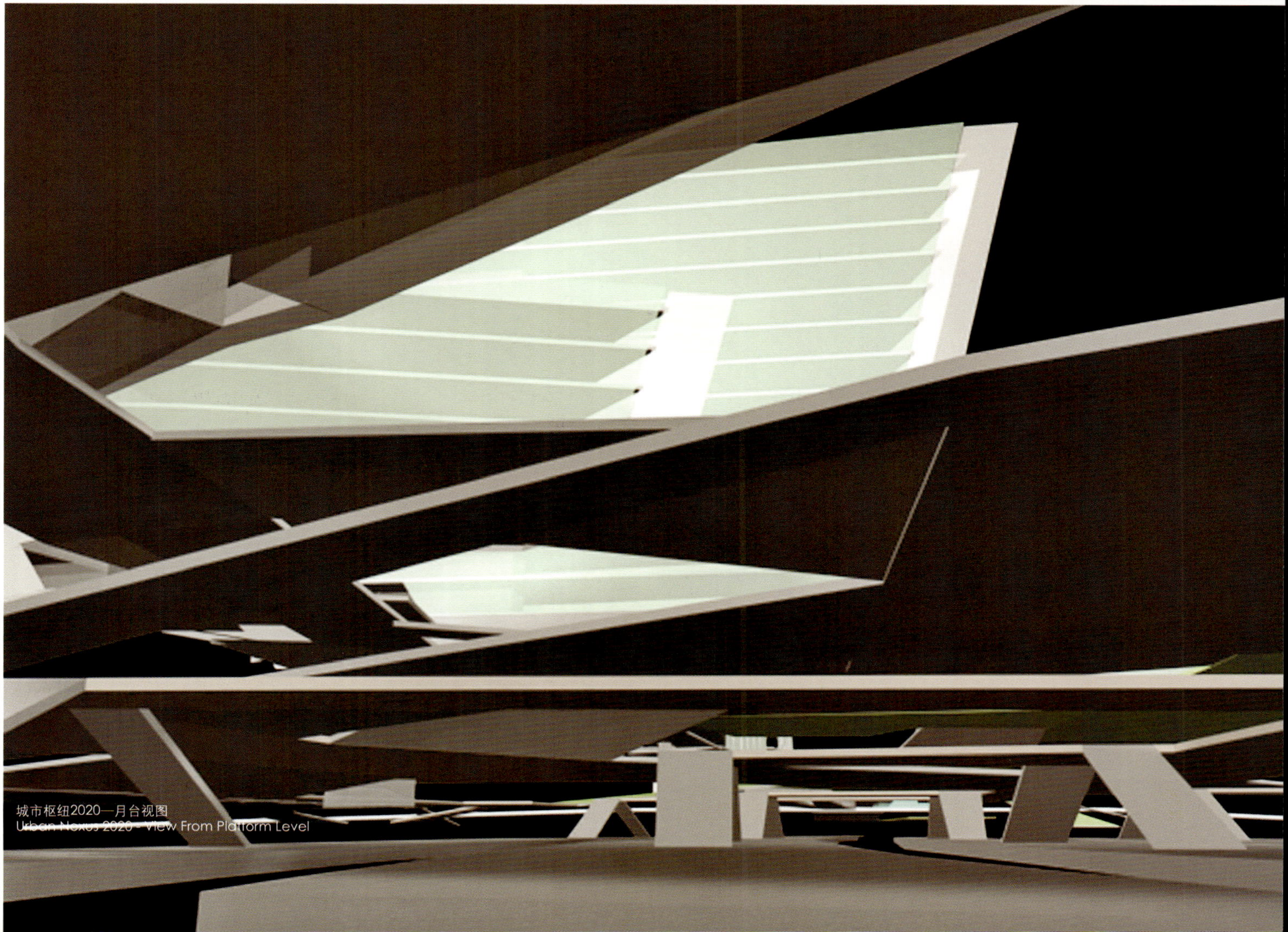
城市枢纽2020—月台视图
Urban Nexus 2020 - View From Platform Level

Chief Architect/Office Name: Student Architect – Bolojan Daniel Calin

Recreation Center

康乐中心——建筑与光的完美结合

The project idea is to analyze and integrate the Trenta main characteristics the: mountains, fog, vapor, smoke and the village. By using different elements and their compositions allow to achieve an adaptable system which maintains also its instant recognisability.

As we already know until now architecture represent a game of shadows and light, game where we report permanent. Our perception was governed entirely by this relation. If we think what happens with an architecture enveloped by haze, we can say that all the perspectives that will result will be distorted like a spherical projection and the fluid boundaries of haze will be in a continuous feedback with the environment. What will happen with the visitor in this haze space? We all know that we find flavor in ambiguity of haze which is full of possibilities, and promises. Fog will condition a reality in continuing transformation and will always require a new reading.

The project will be an interactive system, which have almost the same physically characteristics with haze space

-impossibility of perceiving space as a whole

-wavy space, with no right angles, the range is connected by a clothoid (flat curve whose curvature is proportional to the length of each arc between center and point considered) symbol of passing into an unknown territory.

-the impossibility to refer to the line of horizon

-the interactive system will be activated not only by the exact coordinates of the visitor at that time but also the history route.

Moreover, the structure—curved surfaces connecting different elliptic sections—acts as path of deformation while it can change under the influence of wind, just as mist medium does. So, we can say that it is not relevant to speak in terms of architecture, about indoors/outdoors; it is rather a total space, a space—event, where the visitor should find the way by himself.

The need to provide a community centre will usually be generated either by the need to update, modify or replace an existing facility that has genuinely served a need, or to provide a totally new facility intended to encourage the re-establishment of a sense of community.

The size and functions of program will be designed so that it can be maintained by the inhabitants of area. The program is divided in 2 main programs—first sport and second therapy—spa.

屋顶平面图
Roof Plan

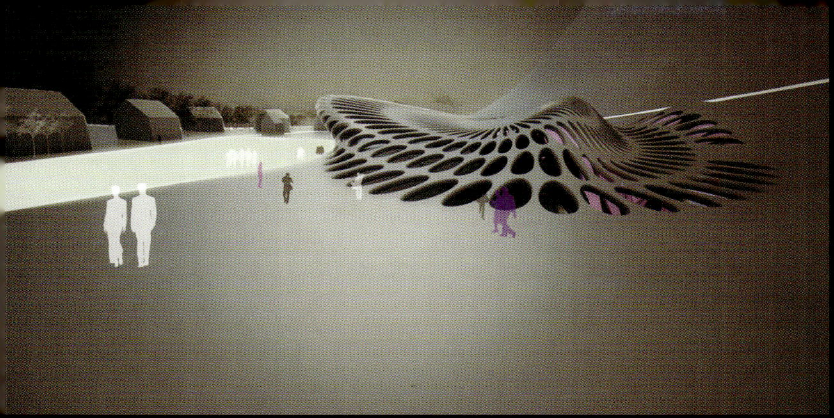

效果图
Render

平面图
Plan

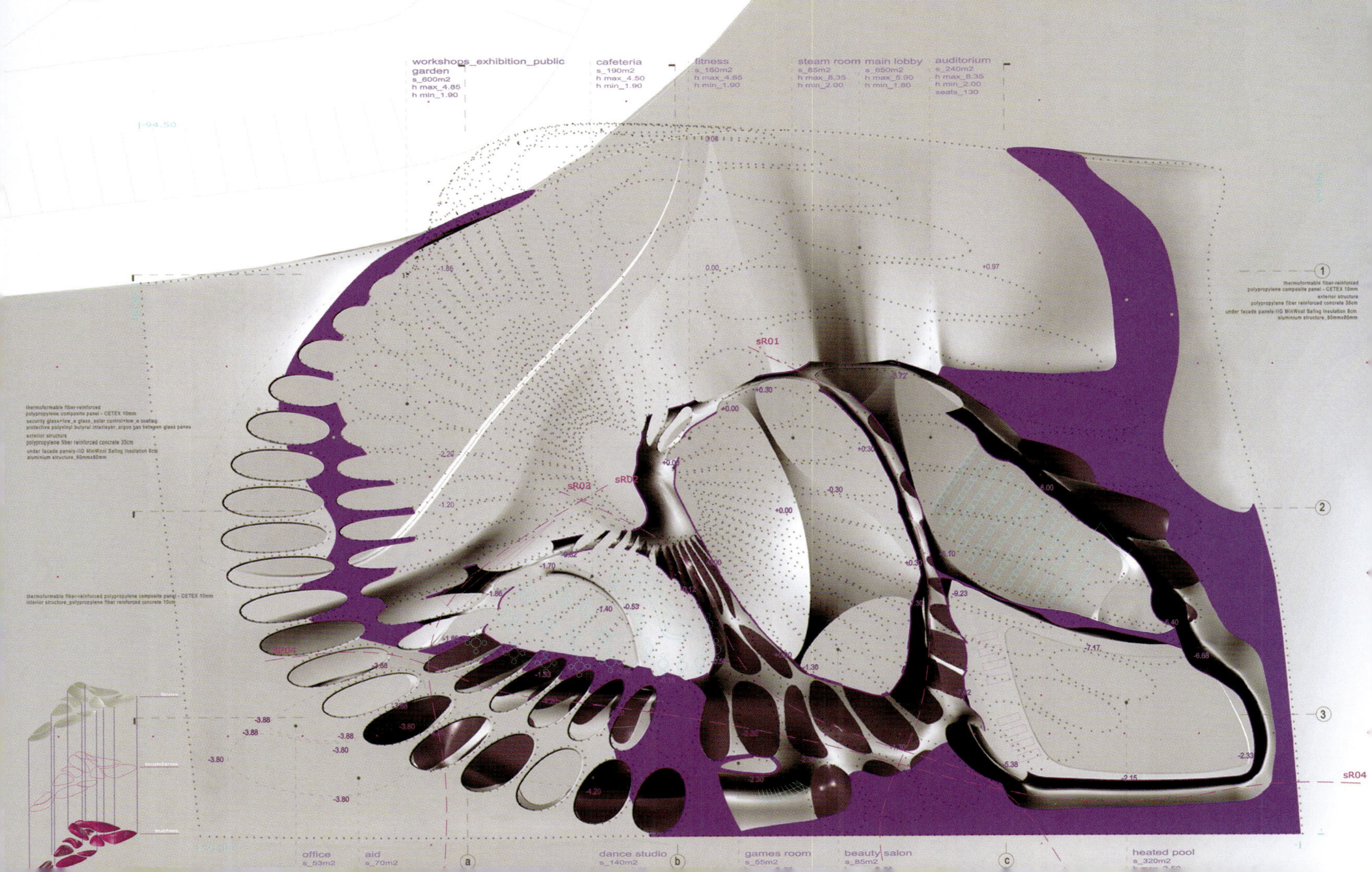

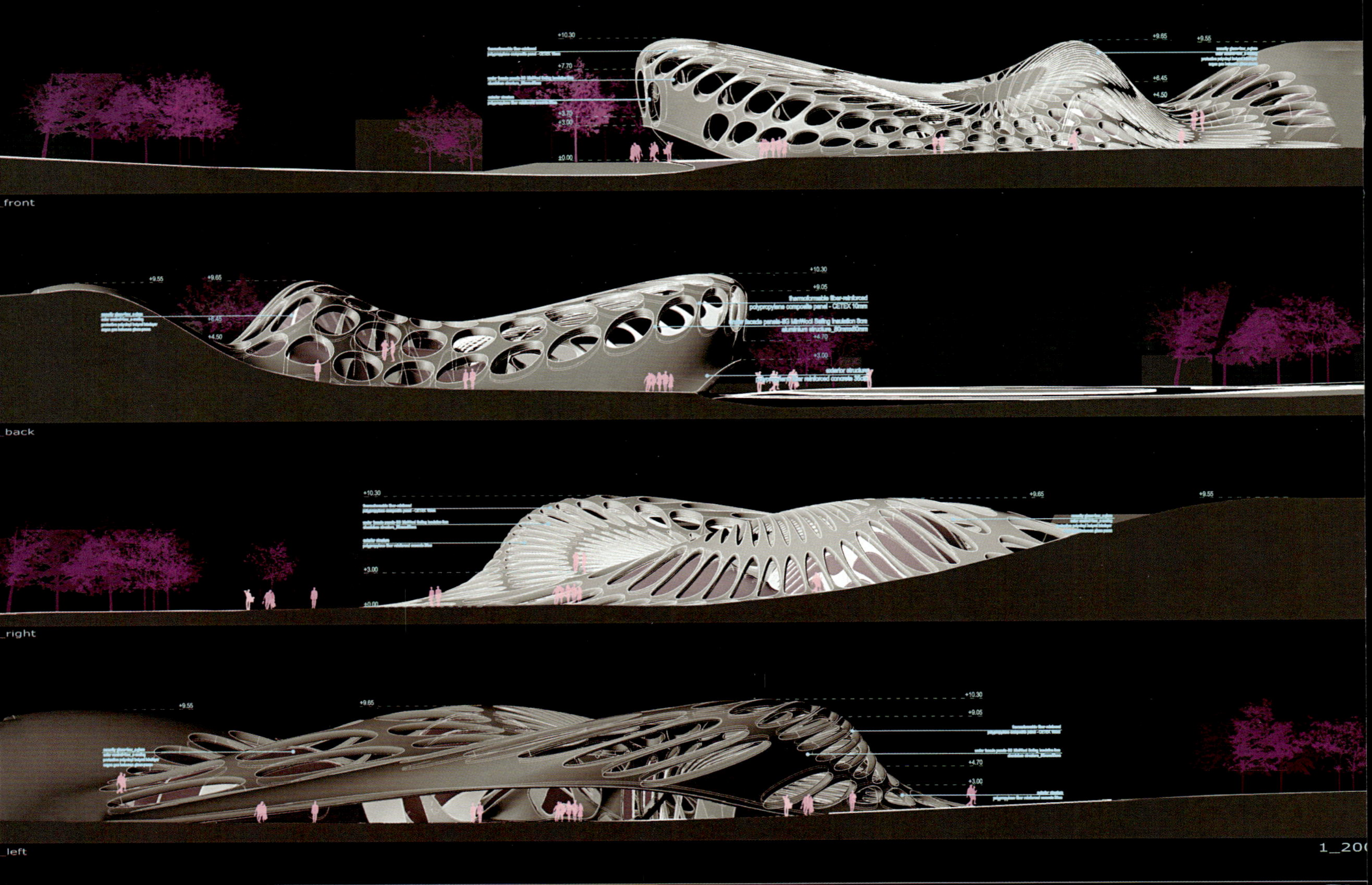

康乐中心正立面
Facade

F.S.V_fog_vapour_smoke_ Recreation center_sport and spa_08_09_univ.dipl.inza.arh.M.Florijancic_assistant:univ.dipl.arch.Mitja Zorc_student Bolojan Daniel

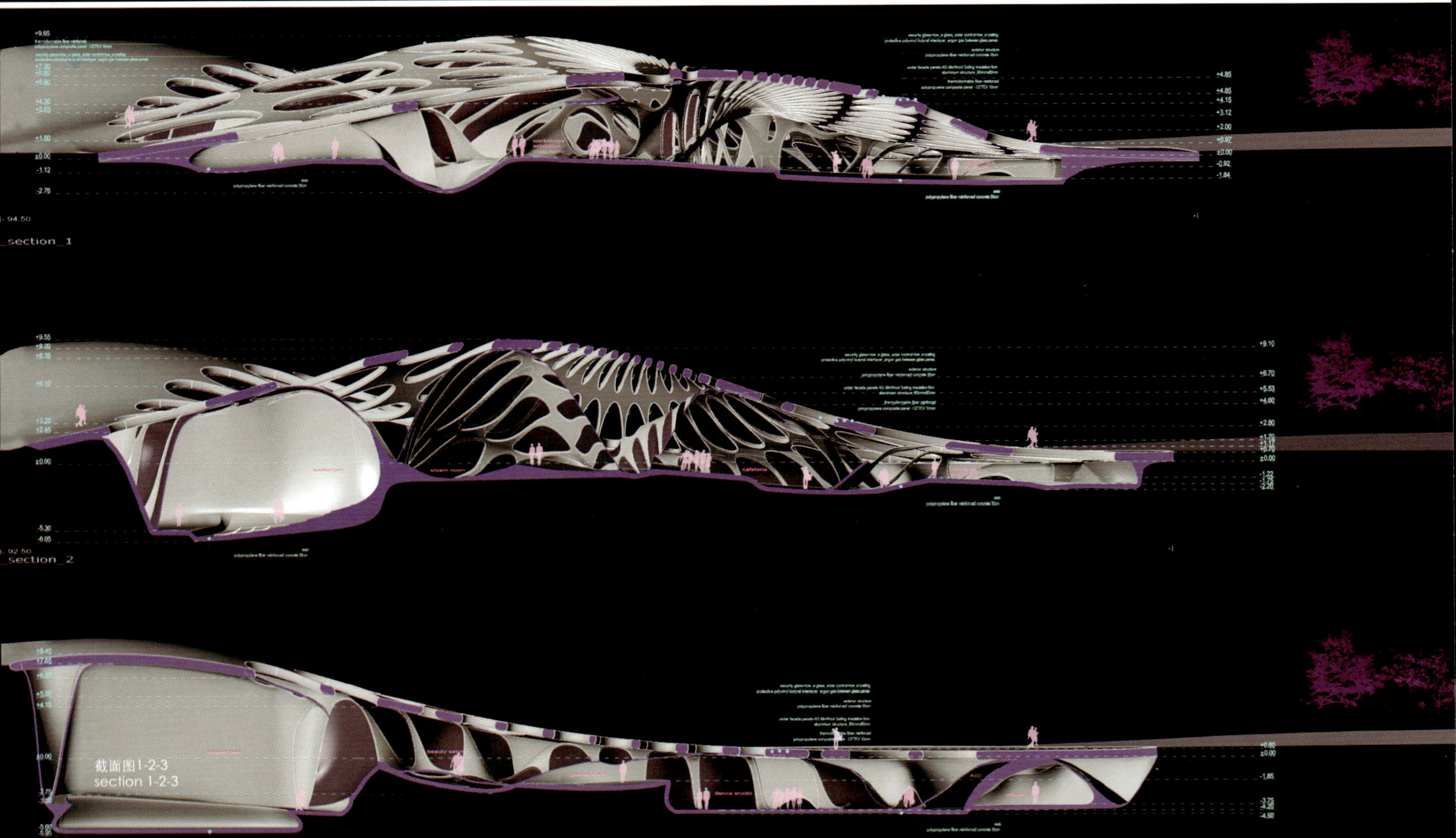

截面图1-2-3
section 1-2-3

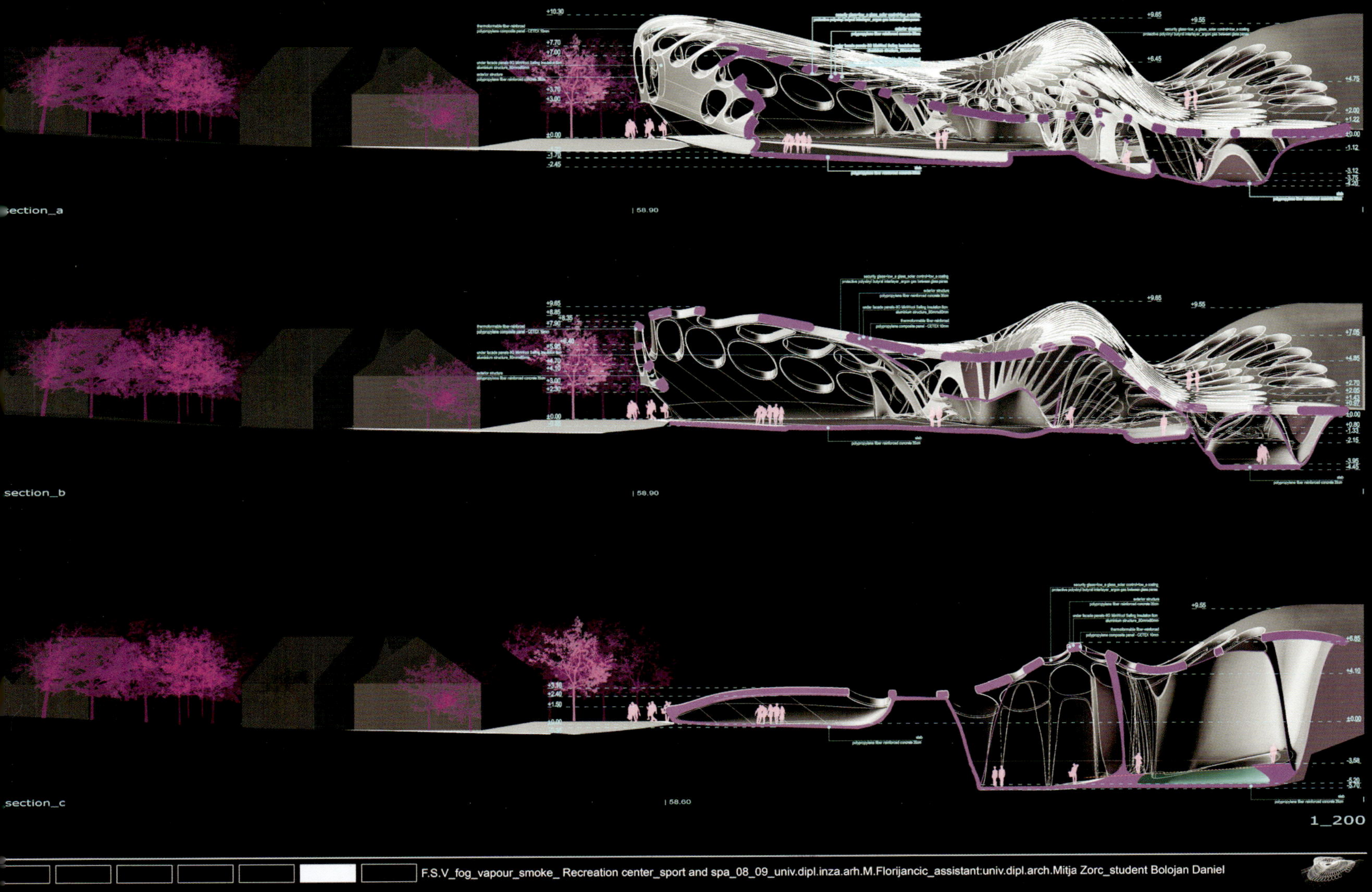

截面图a-b-c
Section a-b-c

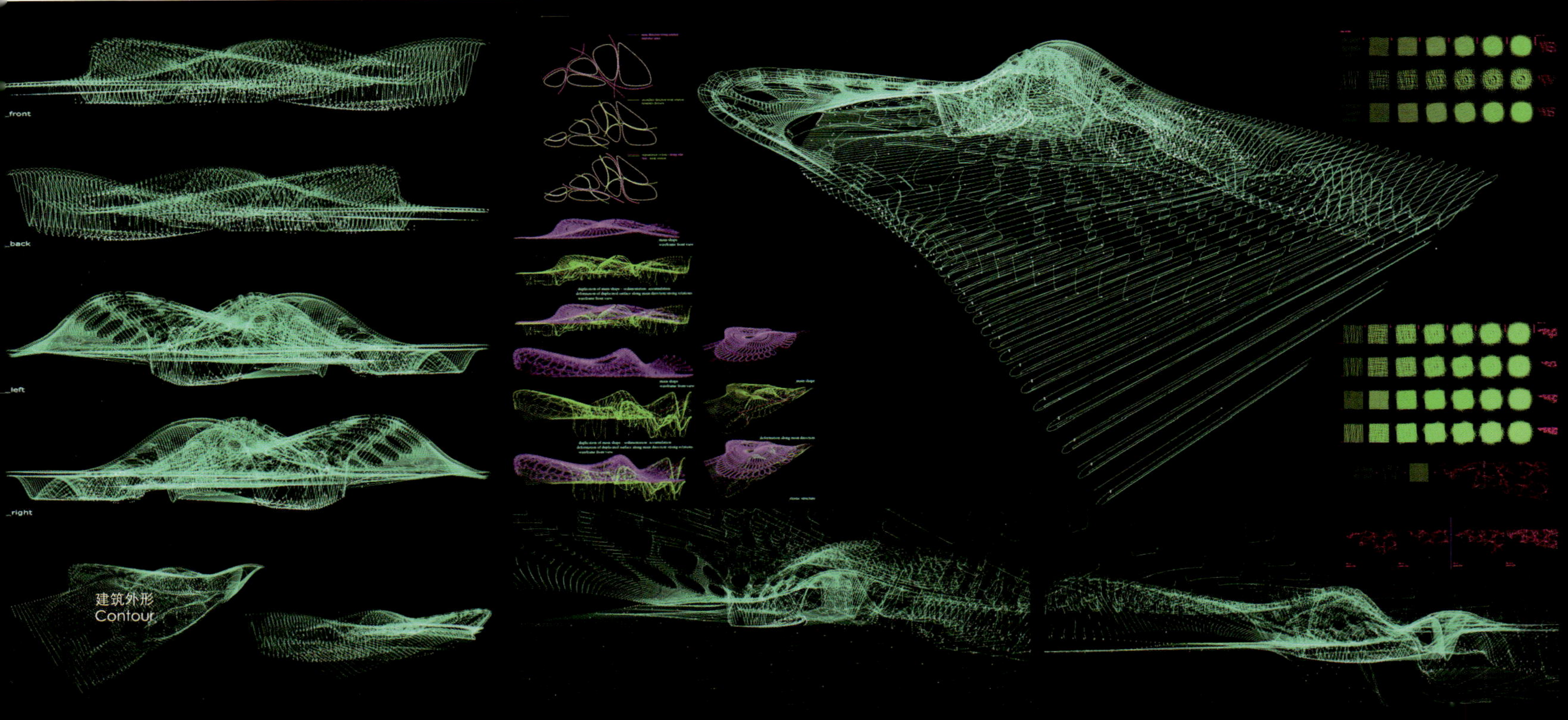

建筑外形
Contour

Rising Core House
增长型核心住宅

For each good thing there is to living in cities, there lurks a dark side—a reality that confront us in this age of rapid urbanization—poverty, unemployment, political instability and social unrest, and environmental degradation. For developing countries, this scenario is exponentially exacerbated to proportions that have spurred many to find ways to address this problem.
Filipino architect Albert Zambrano may just hold the key with the project proposal, Rising Core House (RCH). RCH, as it is conceptualized, is a housing delivery system and a corresponding prototype house design that intends to create more dynamic, humane and sustainable human settlements in cities and urban areas in developing countries like the Philippines.
Geared towards low-income earners, RCH's development process takes on a two-fold strategy: first, it aims to provide a responsive shelter that has the flexibility to accommodate future requirements and adapt to changing conditions; and second, to create livelihood or income-generating opportunities for its residents and the community. These are seen to eventually stave off social unrest and foster political stability.
The main thesis of the project is this: the poor should not remain poor. RCH is projected to be a vehicle of change—a way to harness initiatives, creativity, energies and resourcefulness that will allow the urban poor to improve their conditions over time.
Zambrano's immersion with Manila's urban poor, resulting from years of study, research and planning

透视图
Perspective Views

东方特色
Eastern Character

东南亚特色
Oriental Character

low-income housing has resulted in an advocacy that Zambrano has taken on at the very root of this project. The RCH is a product of evolution from Zambrano's early studies of simply producing the least possible livable dwelling space for low-income housing.

The RCH is designed by combining the characteristics of vernacular architecture with slum development patterns and sustainable building approaches. Each unit can start with a 12 square meter floor area which can expand to over 150 square meters and is conceptualized to function as a shop house. The lower floors are appropriated for income generating activities while the upper floors are used as dwelling. These dwelling spaces, moreover, can generate rental income if owners choose to lease them out.

Collective income of the community is likewise developed with the creation of jobs from new construction, expansion, upgrading, renovation and repairs. Such construction activities then create multiplier effects on other sectors of the economy and stimulate the growth of micro-enterprises (MEs) and small and medium enterprises (SMEs) creating more income generating opportunities for the poor in urban areas.

In a theoretical narrative, Zambrano relates how an individual resident earning a keep from a shop in front of his house can generate enough savings to construct an additional floor. The construction will logically feed other businesses, setting in motion a domino effect that will highly affect the entrepreneurial dynamics and human interaction between members of the community.

The breakdown of barriers between the private and work sphere, and the merging of the two is not a novel concept but its application in light of current realities that beset urban areas can pave the way for a shift in perspective in the way we live, work and interact within the local community.

With strategies combining reduced operational expenses, security of tenure, micro-finance and support for home-based enterprises (HBE) in micro-, small and medium enterprises, Zambrano's RCH proposal could be an effective and holistic approach for poverty reduction and a much needed urban revitalization.

设计特点
Design Features

开发过程
Development Process

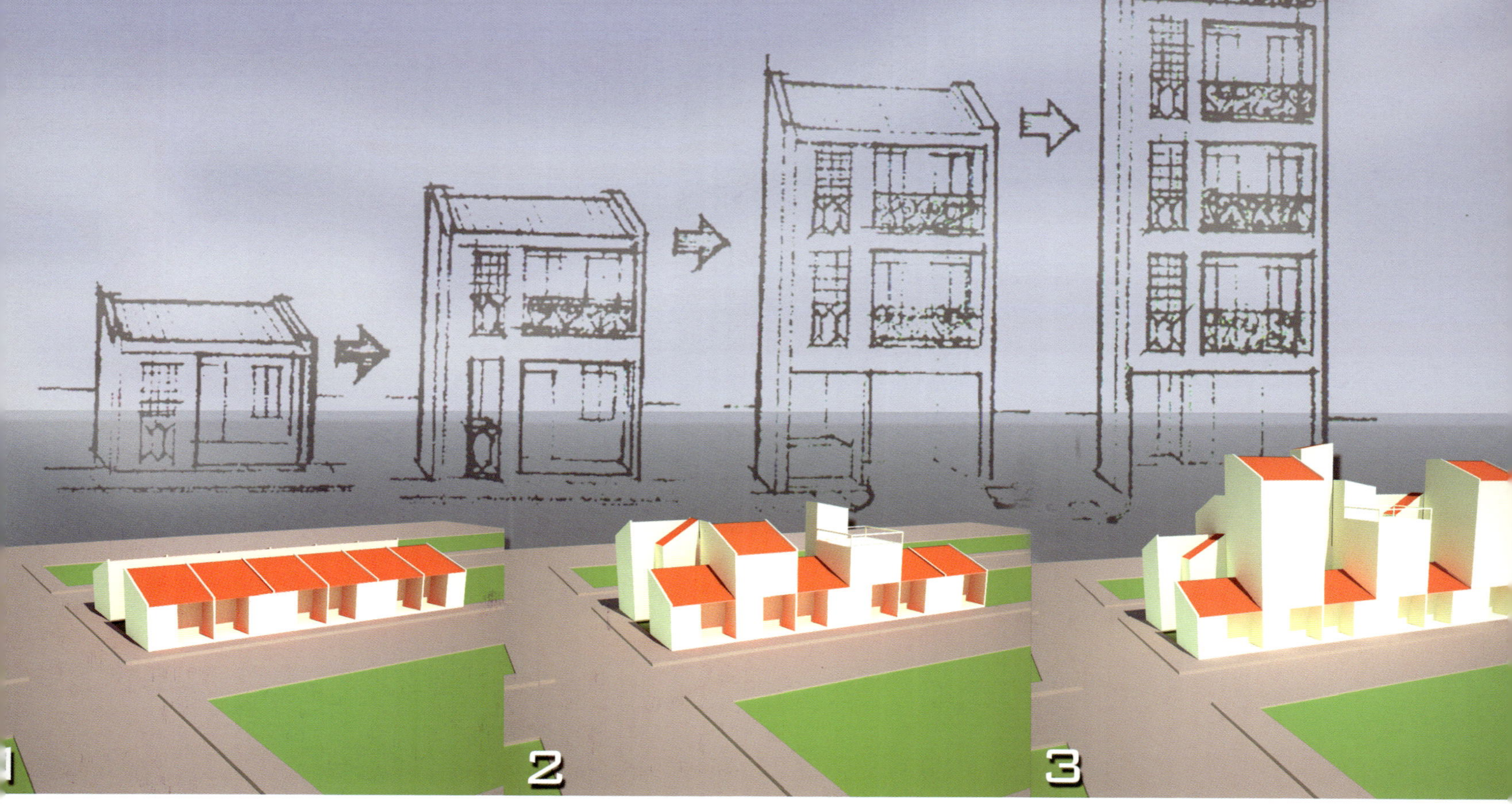

18平单元平面图和等距视图
RCH 18 Plans and Isometric View

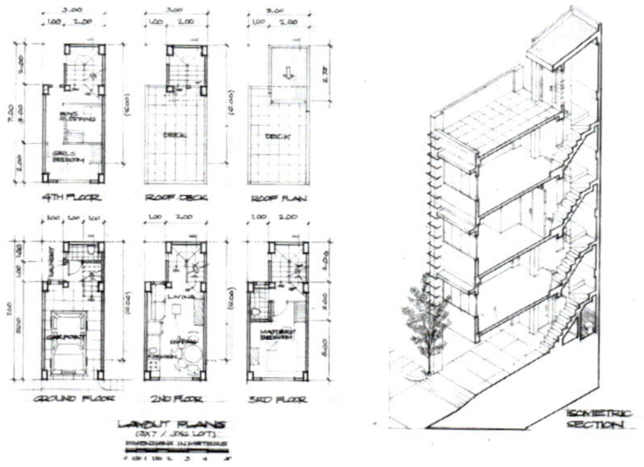

12平单元平面图和等距视图
RCH 12 Plans and Isometric View

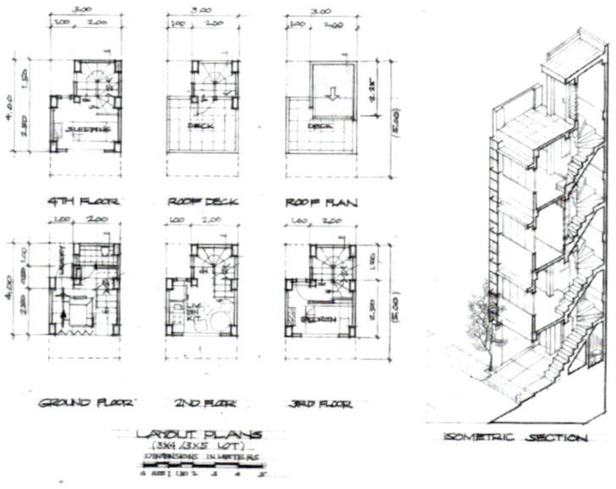

模型图
RCH Scale Model

细长的街景
Narrow Street Scene

外部景致
Exterior Views

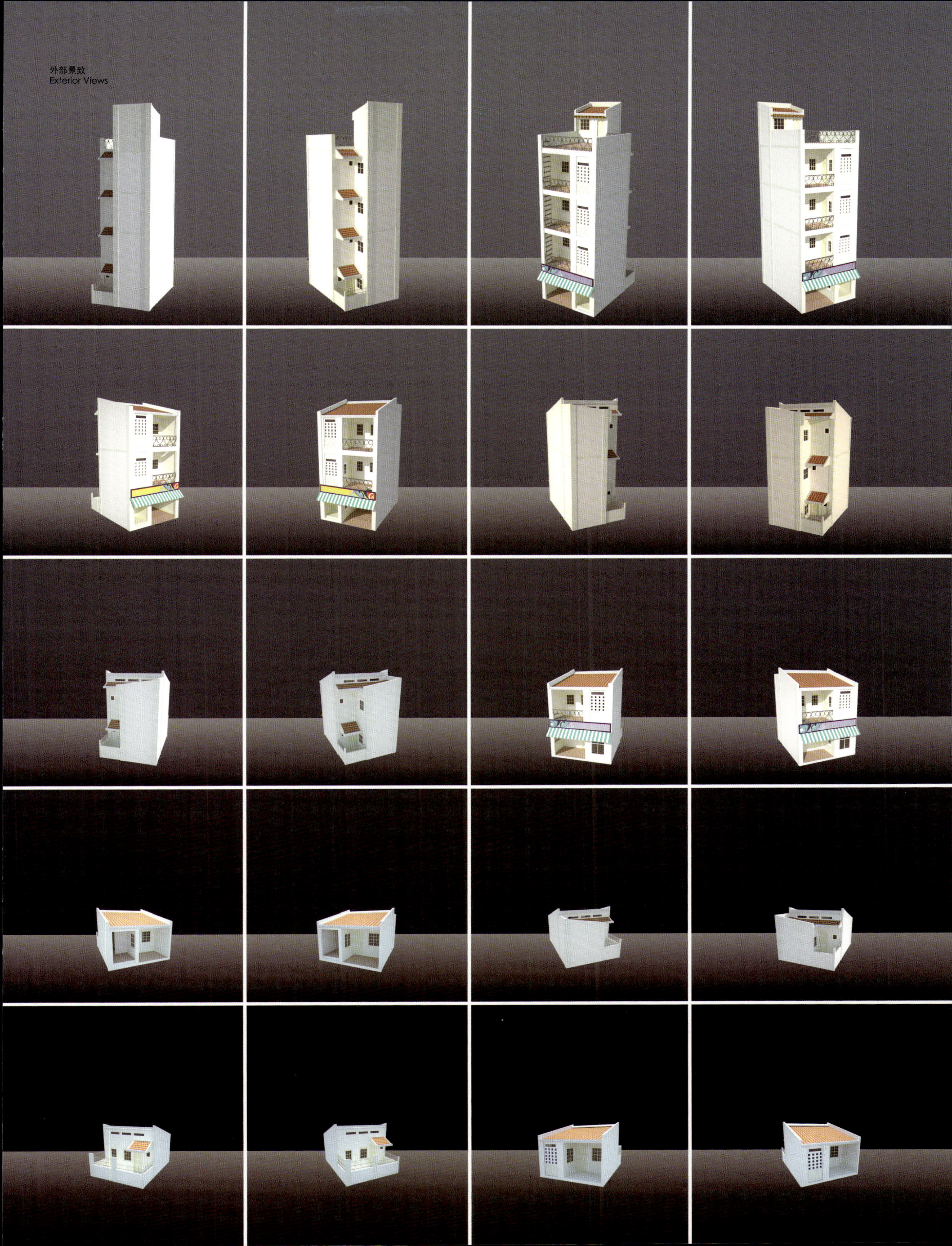

Chief Architect/Office Name: student architect – Bolojan Daniel Calin

RM Library
造型独特的RM图书馆

SITE

The library is situated on Regina Maria Avenue (which connects Unirii Plaza and the Rahova street), which has specific urban tissue for the central area of Bucharest, on a triangle - shaped plot, having valuable neighborhoods like the Church of June 11th and the Romanian Patriarchate.

STUDY TARGET

-Redefining the functional program of a library and its related activities in the context of a IT era scenario

-Develop a valid architectural-spatial scenario, as rescue of a cultural program that tends to disappear from daily life

-Integration of a contemporary intervention in a defined urban tissue

THE MAIN FEATURES OF THE SITE

-combination of commerce and services with collective housing

-train type house—narrow and long yard areas that create intimate and discreet communication with the public space

Discontinuous fronts have the effect of obtaining an continuous transition between public and private space. This transition opens a new continuous perspective, like a drawing of overlapping layers which define the urban tissue. The aim of the new building is to correct the uncontrolled shape of the existing open space and to create a better-defined new one.

The library project developed on the idea of morpho—ecologies design (zantedeschia aethiopica—calla flower growth), open public space which is gradually enveloped in an enclosed public space. This kind of increase can be observed in the case of the so-called train type house, which is something similar to bungalow house, that represent the main site typology.

On both sides of this structure, both sides oriented towards streets, lye the volumes that contain the library, cafeteria and the book store. The north volume works as a buffer that organizes a smooth connection between Regina Maria Avenue and the public square (open public space). The space created in-between these volumes is particularly important because it works as an urban square and in the same time as an open air library, which delivers the means of passive environmental modulation to the area. Below these, at level -1, the multimedia room, and administrative functions are developed.

The growth of the library's envelope is driven by computational processes that derive from specific articulation of the triangular elements in strong relation with structural analysis, spatial, functional, and environmental parameters, and last but not least to the human parameter.

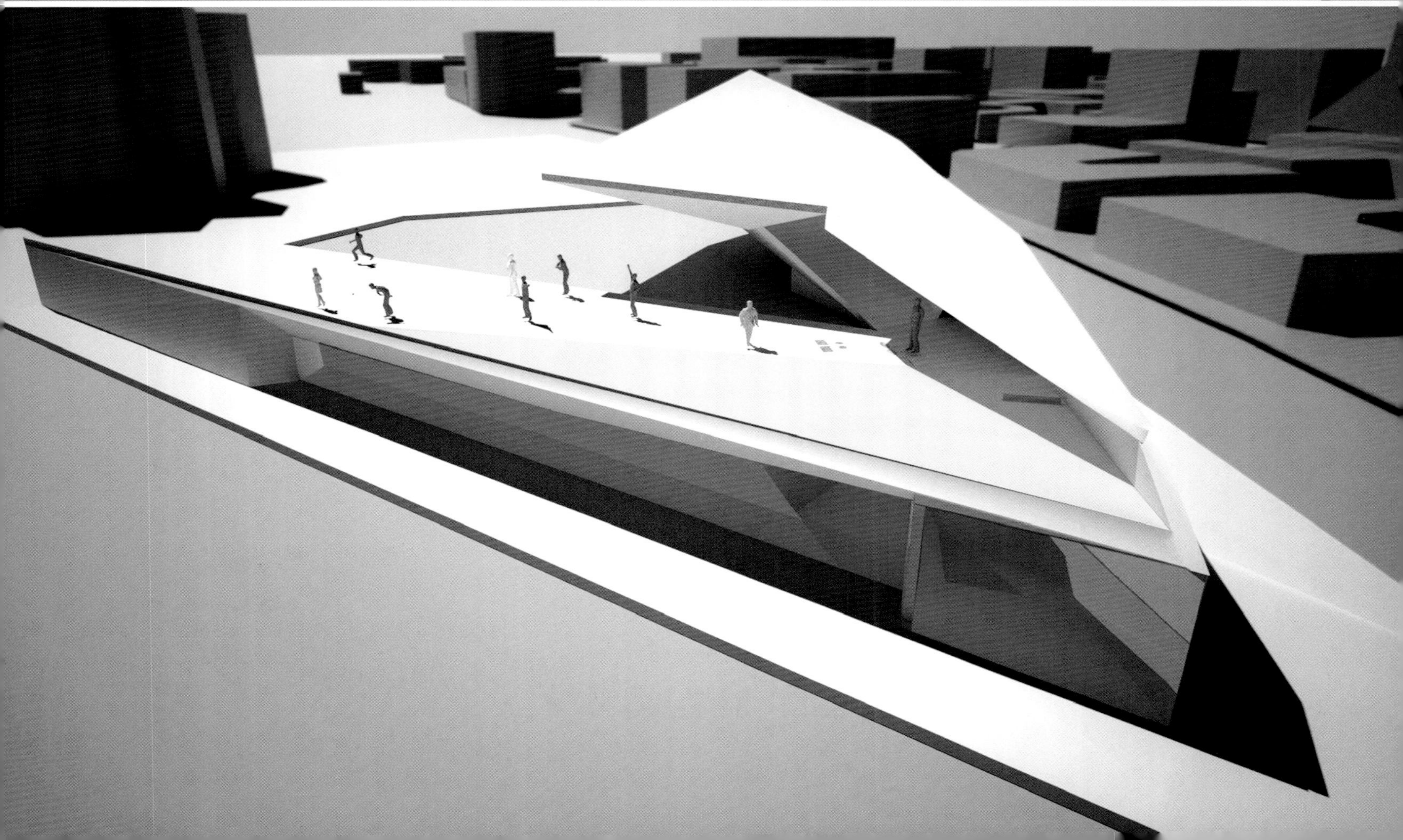

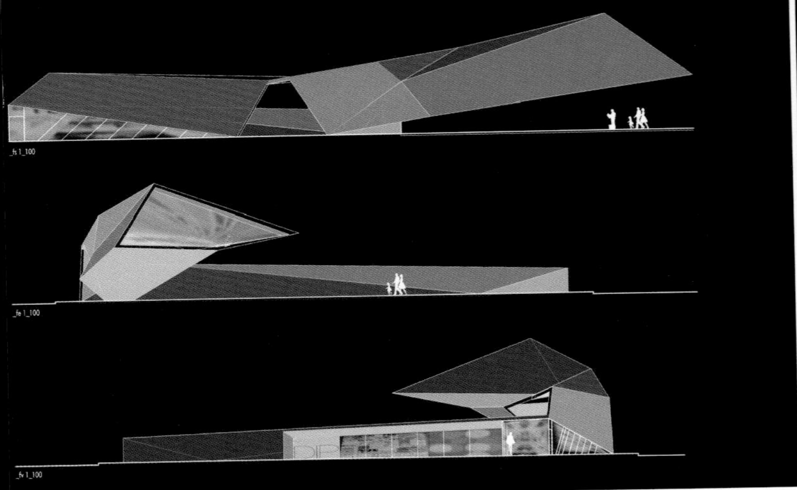

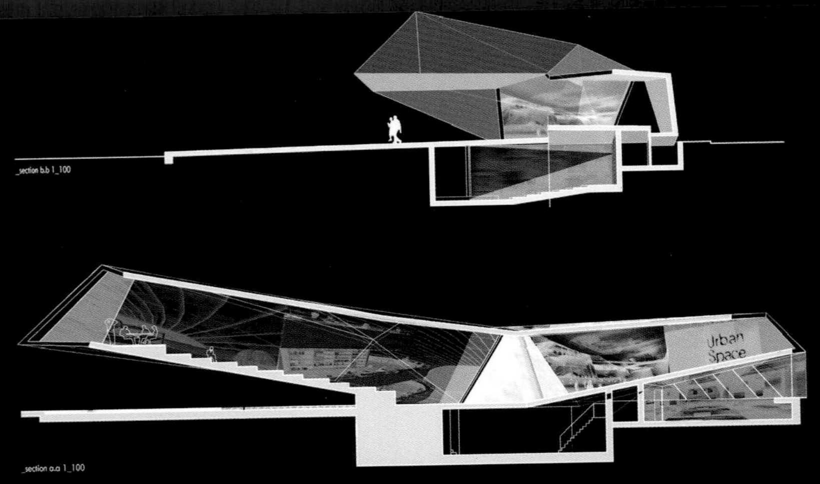

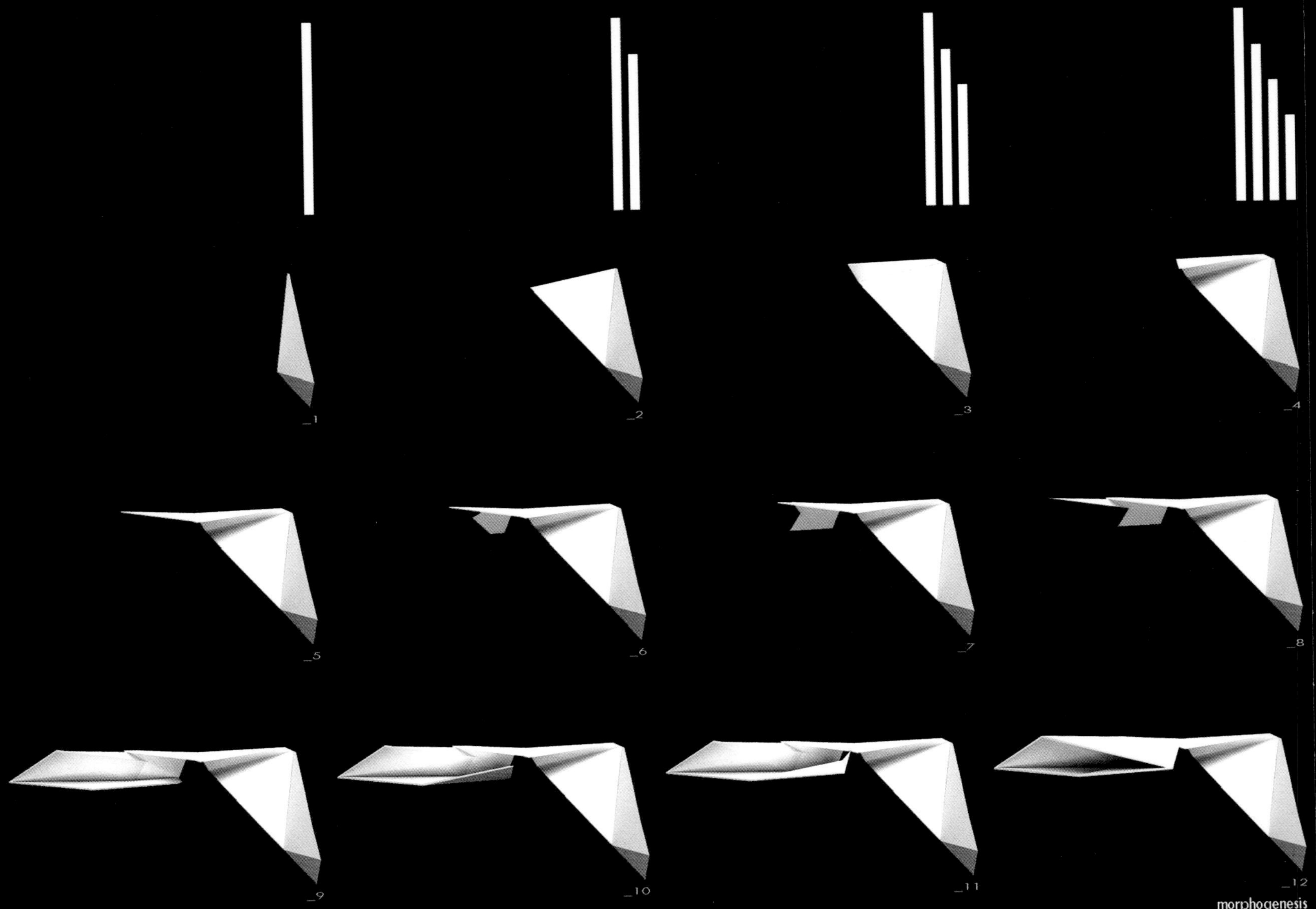

morphogenesis stages

闪蝶型图书馆建筑
Morpho

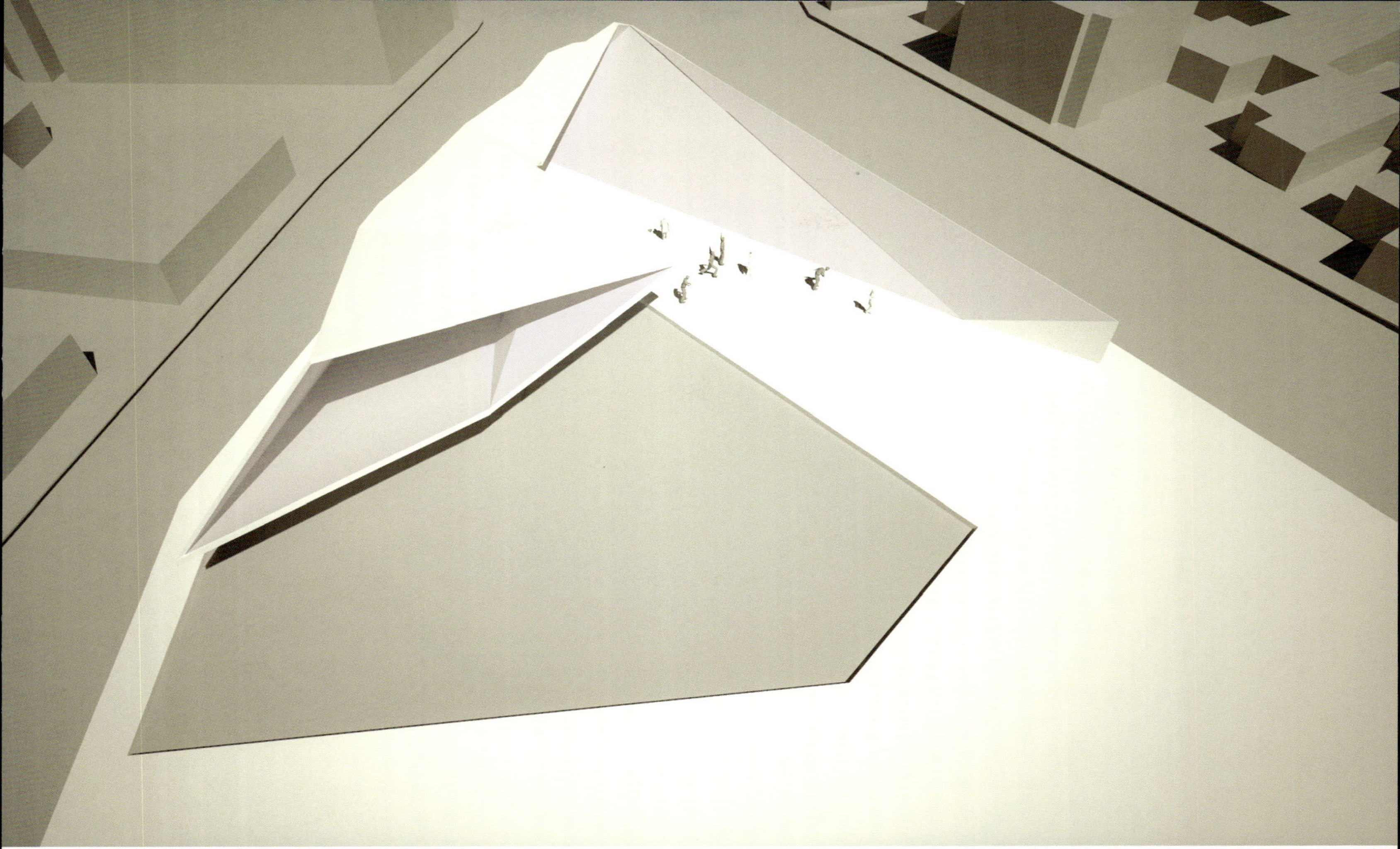

建筑模型
Model

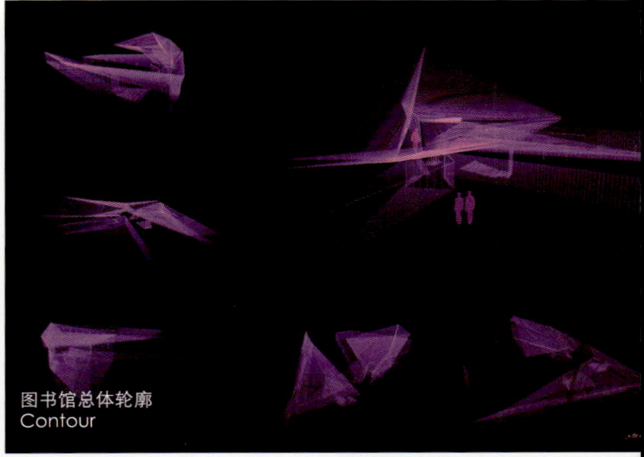

图书馆总体轮廓
Contour

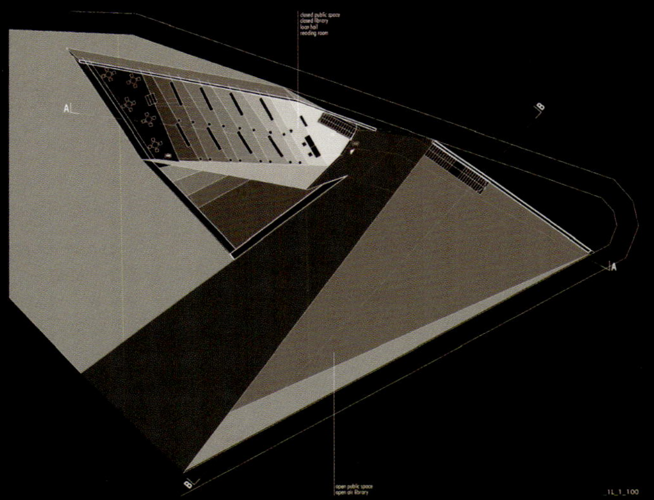

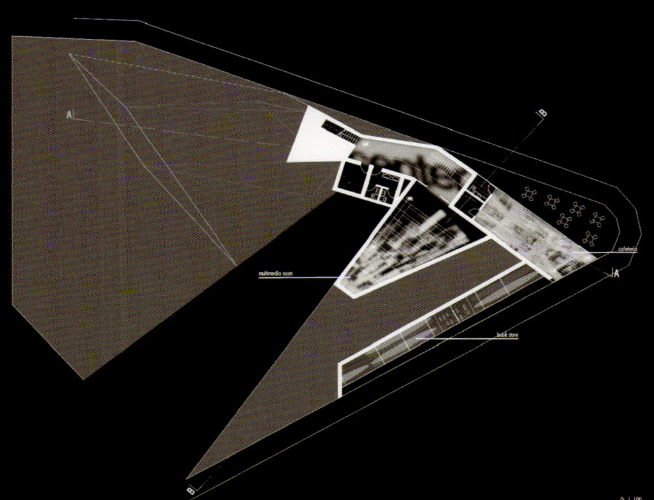

Project Management : Hérault Arnod Architectes

Rossignol International Headquarters
全球总部——建筑对山的赞礼

The image of Rossignol, a longstanding leader in ski equipment, is closely linked to mountains and snow. The design of its international headquarters is a world away from the typical office building, and instead pays homage to nature and peaks, as well as technology, inseparable from high-level sports.

A LANDSCAPED ROOF

The building has been specially designed for Rossignol and is inspired by the fluidity of movement in snow sports, as well as the snowy peaks and glaciers sculpted by the elements. The topography of the roof, which wraps around the building, merges with the natural landscape, its organic wood-clad form reflecting the shapes of the surrounding mountains.

The roof covers three different types of area:

-The production and technical spaces which line the motorway side of the site

-The street, which cuts through the building: welcoming, spectacular and filled with light

-Offices

On the motorway site, the façade creates a kinetic and dynamic vision, reinforced by the repetition of the logo, which emerges gradually. The curved façade becomes the roof over the production units, before rising up to its ridge and descending on the other side across the offices. At this point it is pierced by patio courtyards planted with birch trees which appear to perforate the roof.

The irregular outline of the roof and the office façades allows the possibility of adding extensions as needs arise, without disrupting the harmony and identity of the building. Right from the start, the architecture generates a process of growth. The roof ridge sits above the street, flanked by a glass façade and the high altitude restaurant, the highest point of the building, which echoes the buildings found at the top of ski slopes: with its panoramic outlook, the terrace extends across the roof for al fresco dining, with magnificent views of the mountains, hidden away from the noise of the motorway.

THE ROSSIGNOL "FAMILY"

The Rossignol Group will bring different entities together on this site; these are currently split in different locations but they all contribute to the identity of the Rossignol "family".

Inside, the building functions like a hive, enabling different areas of activity to cross paths, and making them accessible and allowing people to enjoy working and socialising together. The originality of the project is the way it brings in other activities alongside the administrative departments:

- The racing skis production unit, to act as a technological showcase for the brand

- The R&D and Design departments, located at the heart of the building

- Two showrooms to present Rossignol and Quiksilver products

The purpose of bringing these different aspects together is to create an overall synergy and prevent dislocation between design, administration and technology. Each different role - engineering, design, technical, secretarial, sales, etc - will cross paths with the others. To encourage this internal communication, social spaces are dotted throughout the building. The restaurant, situated at the top, right over the street, has been designed as a special place to enjoy the life of the Group, with two large glass façades framing panoramic views of the sky and mountains, one overlooking Vercors, the other Chartreuse. Either

on the terrace or beside a log fire, this space turns lunch breaks into a unique experience.

RESPECT FOR THE ENVIRONMENT

The design of the structure aims for minimal environmental impact. The technical choices made have produced an efficient building with a low energy requirement and good insulation. It is protected from the summer sun by an untreated wooden roofover. Installations will be optimised and the heat given off by the workshop machines will be recovered and passed back through the heating system.

Particular attention has been paid to water management: rainwater is collected, and industrial water and water recovered from the roads is depolluted.

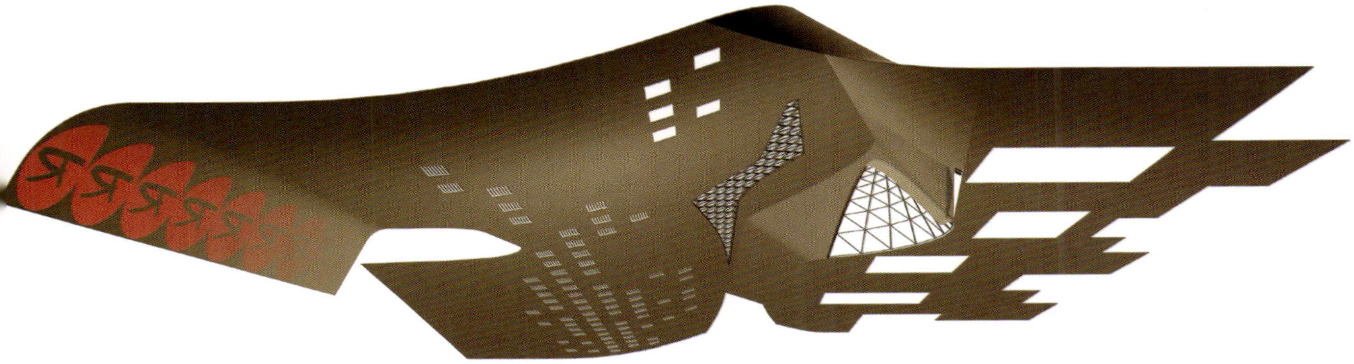

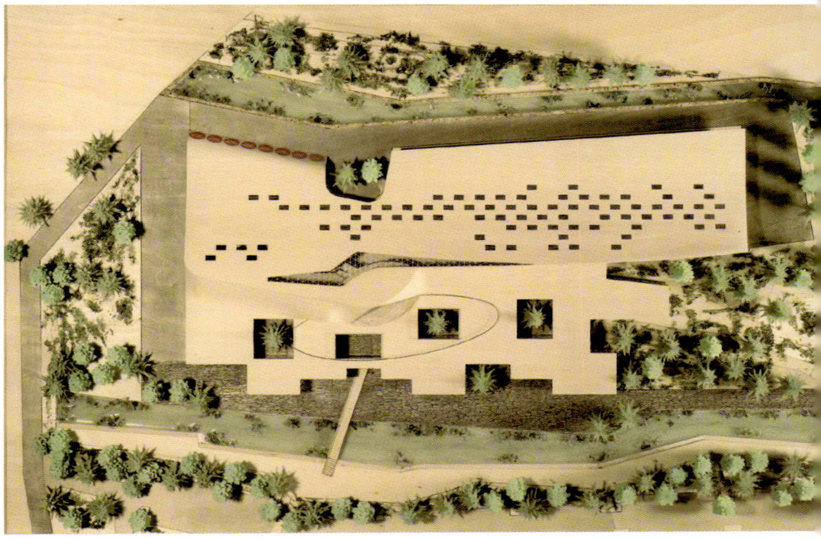

Royal Botanic Garden in Jordan
约旦皇家植物园

The Visitor complex and a (winter garden)
The main existing entrance is from the southern east corner, we want to create a buffering zone for welcoming visitors and to transfer the feeling that you are entering a new area with a new experience. The main gate of the site is located at the neck to make control of the cars and people during the week days and the events inside the botanic garden, the Visitor center is Located on the top of the mountain to have the full visual connection all over the site; it is designed in multi-layered levels by creating an artificial-designed-hill that it will melt the architectural solutions with the existing nature and it will connect the visitor with the body of the Gardens, its major function is visiting, gathering and experiencing with functions like terraces, exhibition halls, shops, library,... .The parking will be underground, and as soon as you will get out of the car you will find your-self in the middle of the visiting center to make the wanted shifting in the experience.
The main composition revolves around the topographical flow of the natural lines at the site; it transforms its mass with ascending elevation as if it was growing from the earth to become a more architectural mass, this transition is controlled by the main datum wall of the composition which in fact is a bridge in the middle of the building. When the composition becomes functional it starts triggering the human senses like sight with

全景俯视图
Bird Eye Main General View

实物模型
Physical Model

its charismatic facade by playing with the human perception of the growing lines, the building also provides a wide range of visual recreational facilities like art and light galleries and a theatre , the acoustics like the- passive- which are represented by the natural reflection of the sound from its surfaces and -active- which interact through the landscape like the murmur of the water features, smell which are represented by the essence of the plant presence in the environment and the scent chambers in the scent factory which is one of the recreational sides of the project.

MASTER PLAN

The royal botanic garden will be serving four main disciplines: conservation, education, scientific research and recreation. The royal botanic garden will be a center for native plant conservation, public education and scientific research where it will have links with other gardens, both in the region and worldwide. The garden will also be an attractive park where visitors can learn about Jordan's heritage in a beautiful setting, as well as a place where these plants will be preserved and displayed to the public. It will be a regional as well as a national public attraction, in a country where there are currently few such amenities.

The royal botanic garden will also have a positive

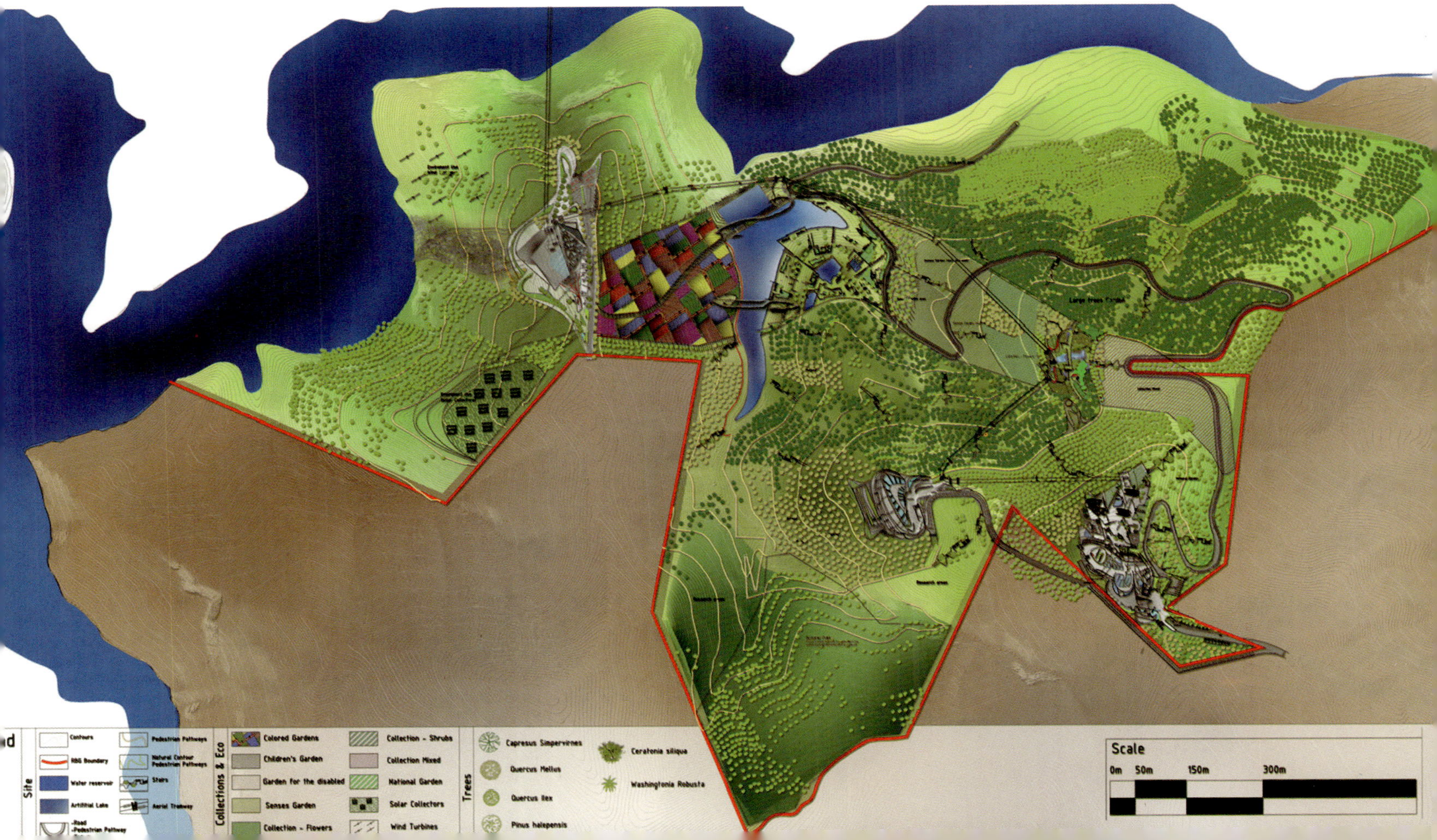

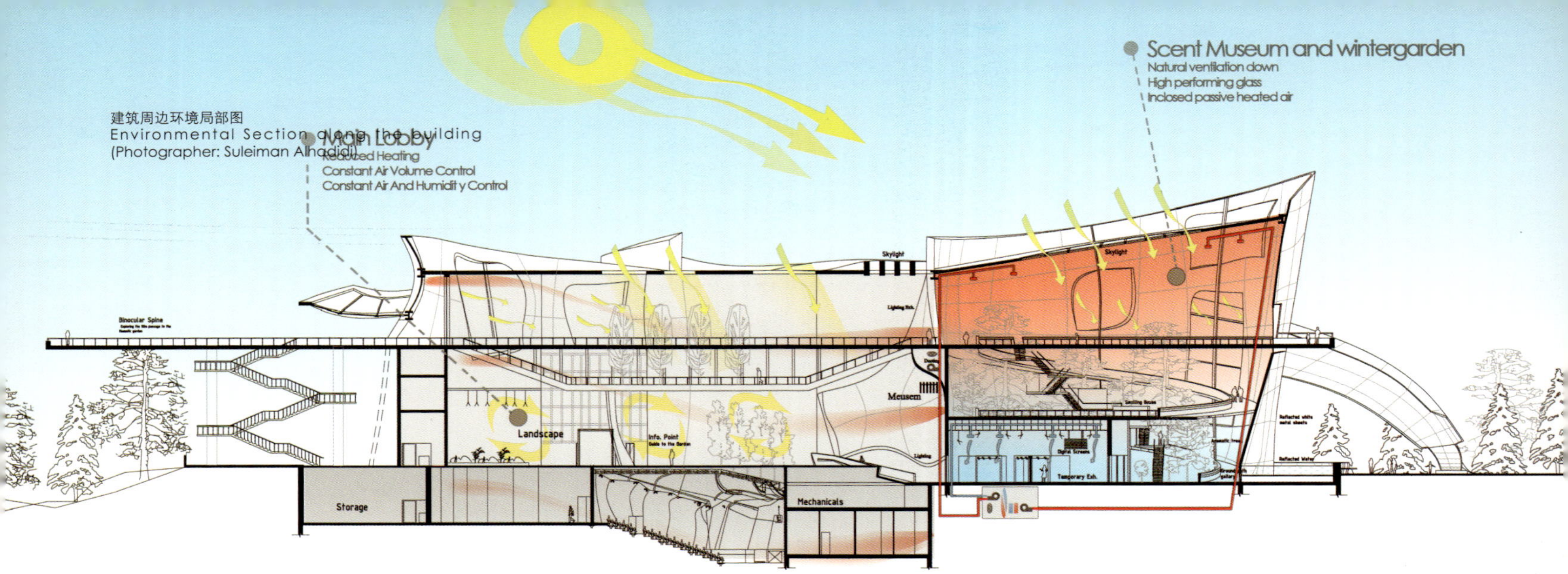

建筑周边环境局部图
Environmental Section along the building
(Photographer: Suleiman Alhadidi)

impact on a beautiful area of Jordan that is suitable for touristic development. Though it's main site at Tal Al Rumman and associated satellite sites throughout the country, the royal botanic garden will conserve the plant species of Jordan and will be a major center of education to achieve a better understanding and appreciation of Jordan's botanical heritage. The mission of the royal botanic garden is the promotion, implementation and management of biological diversity conservation, focusing on native flora, for suitable utilization in; scientific research, genetic recourses, education and ethno-botanical history of medicinal and culinary plants.

A PROSPECT OF THE SITE

The Establishment of royal Botanic garden is part of the biodiversity strategy and action plan, prepared by the ministry of environment to implement the 1992 convention on biological diversity, ratified by Jordan in 1993. The ministry of planning and international cooperation, ministry of agriculture and ministry of water and irrigation are prominent supporters of the garden. The royal botanic garden will be the hub for the conservation of Jordan's biodiversity; this will include establishing natural habitats, theme gardens and a national herbarium through both its main site and associated satellite sites throughout the country. It will be the center of education to achieve a better understanding of Jordan's botanical heritage and will promote the knowledge and appreciation of plants for personal enjoyment, inspiration and scientific research. Furthermore, the royal botanic garden will establish a state of the art research center, with the aim of becoming one of the most credible reference centers in the world.

OBJECTIVES OF THE ROYAL BOTANIC GARDEN

- Rehabilitation of the ecosystem at Tal Al Rumman.
- Raising awareness and knowledge of plants and the importance of biodiversity in the local community.
- Establishing collections of threatened plants and promoting the ex-situ conservation of rare and endangered plants.
- Setting up a national herbarium.

Roles of the Royal Botanic Garden – Jordan:

- Maintaining a collection of living plants.
- Establishing a herbarium collection of specimens.
- Developing genetically representative population in cultivation so that the majority of the genotypes of rare and endangered plants are safe guarded in cultivation.
- Undertaking habitat recreation or restoration projects within the garden where the dynamics of ecosystem functioning can be studied and native plant population restored and managed.
- Assembling a back-up collection of seeds of endangered and threatened species in a seed bank.
- Propagating rare and endangered plants to provide material for population increase, reintroduction and habitat restoration.

Ras Alain Amman城市发展
Urban Development for Ras Alain Amman, Jordan Suleiman Alhadidi

主入口
Visitor Complex WG Main Entrance

屋顶花园
View to the Roof Garden

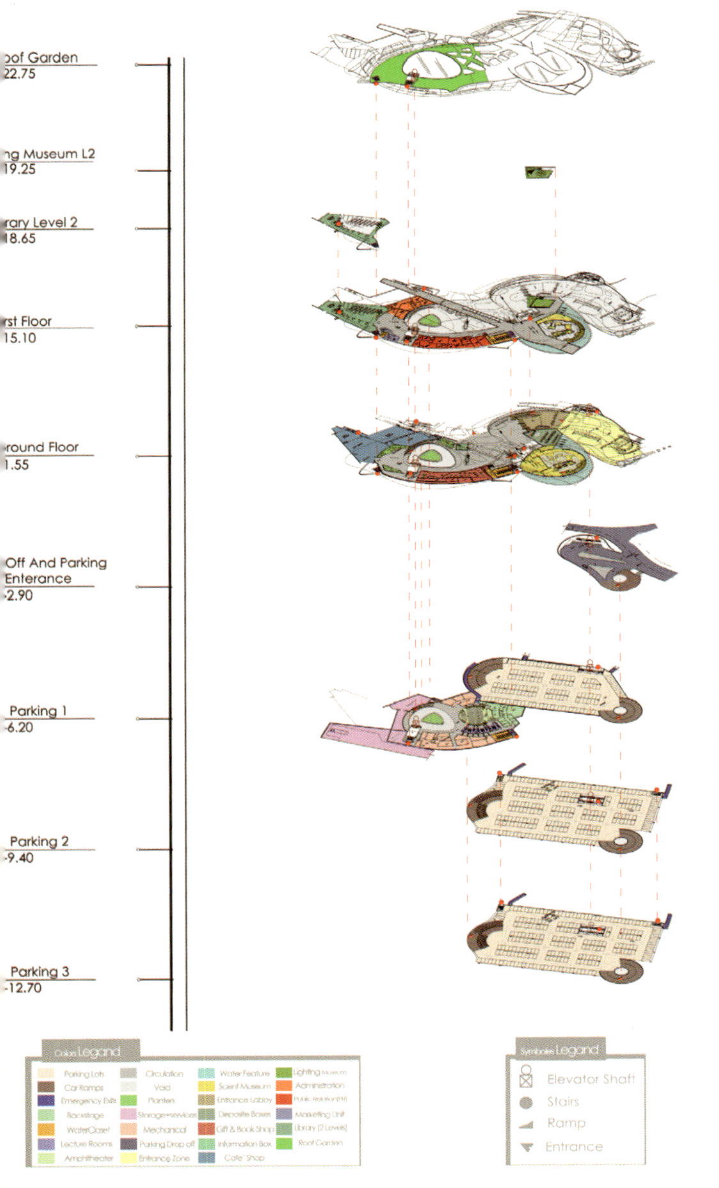

室内主厅
Interior Main Lobby

Project: iCuadro Architects : Massimilano brugia, Valerio Campi,Francesco Bezzi

Shanghai Expo 2010: Italian Pavilion International Contest, 3rd prize winner

上海世博会 2010：意大利馆国际竞赛，三等奖

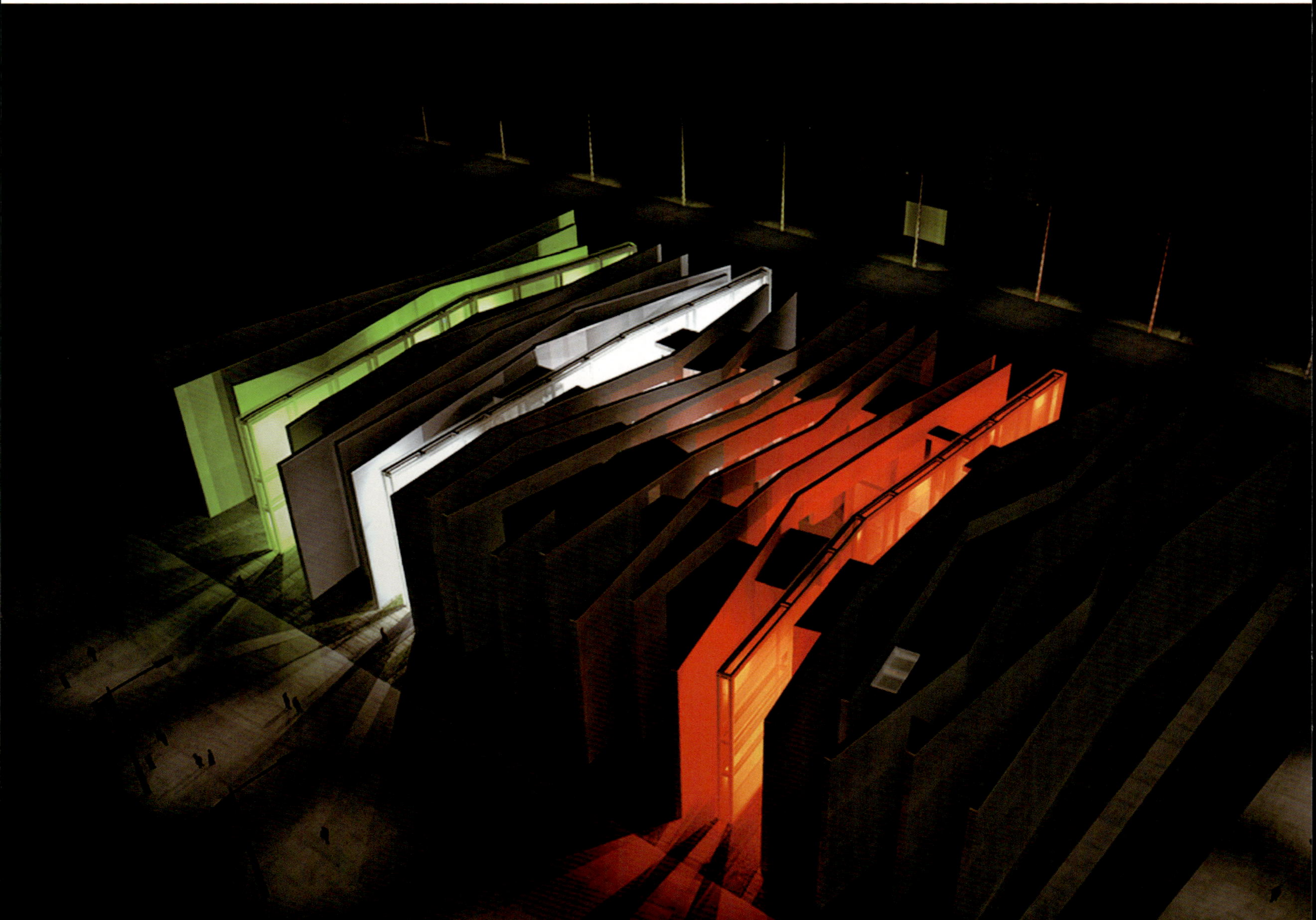

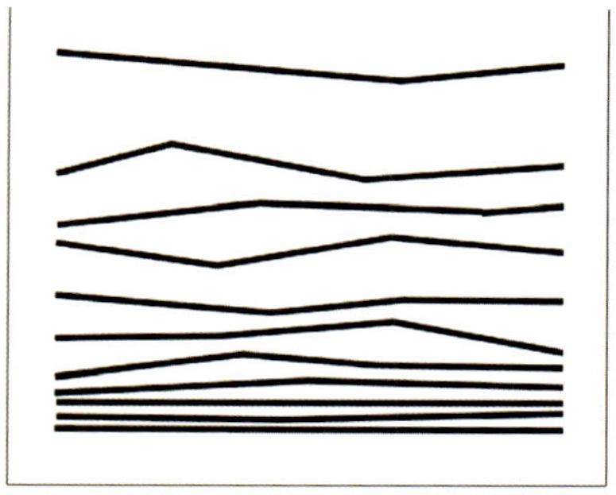

LA STRATIFICAZIONE

+

VENTI DIVERSE REGIONI

+

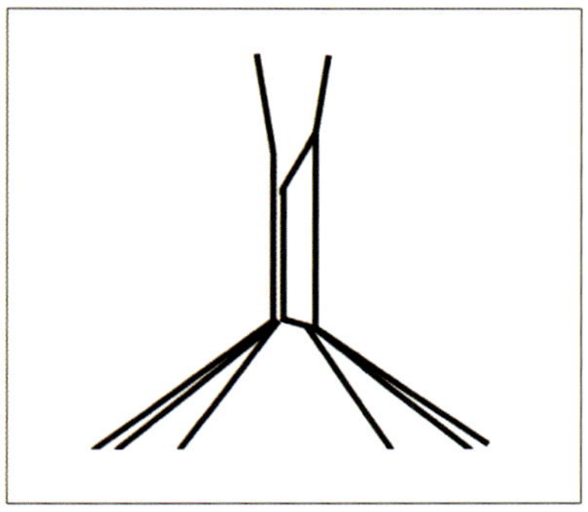

I VICOLI DEL CENTRO

+

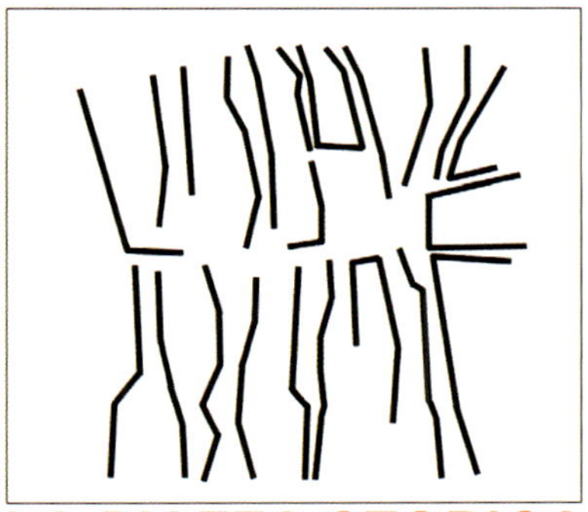

LA PIAZZA STORICA

=

ITALIA 2010

"Better city for better Life". The project presents the Italian style thru a design inspired by the authentic expression of the Italian City: "the historical stratified city".

While big metropolis of the 21st century are having a crisis of sustainability because of wrong growth and environmental alienation, the Italian urban system, which is based on continuity with the historical and the geographical background, is still working well and produces high quality lifestyle.

"The Italian stratified city" keeps resisting to the modern processes of alienation of inhabitants. It still represents a strong human spatial entity that can be perceived psychologically by people. It is connected to the past. Its tissue was built up during the centuries and thru several generations so that life of its inhabitants is strictly related to it.

The Italia Piazza is the center of this entity. It is the core that cannot be replaced. It is the place of "being" and "staying", very important case study, where Art also is expressed. It is a unique "spatial circumstance".

The Italian Pavilion for Shanghai is the representation of this "spatial circumstance which still produces and give the birth to the excellence; It is a prototype that expresses the stratification of the social and urban Italian soul and gives the idea of historical continuity. This continuity is the right approach to the Progress and to the new technologies that in this way can become "sustainable".

The guide line to the project is a volume divided into several planes which represent a timeline as metaphor of the rise of the Italian urban tissue. Twenty planes, as many as the Italian departments, are separated by three window planes where the Made in Italy products are shown. They are the main structure of the pavilion in floor plan.

The interior exposition spaces are influenced by this spatial scan and are organized between the planes as different themes streams. Walking inside the building is like going thru the history.

The building grows around a plaza. Visitors discover this plaza just by walking in as the planes hide it to the exterior. This gives the same feeling you would get by exploring an ancient Italian city or town.

Inside the building everything is material and light. It is a material that protects and creates a "local psychological space". All sides are open to the environment. The planes represent a visual and solar shading.

The inner Piazza is finally a space that helps visitors to relax and meet each other.

BiCuadro Architects is a young award winning, Rome based firm of Massimiliano Brugia, Valerio Campi and Francesco Bezzi.

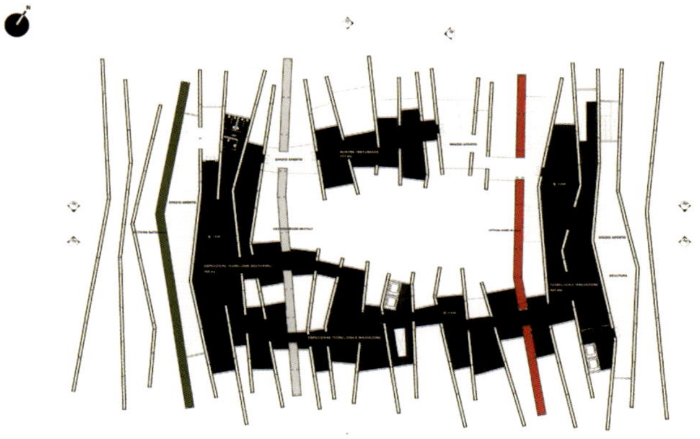

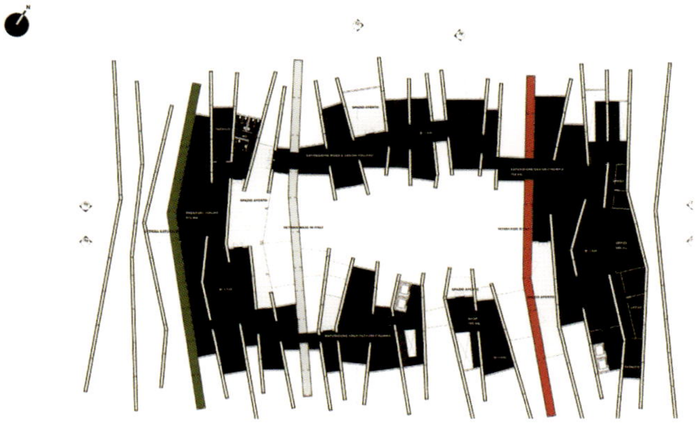

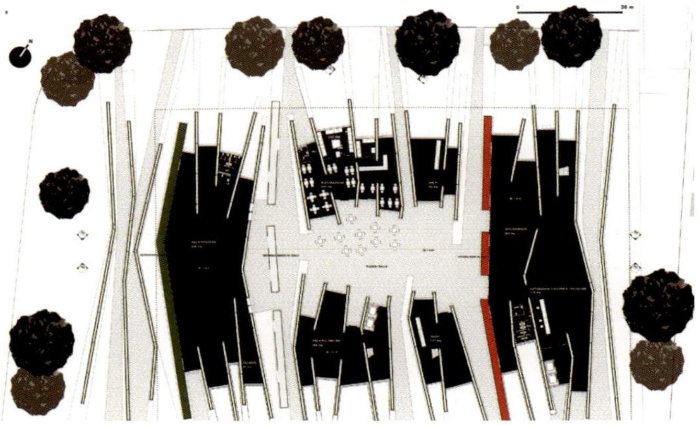

ITALIA
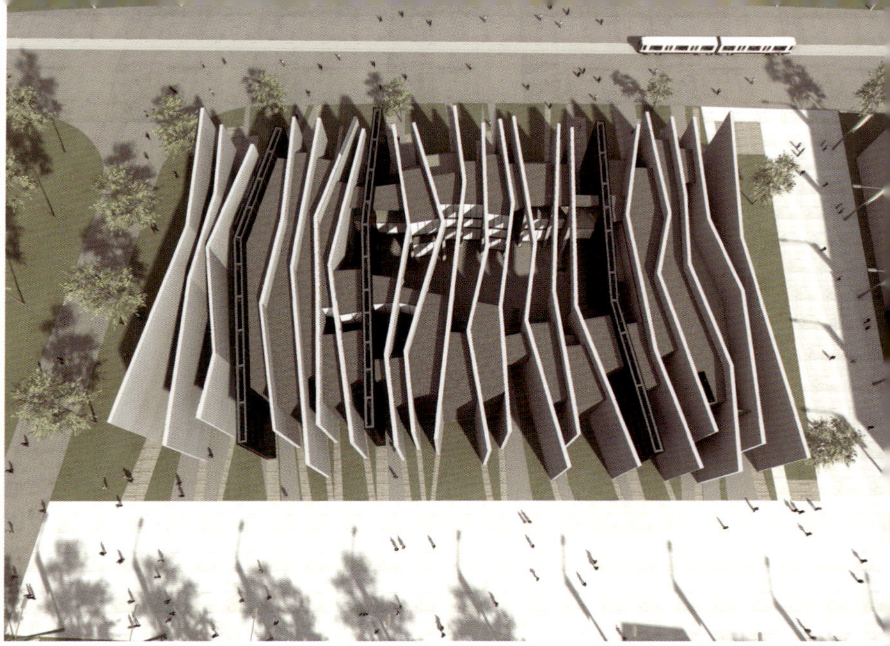

made in italy

Silhouette Istanbul
伊斯坦布尔城市轮廓

Silhouette Istanbul is designed as an iconic building aiming to bring a new look to the integrity of Istanbul and changes in public space.
As located in an area where Europe and Asia harmonically intersect, the structure establishes a dialog with the silhouette of the city considering topographical characteristics of Bosphorus and Istanbul.
In addition, the structure represents an opposite concept against vertical buildings which are designed by a stereotyped architectural approach on Buyukdere Avenue.
The aim of the concept is to create a scene from a contemporary city and to provide a different life style. The conceptual infrastructure would be summarized under several key aspects: the implementation of sensitive systems dealing with energy consumption; designing inner circulations such as semi-open streets, semi-public squares and gardens which support cultural and public activities. Most of the public and cultural areas are dominated by the green in order to experience the harmony between a natural, relaxing city life, and the gorgeous view of Istanbul.
The justification analyses are performed in terms of converting common public areas into gardens and activity areas. The scale comparison between large, well known public places (Place d'Etoile, Collesium in Rome, Palace of Topkapi) and the urban design approach is carefully generated.

The site is located on a hill followed by a valley which comes open to the coasts of Bosphorus. One of the main ideas of the project concentrates on the protection of existing green areas and increasing the capacity of those in order to bring the site into its original topography.
The shopping centre functions placed in the program of the competition is located into the hill. As it is seen in the schematic cross-section, a connection with the general topographical characteristics of the city is established brought at the center of the hill. As the green areas are carved out through the building's site by levels as far as the project long they are considered as the guiding idea of creating public areas.
Basically, private and public uses are both separated and these usages are associated with the site and the environment in the light of local and architectural arrangements.
The different functions which are sited in the competition regulations declared by Zorlu Real Estate are considered as independent elements.

During the concept works, we seek to find out the answer of this question "If every function is designed as an individual tower, what kind of an image we could get?" Buildings having 25mx25m=625m^2 square-shaped substructure are designed in order to observe the dimensional impact. Then, we concentrated on the second question "How are the buildings horizontally established without loosing its functionality?"
If we observe the urban texture on Buyukdere Street, a repeated characteristic is used in every structure. Most of the buildings do not have any original relation with their site. Furthermore, the relation with the city is limited with a cold verticality. The main idea of Silhouette Istanbul is against this redundant approach. Its opposition is based on a relation with the city, and creates various perspectives of this harmonic dialog from Bosphorus and different angles. Additionally, the structure provides an environmental fortune dealing with its original appearance based on the idea of combining several structures.

Proposals are prepared dealing with the connection of ground and underground public transportation, and pedestrian. The park is structured together with the existing and possible bus stops located near the site on the axis of Boshporus Bridge.
Residences, hotels and offices are horizontally placed at flying structure of the building. Shopping center and cultural activities are placed in different levels on ground and underground. They are combined with the green areas representing natural topography of Bosphorus.
The inner circulations of the structure are based on semi-open streets and semi-public squares and gardens opening to the landscape. In addition, gardens and parks are articulated with inner functions in order to create a modern life style.
At the lowest shopping-reaction level, public areas are enriched via concert and conference halls, and cinemas providing opportunities to perform open air organizations.

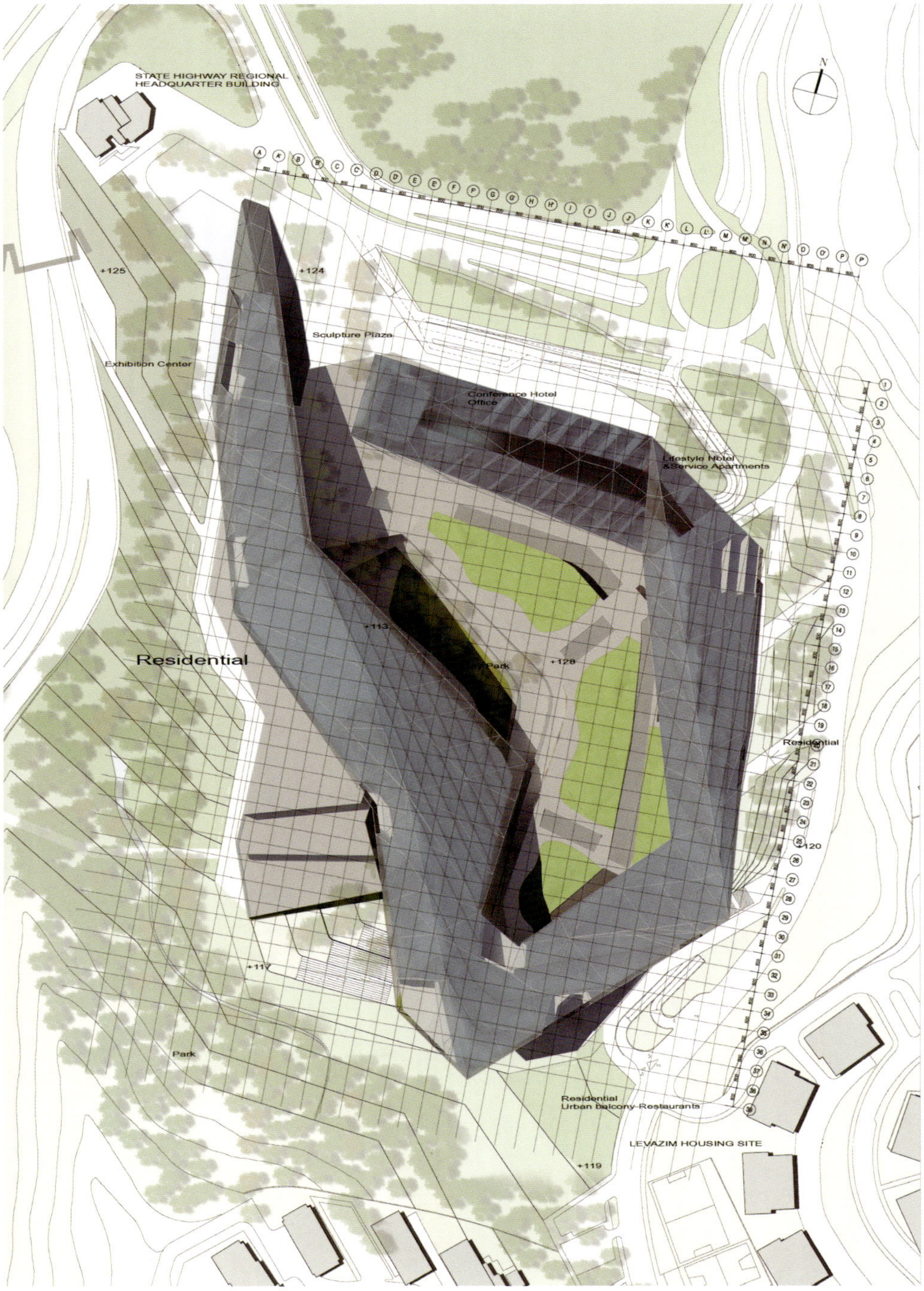

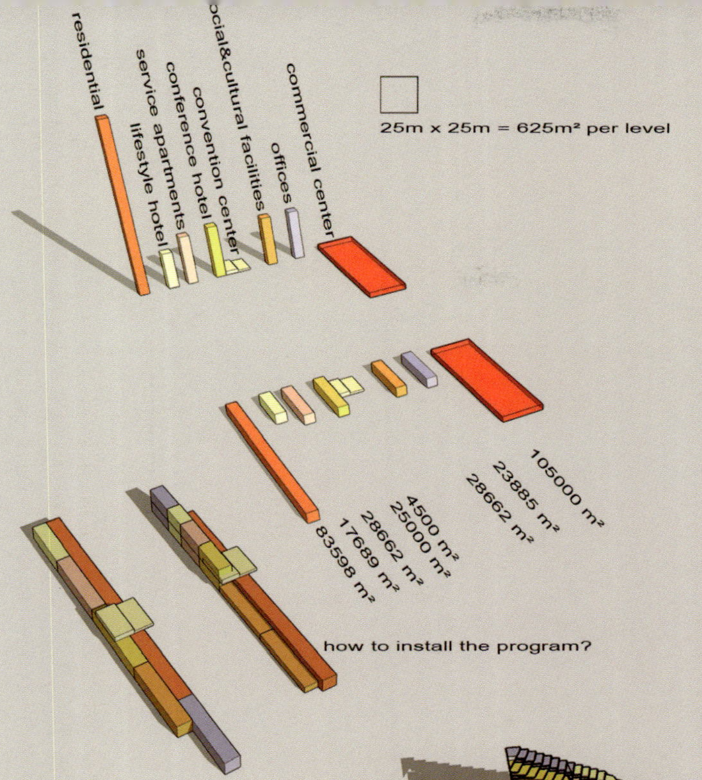

how to install the program?

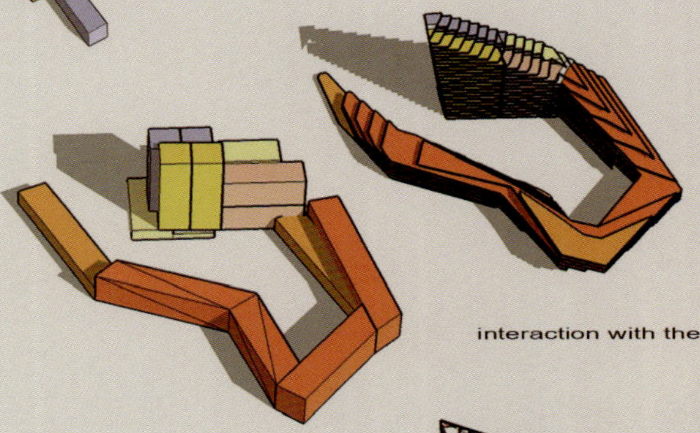
interaction with the site

last formation in levels

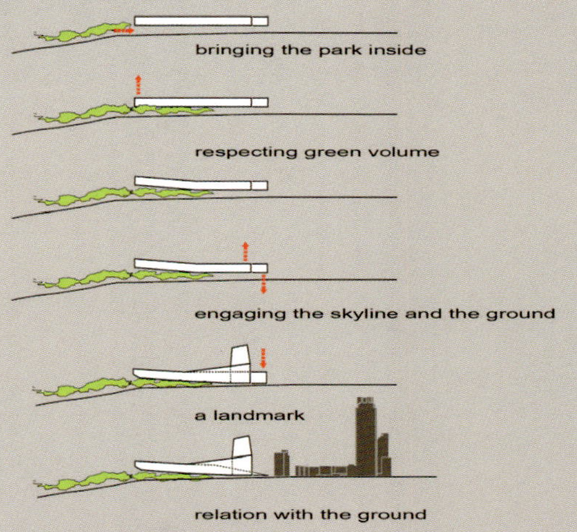

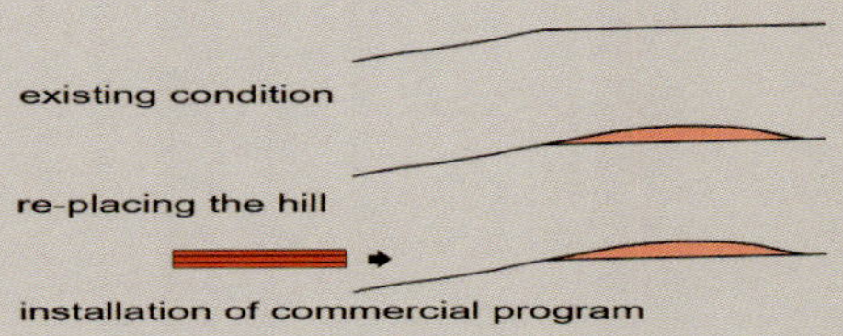

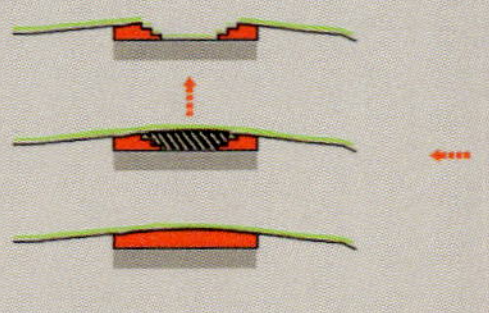
spatial reference

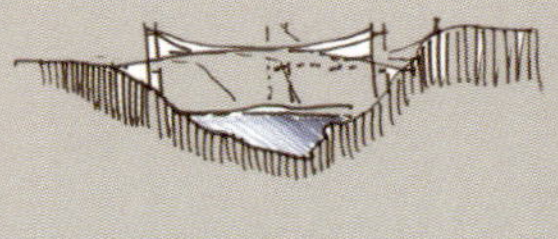
bosphorus

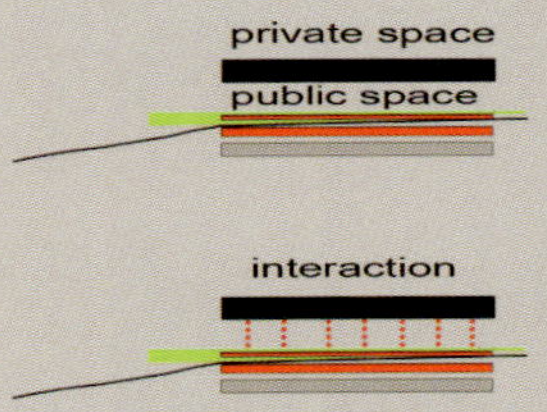

private-public spaces

comparative scale of the site

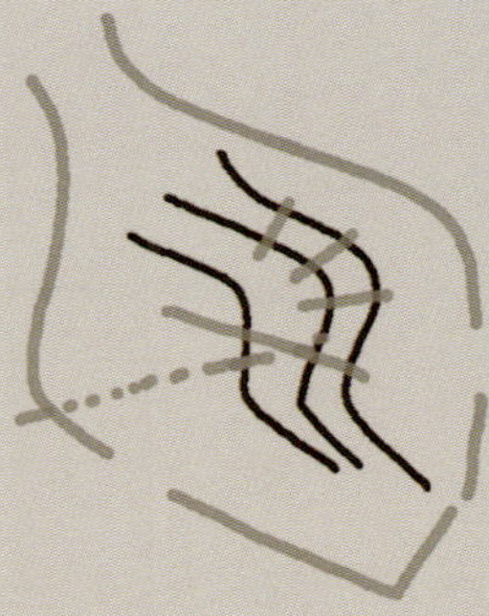
Integration of streets and inner piazza

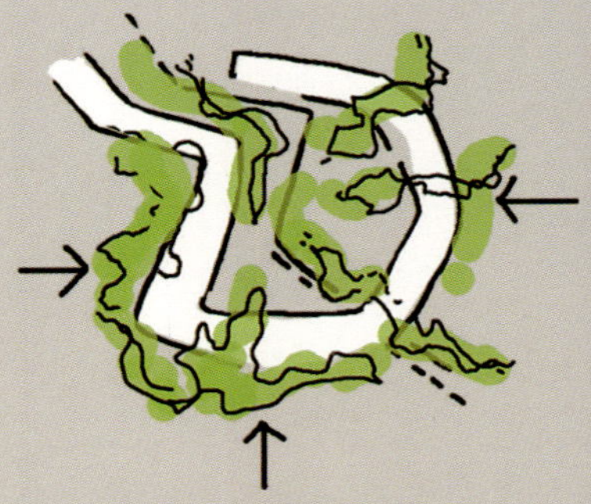
Containing the Landscape In/out + Public

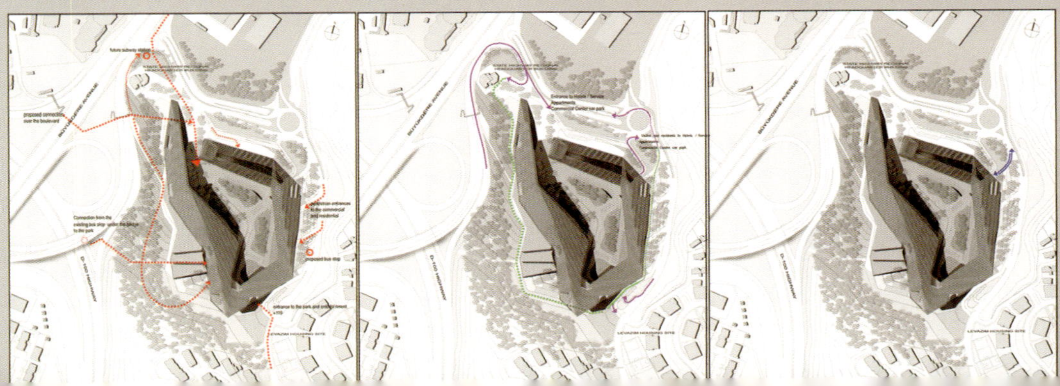

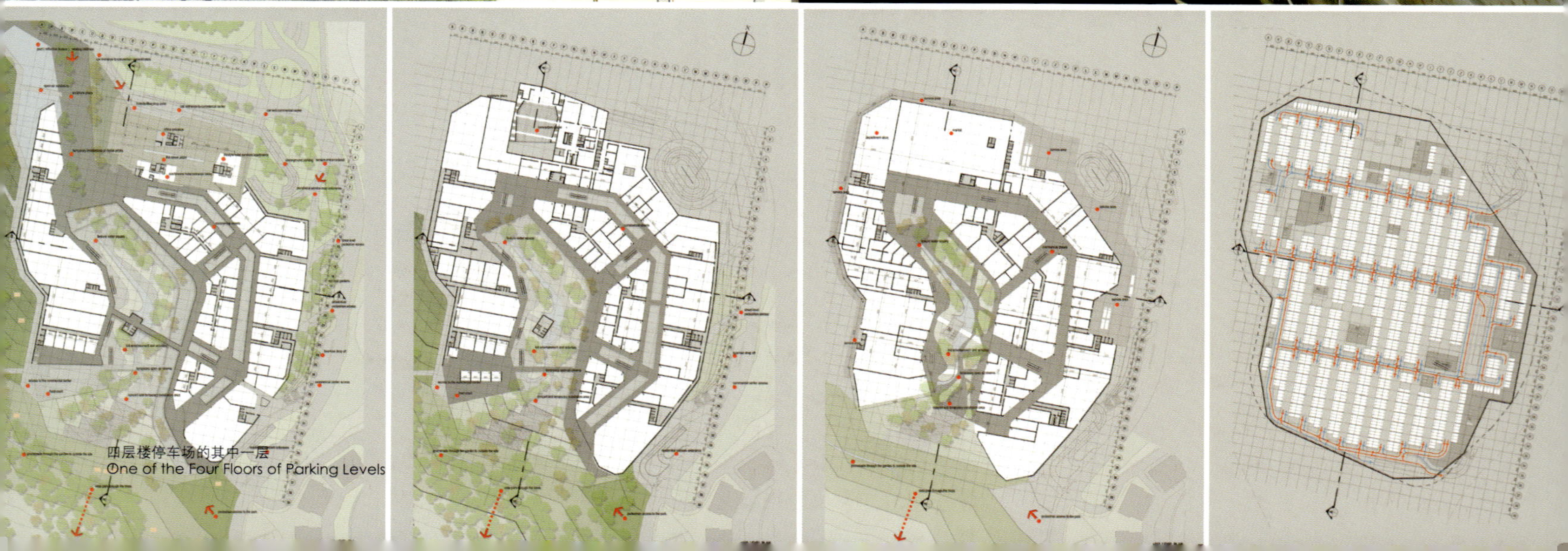

四层楼停车场的其中一层
One of the Four Floors of Parking Levels

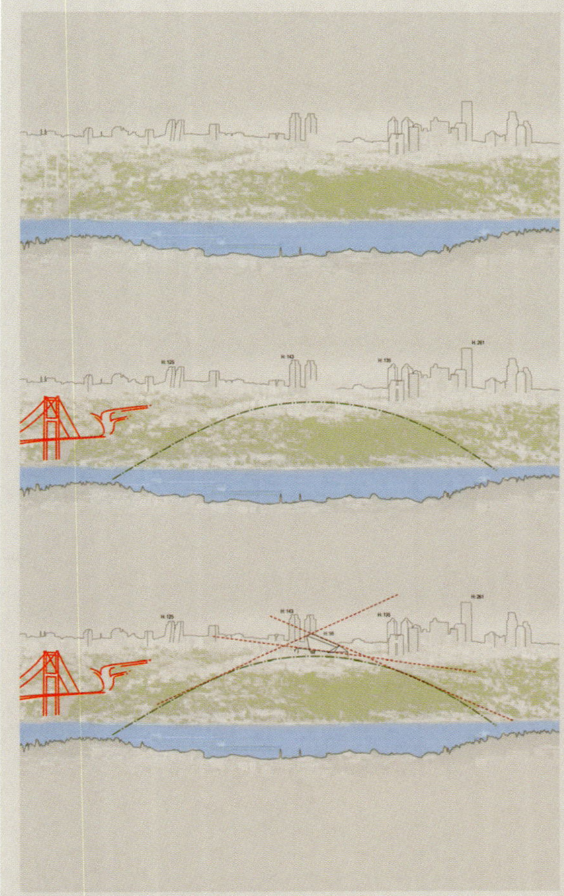

Bosphorus Zorlu Center Relationship

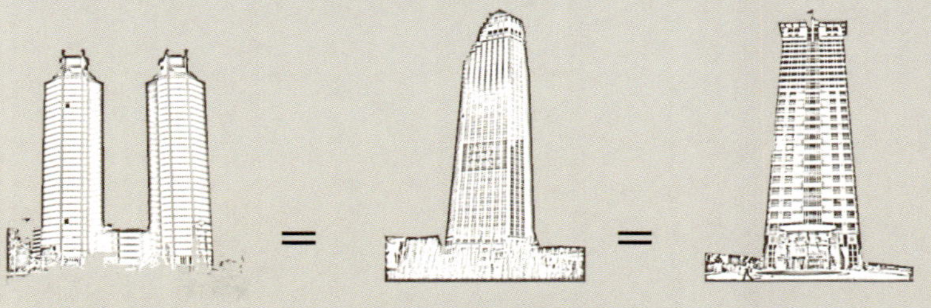

Which one is different?

Is it the same building?

Highrise - Bosphorus Relationship Diagrams

Comparison Chart of Highrise Buildings of Istanbul

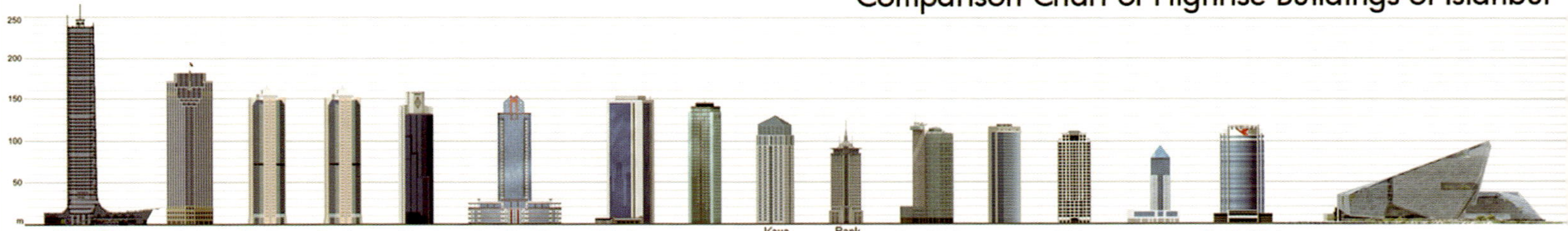

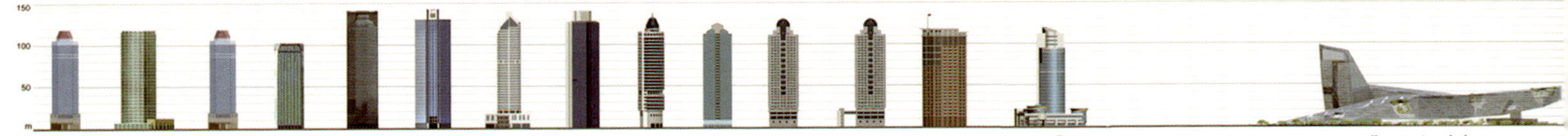

Name	Floors	Use	Antenna	Spire	Roof
Sapphire Tower					
Is Bankasi Tower 1	52	office		195 m	181.2 m
Tekstilkent Plaza 2	44	office			168 m
Tekstilkent Plaza 1	44	office			168 m
Sabanci Center 1	39	office			158 m
Süzer Plaza Ritz-Carlton	34	hotel			154 m
Polat Tower Residence	42	residential			152.5 m
Sun Plaza	38	office		147 m	145 m
Kaya Ramada Plaza Hotel	32	hotel			130.1 m
Bank Ekspres Tower	27	office			124.4 m
Garanti Bank Headquarters	22	office			121.6 m
Zorlu Plaza	30	office			120.4 m
Selenium Residence	30	residential		120.1 m	104.5 m
Polaris Plaza	28	office			120.1 m
Yapi Kredi Bank Headquarters	25	office			120.1 m
silhouette istanbul	**26**	**mixed use**			**98 m**
Is Bankasi Tower 3	36	office			118 m
Istanbul Canyon 1	30	office			118 m
Is Bankasi Tower 2	36	office			118 m
Maya Tower 1	30	office			103 m
TAT Tower 1	34	office			143 m
TAT Tower 2	34	office			143 m
Sisli Elit Residence	35	residential			140 m
Sabanci Center 2	34	office		140 m	130 m
Metrocity Millennium 1	23	office		138 m	134 m
Beybi Giz Plaza	36	office			135.9 m
Metrocity Millennium 2	27	residential		135 m	128.4 m
Metrocity Millennium 3	27	residential		135 m	128.4 m
Tekfen Tower	28	office		135 m	118 m
Kozyatagi Business Center	30	office			119.2 m
silhouette istanbul	**26**	**mixed use**			**98 m**

View from the D-100 Axis towards Bosphoru

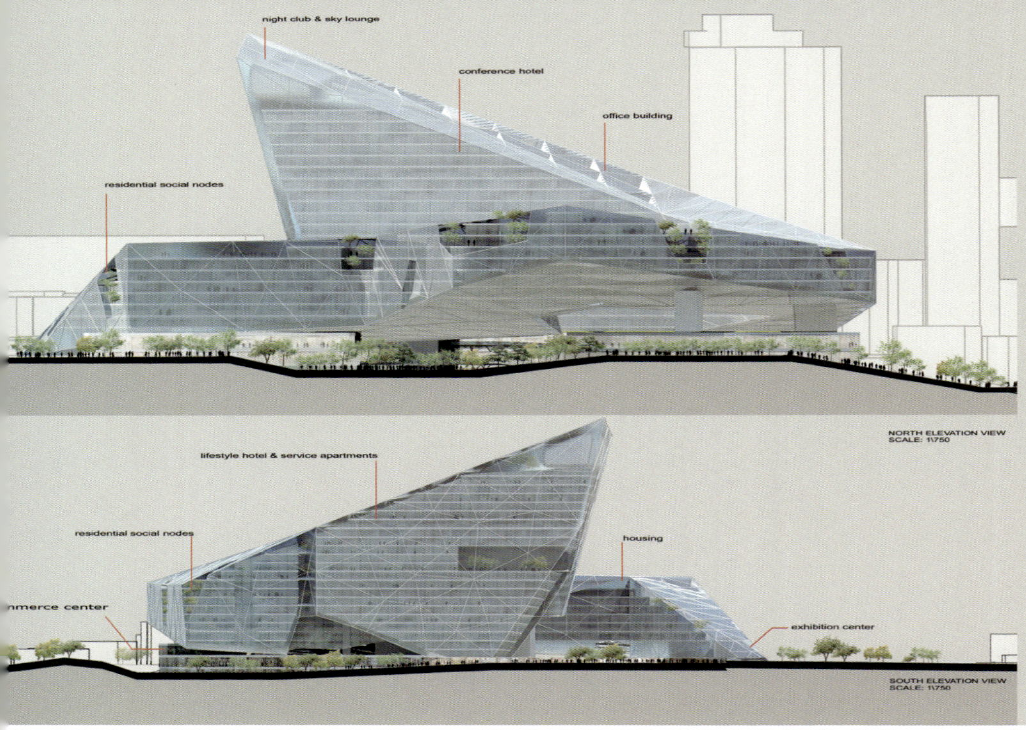

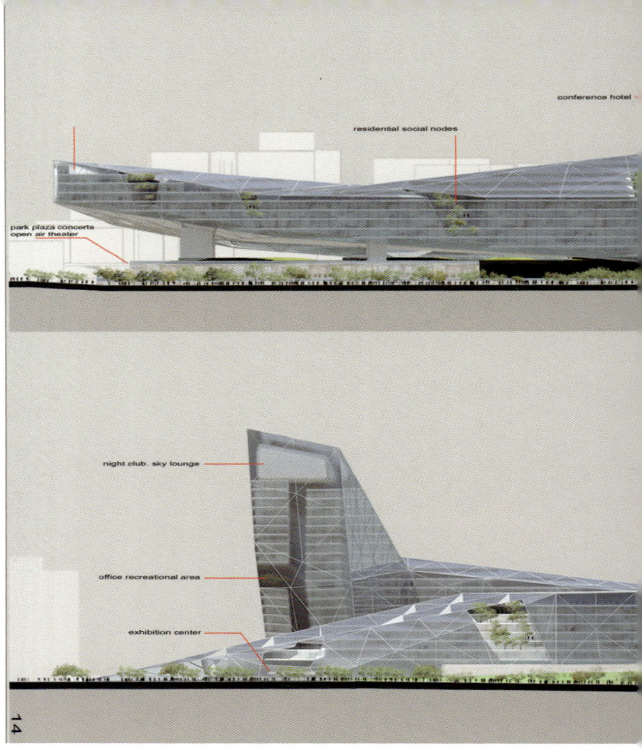

ZORLU PROJECT MECHANICAL INSTALLATION SYSTEMS

Air Conditioning Systems

The thermal energy coefficient for the core is planned to be approx. 0,35 W/m2 according to the aim of providing a high level energy conservative building. The roof and façade cladding materials shall be chosen from the CFC-free (harmless fo hausted air shall be mixed with the aspirated fresh air to provide heat recuperation. It is also planned to save energy by %35 with the heat exchanger fans installed in the Air Handling Units.
The mechanical systems shall be managed through BMS and according to the exterior and interior data the HVAC and lighting systems shall be converted to economized operating modes for the areas out of usage or with the lack of human de be provided with BMS also by maintaining the comfort conditions.
Following the outdoor conditions, the air conditioning shall be provided by the means of fresh air ventilation initially and due to the deficiency in comfort conditions the mechanical systems shall be operated.
Interior and exterior levels of illumination shall be controlled by lighting sensors and artificial lighting will not be allowed as the daylight is adequate.

IAQ

The fresh air quality of the interior areas and closed parking areas will be controlled. The usage of chemicals which might cause pollution within interior sections of the buildings will be minimized during the design phase.
To prevent the mixture with interior fresh air of the building the ventilation of areas such as warehouses, garbage collection areas and cleaning rooms shall be conditioned by direct ventilation.
The VAV systems which provides variable amount of fresh air will be established and CO2 level will be controlled in crowded areas such as cinemas, conference and exhibitions halls. The systems will be stopped and will be in stand-by mode a

W&S

The main aim is to save needless usage of water in the building complex, which the value is equivalent to the primal energy sources and very fast consumed all over the world at the present day. The water consumption in irrigation works will b on the local plants and trees. Rain water will be drained to the cisterns established on the various areas of the plot and all the water demand of the complex shall be supplied except fresh water.

Fire Extinguishing System and Seismic Precautions

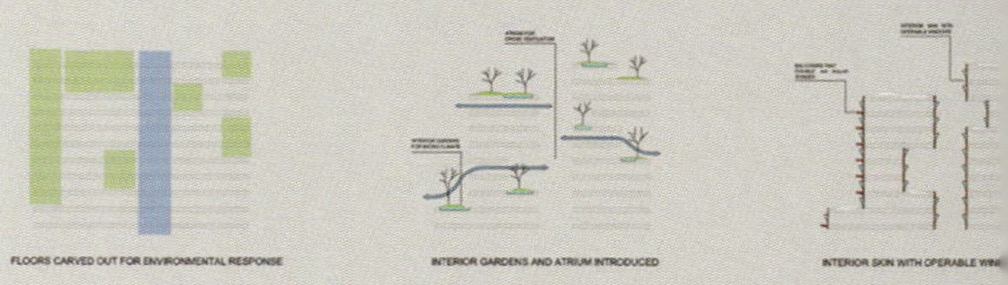

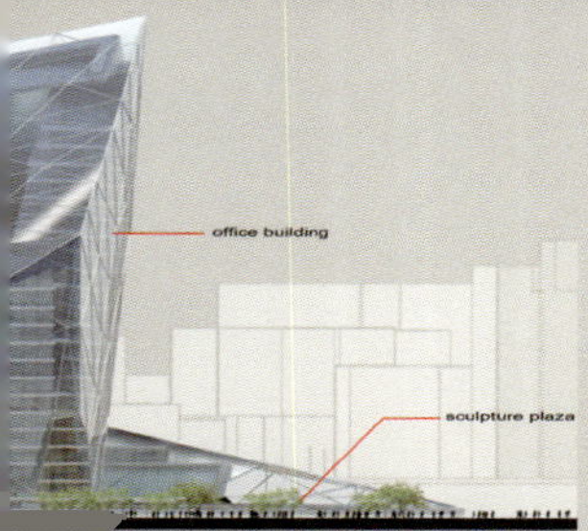

EAST ELEVATION VIEW
SCALE: 1\750

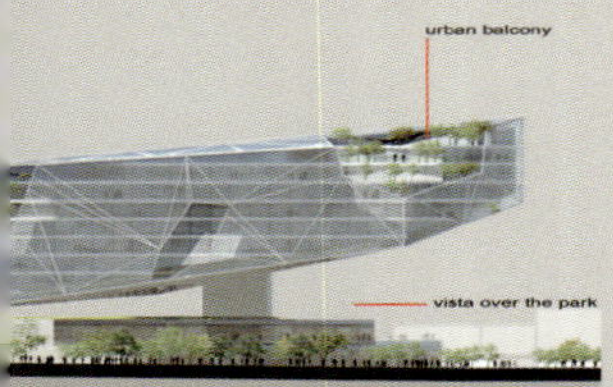

WEST ELEVATION VIEW
SCALE: 1\750

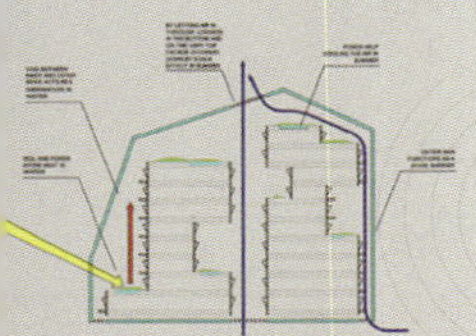

AN EXTERIOR SKIN WRAPS THE BUILDING

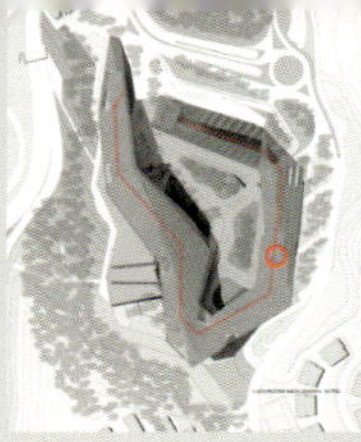

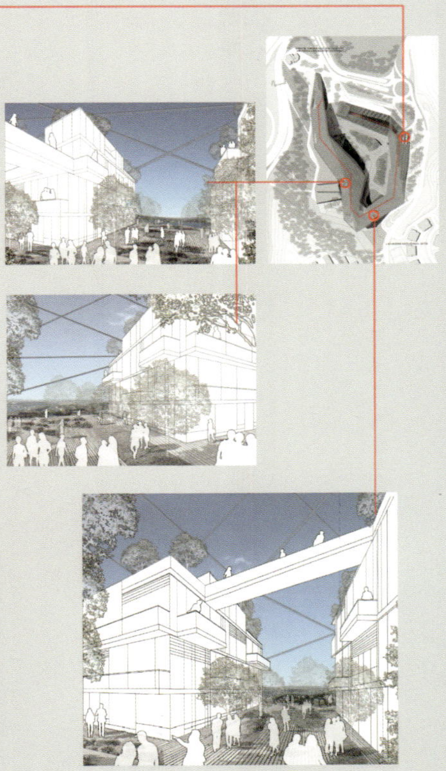

A-A SECTION
SCALE:1\750

Architectural Project Team: Lee Soon Pyo, Kim Jin Kyong, Byun Young Ji, Yu Seung Min, Koo Huyn Jin, Cho Jae Yong

SKMS Institute

SK管理系统研究所

A gentle hill flows into another topography, shaped by the rich branches of the nearby chestnut trees and ginseng farms. The mystical first impression of a chestnut forest with its loosely spaced trees (for better quality, not quantity, of production), the refreshing wetness of the land given by the morning dew, and the soft wind that rustles the young leaves and tree branches, wake up our senses. For the chestnut orchard to look so appealing today, much hard work went into reshaping the rugged hill over a long time. The spirit and memories of those who have cared for the rough environment, and the communication with nature that awakens our senses: It was believed that the former will explain the enterprising spirit, and the latter will free up our philosophical thinking. Our work is looking for the potential of the land and finding ways to build a construction with minimum traces. The program is juxtaposed along the long boundary between the chestnut forest and the newly created site. Based on the simple method that shaped the nearby ginseng farm, we make sure that the building will be part of the scenery, and not a hard edifice. We create an environment where, through this loose framework, a different time can be experienced at a longer pace, where nature is absorbed into architecture. Thus, this architecture becomes part of the environment, part of the continuum with the land, and a framework for thinking, through communication with nature. We hope that the traces we left will, over the years, become part of the land.

OBJECT OF THE WORK

The project was oriented to build create a place where researchers can reside and explore the management systems of the corporation. The gentle hill that has been formed by thinning old chestnut trees flows as an alternative landscape with flourishing branches of the chestnut forest and the ginseng farm. The project began by discovering potentials that the land possessed, subsequently followed by an architectural composition which leaves minimal trace on the premise. By using the integral features of the landscape, specifically the extensive boundary between the chestnut forest and the thinned out land, we were able to juxtapose the architecture with nature. We have been inspired by the loose framework which has been used for the ginseng farm that eventually influenced the approach of the building to exist as a landscape, not as a rigid structure having a compelling presence. Hovering over these loose frameworks, various times can be experienced through the long hauls of breath, creating an environment where the nature can reach into the building. The land

INFORMATION

preserves the memory of hard work and effort of those who cultivated the rough environment into a handsome chestnut forest. We believed that this remembrance will provide support in explaining the corporate vision of our client, in addition to liberalizing the thinking process of researchers through the sense awakening communication with nature. Accordingly, this architecture is an integral fragment of the land and its extended landscape, also functioning as a cognitive framework to communicate with nature. Thus, we patiently wait that the traces we have left behind, over the years, will be embodied into the vegetative landscape.

CHOICE OF MATERIALS

The natural environment surrounding the site itself is the main material of this architectural project. We tried to 'architecturize' the chestnut forest which consists of loosely planted trees that were arranged for better quality than quantity, the moistness of the earth induced by the morning dews, the soft wind that rustles through the baby leaves and branches of the chestnut trees. By expressing the exposed concrete in the most androgynous way, we allowed the building to become a background that sympathizes with the nature. Moreover, the notion of recycling materials which are to be retrieved from the nature has been taken into deep consideration throughout the project. A small amount of wood which has been carefully gathered from its long time owner, the chestnut forest, was made into architectural use along with the soil and stones naturally collected from the surrounding site. The gathered soil and stones were replaced to create a mound for the landscape, in addition to anointing a non-structural gabion texture to the retaining wall of the dormitory building and a small part of the chestnut trees have been generously replanted in the garden. The finishing of the gabion retaining wall is a metaphor of the sedimentary strata. It substitutes the land's former strata by using stones which increase in size directing upwards that evidently presents itself as a non-structural wall. As figurative elements within the traces of our intervention, the materials gathered and utilized will unfold as a medium of materiality, time and memory between the nature and the architecture.

TECHNOLOGICAL-STRUCTURAL CHOICES

We tried to consecutively embody the horizontalness of the land and acknowledge the surrounding landscape. In order to replace the earth pressure that has been cut along the margins between the chestnut tree farm and the thinned out land with self-weight load, we unified the footing of the retaining wall with the 50 centimeter-thick mat footing. Accordingly, the land and architecture adjoin as one, not only by means of physical circumstances but also in respect to the power flow of the land.

USE OF MATERIALS IN RESPECT OF THE CONTEXT

The earth holds on to the memory of the chestnut forest and the scenic walks. As the nature gradually suffuses into the building, we extend the memories of the long walks into the inner space of the edifice. Serving as an ambiguous boundary between nature and building, the exterior and interior, the architecture becomes nature and nature becomes architecture. Consequently, the use of androgynous texture of the panel formwork system is derived from this foundation. Moreover, this also leads to the belief that architecture can once again coexist with nature.

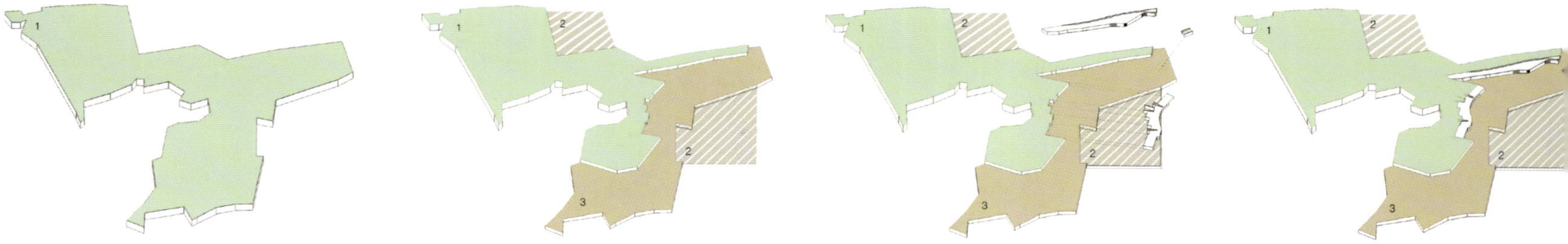

1970's ~ 2006

2006

2007

20

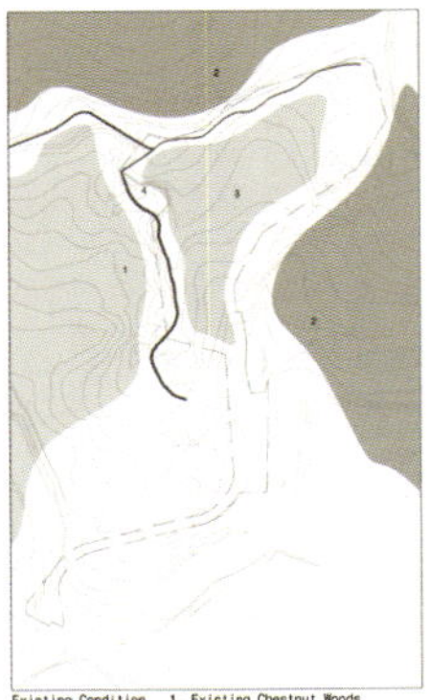

Existing Condition 1. Existing Chestnut Woods
2. Existing Ginseng Fields
3. Nursery Tree Fields
4. Walkway

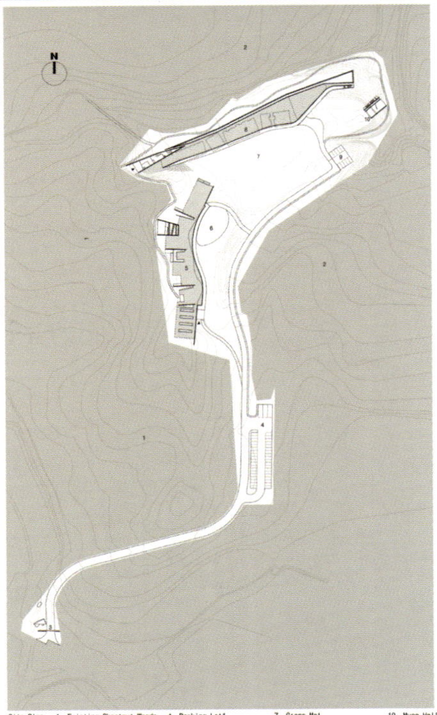

Site Plan 1. Existing Chestnut Woods 2. Existing Ginseng Fields 3. Security 4. Parking Lot1 5. Values Hill(Institute) 6. Mounded Plaza 7. Grass Mat 8. Happiness Hill(Domitory) 9. Parking Lot2 10. Muse Hall

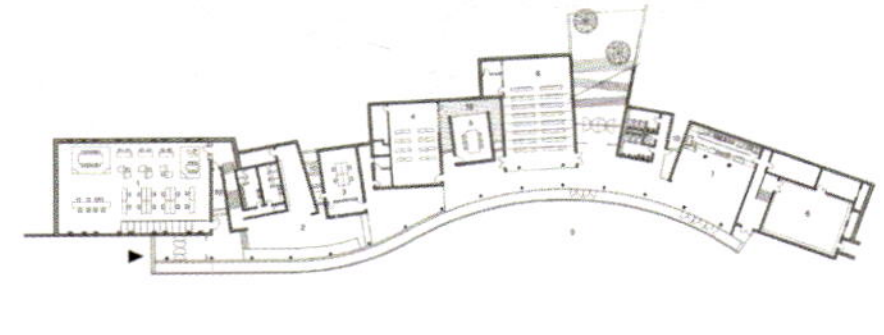

Values Hill (Institute)

1. Reserch office
2. Exhibition
3. Information
4. Can Meeting Room A
5. Can Meeting Room B
6. SUPEX Hall
7. Cafeteria
8. Machine Room 1
9. Mounded Plaza
10. Slant Garden
11. Community Steps
12. Machine Room 2
13. Administration
14. Unit A
15. Unit B
16. VIP Room
17. Refresh Zone
18. Refresh Corridor
19. Sunken
20. Muse Hall
21. Storage

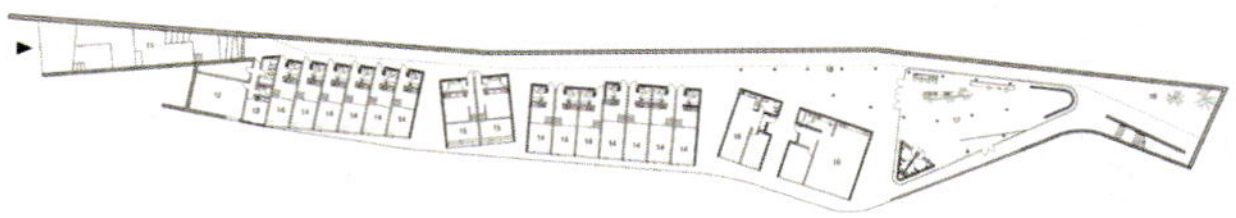

Happiness Hill (Domitory)

Muse Hall

Elevation

Section-1

1. Existing Chestnut Woods
2. Slant Garden
3. Muse Hall
4. Reserch office
5. Exhibition
6. Information
7. Can Meeting Room A
8. Can Meeting Room B
9. Supex Hall
10. Cafeteria
11. Machine Room
12. Unit A
13. Existing Ginseng Fields
14. Administration
15. Unit B
16. VIP Room
17. Refresh Zone

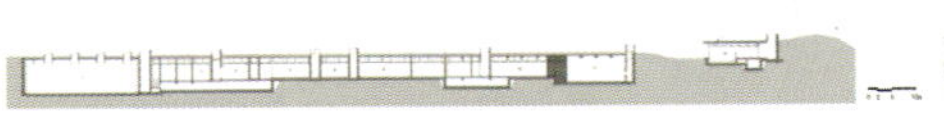

Section-2

Section-3

Section-4

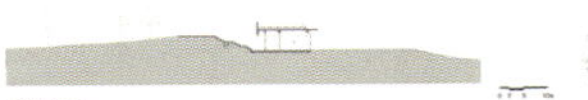

Section-5

Skyscrapers City
摩天大楼建筑群

Skyscrapers started to emerge in the cities like Chicago and New York towards the end of 19th century. Over the century has passed since, and the skyscrapers became the norm for the city centers throughout the world.

The cities citizens' number is higher and higher every year. For the first time in history it was reported in 2008 that more than 50% of population lives in the cities. According to the scientists forecast this percentage will rise to 75% in 2050. Fifties years ago there were only eight metropolises with the number of citizens more than five millions. Today there are 64 cities like that, 26 of them are 10 millions citizens and the record holder city – Tokyo is inhabited by 34 millions people. The problems of overpopulation and the lack of space are getting to be more and more important.

For years we have been trying to arrange our apartments to find as much space as possible and when we decide to live in a present-day city we automatically agree for a lack of this space. The solution can be the high rise building model proposed by us, the one which can completely change the metropolis structure, one which will be practically a separate city.

While developing the model our main inspiration was Manhattan. The world highest density of population is reported to be on the island in the center of New York City. Under the metropolis there will be placed special balloons which will gather warm air 'given back' by the overpopulated city. When the balloons reach the critical mass they will push out the buildings placed above them. What if the giant balloons are placed under the whole Central Park?

The city zoning shapes the form of skyscrapers. The center of the metropolis with the high rise building will be the highest part of our skyscraper-city. The surrounded buildings – a great gathering of culture and business, will create another layer. The next layer will be made of housing estates and tower blocks. The lowest layer will be a construction made of single-family-house estates which usually are to be seen in the large cities suburbs.

Project Team: Alexandros Zomas, Alexia Raisi

Small Library in the Centre of Thessaloniki
萨洛尼卡市中心小型图书馆

The building is located in the centre of Thessaloniki, Greece, near one the most lively squares of the city - Navarinou square. The given plot is very small and narrow and it is part of a city block with high residential buildings. Only one side of the plot overlooks the street while the two other sides are adjacent to the near buildings and on the back it opens to the open air area of the city block.
The building accommodates a three storey library, a small conference room, a small exhibition space and a cafe. The main idea was to develop the building in high by stacking these different functions while giving each of them different morphologies. A very important element of the building was the insertion of the public space inside the building and to accentuate the ancient ruins that are found on the basement of the building. The entrance is drawn back so as to enable the lighting of the ancient ruins but also to make a smoother transition from the public space towards it. The small conference room was designed as a dynamic surface where the inclination of the seats produces the inclination of the surface. It' s placement on the second level along with the development of a basic axis of entrance ensures a fluid free space equally important that functions as a channel of light, air and motion on the level of the ruins. The library unfolds on three levels with internal atrium, where the two first levels are drawn back while the third one thrust itself forward and is protected from the intense lightning by wooden blinds. On the top of the building we find a cafe which exploits the view towards the square with a dynamic form of a light metallic construction. As a public building the library opens itself towards the public life of the city. The structure of the concrete and the definition of the scale with the diversification of the volumes, and the elaboration of the section and the elevation at the same time make the understanding of the building possible. The continuous notional relationship of the inside and outside, of the open and closed spaces describes the project. This project was made by Alexia Raisi and Alexandros Zomas students at the Aristotle University of Thessaloniki department of Architecture Greece.

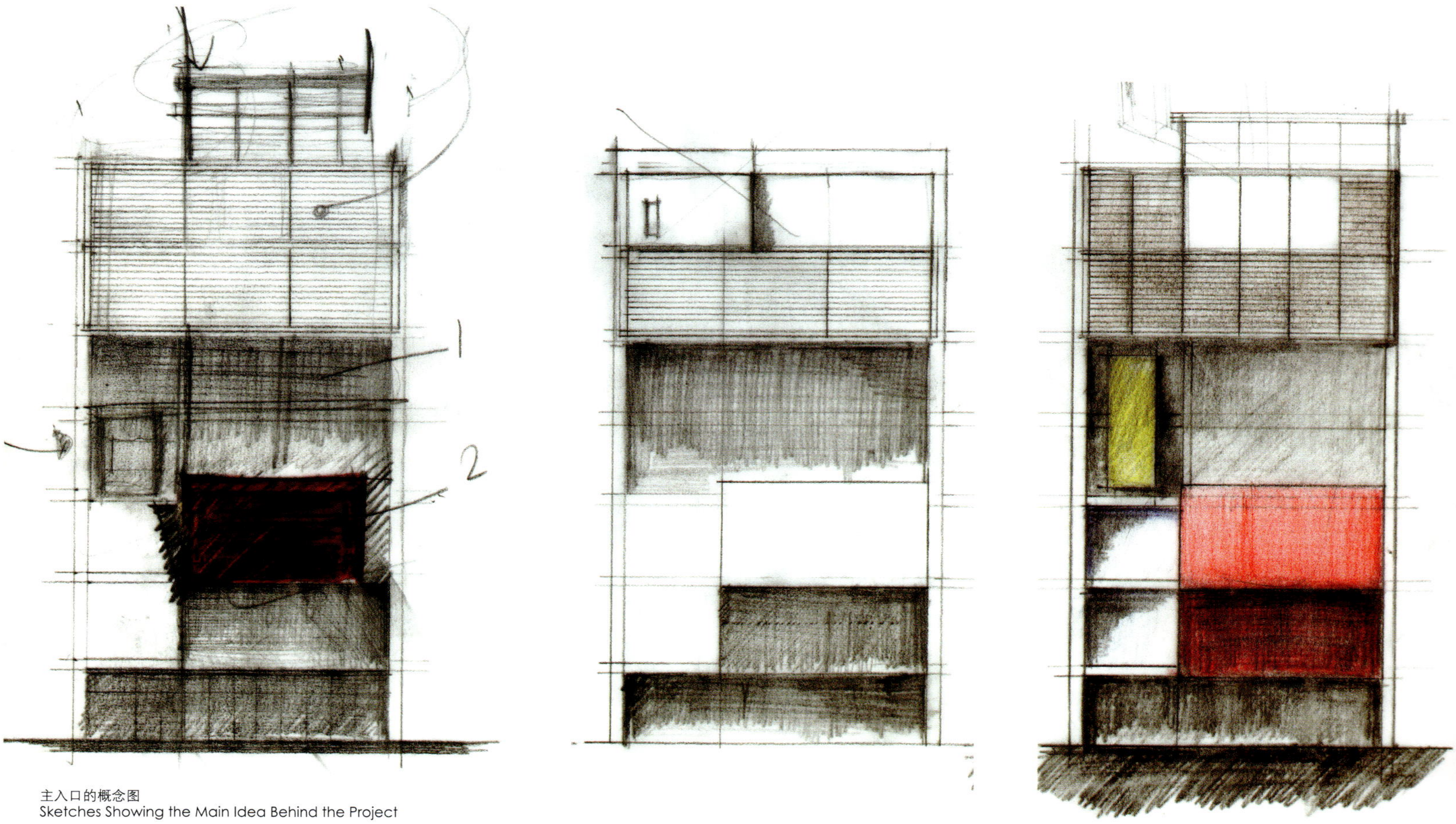

主入口的概念图
Sketches Showing the Main Idea Behind the Project

模型后立面
Model View of Back Elevation

模型截面
Model Section

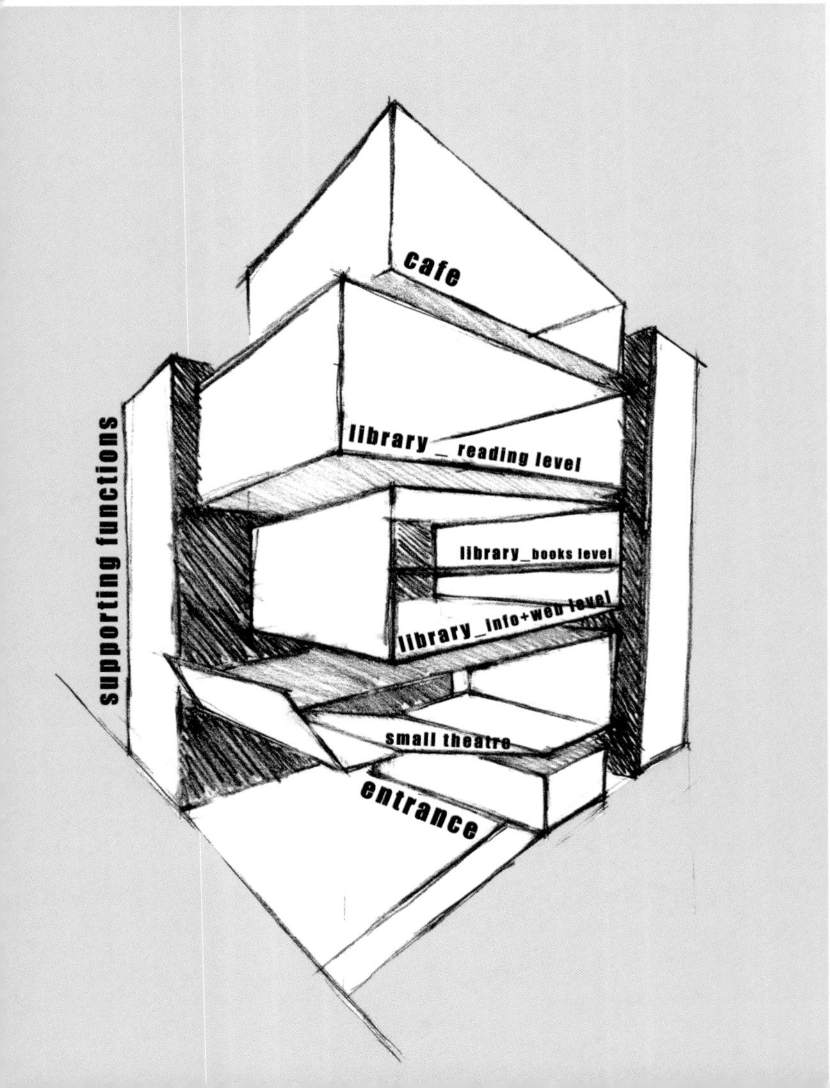

建筑后立面
View of the Back Elevation

临近咖啡馆的开放空间
View of the Open air Space Next to the Cafe

咖啡馆一瞥
View of the Cafe

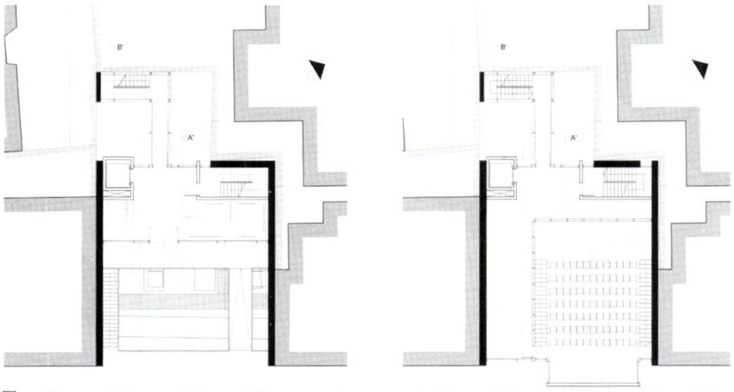

平面图
Floor Plans Level

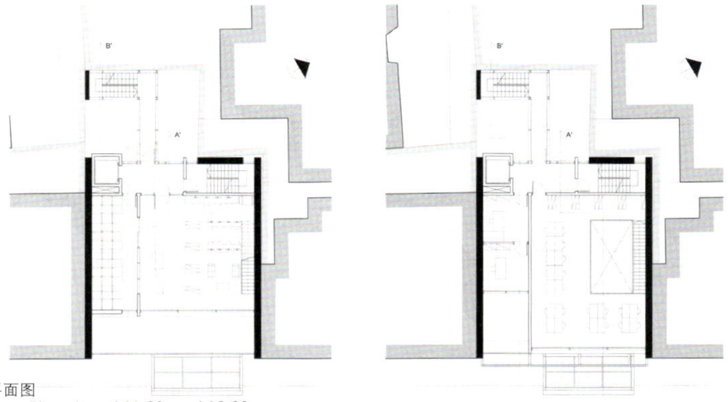

平面图
Floor Plans Level 11.80 and 19.00

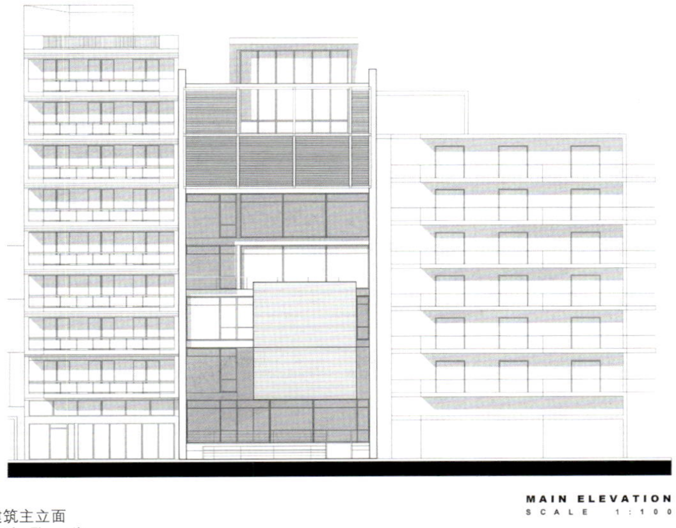

建筑主立面
Main Elevation

截面图AA
Section aa

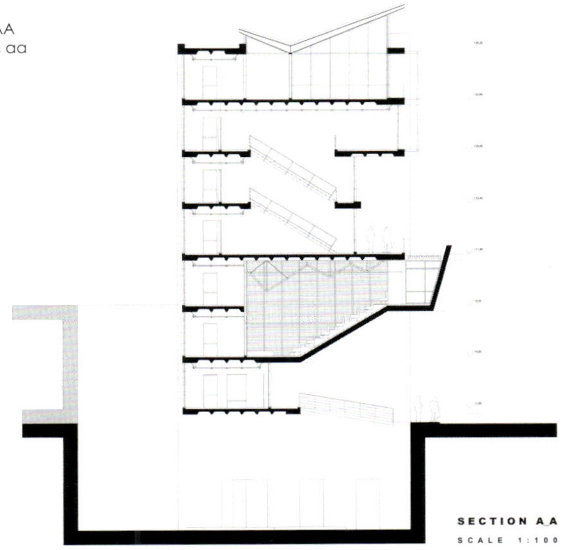

从主入口望向桥梁
View of the Main Entrance Looking Towards the Bridge

Project Team: Georg Driendl, Principal; Franz Driendl, Franz Bergles, Reinhard Weber

Solar Tube — A Symbiosis between Nature and Architecture

太阳能管式住宅——自然与建筑的共生体

The property, where Solar Tube was built, is situated in Döbling, a north-western outskirt of Vienna. Döbling is a quiet, rather wealthy residential area with mostly single-family-homes. The house offers a singular, individual presence among the traditional buildings that are located near it. Tough it also has offended a few locals. "One woman across the street complained that Solar Tube ruined the neighbourhood", Driendl says. Inspiration for the project was provided by a humble tubular skylight, a device that channels sunlight indoors by means of a reflective pipe, rooftop collector and diffuser. The idea was to use the entire house as a solar collector which needs only little heating, cooling or electric illumination. This was the starting point, the root which defined issues of form, structure and choice of materials. The property itself which the family bought is rather narrow but long with a high tree stock. The size of the property totals 1.300 m^2. A narrow, wooded lot, sloping up from the street, heading to the north. Driendl sited the house at the lowest point of the slope, which elevates the living spaces above street level. The location of the house at the bottom of the slope allows people to take full advantage of natural ventilation and the shades which the surrounding trees provide. The natural environment of the property hasn't been changed.

Solar Tube is organised around a fully glazed atrium that accommodates the vertical circulation, connecting all five levels. The entry from the street – which happens to be at the same level as the roof terrace – is to be reached via a ramped bridge directly leading into the living room. There's also a second way to enter the house – by walking over a staircase along the boundary to the other end of the house – ten metres below street level.

Before the actual design process could be started, the opportunities had to be checked. Opportunities are always defined by the characteristics of the location, the plot and the landscape. The sun's position is needed to be taken into account as well as the number of hours of sunlight and a range of other weather aspects before starting the actual design process.

WHY "SOLAR TUBE"?

A "solar tube" in general is a small light-and-heat captor which is usually installed on the roofs of houses. In this case, the entire house serves as a collector, opening to light-and-heat on all sides. The structure is designed to act as a light and heat collector.

The wooded site allowed the architect to use generously glazed elevations, which are sloping or curved in various parts. Since the roof and floors are also partly transparent, the core of the house works like an integrated atrium. As for the glass "tube" that forms the uppermost level, it helps cut heating costs in winter. Apart from its energy-saving virtues, glass also offers intimacy with the surrounding nature. As the façade is nearly

entirely glazed the boundaries between indoors and out are disappearing – an especially pleasant experience during summer, when a leafy tree canopy practically nestles the house among its branches. Therefore one gets the feeling of living in a tree-house. It's a symbiosis between nature and architecture – a concept of openness and transparency.

BUILT IN ONLY 5 MONTHS

Solar Tube was built in only 5 months, from April till August 2001. This fast construction was possible due to the usage of mostly prefabricated or in-stock units and base materials. Used materials were stone, reinforced concrete, steel, wood, glass. Custom designed and manufactured elements do not necessarily produce better results. At the same time, choosing an alternative solution saves a lot of money, energy and time. driendl*architects always have been supporters of fast and efficient construction.

LOW ENERGY CONCEPT

Overheating in summer is avoided by the trees as well as by a special ventilation system that works like a chimney – the atrium allows heat to travel upwards to be expelled at the roof level trough an opening panel. In winter the heating costs can be cut to a minimum because of the compact coat of the building that is designed in a way of absorption and reflection which allows a high amount of sun energy to be used. On the other hand of course the defoliated trees in winter allow the sunlight to shine through the glazed facade and roof and therefore also help cutting the heating costs.

DESIGN AND FUNCTIONALITY

Seeing "Solar tube" from the street is an experience on its own. Passersby are attracted by the remarkable shape of the house as well as by the glass facade, although the facade facing the street is more closed (compared to the sides facing the garden) to give the inhabitants more intimacy.

The integrated atrium that builds the core of the house is open to all sides. Even from the ground level one can see through to the sliding roof. This is possible due to the partly transparent floors on one hand and on the other hand due to the fact that the bedrooms in the uppermost level are reached over a gallery. The interior is also characterized by the spirit of open living. All of the integrated furniture which is understood as part of the architectural process (kitchen, library, integrated cupboards, bathrooms, two story sliding bookshelves the owners can browse while standing on a floor of wood slats that filters overhead light to the main level etc.) was also designed by our office and therefore fits in perfect harmony to the design of the building itself.

Stair House
楼梯住宅

This is a residential building designed for my parents and my uncle's family. The initial program called for two 120 Sq. m. apartments and a small studio for myself. The completed building is being used as planned. The low budget called for an inexpensive site. I encourage my family to buy the current 119 sq. m. site with 3.90 m. width and about 30 m. length because of special characteristics of being deep and narrow which made it available at half the price of comparable lots in the same area of town. I could also use the opportunity of experimenting with my ideas about architecture in our home. I had to sacrifice a few design details because of the limited budget and the low quality of construction work, but the general spirit of design has been kept intact. My early sketches were approved by my family but I had to adjust the construction drawings according to the strict Tehran municipality regulations concerning limitations in the height of the building, surface area and the number of stories. The project was scheduled to complete in 1'5 years but financial problems meant it took longer in reality. My brother designed the reinforced concrete structure and my father acted both the contractor and the mechanical engineer.
The structure is reinforced concrete. The walls are made from bricks with a layer of insulation. The facades are hammered travertine which is easily available in this region. The interior walls are covered by gypsum, bricks (which are cheap and easily available) and rough blocks of travertine which are cut in the site. The interesting point about these blocks is the fact that these are the spare parts of some stone factories which are bought so cheap and cut and formed experimentally in the site with some untrained labors. The floors are mostly covered with limestone with touches of wood. The inside stair cases are a combination of wood and steel. The woods are some how rough and natural.
The site is located in a dense urban context and surrounded by other 4-storyt buildings. The narrow width (3.90 m.) and small area of the site, the number of allowed levels and the total surface area designed by the client were the most influential factors shaping the design. I finally decided on two triplex units located along the length of the site. The public access stairway joins the two units and the studio and the rest of middle part is patio. Because of the slope, there is a one and a half story difference between the two units which provides ample sunlight for the northern block. The whole concept was envisioned in a 3-dimentional manner so as to transform the attention of the viewer upward instead of the width of the units and finally transform a seemingly worthless site to a worthy livable place. Each unit consists of 5 levels interconnected by private staircases and voids thus the sense of confined room resulting from the narrow site give way to airy good-proportioned spaces which look larger than their surface area suggests. Based on the different location of the viewer, there are numerous perspectives to discover and explore the interaction of the levels, the void and the stairs are meant to create active varied spaces. The functional diagram is arranged from public to private spaces, from the lower to higher levels. The

coziness and comfort is created with the careful use of a range of natural and rough materials.
The cold mountainous climate called for a sloping roof and this in turn allowed for the main public staircase to be ended at +1.20. The resulting space gained has been used to provide additional surface for the kitchen and bedrooms.
Despite the complicated spaces inside the units, the facades are simple or some how minimal.

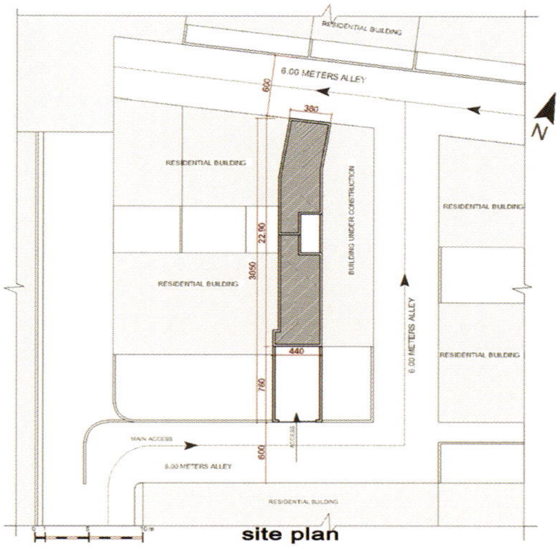

site plan

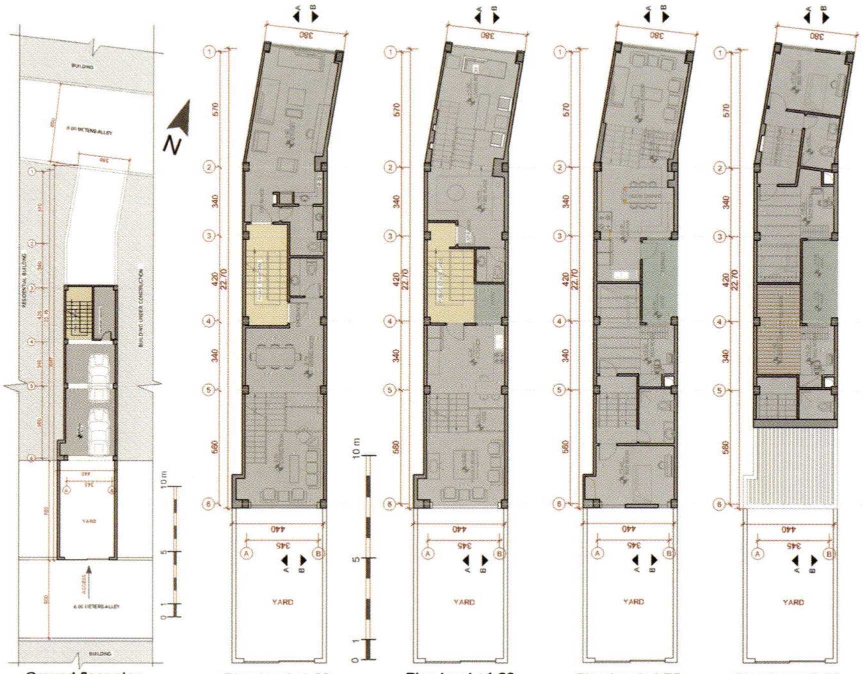

Ground floor plan   Plan level -1.60   Plan level +1.20   Plan level+4.75   Plan level+7.50

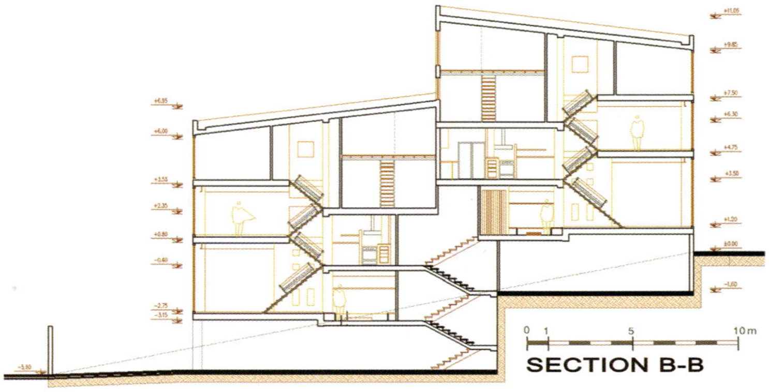
0 1 5 10m
SECTION B-B

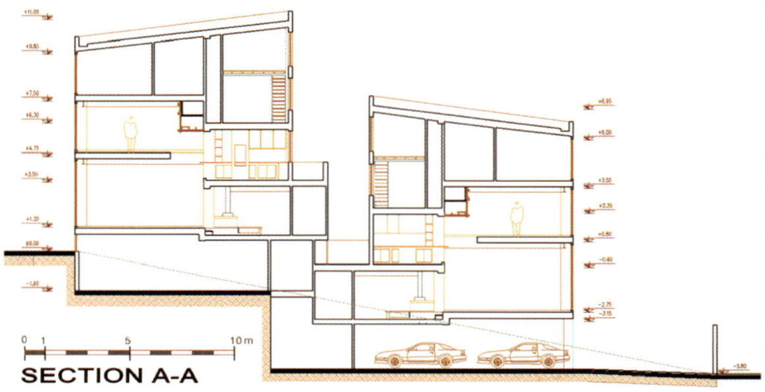

SECTION A-A

Project Team: Kazuyuki Handa

Stone Terrace

石头平台

Rice terraces are machines used to provide and distribute water from the forest to a series of horizontal surfaces. The material and form of this landscape is defined by the artificial system which maximizes the benefits of light, water and wind for agricultural products. The aim of this project is to convert this system for agriculture to new a system for human life. Minimum volume of residence is inserted into the middle of rice terraces with care of continuity of stonewall shapes. A very simple rectangular, flat interior creates various views according to it's relation in shape to the surrounding landscape. Two closed gardens at the sunny side are directly connected to interior and become part of living room in summer. Bathroom has exclusive garden with special view to forest and sky. There are water garden and spacious rural landscape beyond of wide north window, Rain is allowed to drip along the roof and north wall in the expectation that this will create beautiful moss on the stone walls. Water from top of rice terrace and rain create Water garden. Water garden not only gives aesthetic effect to the view from north window but also cools breeze which be taken through north window in summer. In summer, cool air is taken in from the bottom of the north window and warm air in the interior rises along with the slope of the ceiling, where it is exhausted from the top of the south window. A tilted roof casts a shadow into the interior during summer and maximizes sunlight in winter. In this way, Stone Terrace becomes part of a natural circulatory system of water, air and light.

通道
Approach

场地规划图
Site Plan

南面庭院到生活空间的景致
View from South Court Yard to Living Room

住宅成为景观的一部分
House as a Part of Landscape

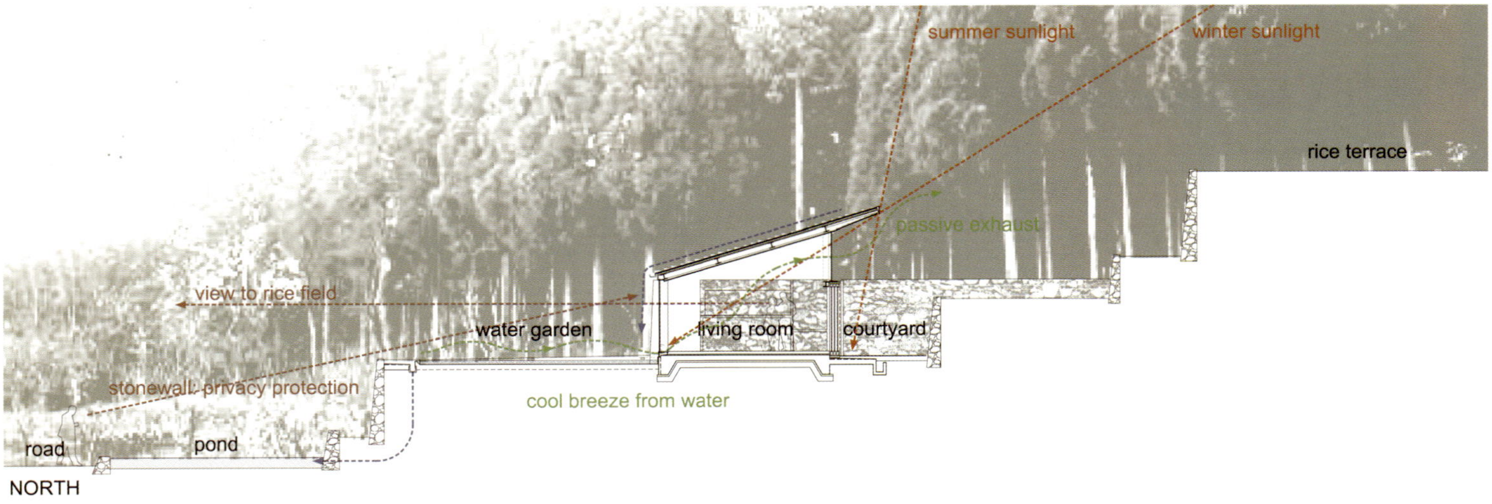

截面图/表现空气、水与光的图表
Section / Diagram of air, Water and Light

卧室
Bed Room

从卧室到南面庭院及石墙的景致
View from Bed Room to South Court Yard and Existing Stone Wall

Project Name: The Consulate Diploma Project Architecture and Town Planning Faculty
Technical University of Lodz, Poland

The Consulate — An Ingenious Building

别出心裁的领事馆建筑

The shape of the building is made on a contrast between vertical concrete walls and soft form of PTFE membrane. These elements emphasize each other and creating minimalistic but spectacular mix that reflect a diplomatic function. Tranquil expression of its facade is "like the reserve of diplomat in the grip of mounting excitement which he manages to keep carefully concealed". The collaboration of mentioned materials existing not only on aesthetic layer but has also sustainable targets. The soft translucent membrane as an element of multi-layer façade allows light to penetrate but diffuse it on the way to interior. Between membrane and internal low emission glazing there is an air ventilating space. The concrete walls, thanks to each thermal capacity play as the part of heat magazine and slowly radiate stored energy.
The additional advantage of this facade is a possibility to use it as an information carrier by putting light projector between layers.

EAST ELEVATION

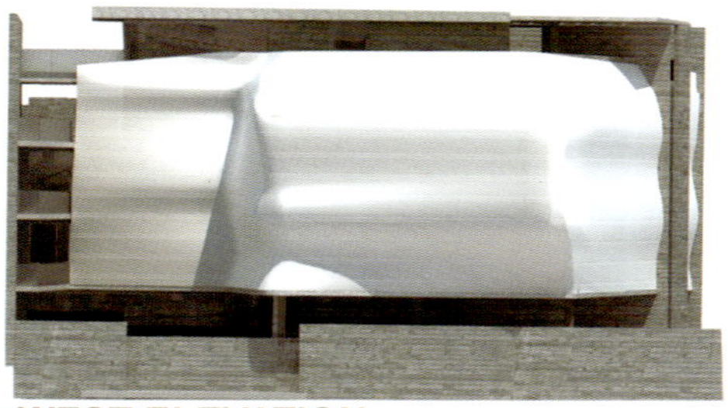

WEST ELEVATION

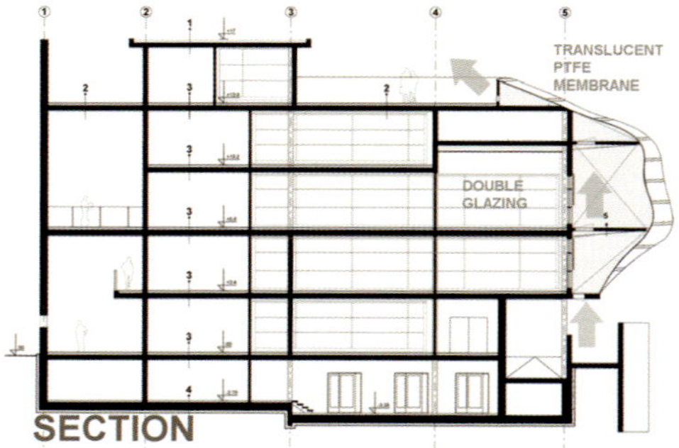

SECTION

ENTRANCE ZONE
CORE
OFFICE ZONE

APARTMENT
OFFICES

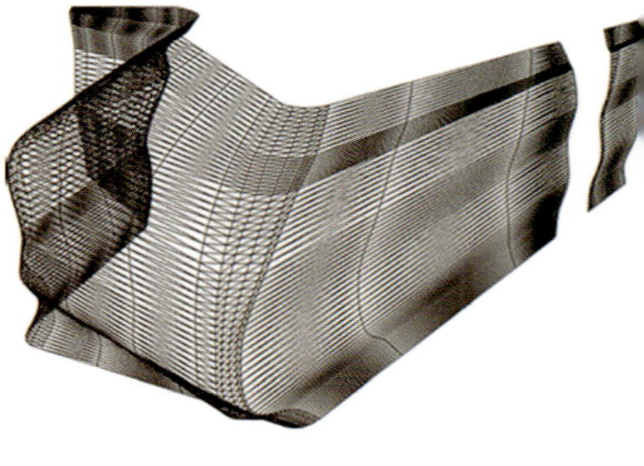

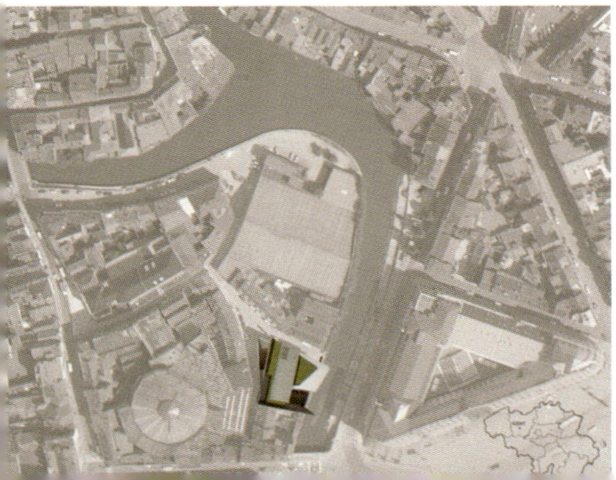

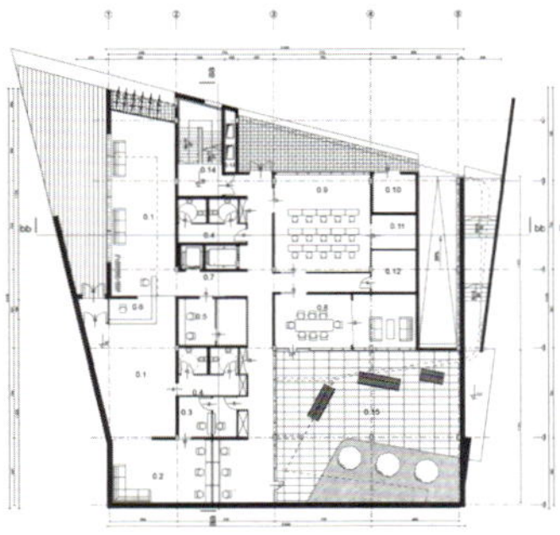

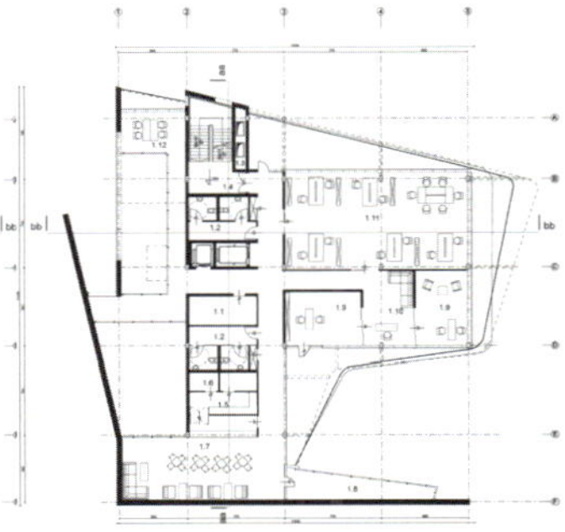

The Lightbox
光线的璀璨交错

The competition-winning project for the Lightbox is a new kind of community cultural centre for Working. It originated from the community and is a flexible and adaptable, a meeting place, a cultural bridge between different groups, an opportunity to present art which goes beyond the walls of the gallery and connects back to the community. The Lightbox opened to the public in September 2007. Since opening it has received a Civic Trust Award and has been shortlisted for the Art Fund Prize. Sandwiched between the Basingstoke Canal and a six lane highway adjacent to the Town Centre, building will be clad in shades of gold and silver facing the roads; and timber on the more Arcadian canal and garden sides.

Architects: Senan Abdelqader

The Mashrabiya House

Mashrabiya住宅

The "Mashrabiya House" is located in the Arab Palestinian village Beit Safafa between Jerusalem and Bethlehem. The house was designed as a contemporary re-interpretation of traditional elements of Arab vernacular architecture, providing at the same time new and imaginative solutions for the transforming social and cultural landscape of the village on the brink of urbanization.

TERRACED LANDSCAPE

The principle of terraces and retaining walls – one of the most typical feature of the traditional Arab landscape - was adopted for steeply sloping building plot, forming an earthbound, partially inhabited 'landscape ground'. The basement itself is carved into the hillside forming a massive stone-clad platform like a deck with sunken courtyards as light wells. This deck accommodates a workshop, studio and gallery and therefore separates working and public functions of the house from the living areas above. At the rear side of the plot this platform folds up to become an enormous, inhabited stone clad retaining wall, which accommodates several apartments.

The landscaped ground forms a stage on which the building itself stands. The heaviness of the ground is contrasted by the lightness of the vertical structure. The floating effect is achieved through the creative interpretation of an Arab "Mashrabiya", a latticed window screen that traditionally formed a threshold between private and public spaces. Here, the wooden screen is re-imagined in the form of a large-scale stone envelop that surrounds the building, combining the "Mashrabiya" motif with stone. Its typical semi-transparent effect is achieved by positioning the stones slightly apart, with irregular space creating an effect of lightness and porousness. The resulting stone envelop is structurally separated by a

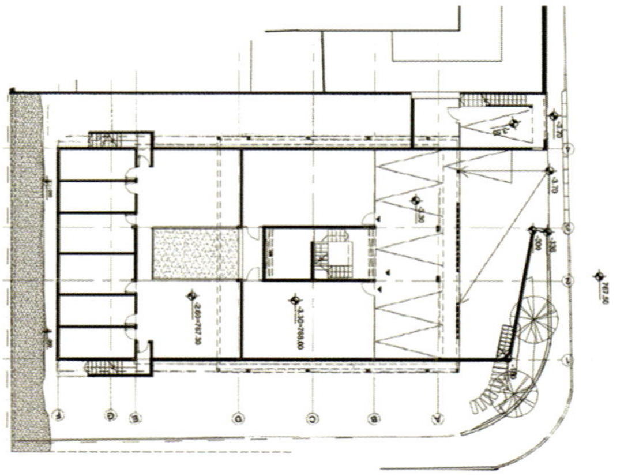

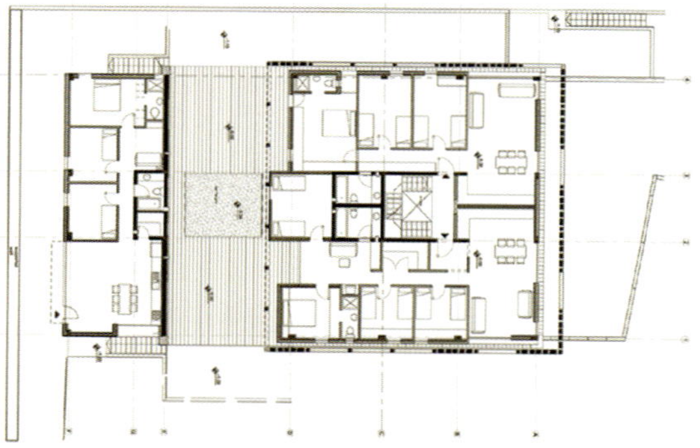

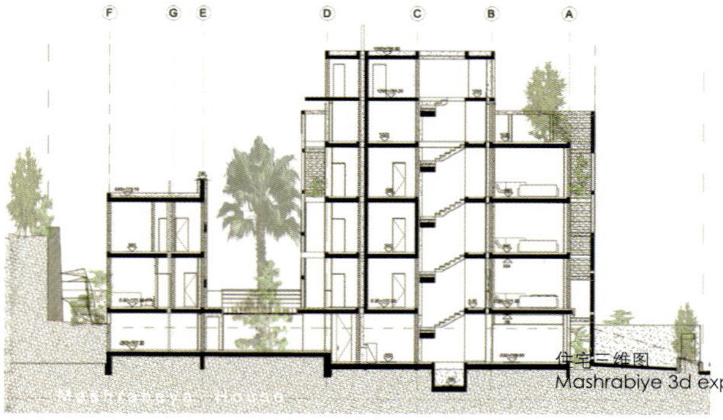

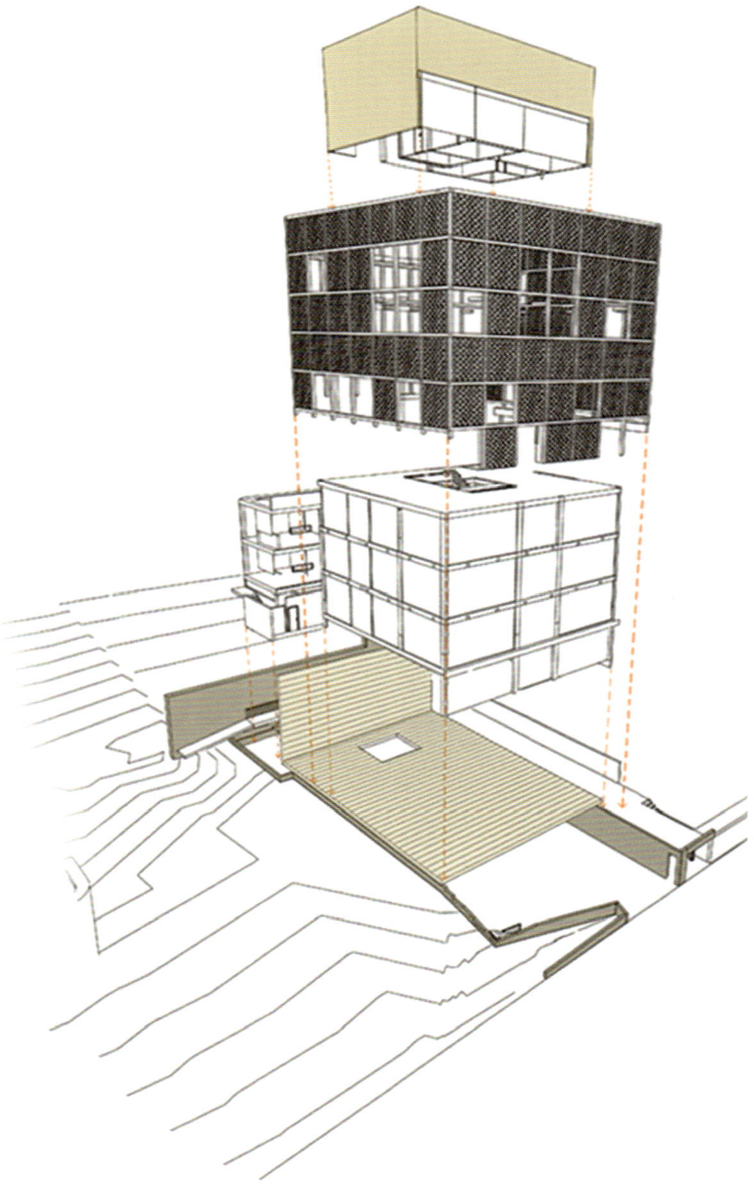

住宅三维图
Mashrabiye 3d explosion

住宅景致
Mashrabiye

住宅景致
Mashrabiye

narrow gap from the actual apartment building behind. The playful arrangement of small and large openings provides views from the interiors out into the landscape while carefully retaining their privacy.

HOSH AND BUSTAN

The concept of open spaces within and around the house is equally informed by sensitivity to traditional Arab elements while insisting on contemporary formal and material solutions. The raised courtyard between the rear retaining wall and the "The Mashrabiya" is reminiscent of a traditional hosh while the "Mashrabiya" at the top of the building becomes a 'garden wall' for a roof top garden following the 'Bustan' motif: a closed garden as a sheltered place of tranquility and privacy.

AN ARABIC INTERPRETATION OF SUSTAINABILITY

Beyond the formal references to Arab vernacular traditions the building developed and tested concepts that could help to inform an agenda of sustainability while maintaining historical cultural continuity: In this way the "Mashrabiya" is not only a traditional threshold between public and private space but also provides an element of climatic control. The stone mass of the outer envelope acts as a climatic buffer: It helps to absorb heat during the day and release heat during the cool Jerusalem nights. Thus it protects the building against solar radiation as well as winter rain and winds. The gaps between the stones ensure a constant flow of fresh air. A further element of passive cooling is the 1m gap between outer and inner envelop. Not only does it ensure constant circulation of fresh air around the building the fact that it remains open towards the top of the building generates a suction effect like that of a chimney: Hot air travels upwards and fresh air is sucked into the gap from below.

Beit Safafa, like most villages in the Jerusalem area, is facing social and cultural challenges, typical for the transformation from a tightly knit village community towards a suburban centre. In this context the building provides an original solution by combining traditional building form with modern urban apartment living. The building is sensitive to the traditional fabric of the village yet open to new residents, as it combines living and working. It is also the home of the new Gallery for Palestinian Cultural Representation.

住宅景致
Mashrabiye

住宅景致
Mashrabiye

住宅景致
Mashrabiye

住宅景致
Mashrabiye

The Museum of Modern Art
现代艺术博物馆

The museum is situated on the inclined relief under ground. Only light elements are lifted from the ground. There are escalators in the museum which help the one to move through the museum. Whole museum is a one system so the viewer can inspect the whole museum only going on the single route.

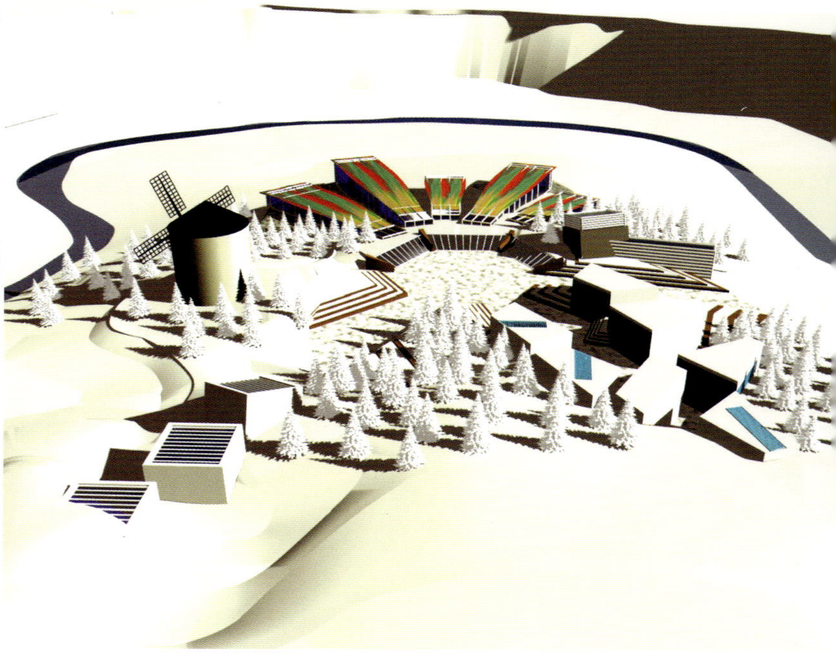

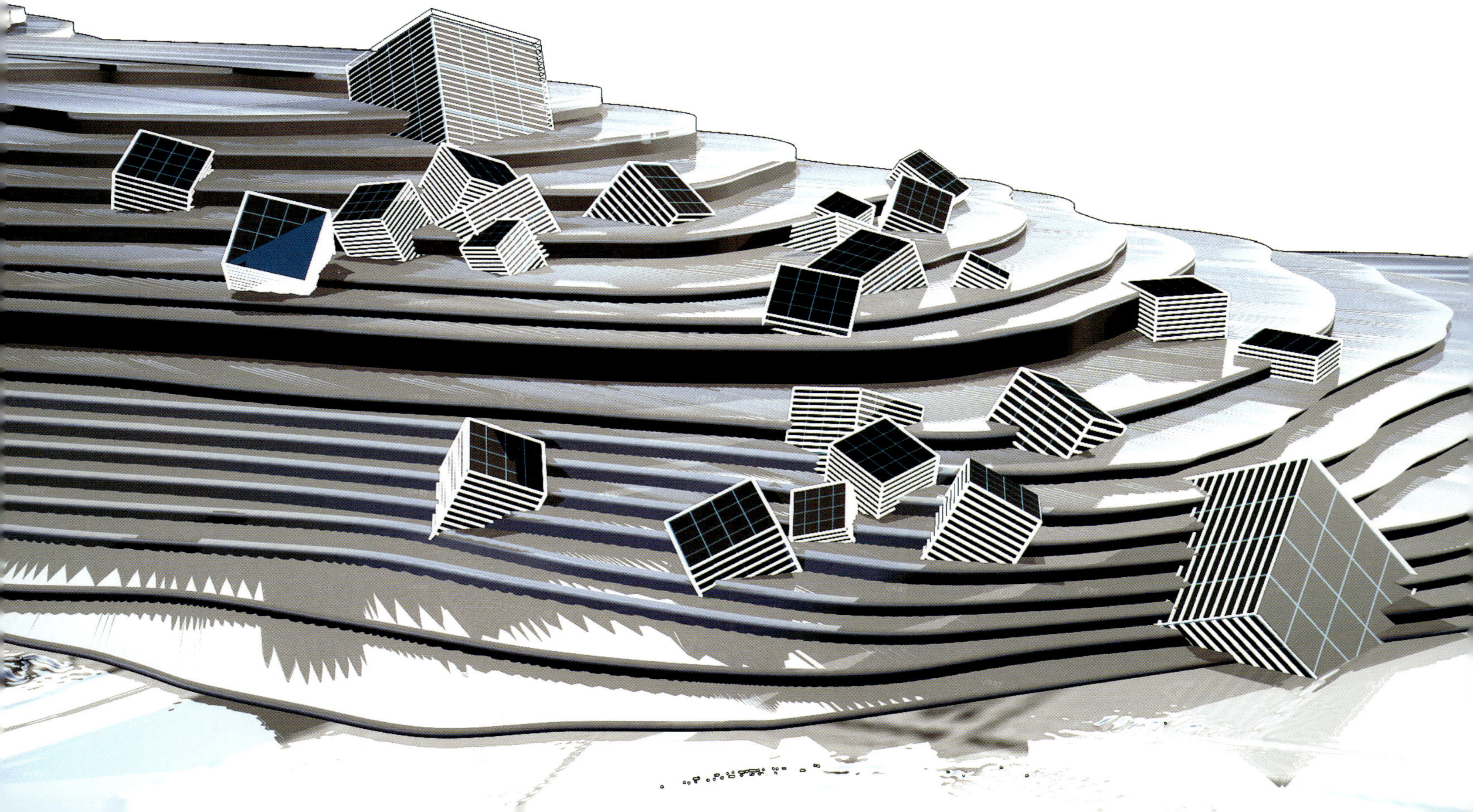

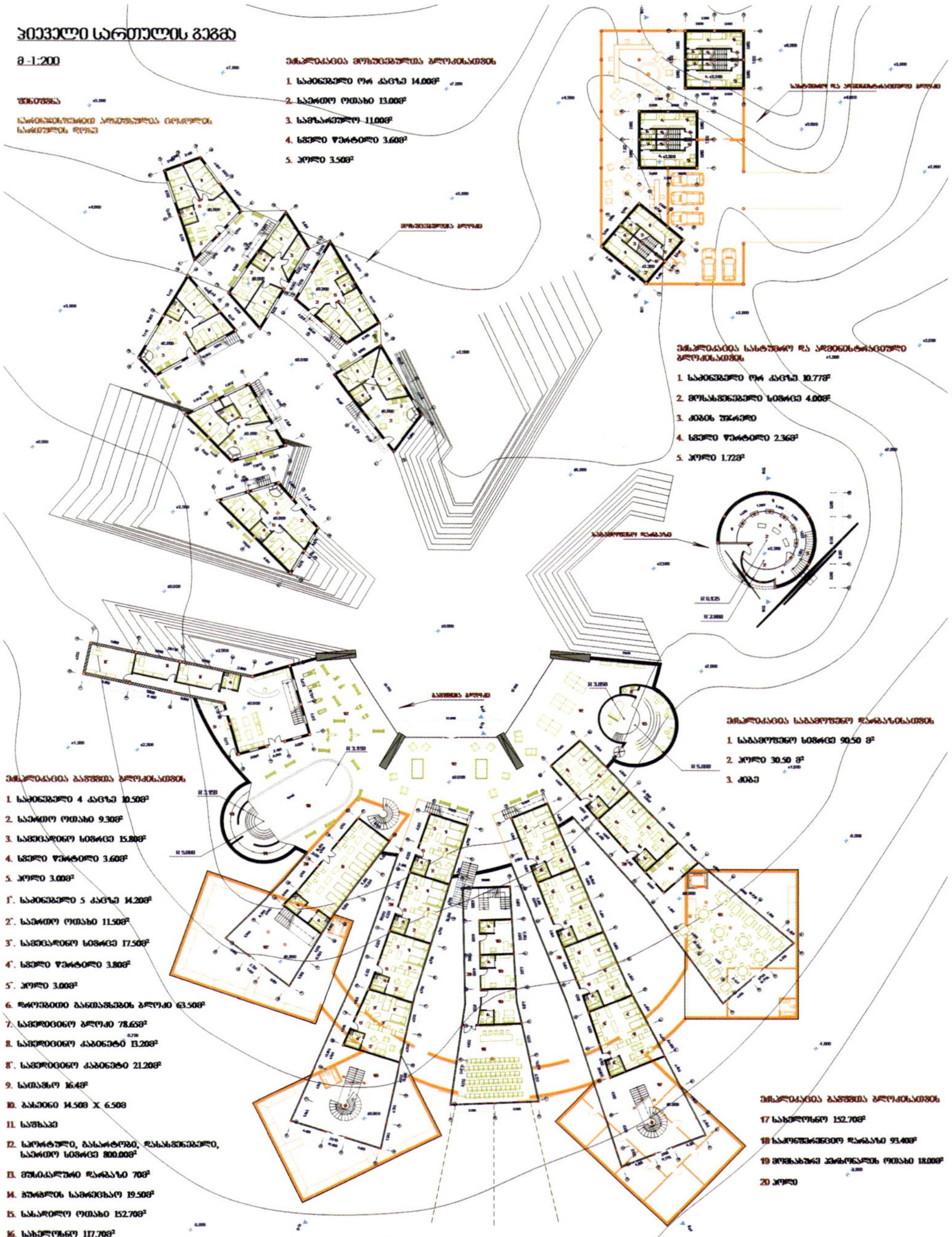

平面图
plan

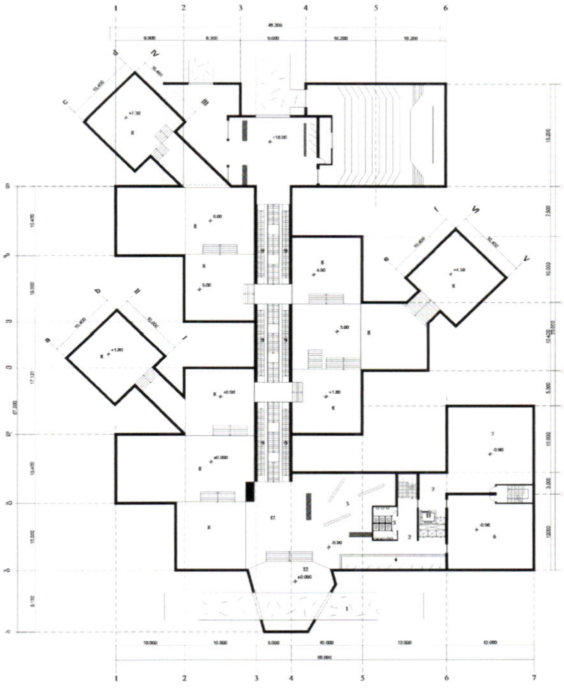

Project Team: Personal Project

The New Venice Biennale Australian Pavilion

澳大利亚威尼斯风情阁

TERRA AUSTRALIS INCOGNITA AND THE CITY OF WATER

Australia was inscribed as 'terra australis incognita' which means 'the unknown land of South' in the Roman times. In my mind, it seems like this description still symbolizes Australia well as a metaphor for the mega-diverse nature. The Outback, the Great Barrier Reef, the Mount Augustus, and the Mount Kosciuszko on the Great Dividing Range are just few of the famous sites among the numerous places that you can feel and experience this diverse and unique nature of Australia. On the other hand Venice, which is nicknamed 'city of water', is a city built on an archipelago of 118 islands formed by around 150 canals. The city is full of rich culture and the charismatic built environment dominates over nature. Thus an architecture that is to be influenced by these contrasting contexts should inherent multiple logics and perceptions.

THE NEW AUSTRALIAN PAVILION

This project aims to create the new Australian Pavilion that architecturally represents Australia's nature and 'sculpt' it into the built environment of Venice. The form of the overall building is pushed and pulled to correlate with the surrounding context and an entry ramp is inserted through the building for direct connection with the adjacent canal. The facade of the building delineates the shapes and colors of Australia's landscape and the operable exterior louvers will change colors as the wind blows. The interior space is designed to be a glowing space full of light with soft shadows created by layered shells, and the coherent use of materials will enable this space to become a neutral backdrop for the art that will be displayed.

AUSTRALIA

AUSTRALIA

Main Gallery / Media Gallery
Entry Lobby / Books / Meeting Room
Office / Staff Facilities
Back- of- House Facilities
SCALE 1:300
volume determined by required progam size
mass is sculpted by site context
connection between entry plaza and canal
INITIAL MASS
MODIFICATION OF FORM
GEOMETRY BREAK
N
SCALE 1:300

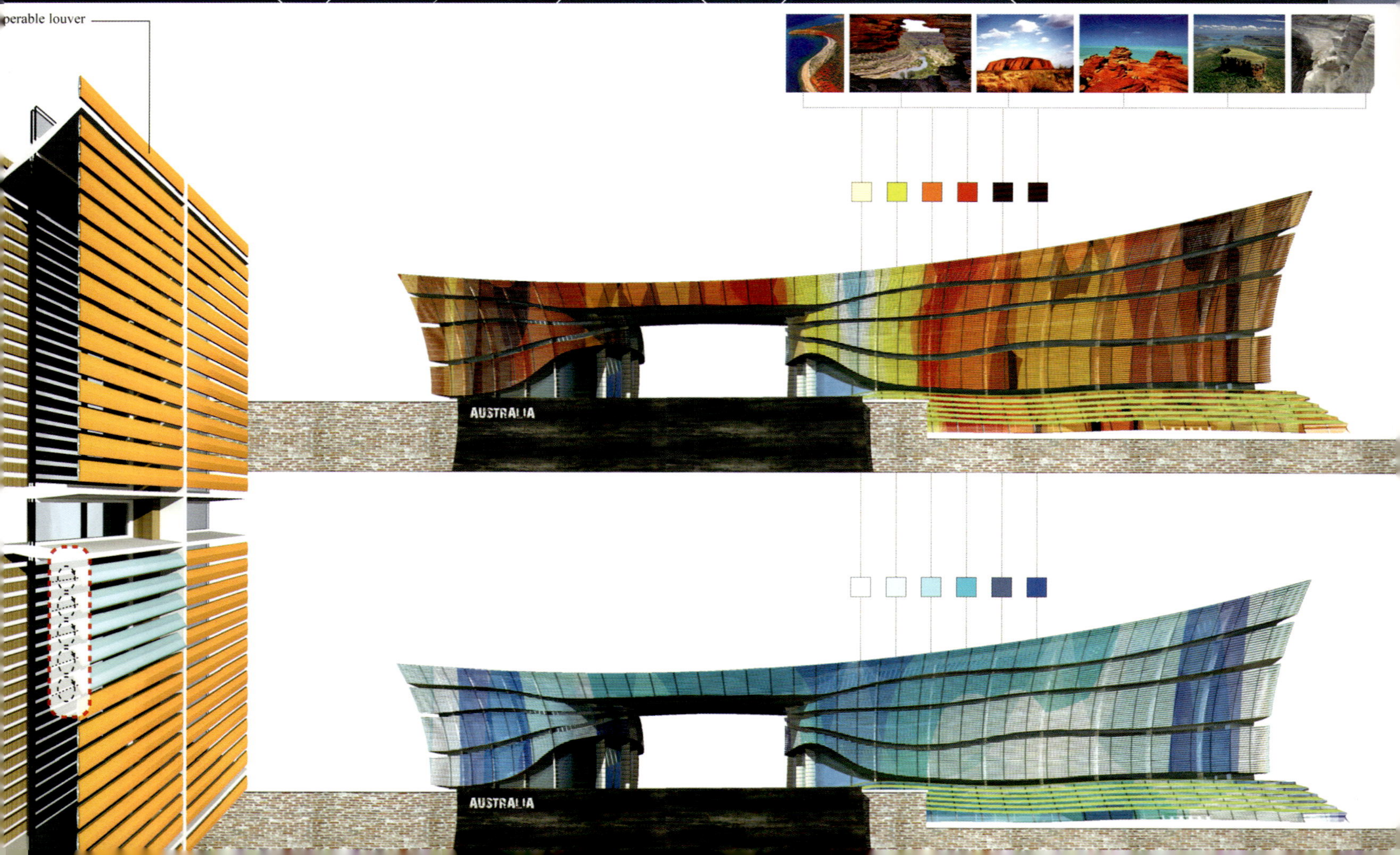
perable louver
AUSTRALIA
AUSTRALIA

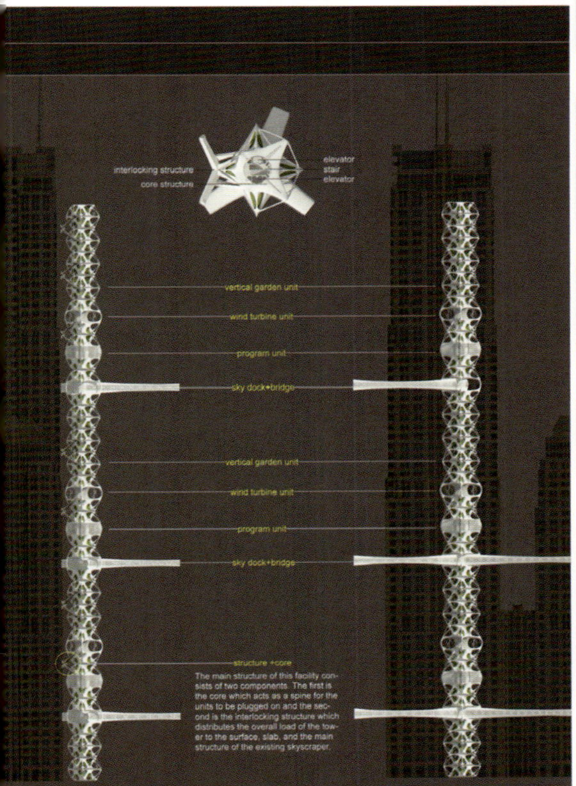
interlocking structure
core structure
elevator
stair
elevator
vertical garden unit
wind turbine unit
program unit
sky dock+bridge
vertical garden unit
wind turbine unit
program unit
sky dock+bridge
structure +core
The main structure of this facility consists of two components. The first is the core which acts as a spine for the units to be plugged on and the second is the interlocking structure which distributes the overall load of the tower to the surface, slab, and the main structure of the existing skyscraper.

Team: L. Altana - F. Di Cintio - D. Fois - R. A. Tavares Saraiva

The Rainbow Tower and Poljane Masterplan

彩虹大厦及Poljane总体规划

We read the competition scheme as a classical zoning by putting different functions in different sites. We found it a bit rigid. Rather, we introduced the concept of an Intertwined system linking each other different functions within the Poljan territory. We planned an open and liquid system with fuzzy borders. Our aim is to accumulate density of functions getting free spaces for public uses. We see Poljan as an advanced district of services (tertiary, cultural, dwelling) immerse in a large green area, a real model of eco-friendly neighbourhood.

In the Watergate and corner city areas this system bring us to carefully mapping the zones and their possible interactions through geo-urban connectors, peatonals, infrastructures and complex "green" corridors synonymous of multidimensional fluidity.
This appears to be a complex system that tries to keep in touch with every parts. An open, auto regenerating and auto recalibrating system.
This system is condensed into the RAINBOW TOWER, a new magnify landmark, a symbol of the laws of nature, a living symbol, full of life in its interior and perceived from the exterior as the new hopeful skyline of Ljubljana. It is called RAINBOW, for different reasons: first of all, the rainbow as event driven by nature which we must learn from. Its Ephemeral, Light and Prismatic aspect reminds us to the sense of the solar energy and how is possible to take advantage of, and what kind of relation exist with sight and chromatics.
Then we have The Bow, like an architectural stereotype linked to a symbolism that over the time has changed.
Starting from representative, celebrative,

View from the ground level garden

nt view from the commercial "plaza"

View from the upper level platfor

contemplative, anti functionalist (like arcs the trionphe, obelisques), to a functionalist structural system; This bow has been recalled here as a new concept of condenser in terms of social, political, functional and ENERGY. It is an accumulator, a store and an attractor. In this new concept, all the old symbolisms are mixed together and are turned into a new one. It is a Unique and singular Element but at the same time is horizontally and vertically multiplied. The new concept of the tower as a container goes toward a richer concept of structural element alive which is now attractor and distributor of energy at the same time. A new eco-symbol for Slovenian democracy in which with more FRESHNESS, it will be possible to start the holistic mechanisms, proper of diversify and complex Contemporary era.

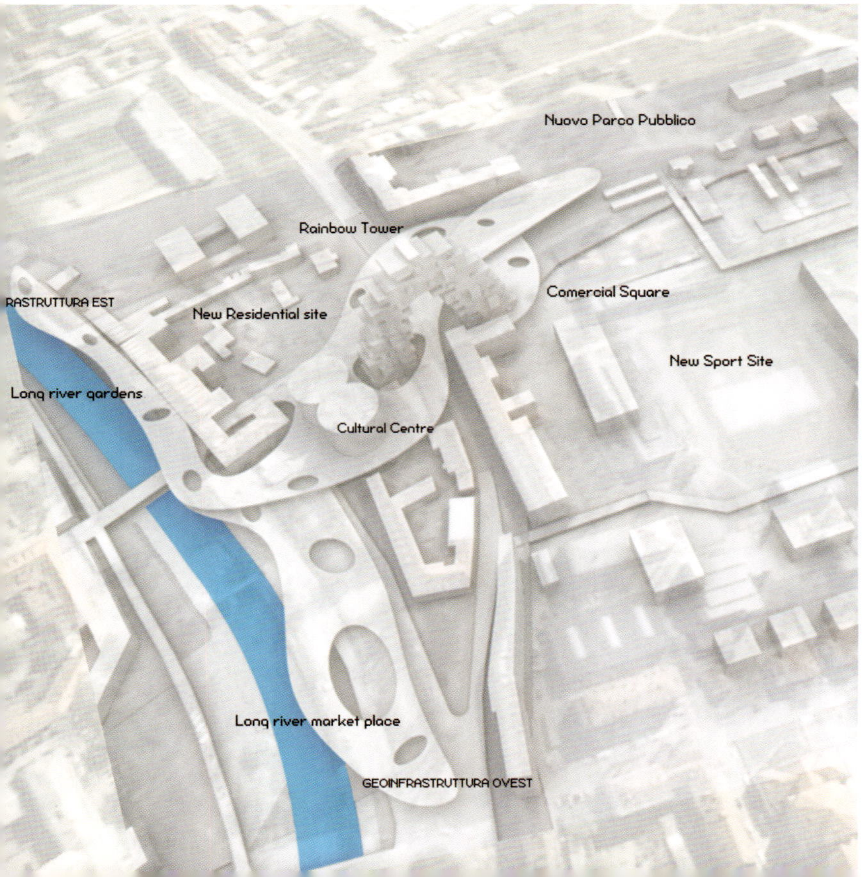

RAINBOW BUILDING

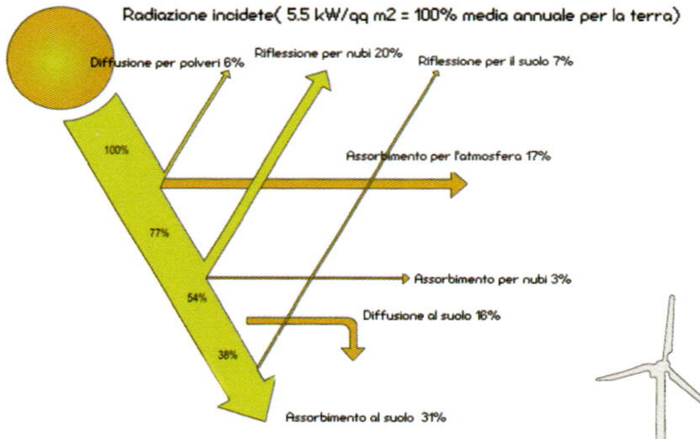

Assorbimento e diffusione solare in funzione del cromatismo

Tecnologia fotovoltaica:

Modulos Fv doppio vetro Vidursol
1,6 x 2,6 m 5mm
..possiilità di serigrafiare
.possibilità del cromatismo fotovoltaico
.vetro translucido secondo il grado di protezione solare
.differenti cellule mono o policristalline. www.vidursolar.es

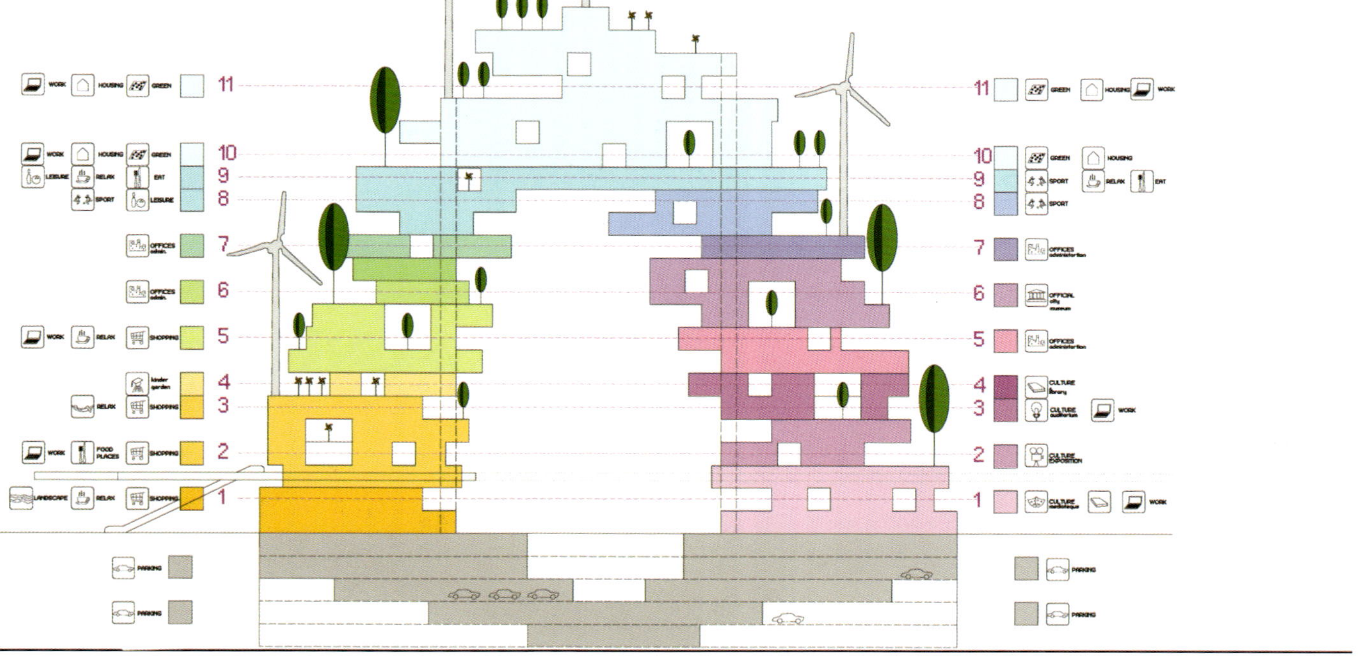

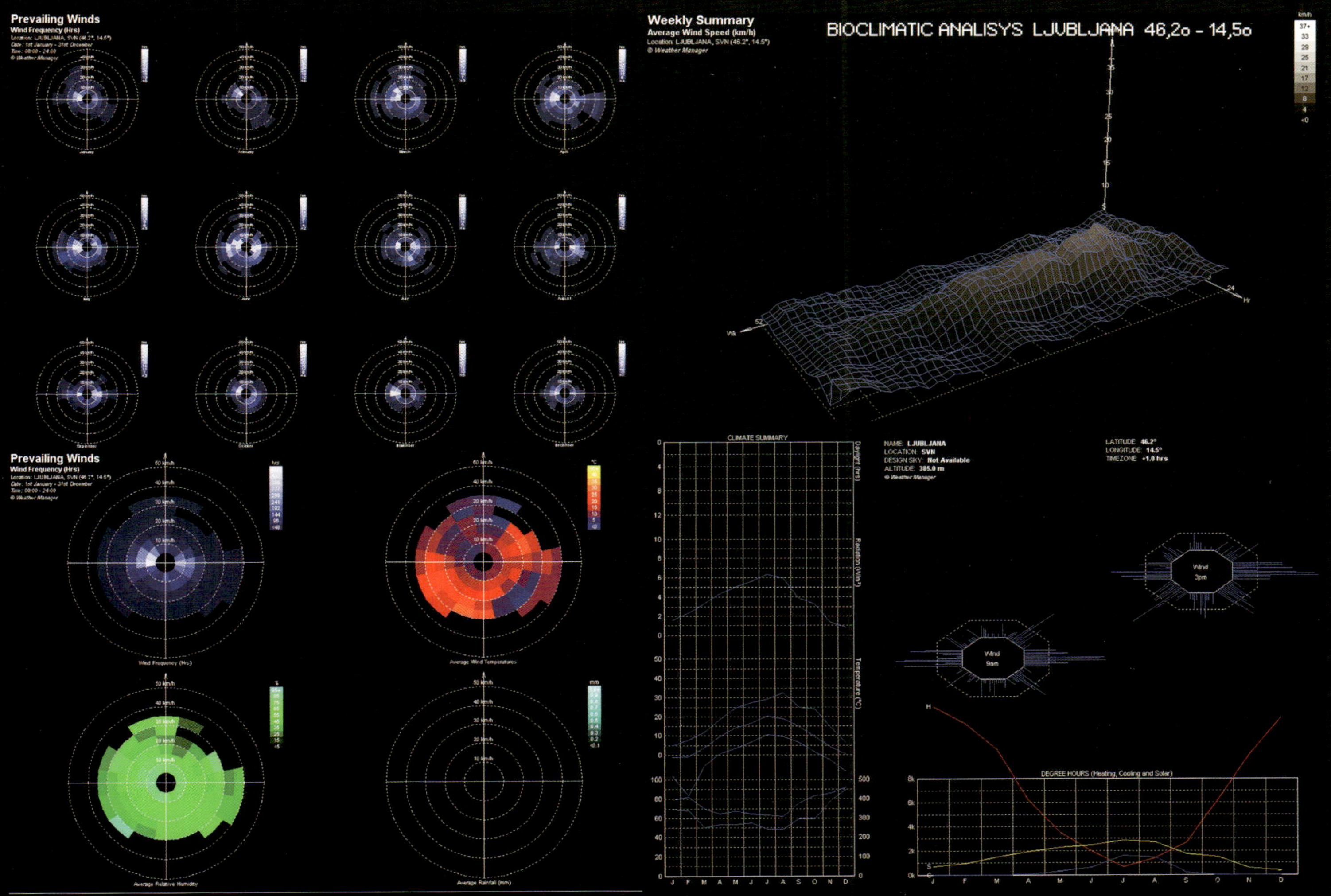

Sustainab.italy - Contemporary ecologies

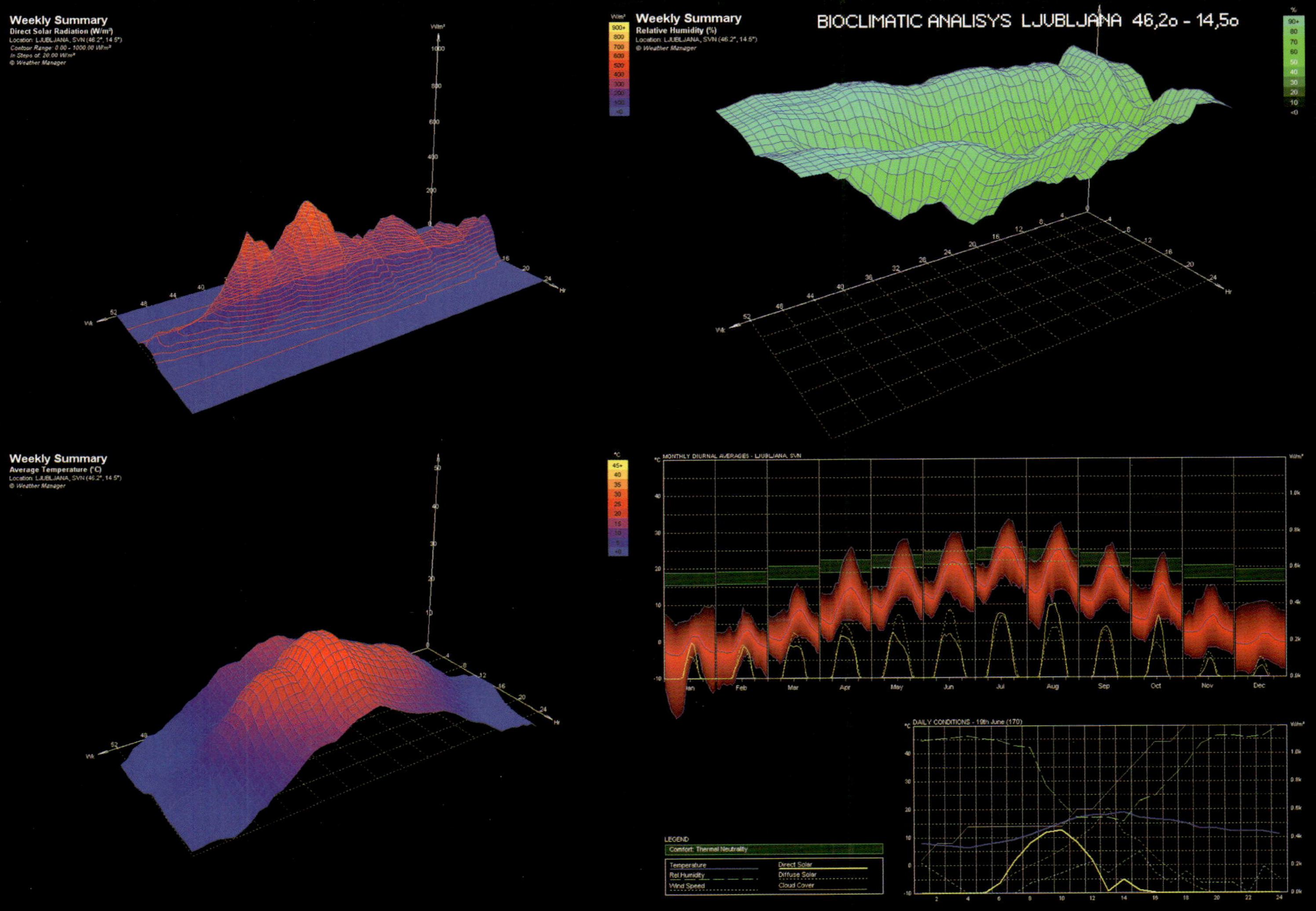

LJUBLJANA INTERTWINED SYSTEMS

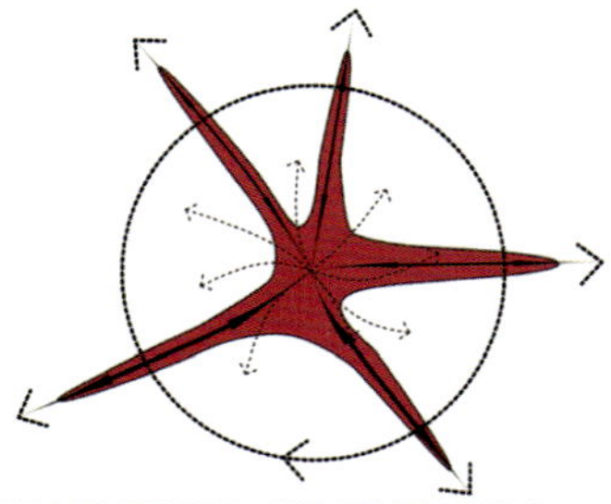

TRASVERSAL CENTRIFUGITY

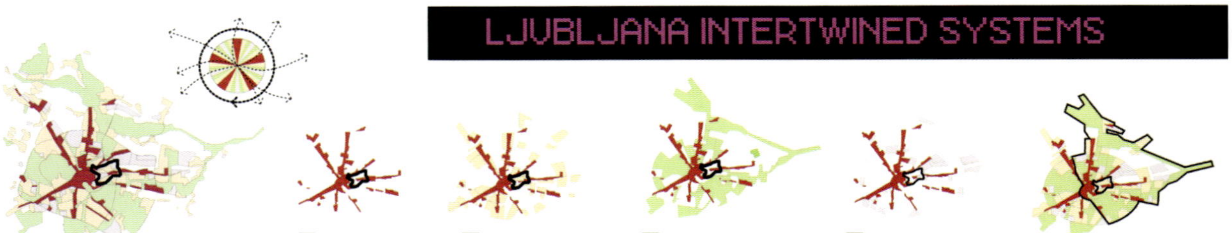

city centre-mixed use area | residential areas | green areas (sport parks, parks, recreation areas) | education and healthcare areas

Ljubljana seems to be a city developing along longitudinal and radial corridors by alternate cords of functions. Its STAR shape works like an hearth beating, receivng inputs and releasing outputs. Generating a tension between the interior factors and exteriors.

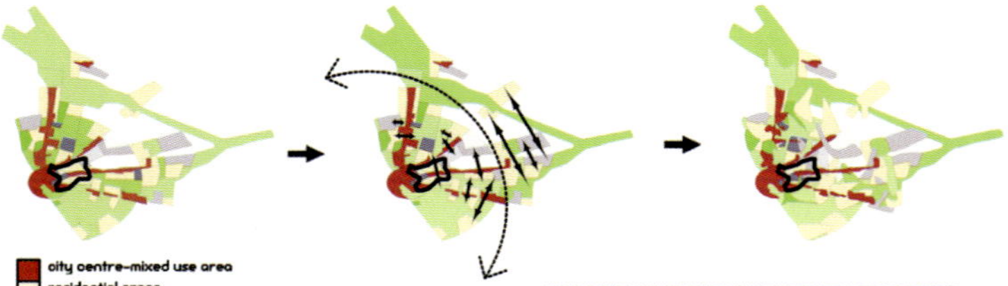

- city centre-mixed use area
- residential areas
- green areas (sport parks, parks, recreation areas)
- education and healthcare areas comercial center and lesure areas manifacturing and special use areas

COULD THE CENTRIFUGE TRANSVERALITY BE A POSIBLE STRENGHT TO lead off a complicity between funcional CORRIDORS and a desirable OVERLAPPING OF A MULTIPLE TRIDIMENSIONAL COMBINATION?

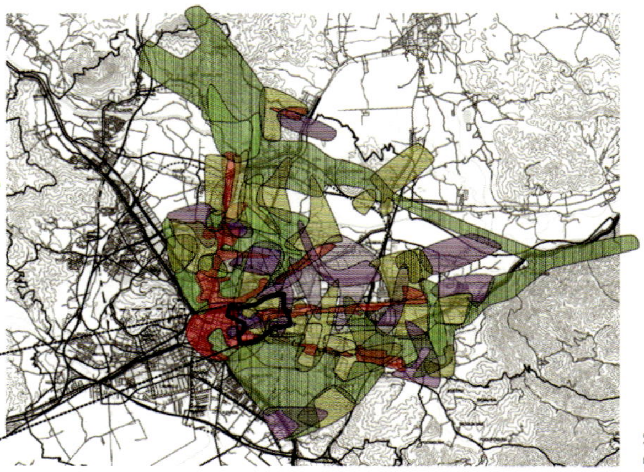

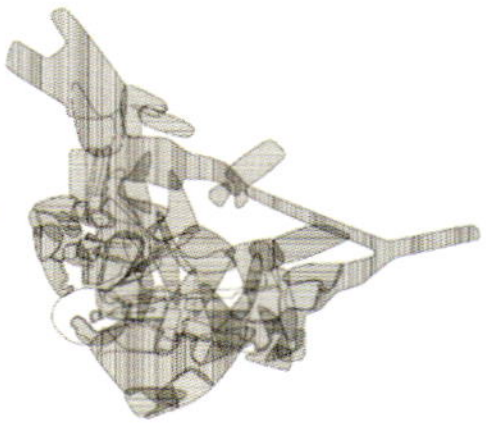

SUPERIMPOSITION OF MULTIPLE COMBINATION

LJUBLJANA HETEREOGENITY: The quality of being made of many different elements, forms, kinds, or individuals: diverseness, diversification, heterogeneousness, miscellaneousness, multifariousness, multiformity, multiplicity, variegation, variety, variousness.

NEEDS INTERTWINED SYSTEMS to be related

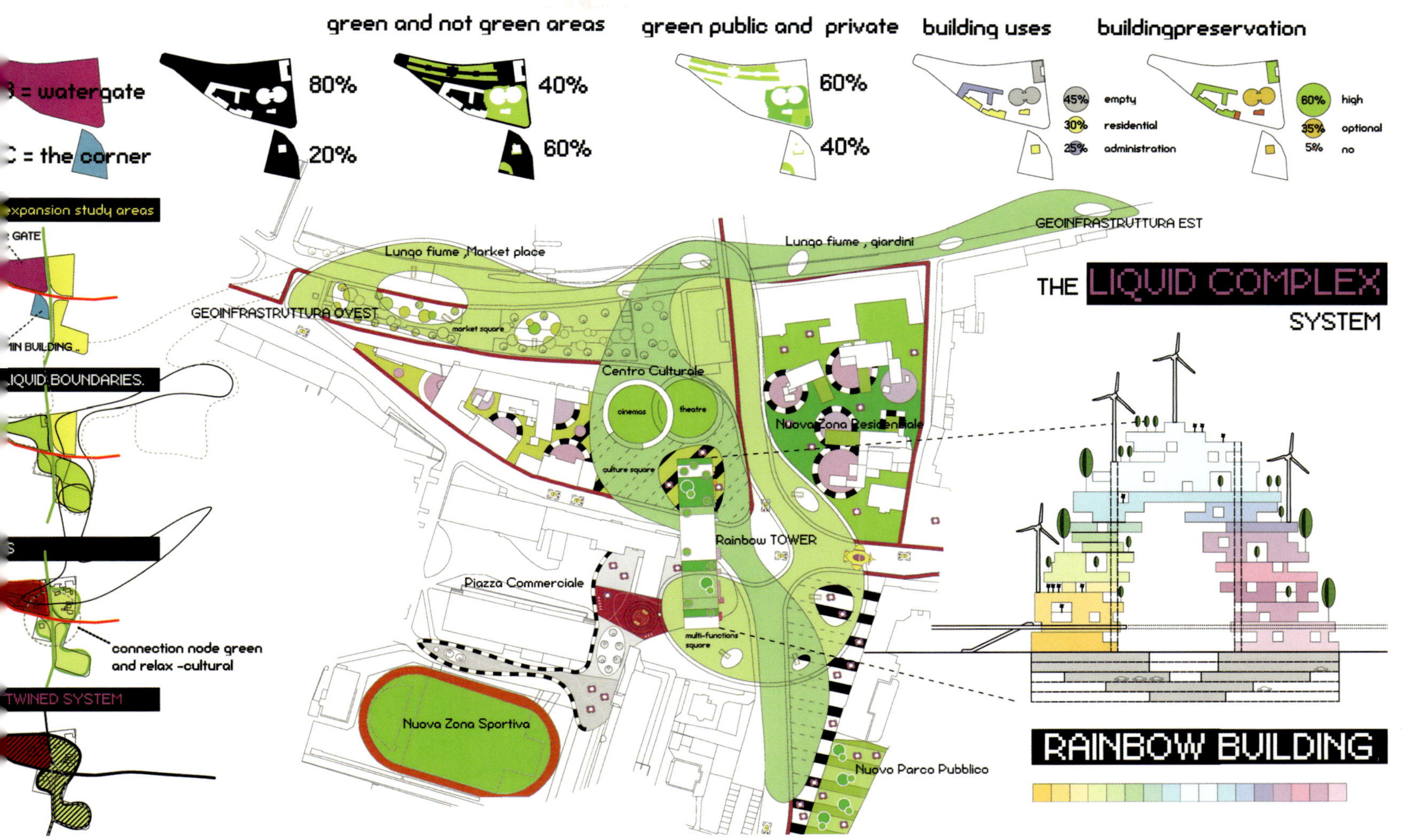
green and not green areas
green public and private
building uses
buildingpreservation
= watergate
= the corner
80%
20%
40%
60%
60%
40%
45% empty
30% residential
25% administration
60% high
35% optional
5% no
xpansion study areas
GATE
IN BUILDING
IQUID BOUNDARIES.
connection node green
and relax -cultural
TWINED SYSTEM
Lungo fiume ,Market place
Lungo fiume , giardini
GEOINFRASTRUTTURA EST
GEOINFRASTRUTTURA OVEST
market square
Centro Culturale
cinemas
theatre
culture square
Nuova Zona Residenziale
Rainbow TOWER
Piazza Commerciale
multi-functions square
Nuova Zona Sportiva
Nuovo Parco Pubblico
THE LIQUID COMPLEX
SYSTEM
RAINBOW BUILDING

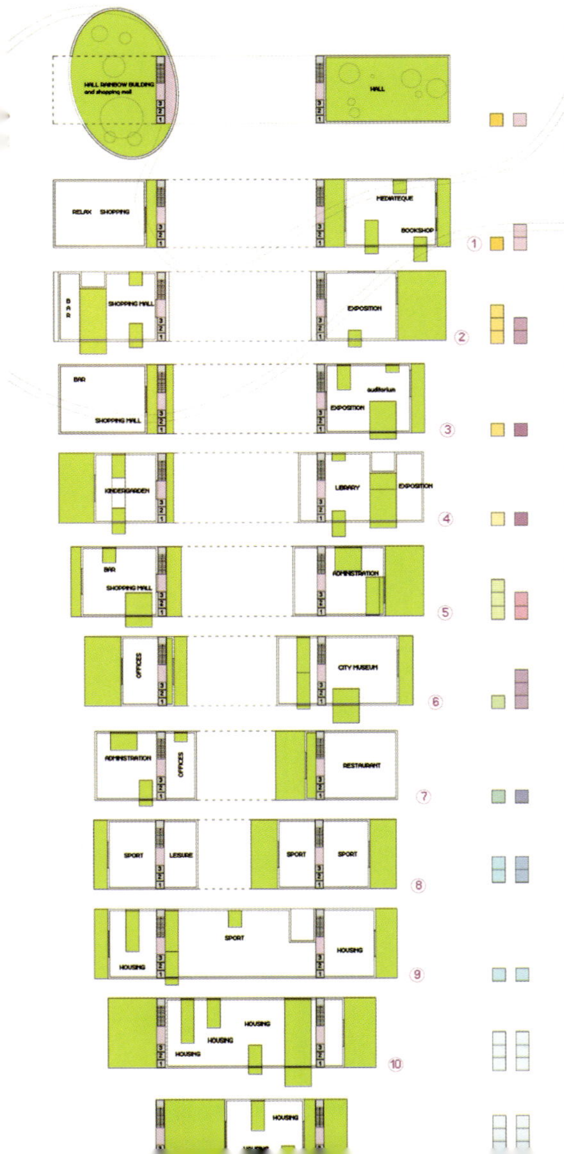

MODEL OF RAINBOW TOWER

Design: LOT-EK, Ada Tolla, Giuseppe Lignano, Principals;Keisuke Nibe, Koki Hashimoto, Judith Tse, Project Architects;

The Special Container Building in the South of SANLITUN

三里屯南面特色集装箱建筑

A new retail complex, with master plan by Hong Kong based The Oval Partnership, is organized like a medieval village with a dense fabric of narrow alleys, low-rise buildings, elevated walkways and bridges connecting all levels. LOT-EK is assigned the north-east section of the "village" to be dedicated to retail, restaurants and event spaces. Other architecture firms commissioned to develop the other buildings include KKAA/Kengo Kuma & Associates, SHoP Architects and Sako Architects. LOT-EK concept is centered on the definition of the internal alleys as multi-level, open-air circulation channels. In each alley, a rhythmic system of scaffolding-like metal frames is wedged between the buildings, adapting to the varying width of the alley's cross-section. The scaffolding-like frames are connected along the side of the buildings by a random system of horizontal metal rods that function as railing and brise-soleil, defining a loggia on the upper levels and generating a tunnel like perspective within the alleys.
The rhythm of the structure is based on the width of ISO shipping containers (8 feet) which are inserted randomly into the facades of the building and jut out into the alleys. At the ground level, the containers function as canopies that hover over the retail stores entrances and house display or other small functions on the interior. At the upper floors the containers are pierced and skewered by the horizontal circulation functioning as entrances to the retail stores and as display windows along the loggias.
The scaffolding-like structure extends the alleys out towards the main street of the Sanlitun area to lure in the passers-by. At every level the containers function as large three-dimensional graphic objects layered with signage and logos. Orange mesh, also pierced by shipping containers, wraps the external perimeter of the entire north-east section adding privacy and sun refraction along the outer façades.

人行小道
Sanlitun South, Pedestrian Alley

Design Team: Andres Remy, Flavia Bellani, Marcos Pozzo, Paula Mancini, Laura Rodriguez Segat. Leandra Rodriguez Llebana.

The Waterfall House

建筑与水的灵动组合——瀑住宅

The waterfall house was conceived by the request of a young business man in a closed neighborhood. The particularities and requisites of the client, along with the lot's characteristics, established the path to follow in the design process.

The borders of the small lot we are referring to, seem to disappear in the unmistakable visual: the lake.

The surrounding houses, the sun's path and the visual were needed to understand the emplacement.

At the back of the lot, at the ground level the lake has a shy presence. So it was considered to raise up to get a better look at the lake. This is the reason why it was decided to place the social area in the upper level, conquering the panoramic views to the lake during the day.

It is there where the water shows up. Unexpected at that height, it embraces the architecture and paints it with reflections, bringing life and movement.

Appearing and disappearing to the eyes of the user, relaxing with it's sound, dazzling with reflects, going through a path that stimulates the senses.

The shallow pool expands the house limits, making them infinite. It erases the line between natural and artificial, inside and outside.

The elevated water fall, as a waterfall, is over the swimming pool in the lower floor, refreshing the ambient and providing privacy at the same time. The water fall also filters the northern light makes it less harmful.

The waterfall is the icon of this house; this is why it was after it.

About the functionality of the house, we designed flexible spaces, so it could be adapted as much as to the socially active life of our client, as to the needs of a future family. The flexibility allows isolating the private areas in the lower floor from the public areas in the first floor and part of the lower floor.

As a way to connect the street life with the inside of the house, was designed a mixed use room, used as a garage, playroom or to host parties.

The vertical circulation in triple height baths the inside with cenital lights, joining the different levels.

The volumetric game outside and inside vanishes at the back, opening the spaces to the lake.

The balance between full and empty tries to avoid the adverse effects big windows may cause to thermal isolation.

In contrast, the front presents a closed design holding the service areas and giving privacy to the house.

映衬在瀑布下的起居室
View to Livingroom and Waterfall

从临街的车库门中望去
View from the Street-garage Door Open

从游泳池望向露台
View to the Pool from Terrace

夜晚，瀑布格外迷人
Waterfall at Night

潺潺的瀑布水流
Water Falling

日落与水面的优雅组合
Sunset and Water

二楼的起居室
Living Room First Floor

位于二楼的壁炉
Fireplace First Floor

厨房
Kitchen

掩映在瀑布下的多功能空间
Mixed-use Room Beneath Waterfall Lower Floor

从套房望向窗外低处的景致
Suite-View to Cieling Window Lower Floor

二楼的入口
View to the Entry from First Floor

鱼缸沿着楼梯延伸到顶部
Fish Tank on First Floor from Stairs With Cenital Light

入口和楼梯延伸出的玻璃桥
View to the Entry and Glass Bridge from the Stair

Project Team: Giuseppina Forte, Joao Sobral, Erika Trabucco with Emanuela Valle

The Women's Health Centre

布基纳法索女性健康中心

The CBF [Centre pour le Bien-être des Femmes] Women's Health Centre in Burkina Faso was created between 2005 and 2007 by AIDOS, an Italian NGO fighting for Women's Rights in Developing Countries.

The AIDOS project, financed by the Democratici di Sinistra Political Party and with a contribution from the European Commission, is just one of the group's international programs focused on contrasting the diffusion of Female Genital Mutilation [FGM].

The social/health-services program developed by AIDOS, together with its local partners was focused on providing the educational services, information and awareness about women's sexual and reproductive rights in Sector 27 of Ouagadougou, a peripheral urban area settled by the rural population.

The social program called for the realization of a building complex capable of hosting a variety of activities in very harsh circumstances. The architectural project represents the response to this condition.

Completed in 15 months by a local builder, under the direct supervision of FARE studio, the CBF is functionally and cost-effective answer to the needs expressed by AIDOS, while simultaneously and primarily representing a centre of aggregation and identity for the entire local community.

The technological and typological responses offered by the project, on par with its social programs, represents an innovative approach to traditional local building practices, presented as the natural formal expression of the changes and new approaches promoted by AIDOS.

The project privileges an integrated approach to interactions between built space and climatic-environmental conditions [building orientation and layout, control of resource consumption, the use of natural vegetation, the selection of building technologies], based on considerations of sustainability and appropriateness.

The project is based on the separation of the primary activities performed by the CBF into two distinct, though closely related buildings: a Training Centre dedicated to activities of awareness-building and the administration and management of the CBF, and a Consultancy Centre, used for medical visits, legal assistance and psychological counseling.

The two main buildings are set atop a single structural element: a raised platform that creates a true artificial plane that supports various buildings used for different purposes. The raising of the platform above the ground ensures hygienic/climatic conditions that are extraneous to local culture and practices of building [protection against dust, mud and humidity].

The two main buildings are protected against rainfall and, above all, direct sunlight, by a light, waterproof PVC recyclable velarium, supported by an independent structure of steel "trees". This sloping tarpaulin is part of a system that collects and stores rainwater, which is used to irrigate the garden.

The volumes that contain the various rooms are independent of the roof structure and freely

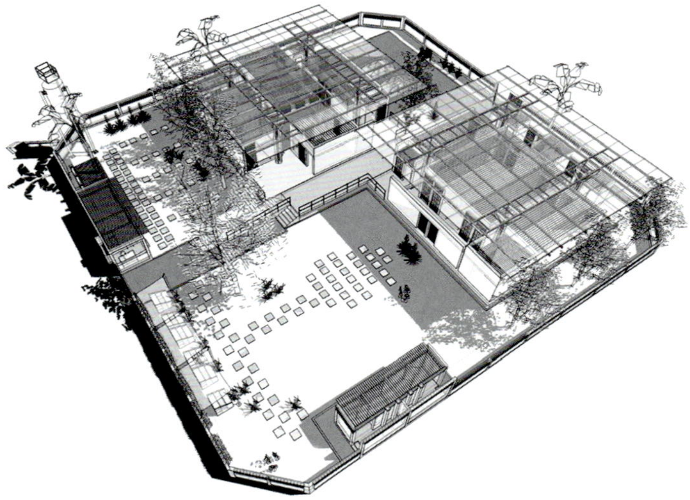

花园
Garden (Photographer: Sheila Mckinnon)

placed atop the platform. They are articulated around a series of shaded and ventilated patios that ensure privacy from the exterior. The modular configuration of the structure allows for future expansion, preserving the general framework of the building.
The building walls are constructed using compressed dry stacked clay bricks, BTC [briques en terre comprimée], made on site using a rough mixture of earth, cement and water. The bricks were baked in the hot sun, with no energy consumption, thus limiting the environmental impact of the material. The choice to use these bricks is based on their temperature and humidity reduction characteristics, enhanced here by their protection against contact with water, perhaps the only serious limitation they pose.
The choice of using this technology represents the desire to introduce alternative and sustainable technologies within a context that is tied to standardized, though not always optimal building practices, and to the importance of importing foreign materials.
The buildings are covered by corrugated aluminum and translucent decking, which allows light to filter into the interior, reducing the need for artificial illumination.
The space between the steel roof and the velarium, the open cavity underneath the platform, together with the exterior openings fitted with operable glass fins, help to improve the natural ventilation of interior spaces, drastically reducing the need for mechanical air conditioning.
The exterior space, similar to the interior, is designed as an open area to be used by the entire community. It is a space of sharing and of information, used to present the themes dealt with by the CBF. Indirect and informal communication is also favored by the organization of small events and public discussions. The garden is a micro environment that surrounds the buildings taking advantage of the shade provided by the building and trees and the humidity produced by the plants. A layer of grass covers the earth, reducing the effects of erosion, while various species from Western and Sub-Saharan Africa have been planted with the twofold intention of creating shade and promoting the return of autochthonous vegetation.
Temperature control, perhaps the most significant climatic issue, has strongly influenced the overall design. The adopted strategy includes also:

- carefully studied [beneficial] building orientation, reducing the effect of hot wind and taking advantage of mutual over shading
- the shading of heavy material against direct exposure to the sun
- extensive use of operable windows
- the separation of enclosed areas by transitional spaces, such as verandas or patios.

Being water and power are not available in the area; the Centre is fully independent, integrating the systematic control of consumption and the self-production of resources, possibly renewable.
Water is provided by a newly drilled and dedicated well.
In order to achieve independence from external sources, photovoltaic cells have been installed along the perimeter wall, reducing the use of the electrical generator.
All this affects both personal behavior and collective responsibility: elimination of mechanical air conditioning [limited to medical rooms in order to assure filtered air] is probably the major achievement of the project in terms of environmental sustainability.
The outside walls, with no openings, are finished with a plaster coat, painted with bright colors. The local NGO's slogan, translated in 5 languages, completes the decoration of the walls. Each wall becomes a canvas, presenting the social message in an informal way.
Internal finishes such as the colorful mosaic tiling have been realized with the aim of combining a pleasing appearance with ease of construction, using locally available materials in a different, playful way.

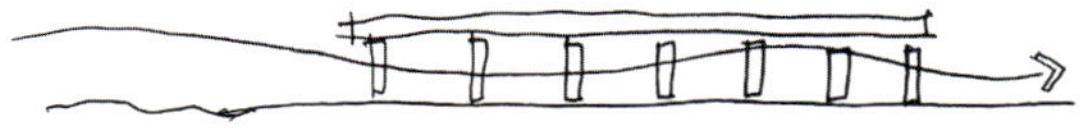

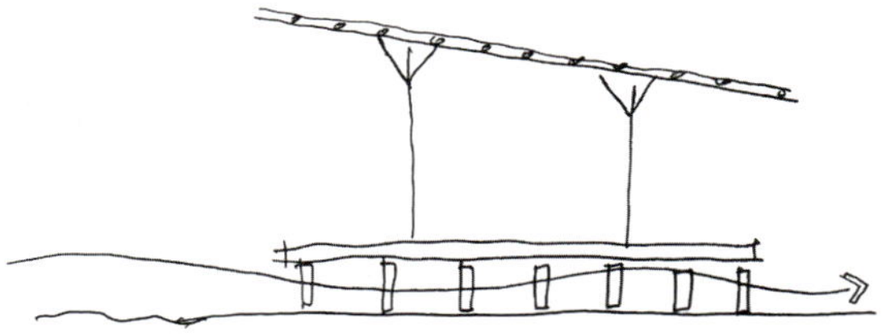

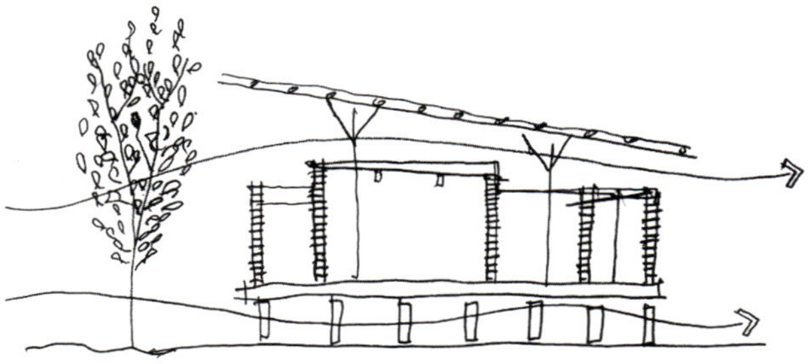

培训室
Training Hall

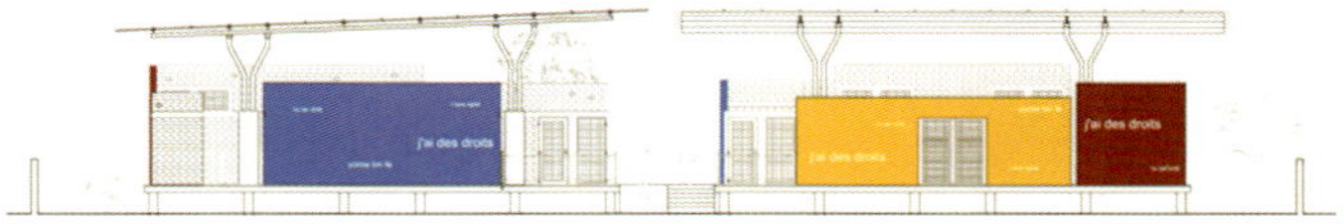

West elevation

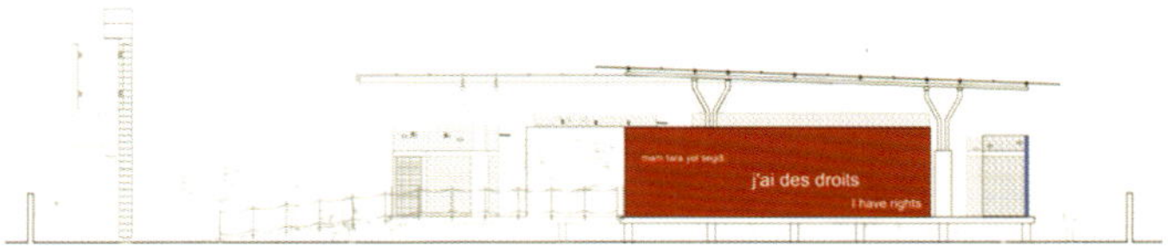

South elevation

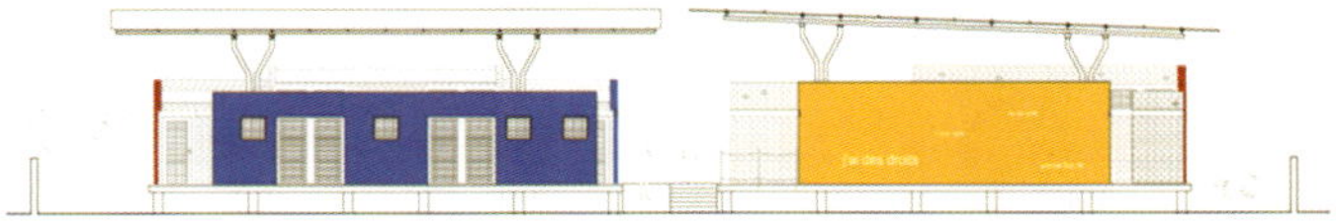

East elevation

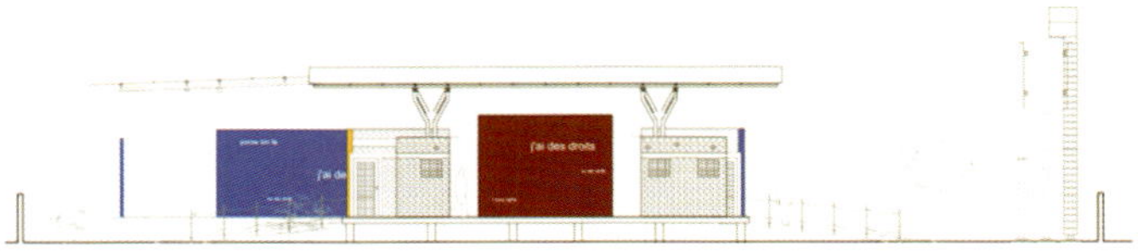

North elevation

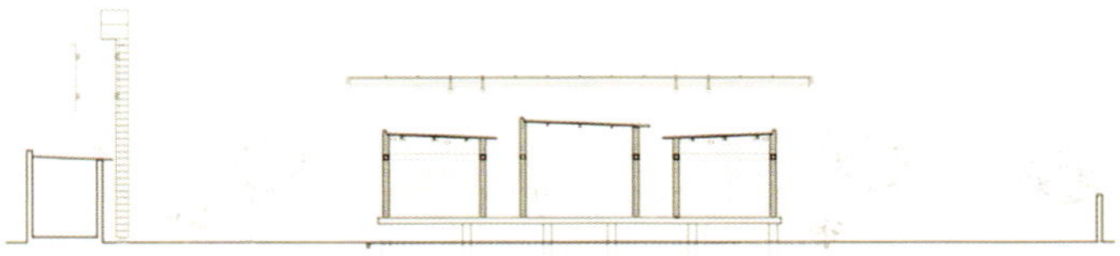

AA section

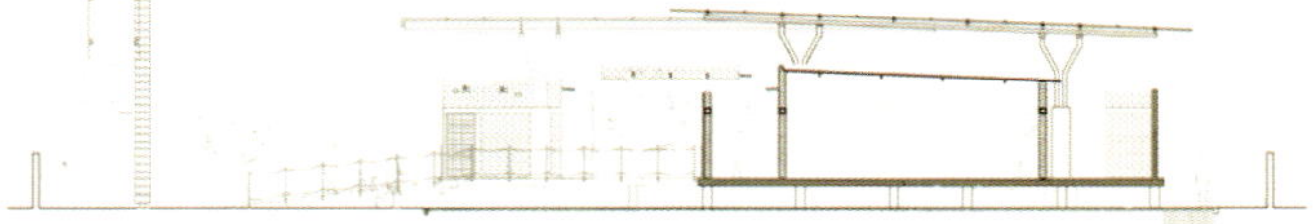

BB Section

The Yoga Studio
瑜珈室

CLIENT
The owners, both practicing Buddhists and avid modernists, were interested in creating a weekend retreat from Washington, DC in the Shenandoah Valley. The natural setting with distant views and nearby rock outcropping were maintained for enjoyment while meditating. The clients' conviction for energy efficiency and minimal waste or consumption in producing, delivering, installing, and maintaining materials and products inspired a sustainable project. The team of client and architect agreed on the importance of local craftsmanship in nurturing local culture -- the other half of sustainability. The project, designed by Carter + Burton Architecture, is part of the LEED for Homes Pilot program and received Gold certification. Southface was the LEED for Homes provider for the project.

DESIGN
County restrictions on size for a Studio in Clarke County warranted a design which feels big for spaciousness, views and light while maximizing efficiency. An organic shape felt right which stretches out to take in as much southern light as possible while still being compact. This outbuilding fits with the site while maintaining a modern purity of form and space rarely seen in this rural setting. The structure's circulation features doorways at each end. A bermed entrance to the east and the western end sits high with a deck on the view side feeling like a tree house. The walls and ceiling curve to provide an energy efficient and site responsive design which respects the allotted space requirements. The simplicity of the structure required an attention to detail and materials that are achieved with the customization of most elements in the project.

SITE
The mountainside site proved challenging for design and construction. The chosen site rests behind a craggy stone ridge on a 5 acre lot an hour west of Washington DC. The placement of the Studio 100 feet from the clients' Main House allows for some independence while staying connected enough for shared functions and systems, reducing the overall impact on the site. Care was taken in the design of the landscape around the house to allow for the forest to return to its natural state as quickly as possible.

SYSTEMS
A geoexchange system provides efficient space heating and cooling and all of the Studio's hot water needs. The system consists of a ground-coupled heat transfer loop (geoexchanger) connected to a liquid-to-liquid heat pump and a liquid-to-air heat pump. The loop employs vertical and horizontal ground tubing runs, and is sized to heat and cool both the Studio and the Main House.
Heat pumps are limited in their ability to raise and maintain temperature in domestic hot water storage tanks. The mechanical engineer was concerned this risks growth of biological organisms

北立面的开放面较少，双层的钢铁边缘能够为建筑很好地防护恶劣的天气的影响
Less Openings and Durable Steel Siding Protects the North Side of the Studio from Storms

绿色屋顶可以采集65%的降水
The Green Roof Collects about 65% Rainwater

东入口和西面的平台都比较小，衬托出起居室的宽敞
The Eastern Entrance and Western Deck are Small so the Main Living Space Feels Larger Inside

瑜珈室西南面的道路
South-western Approach from the Rocky Outcropping Below

从南边望去：瑜珈室坐落在优美的山景中
Southern View of the Yoga Studio Nestled Into the Hillside

in the tank. For this reason, a thermal storage system (TSS) using an aqueous heat transfer media (HTM) was employed.

The liquid-to-air heat pump provides space heating and cooling via a forced-air system. A heat recovery ventilator (HRV), along with high-purity air filtrationan air filter ensures the indoor air is kept exceptionally clean.

MATERIALS

This project tested the limits of construction technologies for the area and its tradesmen. The following items created unique spaces, details and functional attributes:

The design team considered S.I.P.S. for many reasons including structural strength with less material, high R-value, low air infiltration, and use of recycled materials in the OSB and expanded polystyrene bead foam. All these factors helped to pilot a unique design strategy for the Yoga Studio, where the client can now lead others in yogic practices in a material and energy efficient space.

The project also employed many environmentally preferable products, as recommended by LEED, to satisfy the clients' desire for a "healthy" interior environment and sustainable materials. These materials include:

Poplar boards from a sawmill 3miles away were used to board-form the concrete foundation walls. Soy oil was used as a natural release agent before the boards were air dried and planed for reuse as flooring and curved wall panels inside the Studio. No VOC-laden carpet or paint was used in the house.

Stainless steel ceiling panels were used indoors and on the porch ceilings for indirect reflected lighting and an airy feel. Colored glass tile lines the bathroom walls and ceiling while a radiant terrazzo ground concrete floor with cast in place floor lights covers the bathroom and entry hall areas which creates a larger feel. Custom steel handrail at the loft with resin and grass panels serves dual purpose as safety feature and an art element. Custom wood windows were fabricated 10 miles from the site by the builder and are low-e with a solar heat gain coefficient high enough to allow for some of the sun's heat to be absorbed by the building as a passive solar technique in the wintertime.

CULTURE

The simplicity of the diminutive structure required more attention to detail and materials to maximize the perception of a large space. There are some key features that are worth highlighting about the project:

After using the 22" wide poplar boards from a sawmill 3 miles away for formwork of the foundation, the builder and his team planed the wood for use inside the house as flooring and wall panels.

The builder and his team spent weeks in his shop constructing all the wood windows and doors.

The doors have custom steel handles sized to be comfortable for a hand grip with a narrow profile for the tight spaces.

Another key feature of the Studio is that it doubles as a bunkhouse. There are beds built into the floor diaphragm with trap doors hiding them during the day. Bunk beds built out of a steel frame with custom mattresses are located in the loft space.

A custom steel and wood ship's ladder saves floor space while allowing natural light to blend to all corners and doubles as access the loft and as a sculptural piece for the Studio space.

All the cabinetry in the Studio was built by local craftspeople.

Materials which are rooted to the region, such as the maintenance free galvanized corrugated metal siding creates a rural outbuilding feel. The grain of the wood extracted from the region is left exposed inside the Studio highlighting the sense of place.

All the subcontractors working on the site were from within 30 miles of the site. The benefits of utilizing local craftspeople as well as experimenting with and using environmentally responsible materials in inventive ways reinforces sustainable concepts and creates and enduring sense of place.

瑜珈室与环境相互融合
The Yoga Studio Fits Comfortably With the Site

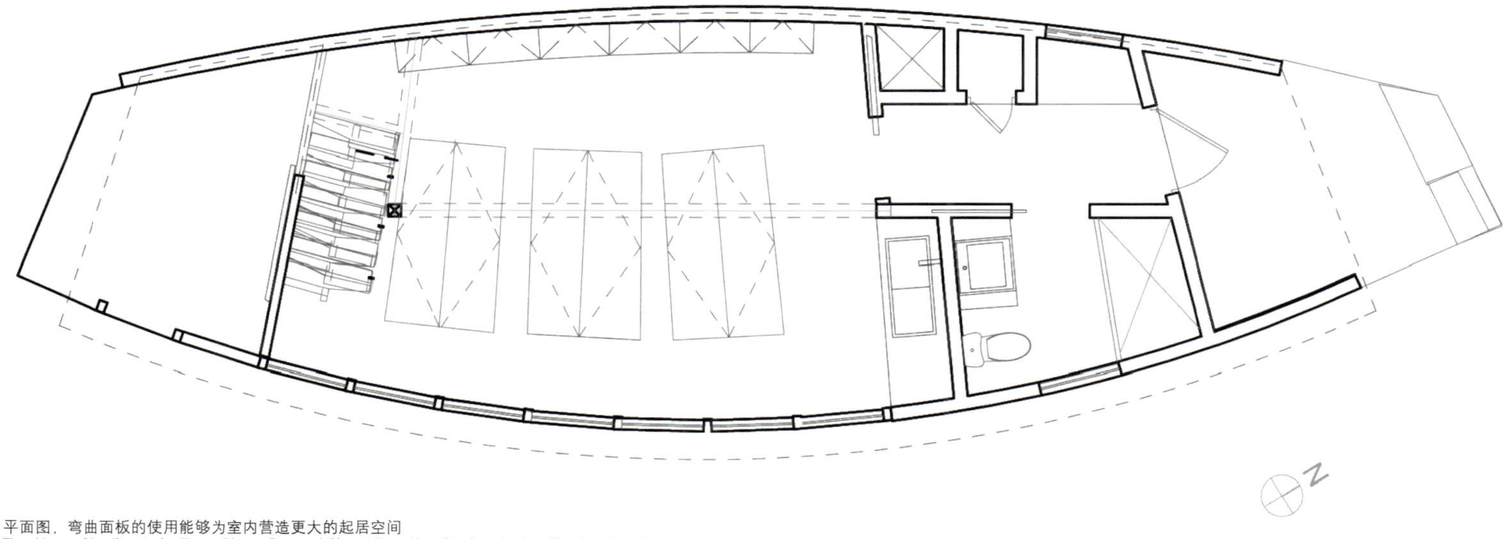

平面图，弯曲面板的使用能够为室内营造更大的起居空间
The Yoga Studio Main Floor Plan. Curved Sips Allow the Design to be Pinched at The Ends and Roomier for the Main Living Space

订制的长凳能够储存物品，楼梯也很好地节省了空间和材料
Custom Carbon Fiber Benches Support the In-floor Storage Covers and a Custom Ship-ladder Saves Space and Material

从顶楼往下看，地板的活门开启时的床铺
View of Open In-floor Bunk from Above

室内景致成为了窗外自然风景的背景
The Interior was to be a Simple Backdrop of Materials so the Art is the Natural View Out the Windows

窗子、地毯和门都是在附近手工制成的
The Windows, Cabinets and Doors were Hand-Built Less than 5 Miles from the Site

墙壁和天花板与光线的组合扩充了浴室的空间感
Blue Mosaic Tile on the Walls and Ceiling and In-floor Lights Make a Small Bathroom Feel Larger

室内的天花板和墙面的设计，LED灯光为顶楼提供较好的照明
Beeswax on Canvas Cover the End Walls and Recycled Poplar Boards (from the Concrete Forms) Cover the North Wall and the Floor. Led Lights Illuminate Under the Loft

Tokyo Matsuya Unity
东京松屋单元住宅

This design rediscovers the traditional techniques and decorative use of Japanese paper from the Edo period. The thick paper, originally imported from China, is patterned with ground oyster shell and mica powders and is commonly used to cover sliding doors. 'Tokyo Matsuya' is a well-established company that has continued to hand-make traditional fusuma paper and Japanese traditional paper, using traditional methods. The company has also been engraving wooden blocks since the Edo period and have been involved in wholesale publishing since that time.

At he lower level of the building, 'seeing, knowing, and buying' is communicated upon entering the show room, and at the higher level, 'living, using, and experiencing' is conveyed for multiple housing complexes. Japanese paper from the Edo period lined sliding doors and wall spiced up in modern taste. Sliding doors can be opened to refine to space. Shoji(paper sliding door) applied to each home. Natural light shines through.

The above also relates to hanging gardens, lighting, and ventilation at the middle level according to an old saying, 'a sense of homogeneity with nature is a characteristic of Japanese traditional buildings' where communal areas are linked and consolidated homogeneously through the architecture. Even in the interior it is possible to sense the light, wind, and natural scenery of the external spaces of this high-rise building.

White wall is painted with the Japanese traditional color haijiro(ash white). Even though the appearance is modern, Japanese motifs such as louvers reminiscent of the sudare(reed screen) are adopt.

正面局布图
Facade Detail

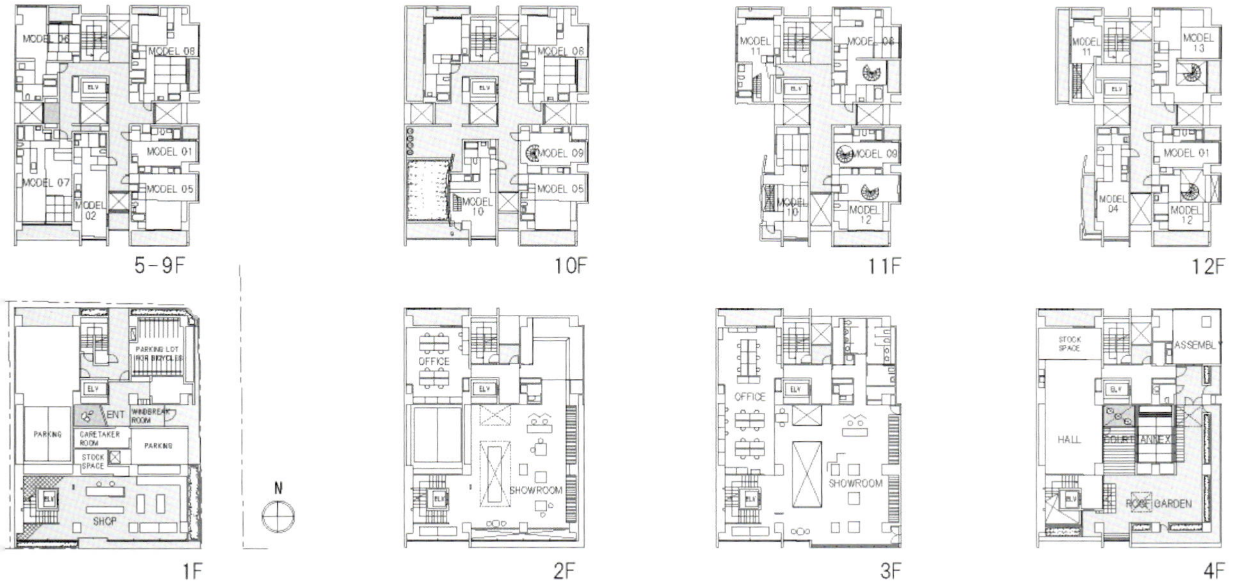

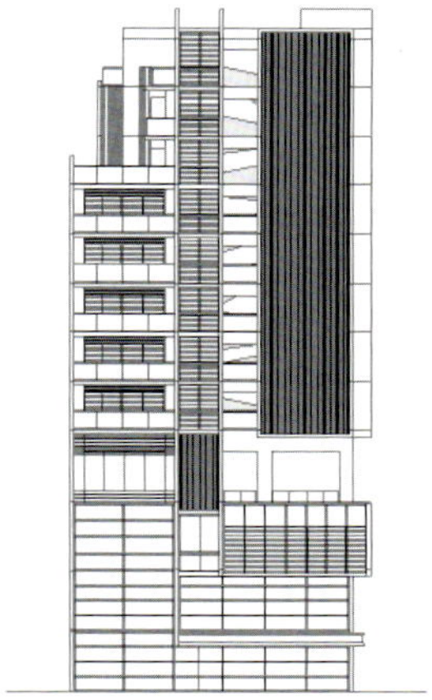

TOKYO MATSUYA UNITY

Principal Use: Shop Office Showroom Residence

Site Area: 452.97㎡

Bulding Area: 409.21㎡

Total Floor Area: 3703.28㎡

Stracture: Reinforced Concrete

Scale in Building: 12stories

COMPOSITION

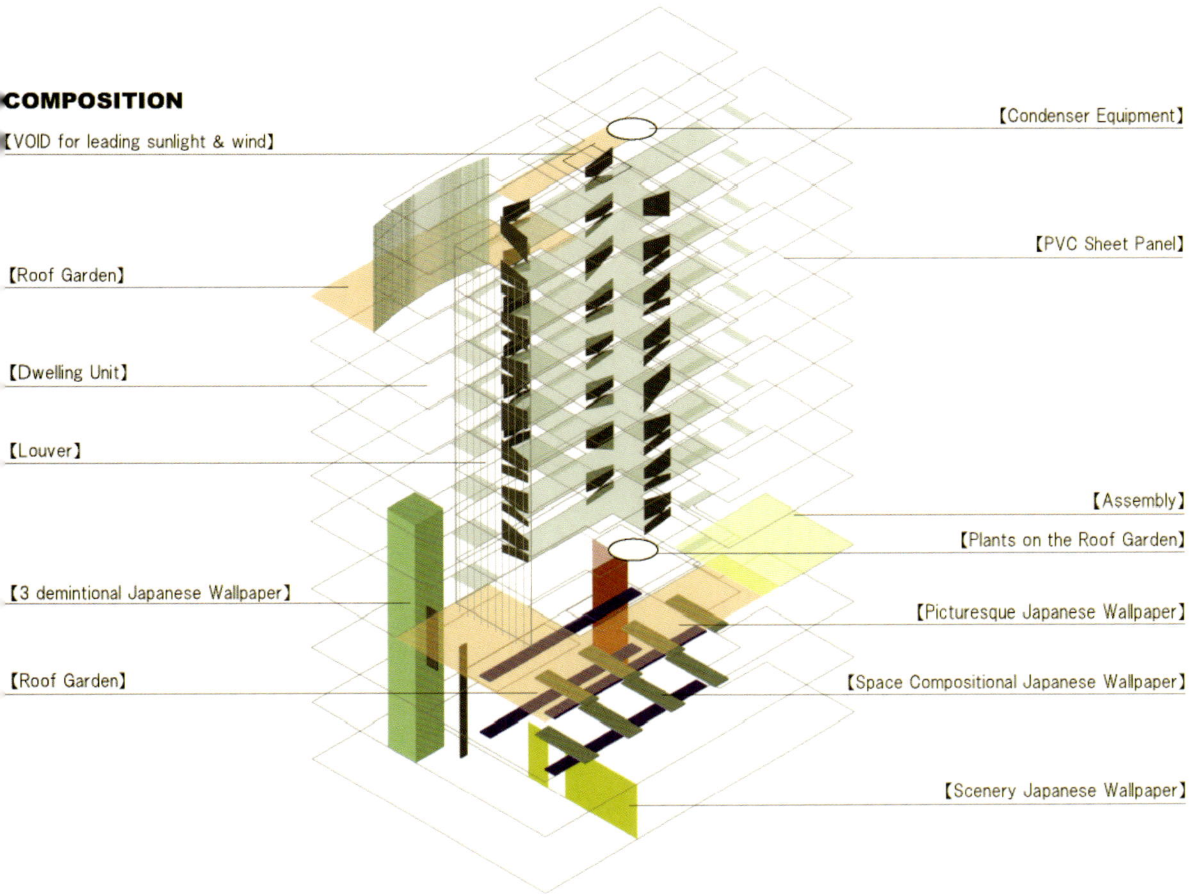

SECTION 1:400

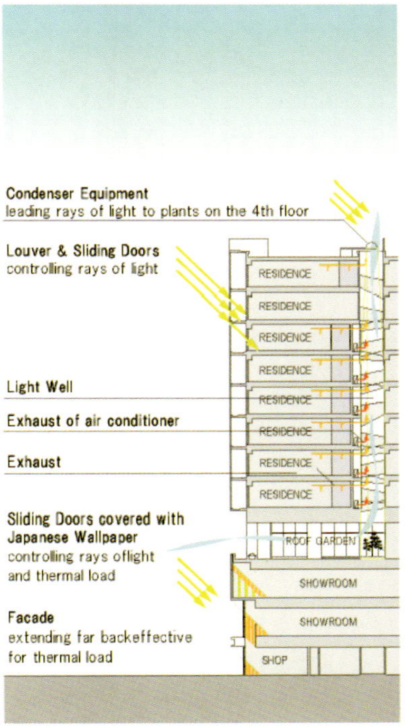

CONDENSER EQUIPMENT SIMULATION

▲ condenser equipment above the well-A ▲ well-A

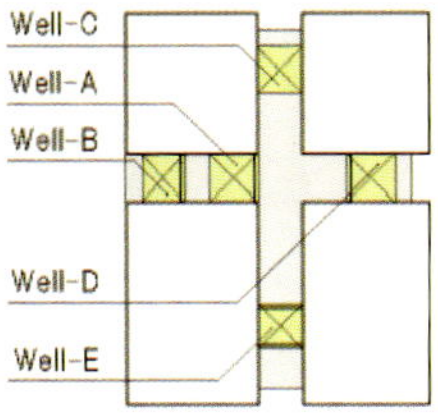

Well on the Residential Floor(5-12F)

Cross shaped corridor : Horizontal wind way Well (Void) : Vertical wind way

WIND SIMULATION

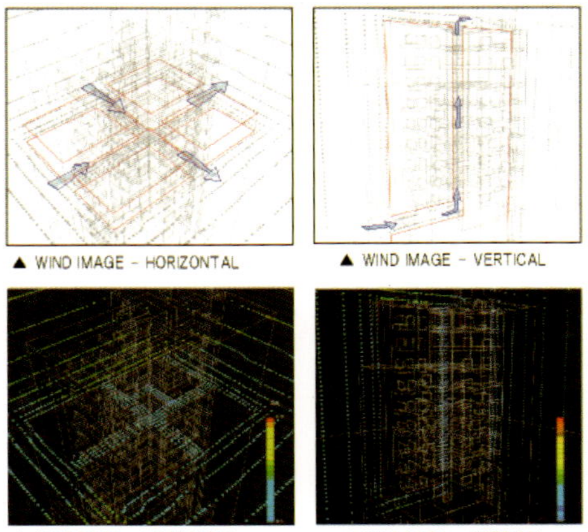

▲ WIND IMAGE - HORIZONTAL ▲ WIND IMAGE - VERTICAL

▲ WIND DATA - HORIZONTAL ▲ WIND DATA - VERTICAL

Wind blows horizontally in the cross-shaped corridor on the residencial floor. In the urban life, it is the place to feel refreshing air. Ascending current of the well makes exhaust out efficiently.

LIGHT SIMULATION

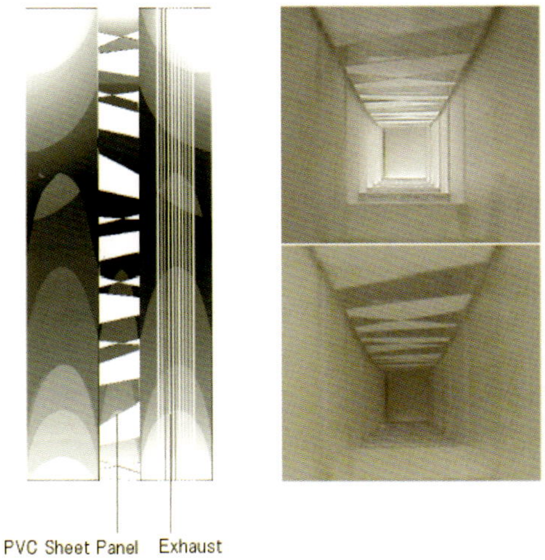

单元住宅内部装饰
Dwelling Unit Interior

单元住宅内部装饰
Dwelling Unit Interior

二楼展厅内部装饰
2F Showroom Interior

二楼展厅内部装饰
2F Showroom Interior

单元住宅内部装饰
Dwelling Unit Interior

单元住宅内部装饰
Dwelling Unit Interior

二楼展厅内部装饰
2F Showroom Interior

Project Team: Gianluca Evels, Stefania Papitto with Lorenza Giavarini, Matteo Rossetti, Luigi Valente

Touching Water
水上建筑

Kotka was born as harbour town for the wood and paper industry. The actual urban configuration has grown around this productive activity, with two main centres; Kotkansaari and Karhula. Nowadays the town's main character is still strictly connected with the sea and harbour activity.

The big emptiness in the competition's area gives the opportunity of designing a proper 'city's piece', on which we can express and realize a contemporary idea of living.

The project proposes a way of designing the city not as a succeeding or overlapping of single objects or architectural episodes, but instauring complex spatial and architectural relations according to a "systemic" proportion of the city. In this "systemic" idea, the varying architectural and landscaping elements integrate themselves in a proper idea of net, as alternative of a hierarchical and centripetal city or only made of architectural emergencies.

The project promotes the accessibility to any space and function, underlining the connections and free routes between the different parts and the rest of the city, and above all a new urban

interface between land and sea.
The entire project is concentrated on this dialectic concept's reality of the limit between land and water: the cost line is contemporarily the start and the end point of the city's territory.
It is the place where the limit of the vastness of the Nature and the sense of the anthropology's made up order imposes itself.
This city's part marks mediation between the nature's infinity and the basic human will of building 'inside' spaces.

THE EUROPEAN 9 THEME AND THE PROJECT STRUCTURE

In the competition's area the project researches a clear and precise formal and architectural connotation answering to the program's requests and to the desire of the town to create a special district with a marina on the shore of Hovinsaari following some programmatic essential lines.
The context as starting off
The project proposes a contest's reading that starts from the pre existent urban plot as that of the Cartesian plot of the central area quarters and that of east part of the competition site of the Polish Docks and the residential part of the west side.
According to the competition's program, the theme of the floating houses is met- not only with a pure fulfilment of the quantities and volumes proposed but building a proper marine village that evocates a feeling of infinity with the expansion to the water.

THE WATERFRONT

The waterfront theme of this part of the city is met with a continuous functional and spatial variation of the coastal line: on the west side, in front of the residential blocks, there will be a system of public spaces immersed in the green and made of small platforms and bridges that create a swimming area presenting small kiosks and parking areas fairly distributed.
The square in the urban west culture is generally the collective space where material culture and symbolic projections, ritual habits and social behaves all linked together. The presence of some green in the square declares the intention of recovering the sense of nature in a public space, very important in order to 'humanize' a pedestrian contemporary space.
The waterfront system ends with the marine with fixed and floating jetties for boats, services for pleasure boat owners etc.

THE 2 PRIMARY NATURAL ELEMENTS

Substantially the project is articulated in four main systems connected between them as an "architectural network". These systems are summed up by two great categories

WATER

Floating house system: a marine village on which the special house typology on the water articulated with bridges of reserved mooring jetties, allows to articulate a complex space strongly linked to the water element. The interstitial spaces between the floating houses work as water gardens that concern the residences the virtual perimeter of the small basins.
Polish dock system: The memory of the geometries and alignments of the existing harbour's facilities is confirmed by the presence of some blocks with strictly urban functions, like restaurants, café, shops and warehouses and receptive activities connected with the small harbour.

EARTH

Sliding urban system: the long residential stick architecturally articulates itself in a juxtaposition of autonomous blocks, some of which 'slide' compared to the edge of the theorical fa□ade, allowing in this way to articulate the long volume in a serious of smaller and more of a human scale episodes.
Patio housing system: the residential system of this small houses with patio-garden is constituted by the complex and vary aggregation of simple architectural units: the 6x3 volumes that articulate the house on two levels, the surrounding walls of the patio-gardens with full modular elements or realized with horizontal fillets, the wooden platforms of the gardens, the small street that enter the residences.
All the four systems are interconnected by a system of routes, streets and cycling and pedestrian paths and a system of public spaces that have the role of hinge-connection between the systems.
With the public spaces it has been thought a system of green areas as well, very specialized, that connect the existent green system too, either in the surroundings of the residential blocks or on a bigger scale, to the green areas of the territory and the city in order to improve the sense of ownership that this part of the city aims to have.
The project's idea is based on a technical-spatial organization able to welcome flexibility variants and the different needs of various clients, and able to offer a high standard architecture and liveability in a special context as the living on the water can offer.

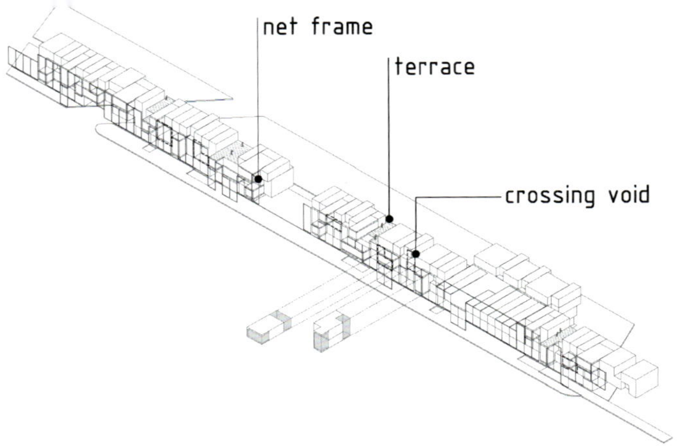

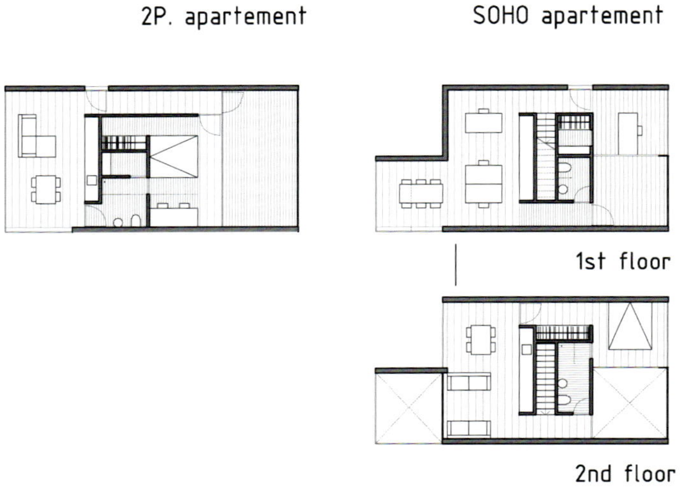

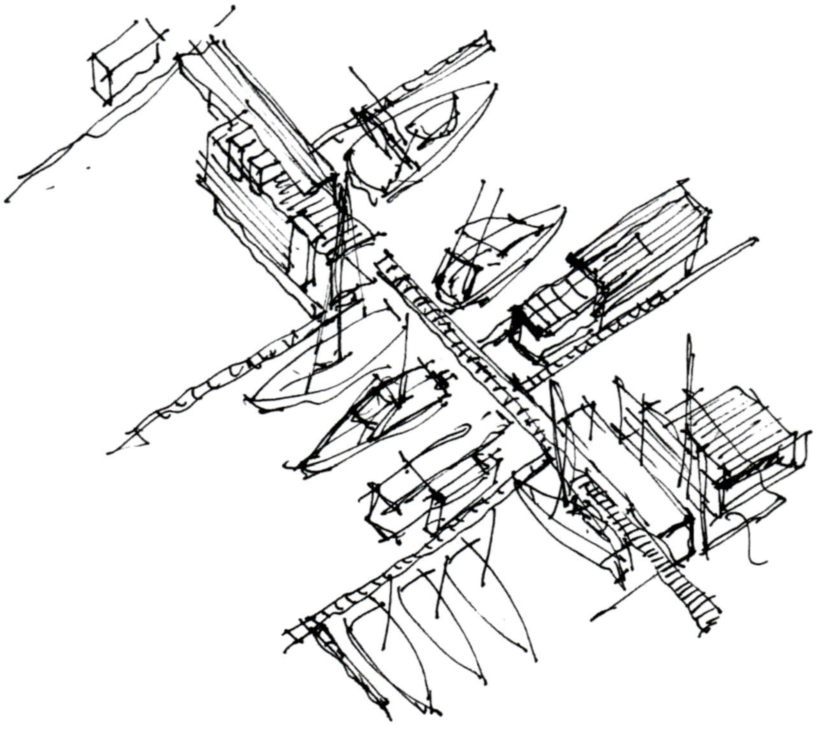

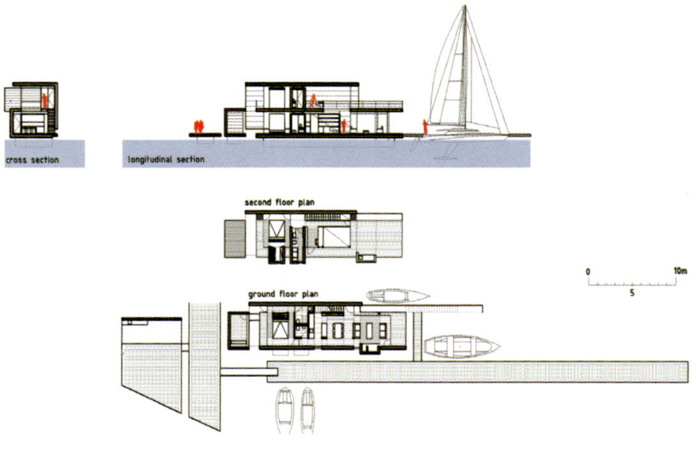

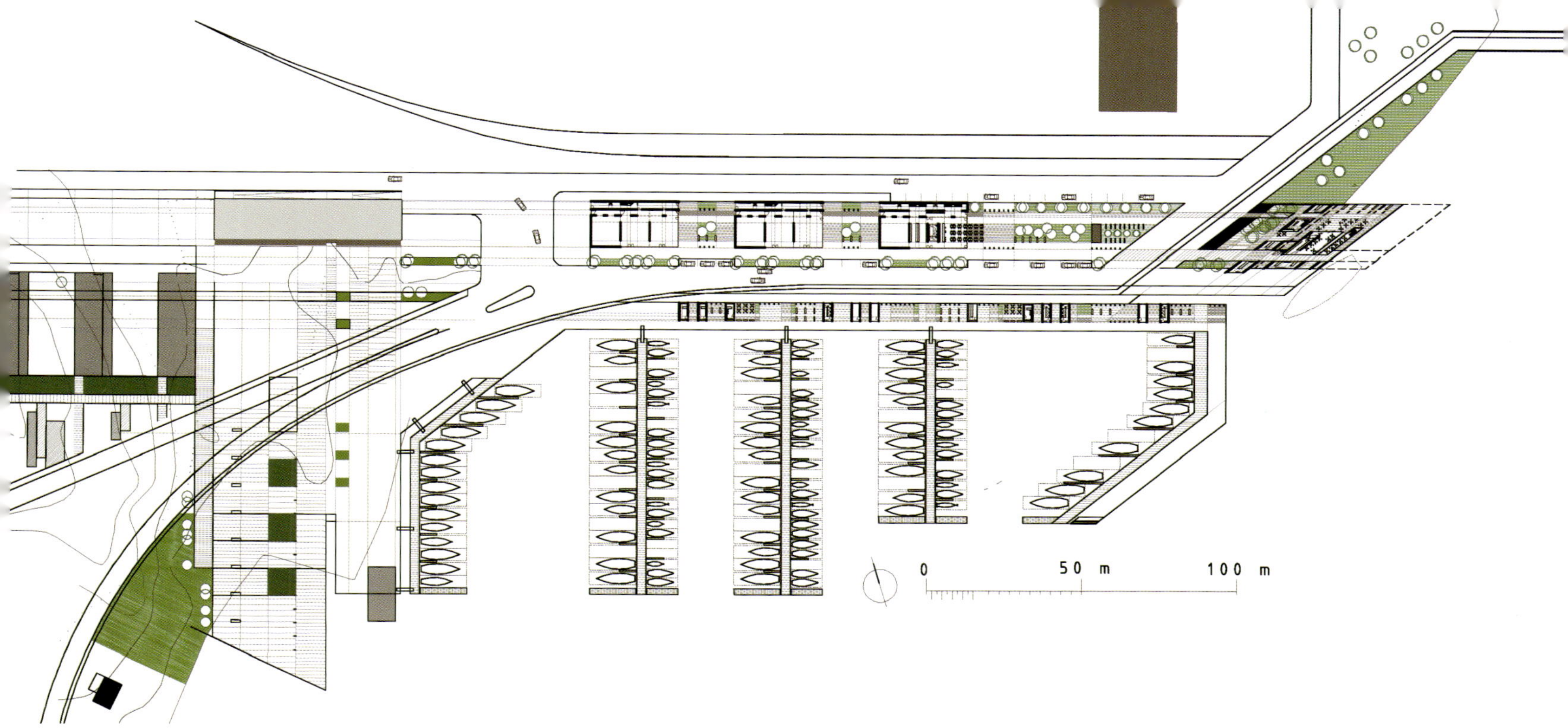
0
50 m
100 m

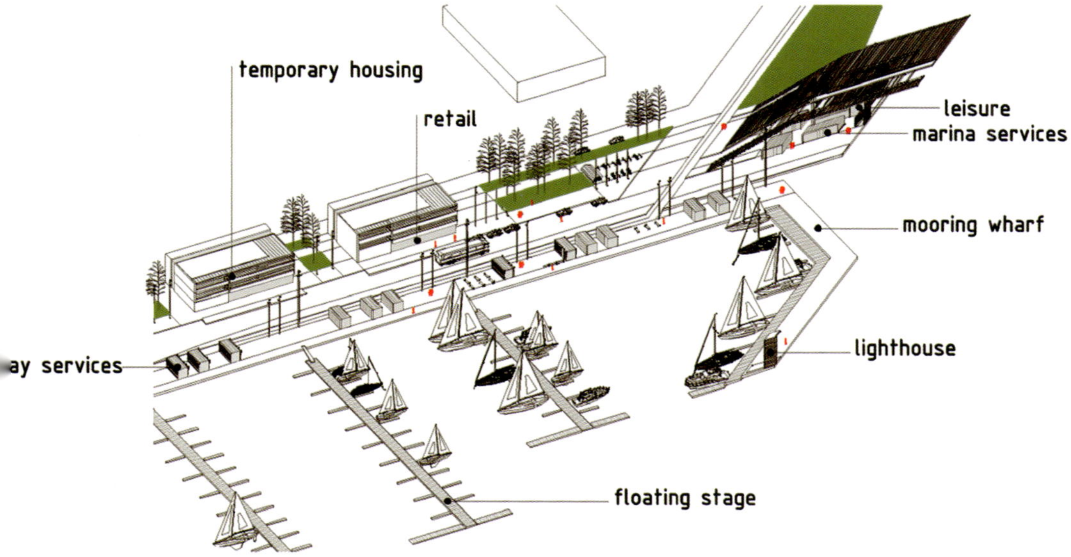
temporary housing
retail
leisure
marina services
mooring wharf
ay services
lighthouse
floating stage

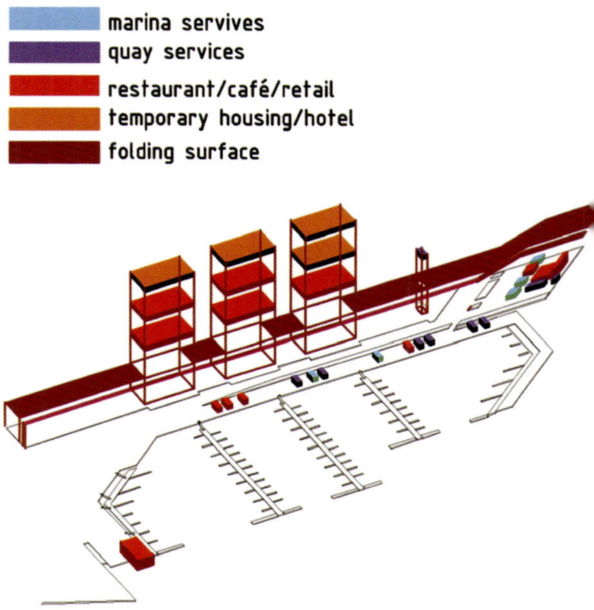
marina servives
quay services
restaurant/café/retail
temporary housing/hotel
folding surface

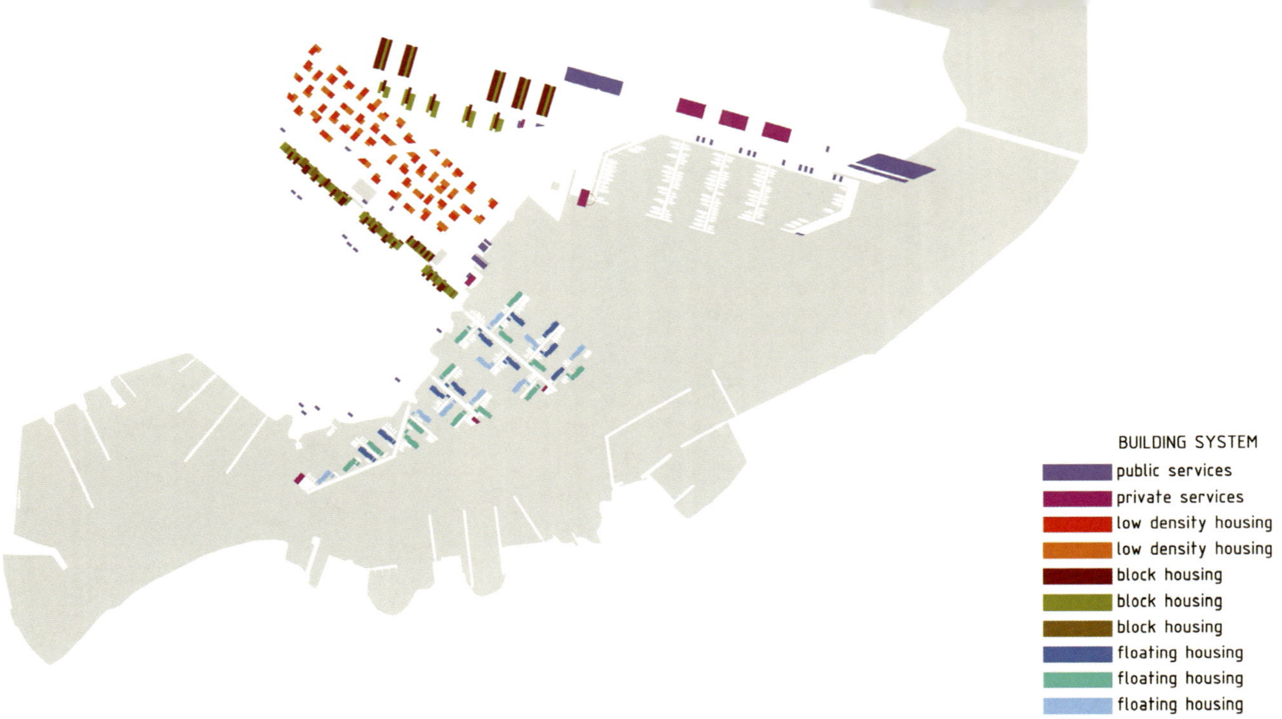
BUILDING SYSTEM
public services
private services
low density housing
low density housing
block housing
block housing
block housing
floating housing
floating housing
floating housing

GREEN SYSTEM
forest
park
urban park
residential green
playgroung areas
winter ice/summer water

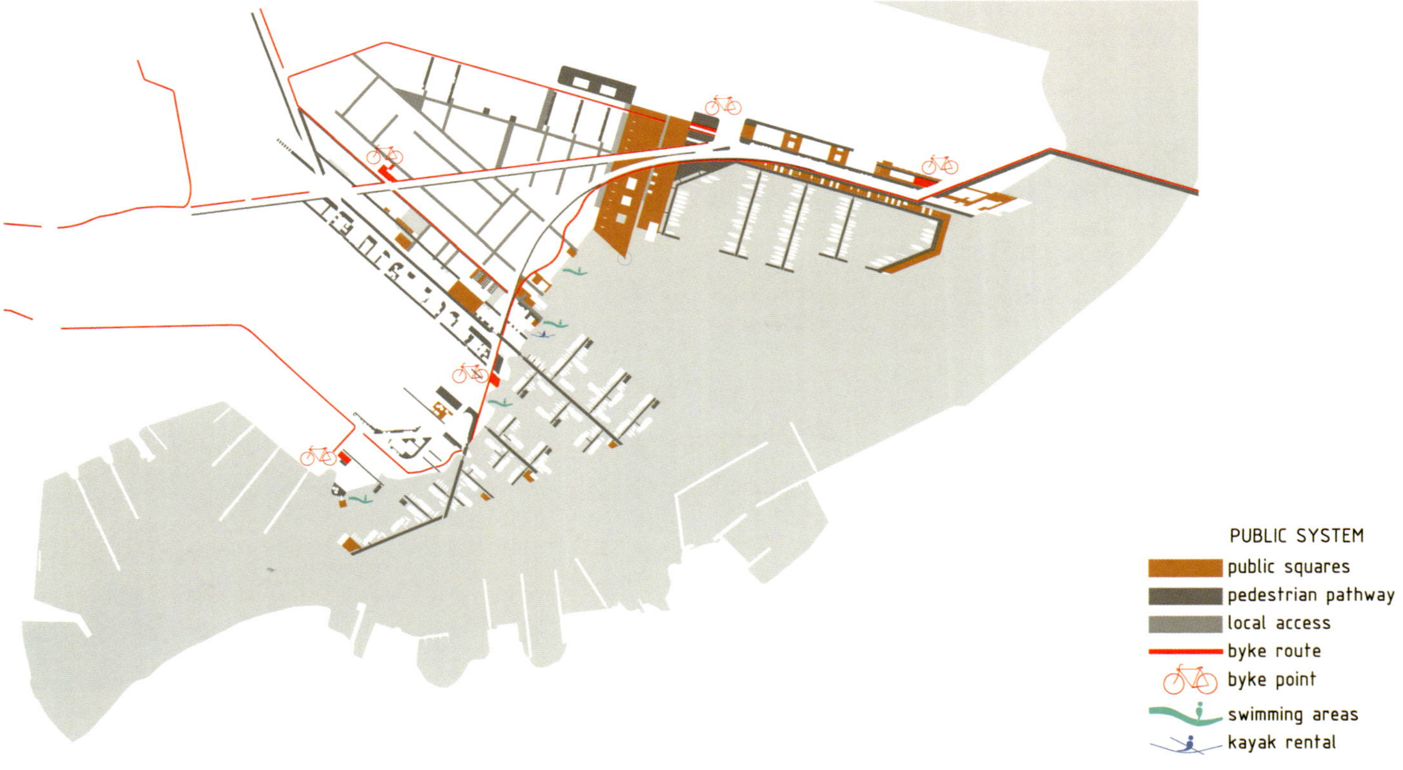
PUBLIC SYSTEM
public squares
pedestrian pathway
local access
byke route
byke point
swimming areas
kayak rental

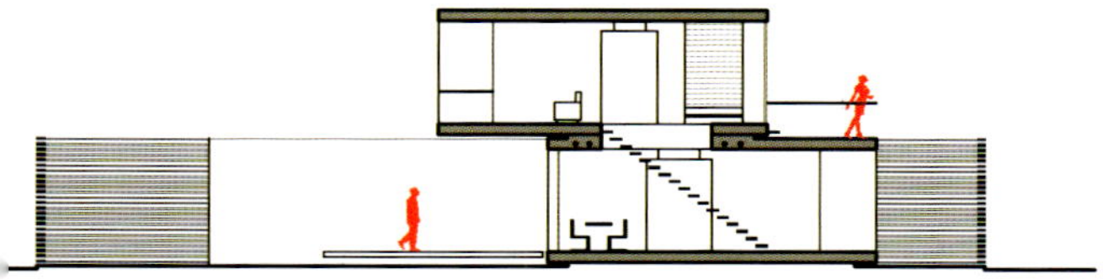
longitudinal section

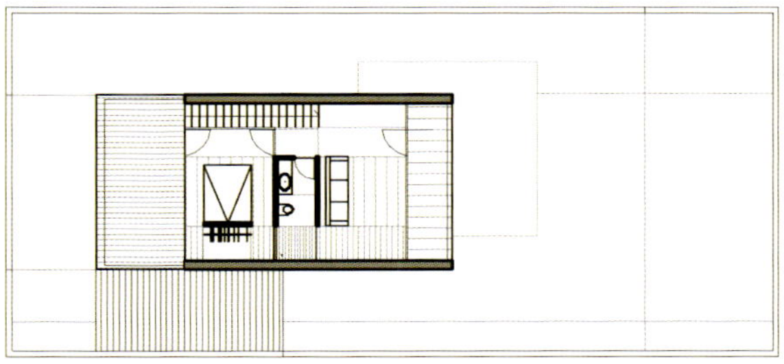
first floor plan

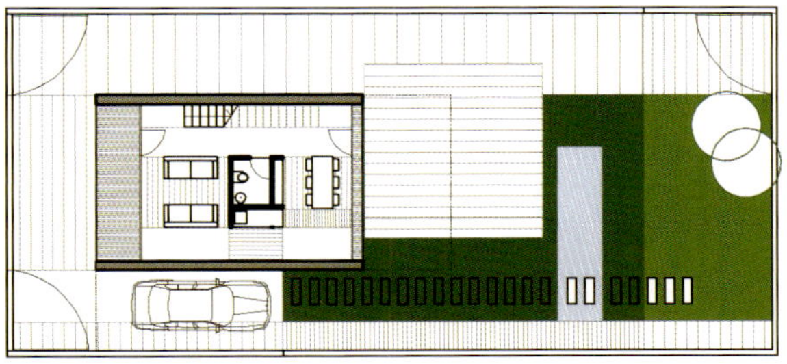
ground floor plan

0 5m 10m

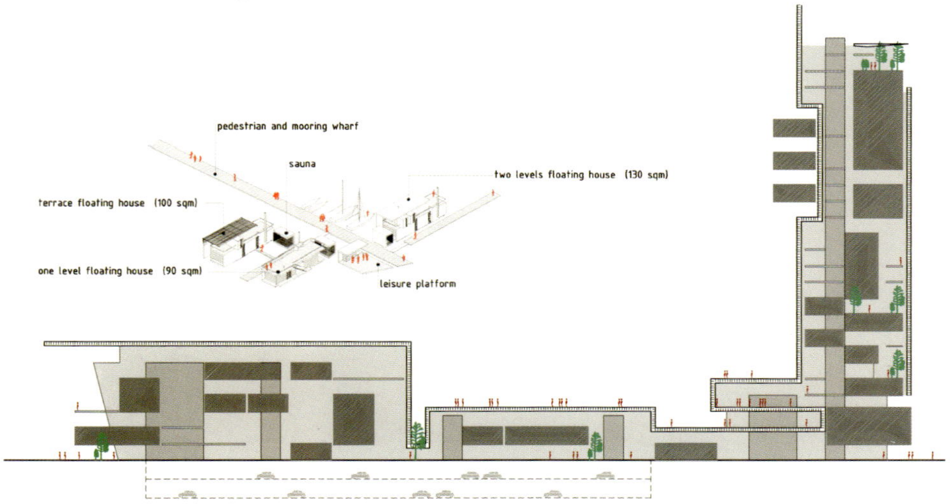

elevation

0 50 mt

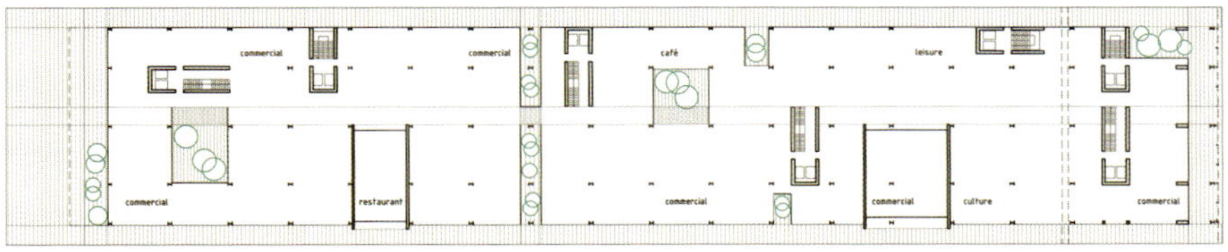

groundfloor plan

9th floor plan

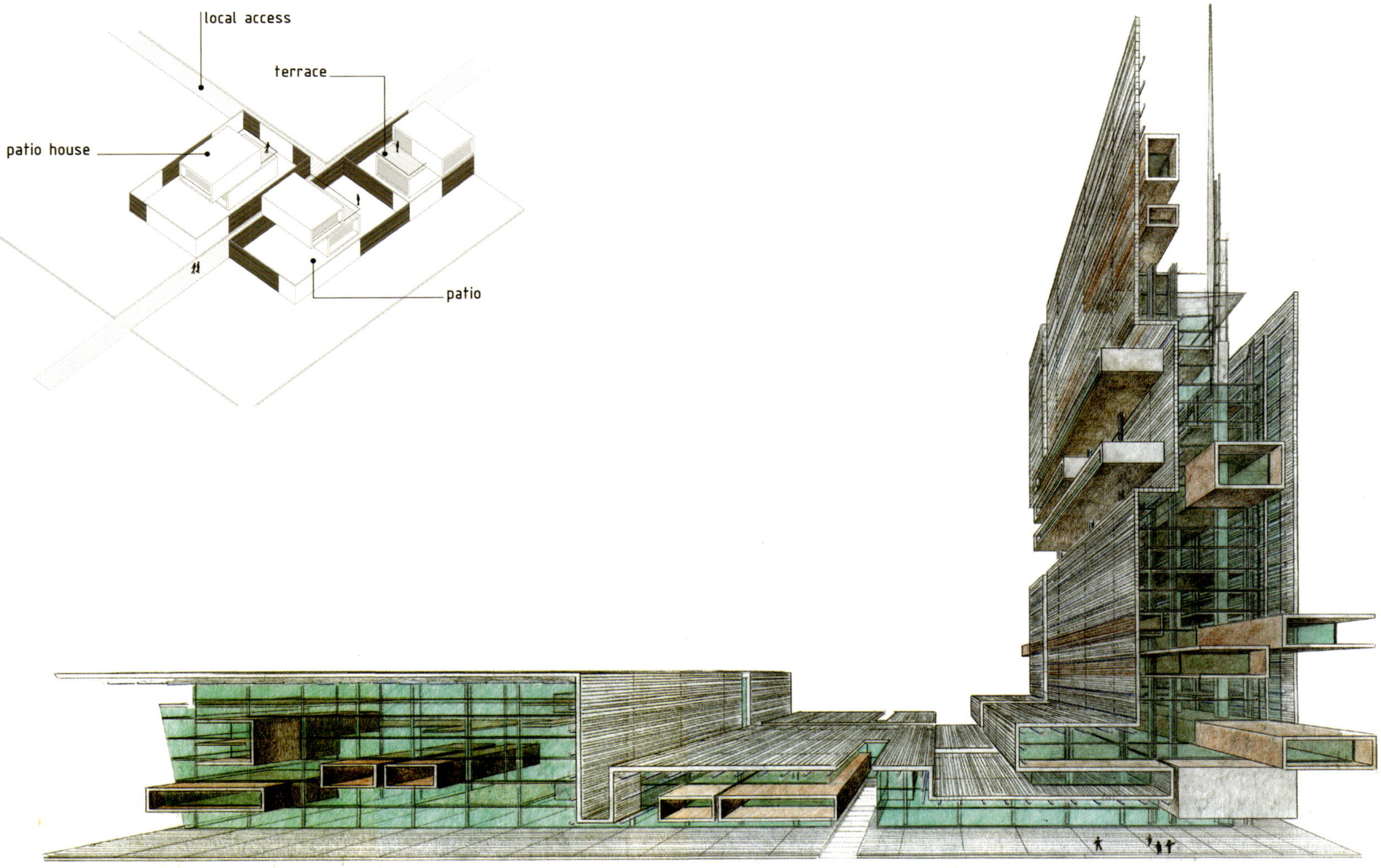

Project Team: Ioannis Perisoratis

Tracing Perception
认知研究

The main goal of the project is the development of a methodology to analyze the impact of technology on the perception of the contemporary urban space. More specifically it focuses on the study of the use of the cellular phone in the public space. At first level it tries to visualize the dynamical change of perception through the analysis of people's movement patterns and behavior.
The area of interest is the area of Liverpool Street station in London which is characterized by the diversity and flow of people through the day. In this area a route has been selected in order to investigate the movement of people using cell phones through the city. The selection of the route is based on the diversity and combination of the public spaces. This analysis, through the development of a mapping technique, is extracted from the visualization of a dynamic field of temporal privatizations.
The next step of the methodology is the extraction of spatial qualities out of the results of the mapping process. This is happening gradually through the development of a translation procedure. This translation is extracting a deformation of the public space itself. It is my intension now to investigate this deformation and the spatial relations it can create. Therefore the deformation is translated into a three dimensional steel truss. The truss has specific geometric attributes which create various spatial relations. It is the moment where the program of the media experience urban park is emerging to fill up these relations. The ethereal character of the spatial relations is underlined with the use of wood as a softer material which is partially embracing the steel truss in order to define exactly the spaces and their relations. The final result is a semi-public space which is engaging a passing by space with the media applications.

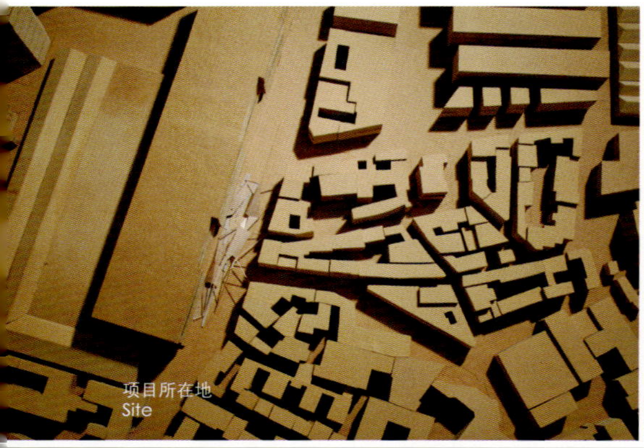

项目所在地
Site

鸟瞰图
Bird's Eye View

映射空间
Projection Space

展览现场
Gallery View

夜景图
Night View

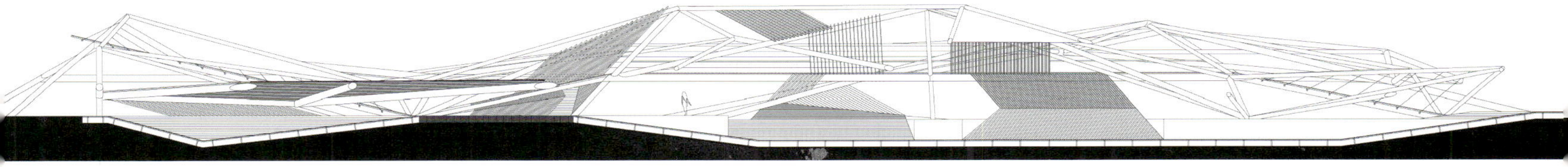
截面图
Section

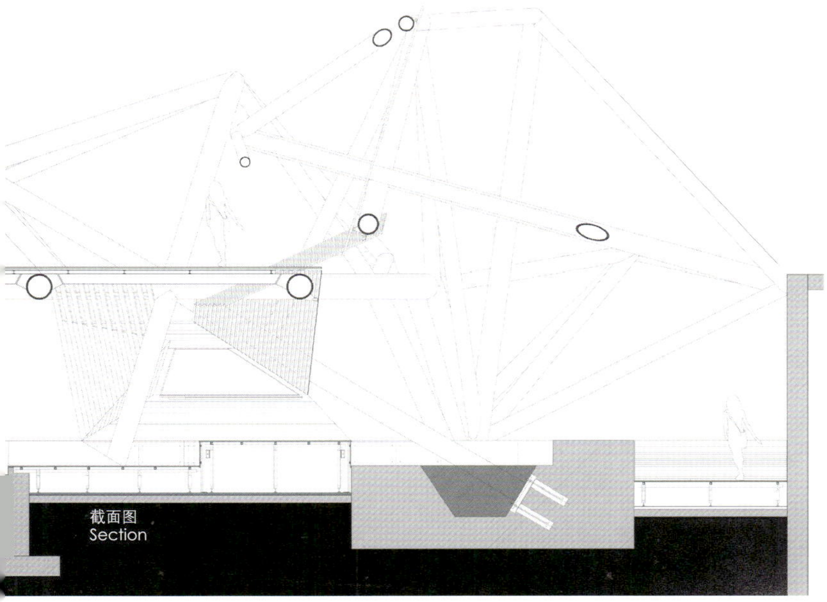
截面图
Section

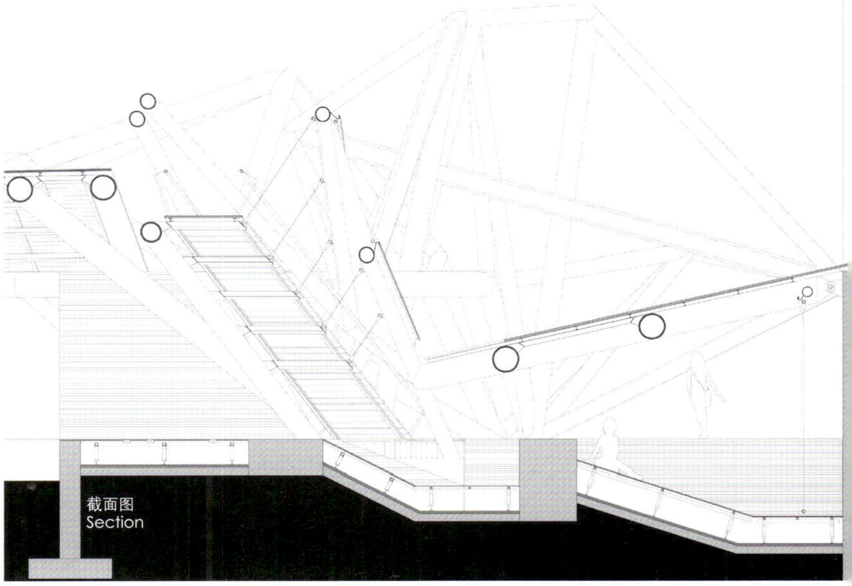
截面图
Section

映射空间
Projection Space

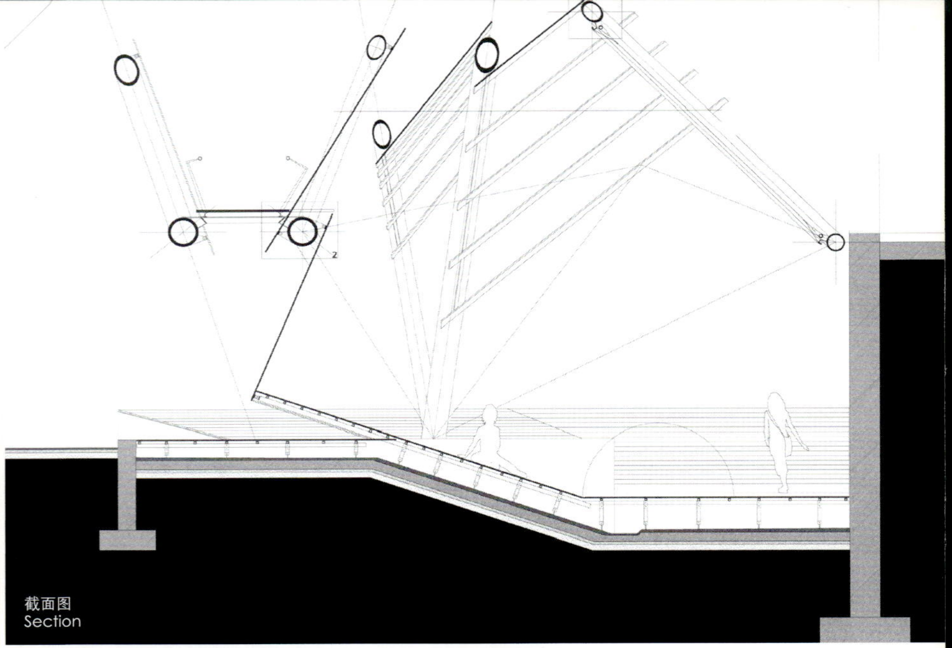

截面图
Section

连接细部模型
Joint Detail Model

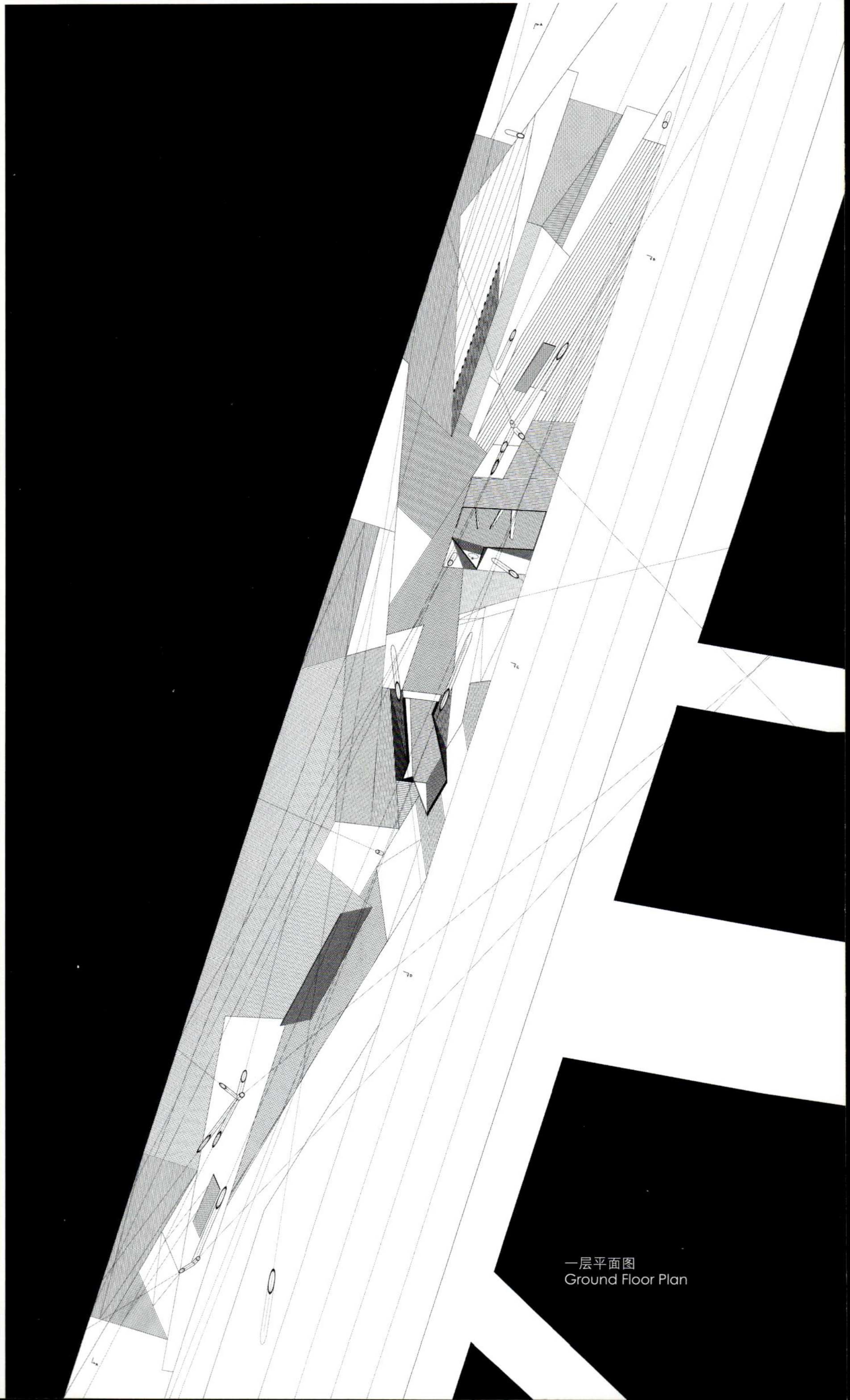

一层平面图
Ground Floor Plan

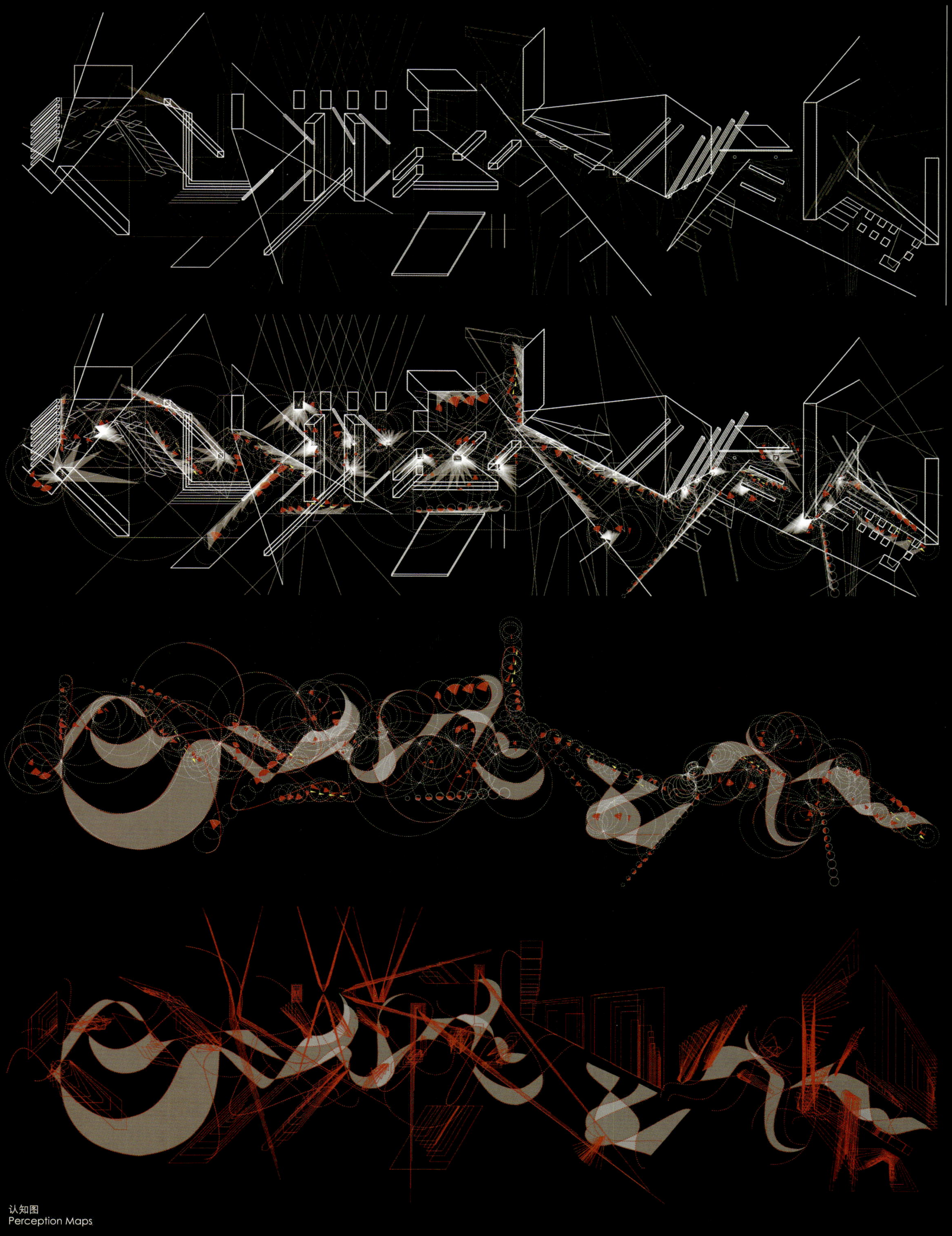

认知图
Perception Maps

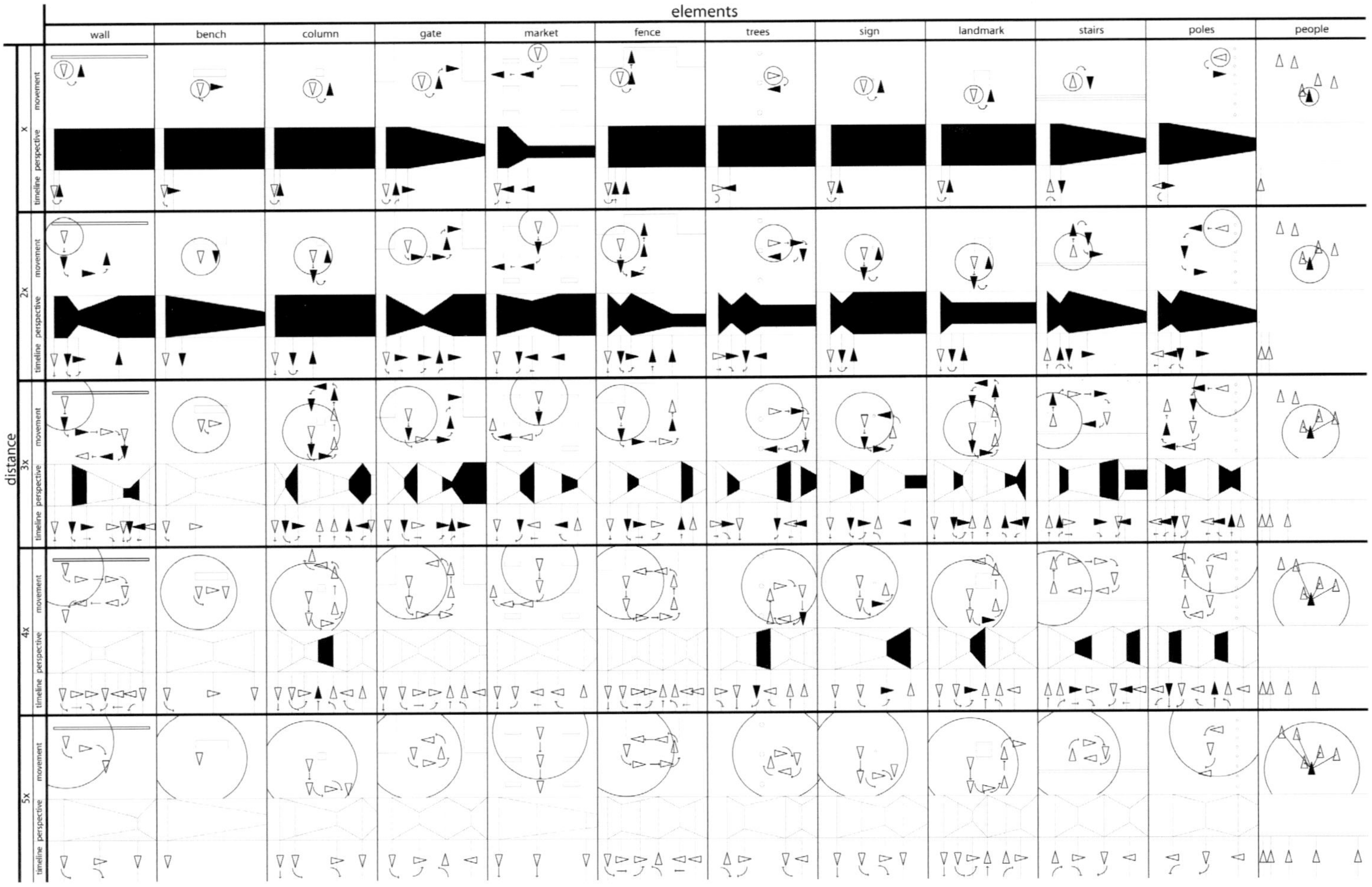

行动矩阵
Movement Matrix

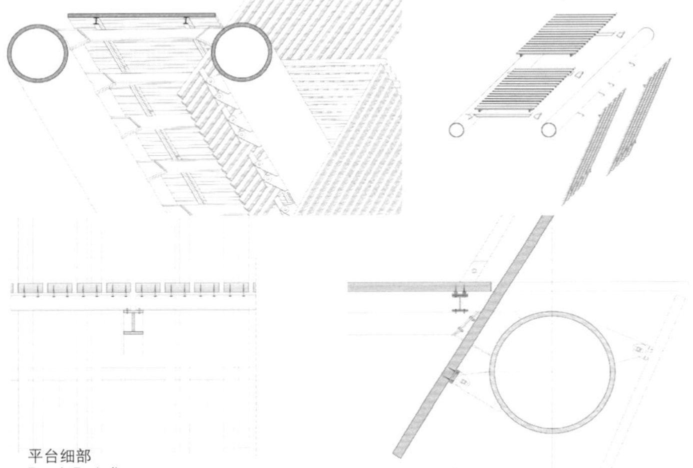

平台细部
Deck Detail

连接细部
Joint Detail

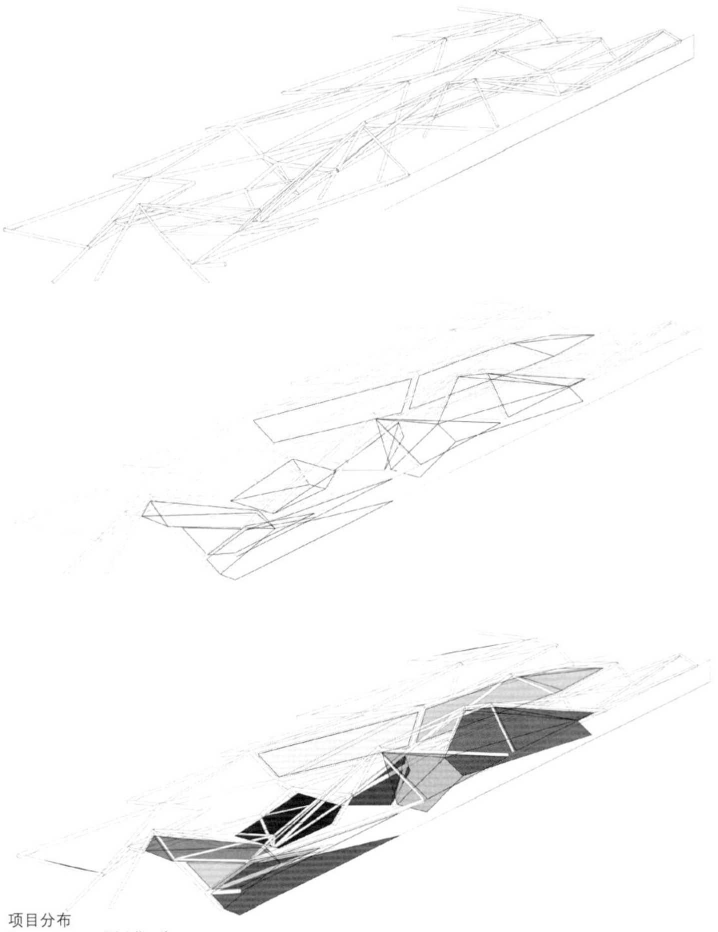

项目分布
Programme Distribution

几何发展
Geometry Generation

Project Team: Juan Lacape / Monica Kurzel / Pablo Roman

Tranquil Multifamily Residence on the Hillside

宁静惬意的山间群居住宅

A steeply sloped site in the mountainous skirts of Guatemala City's valley rested undeveloped while having impressive views. Guatemala is a seismic zone 4 area, and this was a factor for the site to remain undeveloped.

Structures with long spans were designed to rest on top of concrete piles poured into the steep slope in order to be able to develop a residential solution that is cantilevered through the air.

All habitats in the project have long glass expanses, all the walls are light gauge steel sandwich panels, and a wooden Prodema rainscreen was applied on top of some panels, along with aluminum composite panels.

Structural steel has been left exposed in order to enrich the interior spaces.

The habitats were designed for a specific market segment (affluent professionals in their thirties), and thus a specific lifestyle product was developed.

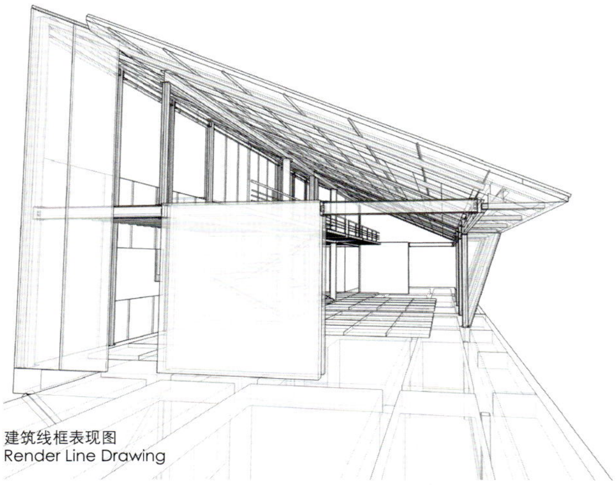

建筑线框表现图
Render Line Drawing

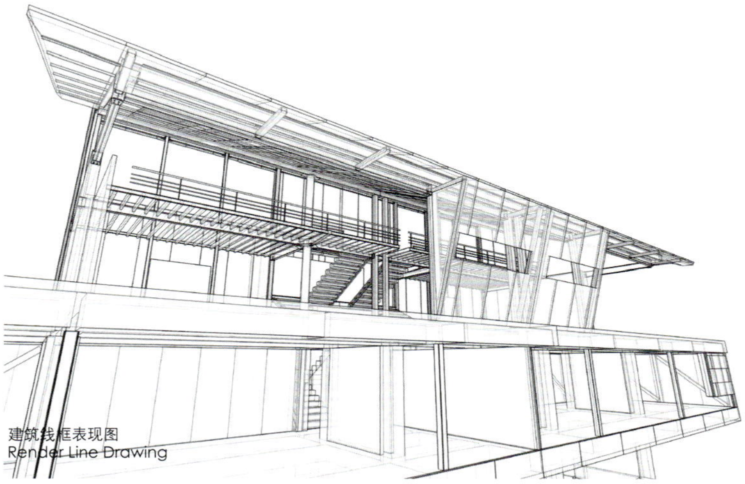

建筑线框表现图
Render Line Drawing

建筑外观
Exterior

室内整体布局
Interior Living

室内整体布局
Interior Living

地板局部
Detail

室内表现图
Render Interior

室内表现图
Render Interior

室内局部
Detail

室内局部
Detail

室内整体布局
Interior Living

Chief Architect/Office Name: Piotr Cegielko, E&L Architects sp. z o.o.

Tulipan House

Tulipan住宅

Travelling along the streets of downtown Mokotow -Warsaw's new business district – passing by modern, tall facades, all of a sudden we fall into a breach in them shining walls of contemporary architecture. You are going to be struck by its low rise and apparent low volume. It only takes to go there personally to fully appreciate what Tulipan House really is about...

Existence of Tulipan House proves that the economical budget, flexible design, efficient structure and rapid execution are not necessarily in contradiction with high quality architecture. Indeed, the deeper you dive into the Tulipan House's innards, the more obvious become subtle qualities that the building has up its sleeve. There's that spectacular atrium space, of course - with open galleries, connecting all four levels into one, big, communal space, spreading through transparent, glass barrier onto the representative courtyard in front of the building. It is not the front facade, however, that is of the essence - it is the area to be rented to tenants: carefully chosen height and depth of future offices, allowing the natural lightning and optimising circulation around personal workspace. Modest form, logical dimensions and practised proportions – all summed up to the friendly, human scale of the whole edifice.

When the building is practically a product put to sale, the architect must walk a very narrow path between the responsibility for commercial success and faithfulness to the basic commandments of his craftsmanship: building should not only be solid and practical, but also...beautiful!

Tulipan House remains the pleasant surprise. There were those, who disbelieved that such a building would be a success in given time and place. The same people are now willing to recognise it as the exception that proves the rule!

Universidade Agostinho Neto Campus Plan

安哥拉Agostinho Neto大学校园规划

This university masterplan for a 17,000 student, 2000 hectare campus is conceived around principles promoting low maintenance sustainable urbanism for the developing world.

The first principle employed was to concentrate development to only a portion of the site, leaving existing vegetation and water courses in place and using the landscape to promote airflow. The main campus is defined by an oval roadway which sits between the watercourses and defines the limits of development. Within this ringroad is a hierarchy of spaces and routes organized in a pinwheel system of streets and shaded courtyard clusters centered about a library and plaza.

The second principle uses building orientation and envelope to promote natural ventilation and cooling. Buildings are naturally ventilated through an optimally oriented design of single loaded classroom bars. An engineered roof shading system is calibrated to minimize solar gain and act as an airfoil to promote cross ventilation through the bars by maximizing pressure differentiation.

中央图书馆
Central Library

校园整体扩建图
Full Campus Buildout

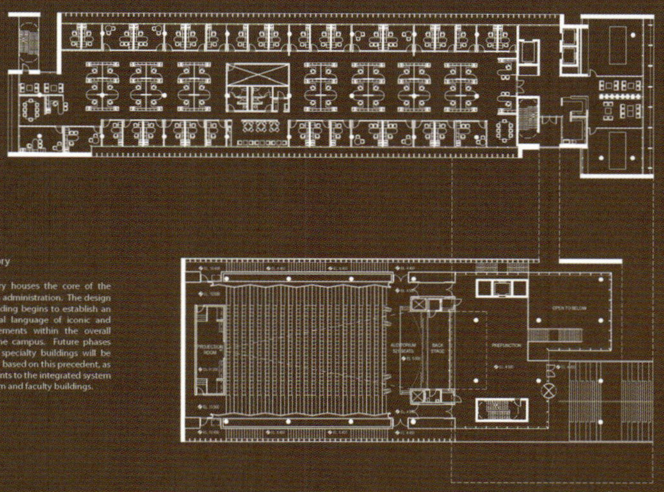

Rectory First Floor Plan

The Rectory

The Rectory houses the core of the university's administration. The design of this building begins to establish an architectural language of iconic and unique elements within the overall fabric of the campus. Future phases and other specialty buildings will be developed based on this precedent, as counterpoints to the integrated system of classroom and faculty buildings.

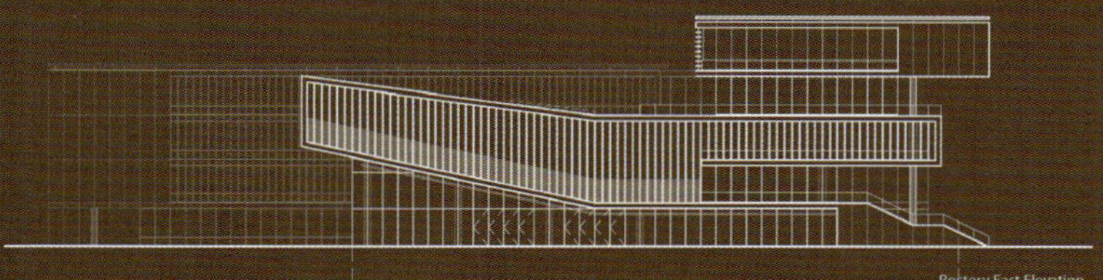

Rectory East Elevation

行政楼
Rectory

教学楼
Classroom Buildings

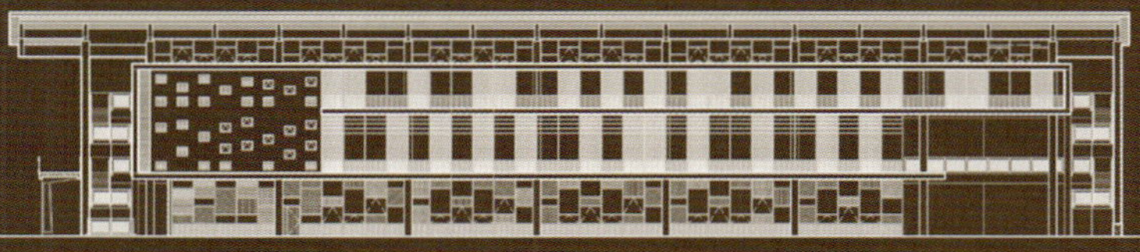

Classroom / Lab South Elevation

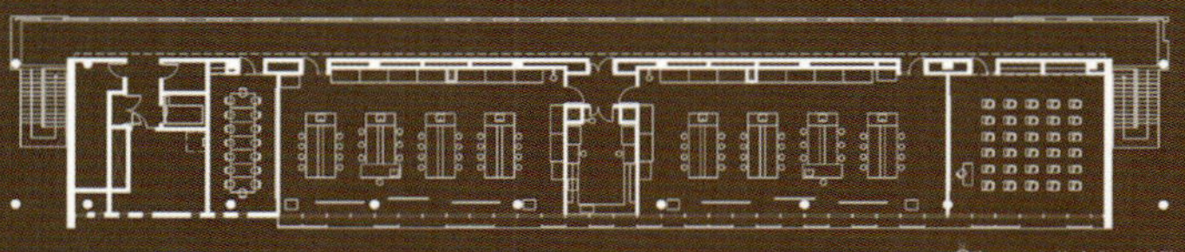

Classroom / Lab Second Floor Plan

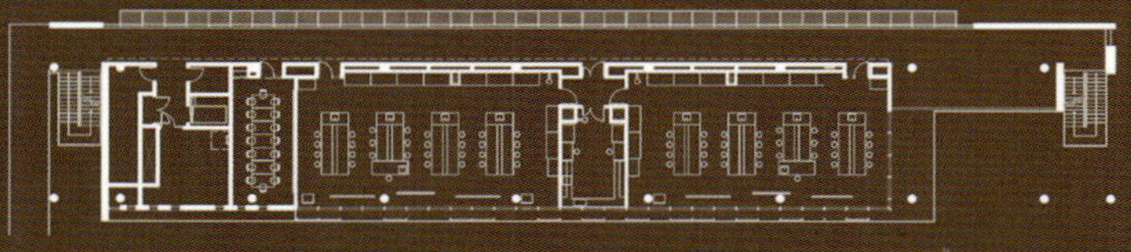

Classroom / Lab First Floor Plan

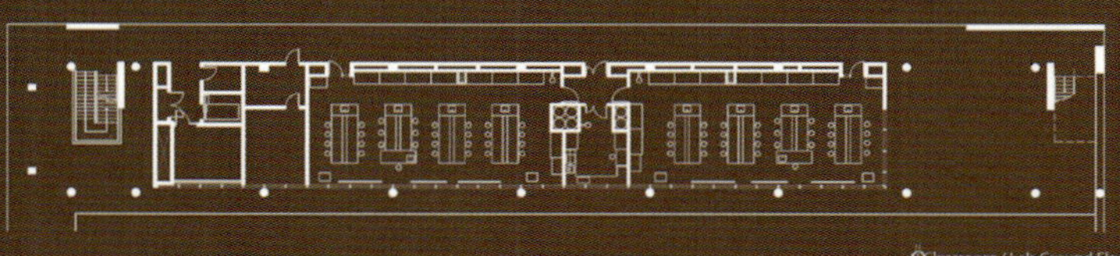

Classroom / Lab Ground Floor Plan

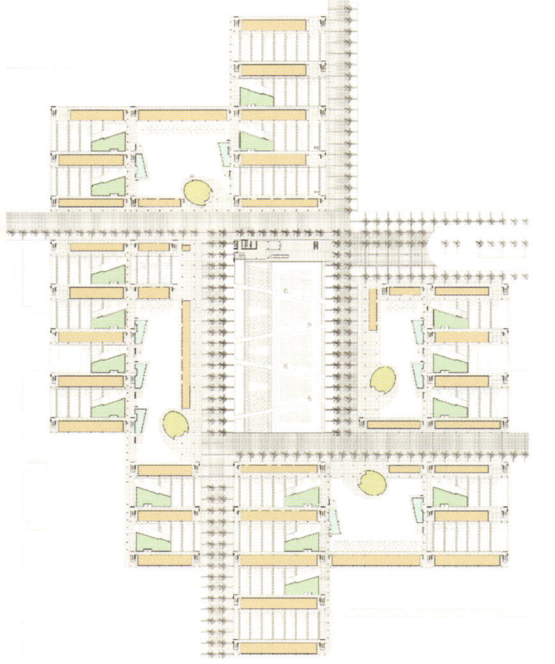

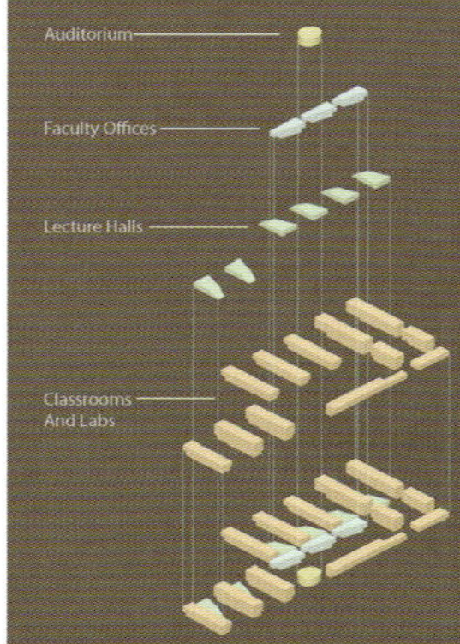

Faculty Organization

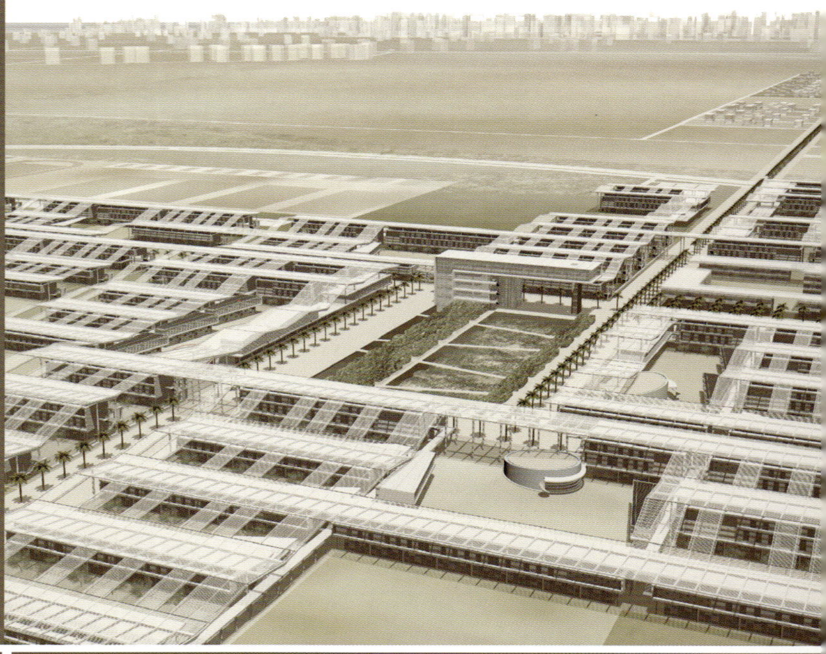

The Core Campus

The university is globally structured with autonomous faculties, each with their respective departments, sectors, and sections. This is the case with Mathematics, Chemistry, Physics and Computer Science. These areas of study are consolidated into core curriculum departments.

The faculties are composed of four fundamental architectural components: faculty offices, lecture halls, auditoriums, and classrooms/labs. Each has been given a distinct architectural character and composition.

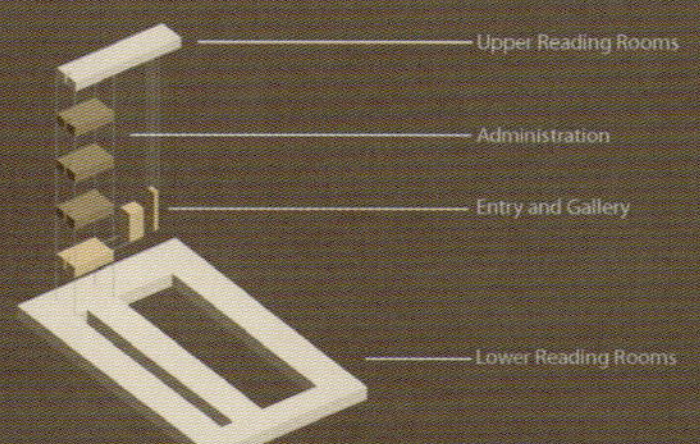

The Central Library

The Central Library, along with its sunken garden and reading rooms, is the organizational landmark for the campus. It's prominent form and central location becomes the defining character of the core campus and the organizational point of reference for all of the campus' faculties.

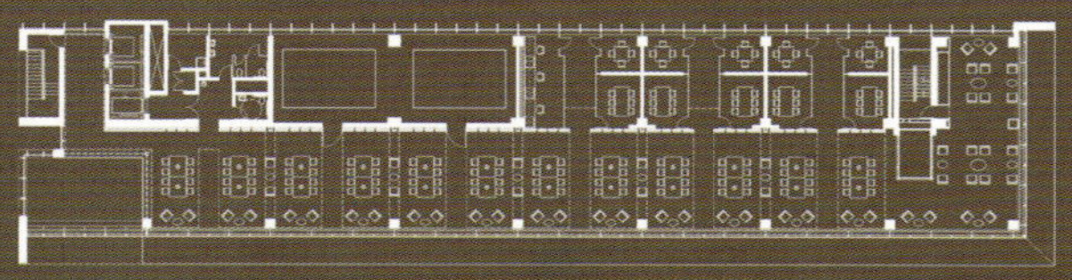

Library Fourth Floor Plan

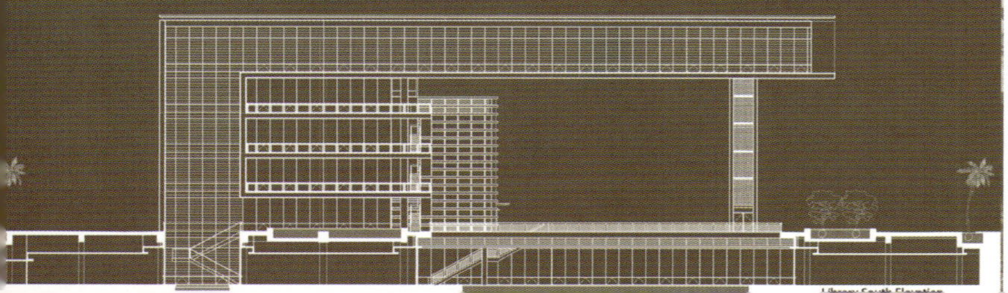

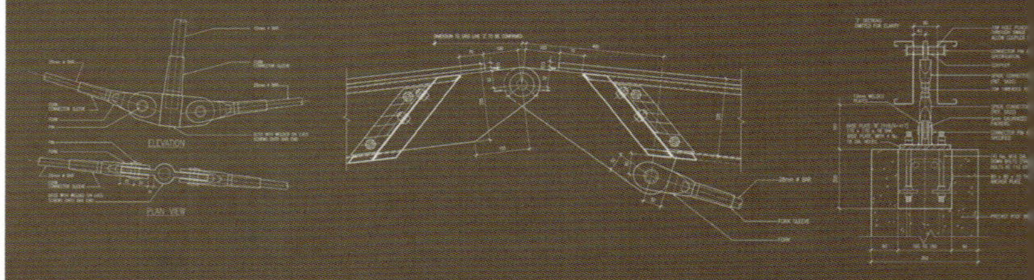

Fluid Dynamic Study for Roof Optimization

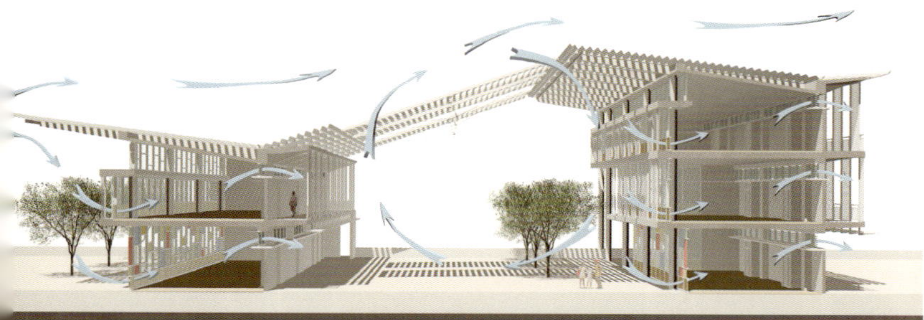

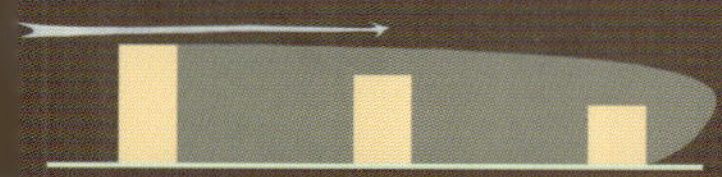

Wind Shadowing

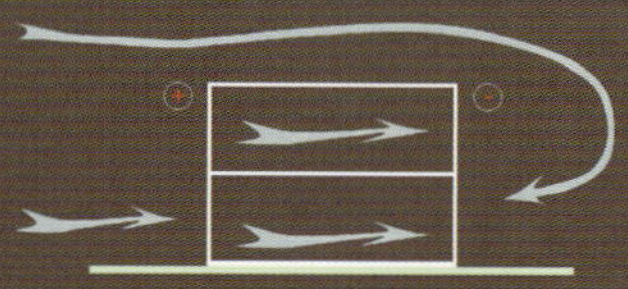

Cross Ventilation Strategy

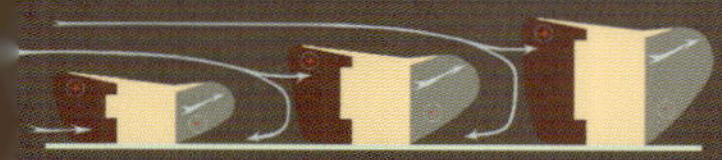

Pressure Differentiation

A Computational Fluid Dynamics Study was done in order to find the most beneficial roof shape and orientation for airflow. Massing hierarchy and orientation prevents the wind shadowing effect. This in turn creates a pressure differentiation across each building which optimizes the cross ventilation of the campus buildings.

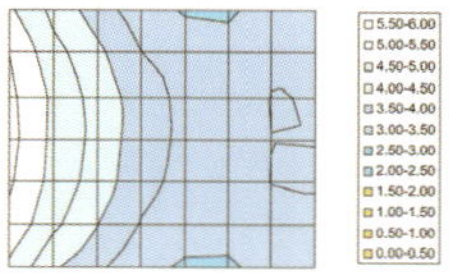

Daylight Penetral Plan

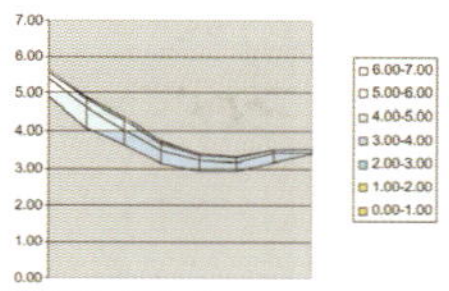

Daylight Penetral Section

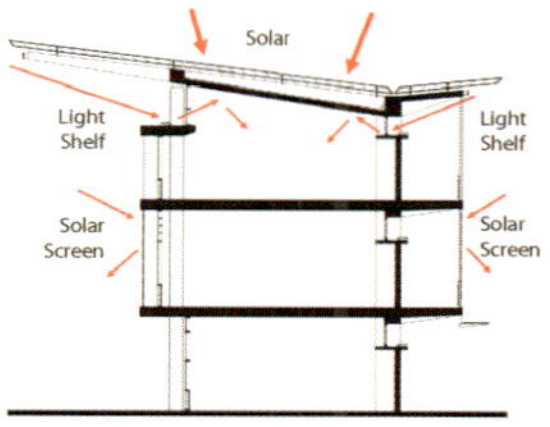

Daylight Control Components

Natural Light is both harnessed and controlled through an assorted system of screening devices and light shelves. Careful analysis throughout the design process organized these elements into a well balanced system that allows for both use of and protection from the harsh sunlight of the region.

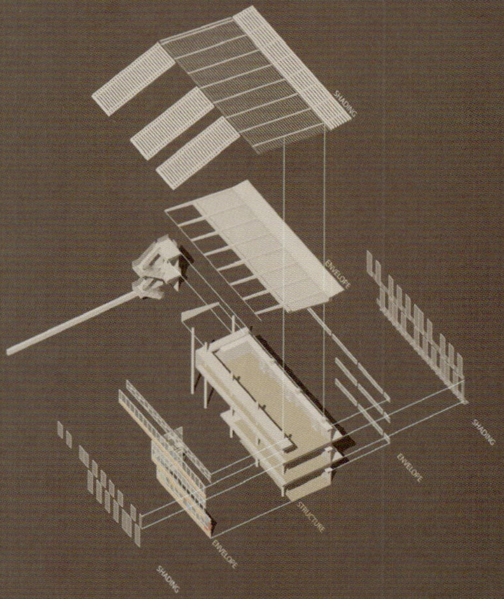

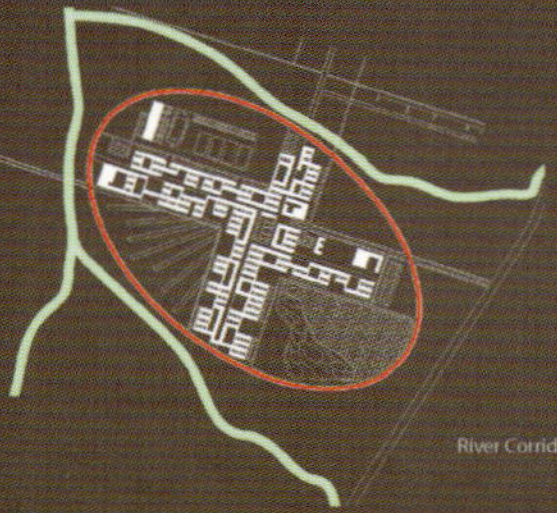

River Corridors

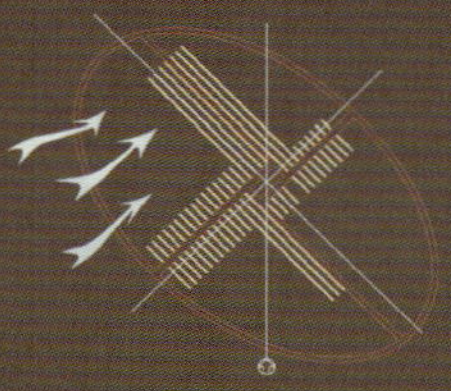

Optimal Wind Orientation Study

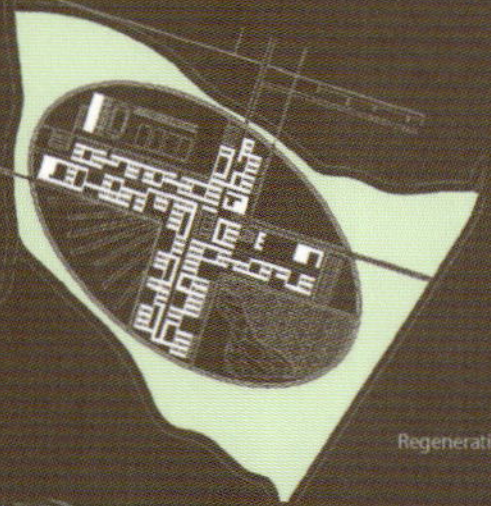

Regenerative Woodlands

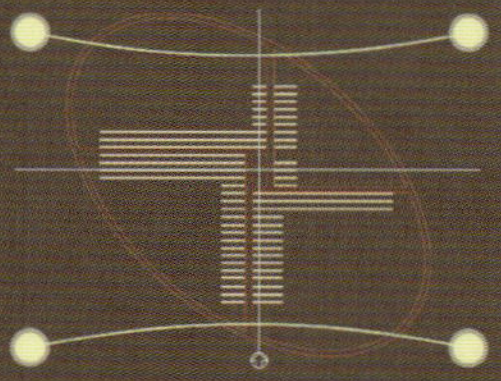

Optimal Sun Orientation Study

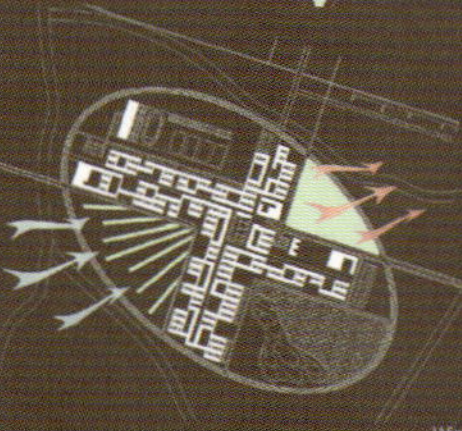

Wind Harvesting

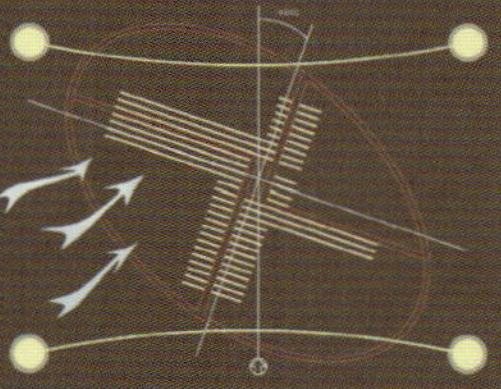

Final Campus Orientation

The university preserves existing river corridors running through the site. These heavily vegitated areas have a distinct character related to the increased availability of water as a resource. Seasonal flash flooding not only replenishes the habitat, but also imposes a natural boundary for the establishment of the ring road surrounding the academic campus.

Natural woodlands at the perimeter of the site are preserved as park area. Fencing of this area provides the opportunity for natural succession and regeneration to take place. The aim is to create a clearly defined landscape character. 'natural' outside the ring road and 'created' within the ring road.

Due to the prevailing south-westerly winds of the site, the south-western quadrant will have a modified topography to enhance the flow of these winds through the interior of the campus. Twenty meter wide tree belts will help to funnel, cool, and remove dust from the winds. Warmer, rising air in the north-western savannah quadrant works in conjunction with the wind channel to help drag air through the site.

In order to achieve the optimal orientation for the layout of the campus two main factors were necessary to consider: prevailing winds and sun paths. The optimal orientation with respect to the wind would be perpendicular to the prevailing direction, in order to maximize airflow throughout the campus. The optimal orientation with respect to the sun would be orthoganally, east-west, in order to maximize the potential for shading of spaces. Collapsing both of these considerations into an orientation of 19 degrees off the north-south axis provides a fundamental basis for massing that will ideally accommodate both desires.

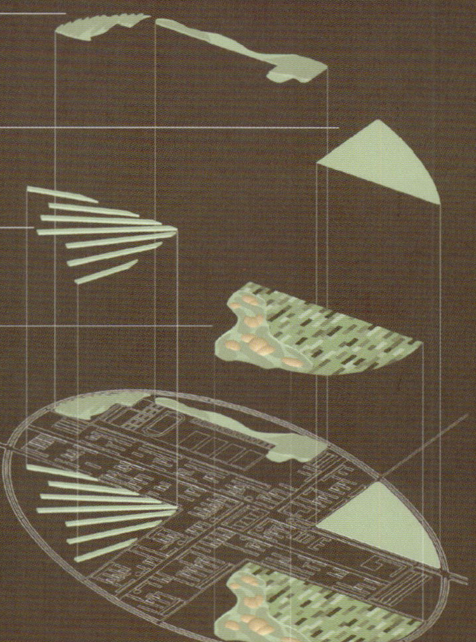

Multi-Functional Landscape

生态设计
Sustainable Concepts

Team: Giancarlo Alhadeff, FAIA Designer

Villa in Ojai

Ojai别墅

Located on an open hillside landscape just outside Los Angeles, this house conforms in spirit to the needs of the new family: its flexible plan permits configurations of privacy and gathering without sacrificing visual unity. Its essential lines contribute to rich combinations of panorama and closure. Views through spaces and out onto the landscape invite in the natural world. The vigorously open space does not preclude a sense of intimacy. The clients spend most of their time balancing between two European cities. This Californian meeting point is not just home, then, but a destination for two merged families and their three children. Their desire, for a home that feels like a gathering of pavilions, provided an opportunity to link Katsura-inspired spaces along defined corridors. The natural incline of the site contributes to the visual interplay of roof-line and disappearing walls, giving the structure movement within the defined plan.

别墅景致
Olivium Villas

别墅景致
Olivium Villas

东立面
Villa in Ojai East Elevation

1层平面图
Villa in Ojai 1F Plan

总平面图
Villa in Ojai GF Plan

别墅景致
Olivium Villas

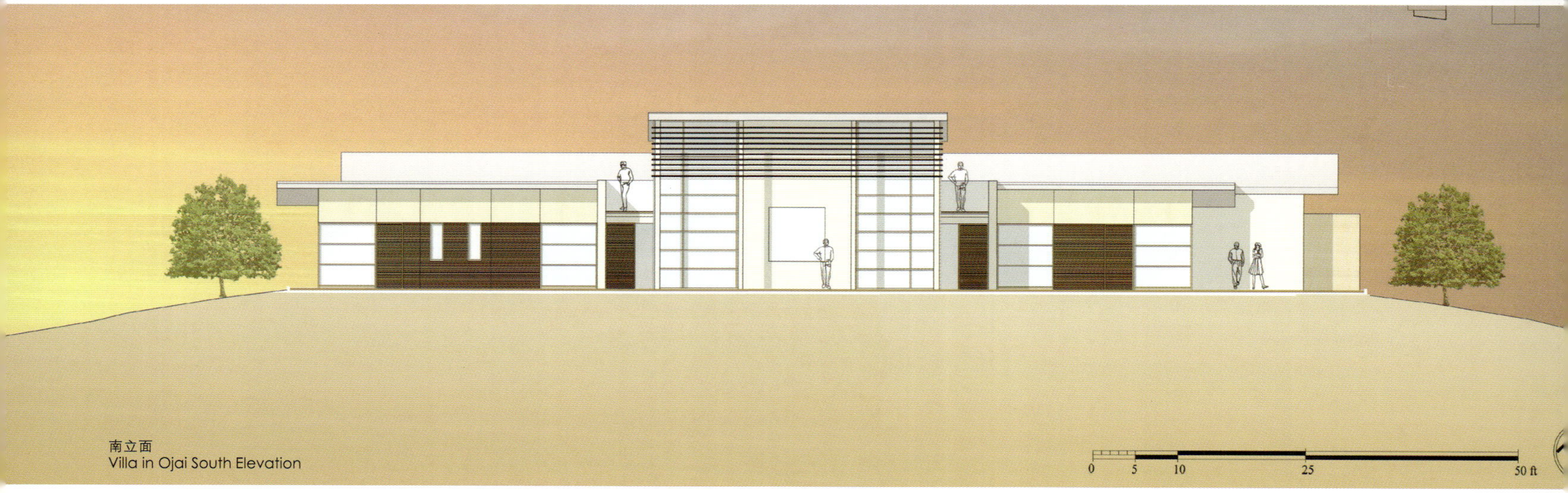

南立面
Villa in Ojai South Elevation

透视图
Villa in Ojai Rendering

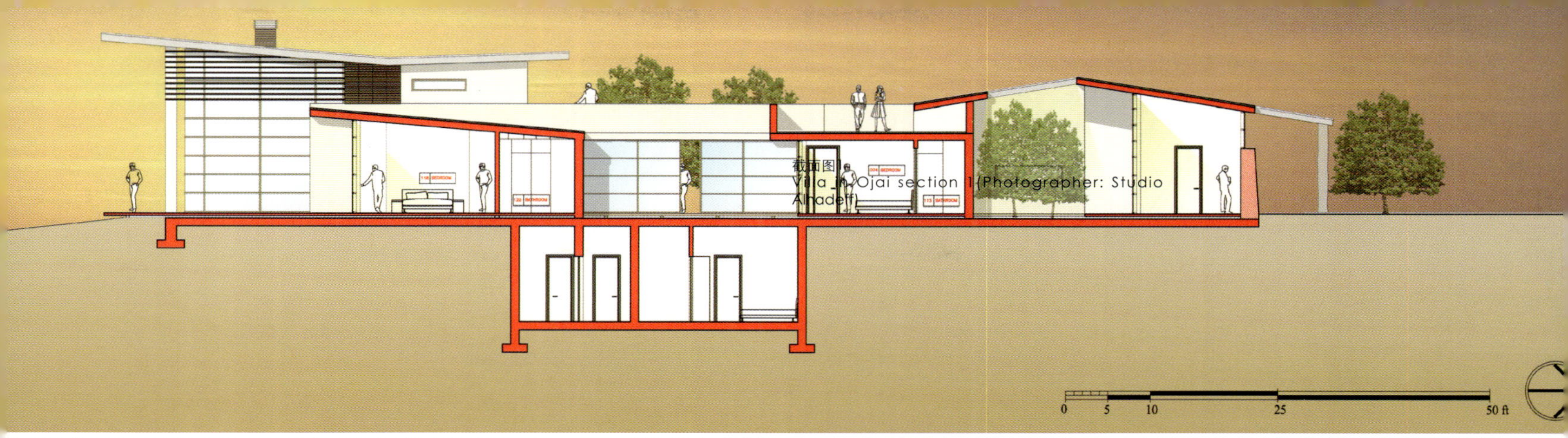

截面图1
Villa_in_Ojai section 1 (Photographer: Studio Alhadeff)

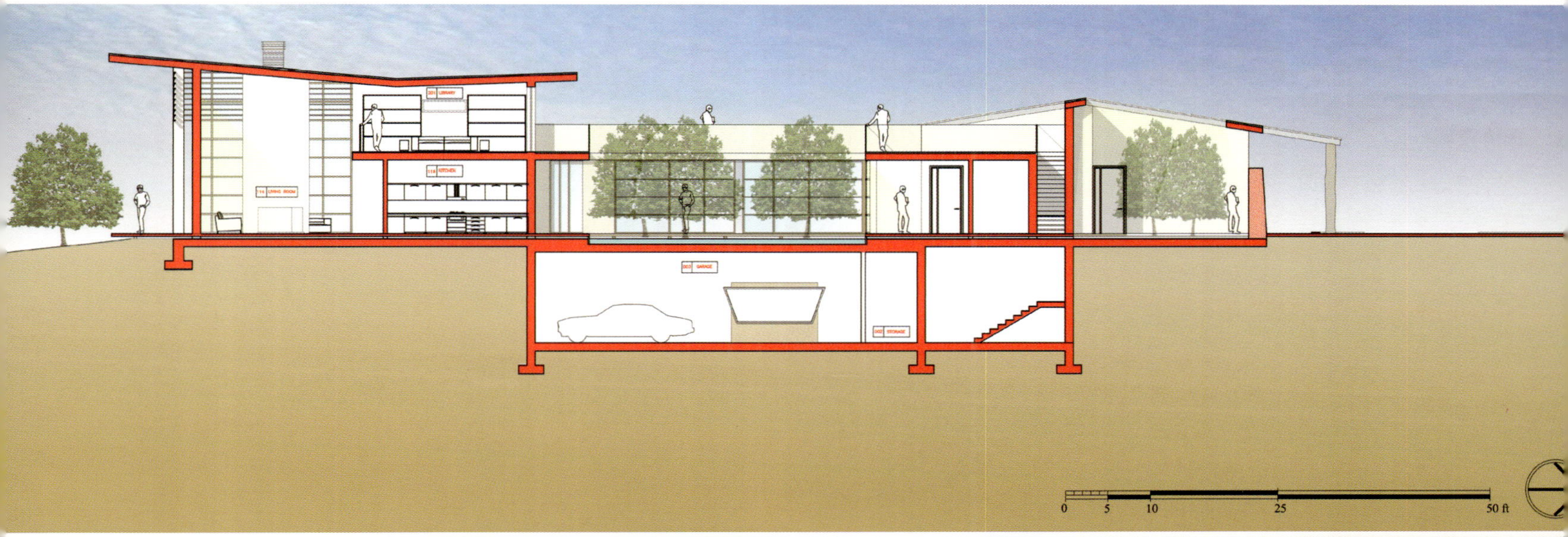

别墅景致
Olivium Villas

别墅景致
Olivium Villas

别墅景致
Olivium Villas

Project Team: Evgeniya Yatsyuk

Water — Touristic Complex in Kaliningrad Center
城市中心旅游开发区建设

The main idea of the project is to create a development which is used to be in the central part, before war started using old red lines according with old drafts and documents related to period before war. A new view is going to be created using new technologies; new materials and using glass-let's say a contemporary view on architecture before war period. In the north, site is bounded with general Karbushev embankment, from the west side with Oktiabriaskaia street, from the south Solnechnui boulevard, from the east general Pavlov street. New quarters will include small private hotel complexes, living area, spaces for artists. Cafés, restaurant, bars, souvenir shops, tourist centers, small museums, and galleries are going to locate on the ground floor.
On the bank of the river Pregol will be situated fishing markets-places Kaliningrad used to be proud of. Infrastructure is going to be pleasant not only for people who live in Kaliningrad but also for tourists from abroad. Another pleasant thing of creating old development is to avert easy view to existing development of 1970, which are nowadays organizes surrounding area. Marina is planned on the site connected with a river Pregol and will be in a sense of transport a connection- in this way an opportunity would appear to use yachts and other water transport for communication.
A hotel complex surrounds water space with open view on the cathedral and water. The hotel territory is represented with the 9 volumes, common space underground where restaurants are located and managerial rooms unite buildings. Pedestrian Boulevard visually connects cathedral on the Knaiphov island and Evangelicheskaia church on Lomse and divides both zones: hotel and living area.
Living complex has inner courtyard for people living in this complex only. Access is limited because the courtyard is situated on -3. 000 mark, while pedestrian Boulevard runs on the 0. 000 mark though living and public spaces are in different levels. The lowest part in the living complex comprises shops, library, sports area, small cinema. Underground level is planned as a parking space for all people living in the complex. On the 5-6 level 2 story apartments are located. Thus a result of all work was a creation a new image of a town which is used to be here not copying it but creating new one with a memory of the past.

HISTORICAL MAP

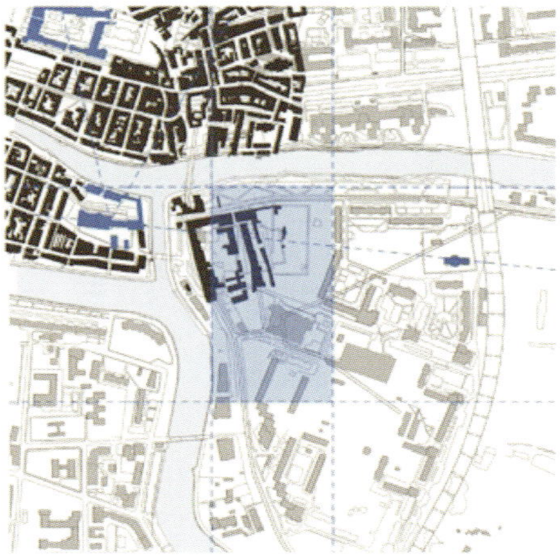

EXISTING MAP

PROPOSAL

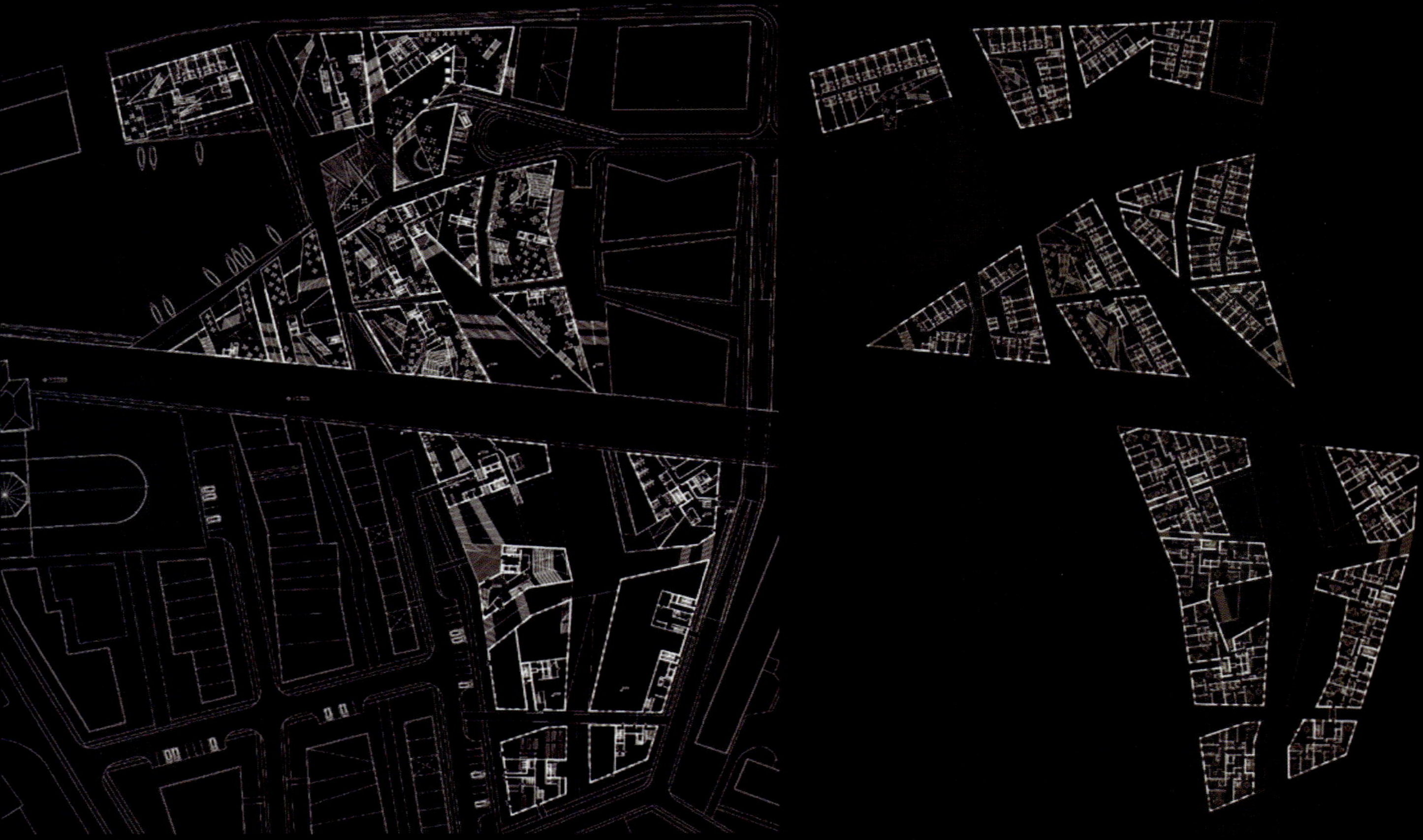

MODEL

FUNCTIONAL SCHEME

FLOOR SCHEME

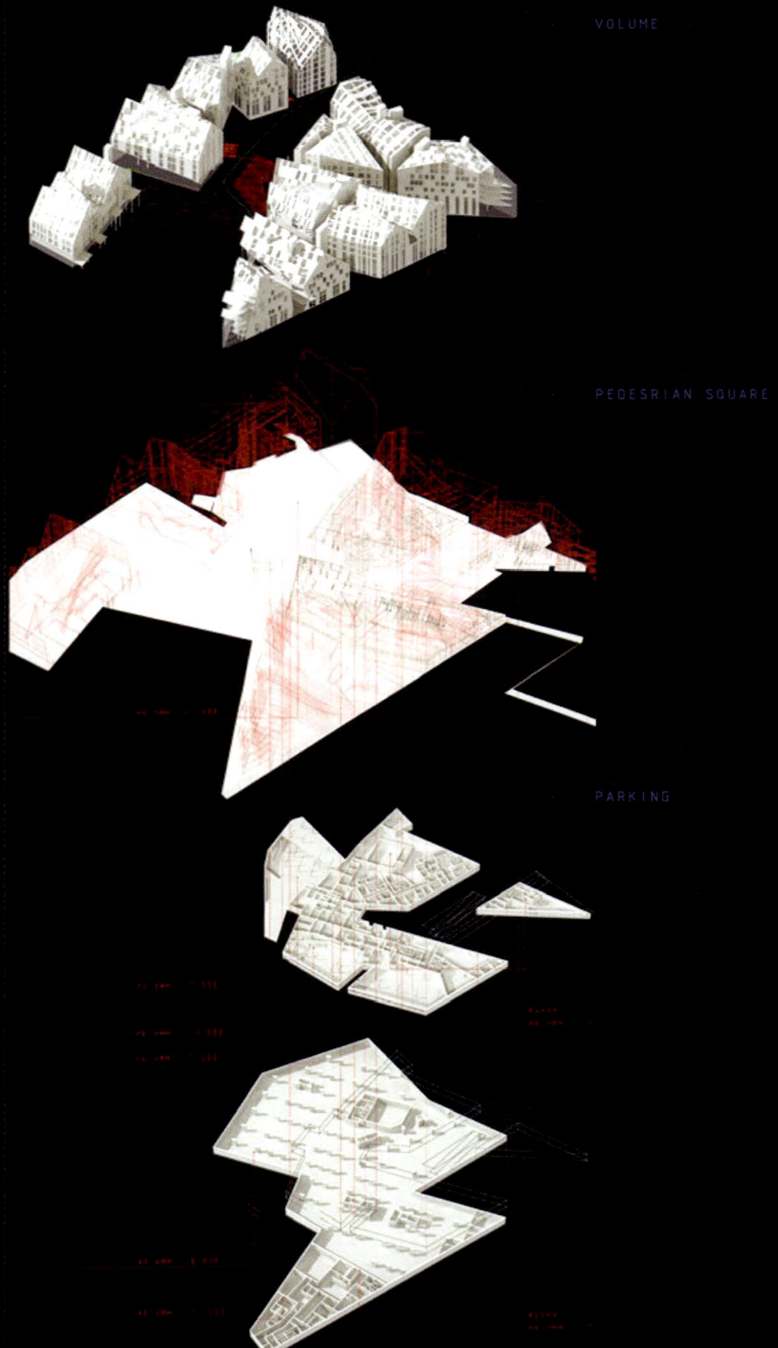

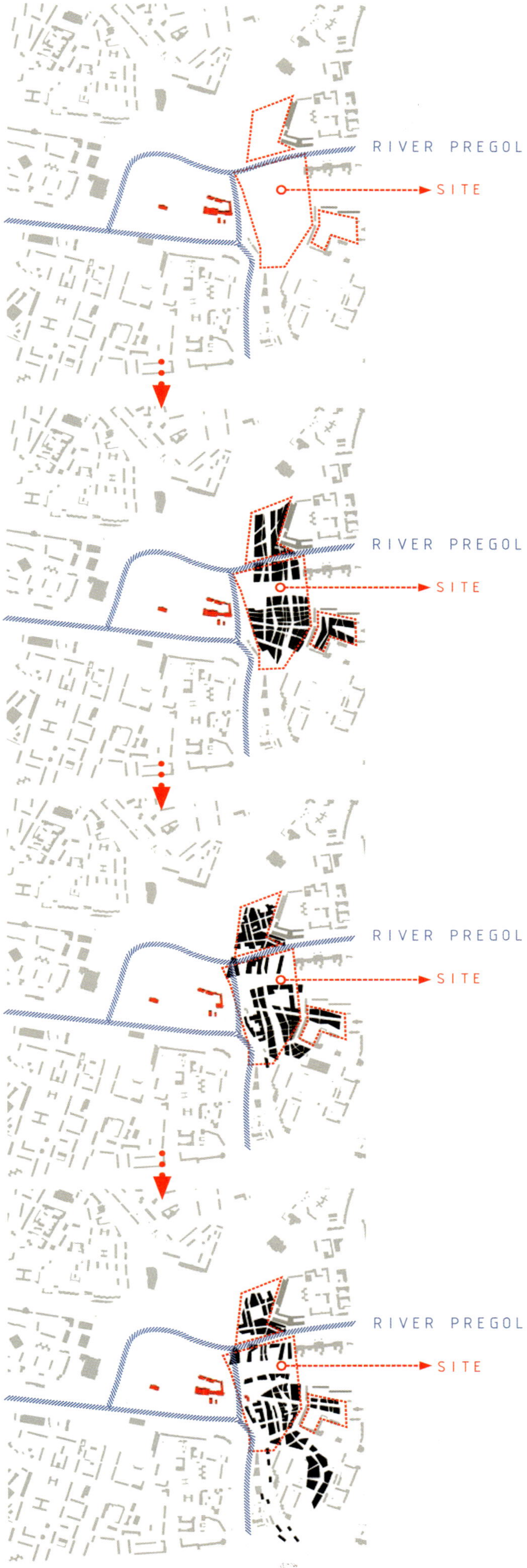
RIVER PREGOL
SITE
RIVER PREGOL
SITE
RIVER PREGOL
SITE
RIVER PREGOL
SITE

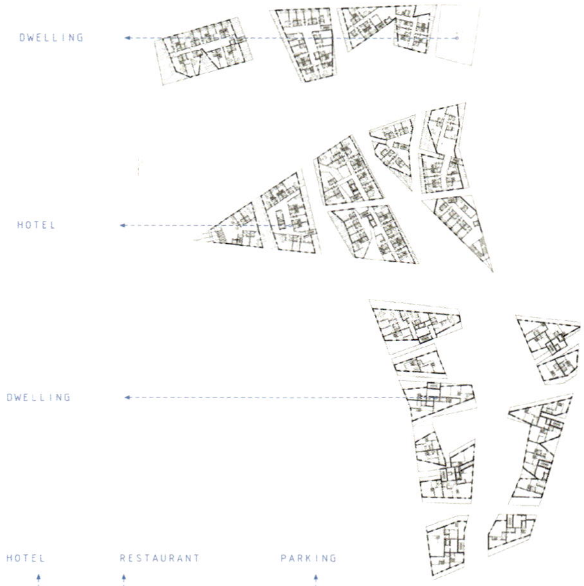
DWELLING
HOTEL
DWELLING

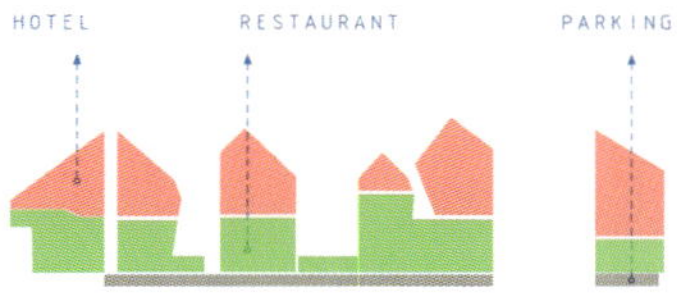
HOTEL
RESTAURANT
PARKING

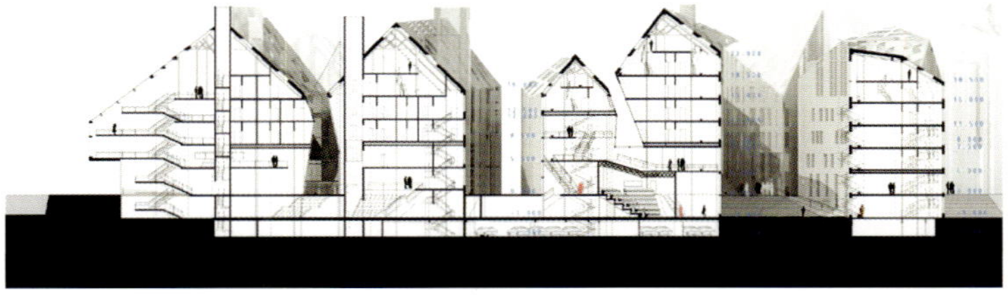

Chief Architect/Office Name: Yannos Yanniotis & Associates

Winery in Santorini

圣托里尼葡萄酒厂

CYCLADIC ARCHITECTURE

The Cyclades -islands in the central, eastern and south-eastern Aegean- are protected by a planning regime which determines the morphology of their architecture. Small single or two-floor structures which blend into the natural terrain arranged around courtyards protect the privacy of residents during the hot summer months and also provide opportunities for communication. The colour of the buildings, as prescribed by law, is white. The only exceptions are the colours of the doors and windows where there are three or four basic colours to choose from (light blue, grey, green, red-terracotta). Amid the barren environment of the islands with their harsh terrain and minimal vegetation, these small structures provide the characteristic visual plasticity, particularly when they are integrated into the fabric of villages that have narrow lanes no wider than 2 m and external entrance steps.

THE COMPLEX AT BOUTARIS WINERY

The Boutaris Winery is located in Santorini – arguably the most unique island of the Cyclades– on the road to Faros, after Akrotiri. The Winery, with its distinctive white dome, was designed in 1987 and went into operation in 1989, playing a leading role in the growth of wine-making on the island. It is open to the public and offers visitors complete guided tours.

ARCHITECTURAL COMPOSITION

The complex, surrounded by vineyards, is situated almost next to the very edge of the crater, and includes a state-of-the-art vinification centre, a tank area, an administration building, a sales department and shop, and a building known as the 'Dome': an indoor amphitheatre laid out in the form of concentric segments of a circle.

This corner plot has a long façade along the road leading from Megalohori to Akrotiri. Passenger vehicles and lorries enter and exit the complex from this road on the north and west respectively. The main entrance for the public to the administration building is also located there. The buildings are laid out around a central courtyard with the management building and winery on a N-NE orientation, the tank on an E orientation, the Dome on a S orientation and the accommodation facilities (to be developed in the future) on a W orientation.

The courtyard is the central reference point in the entire structure and it forms a link to all individual functions. This area is protected against the harsh northerly and north-easterly winds and thus the courtyard acts as a linking tool and helps reproduce the traditional standards of architectural composition and communication.

The building structures are plain, single-floor constructions with flat roofs in the case of the administration building and the tanks, while the roofs of the winery and the accommodation facilities are domed. This ensures a contrast between the stark, plain volumes of the outer, perimetric view of the complex and its interior, with the gentle landscaping of the courtyard and the buildings radiating off it. The geometry of the Dome, which is the distinctive compositional element here, was determined by the amphitheatre-like internal layout. It is a building inscribed on a circle, with concentric circular indents housing rooms where visitors can sample Boutaris' range of wines from Santorini after watching a sound and image show. The domed

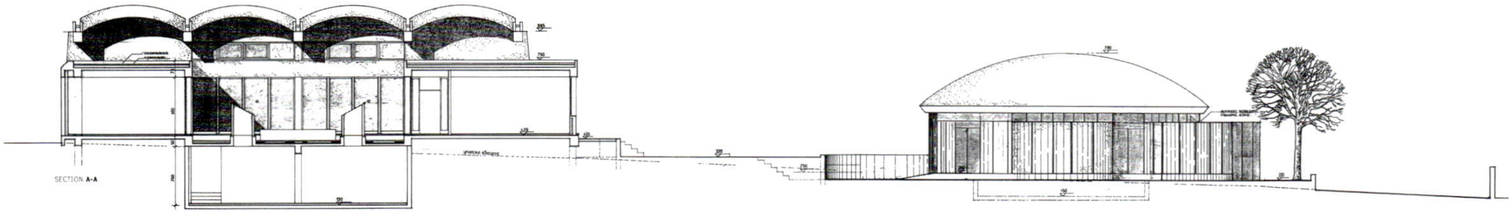
SECTION A-A

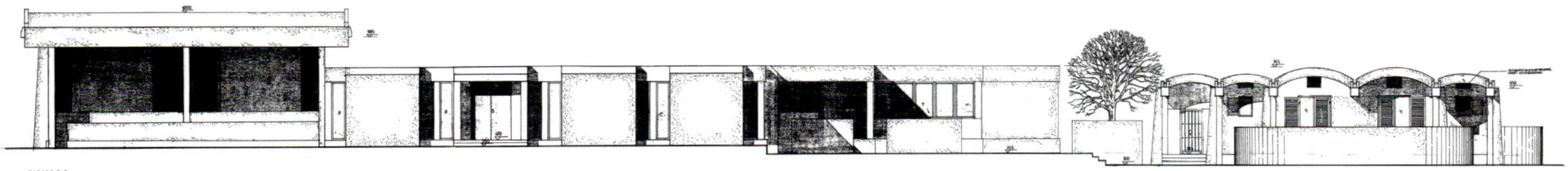
FACADE B-B

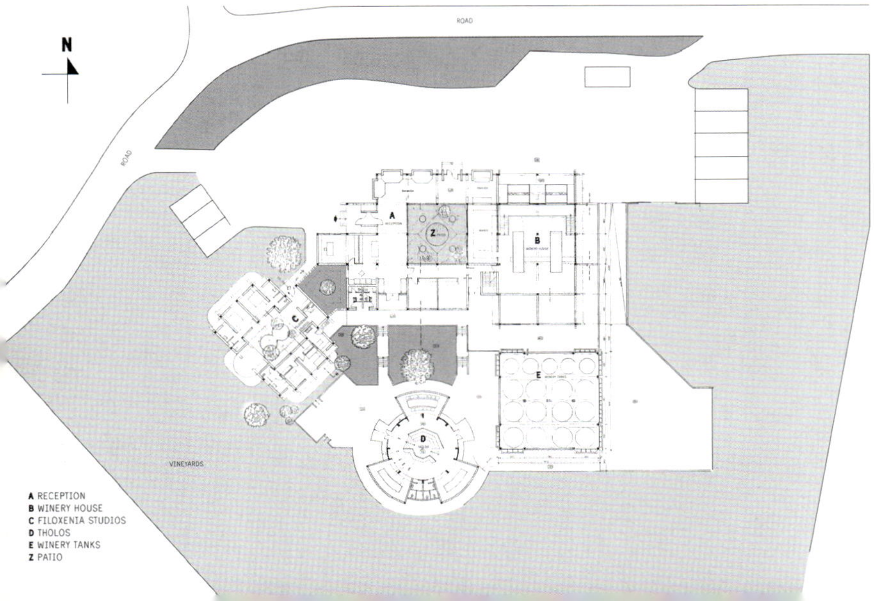
ROAD
N
ROAD
VINEYARDS
A RECEPTION
B WINERY HOUSE
C FILOXENIA STUDIOS
D THOLOS
E WINERY TANKS
Z PATIO

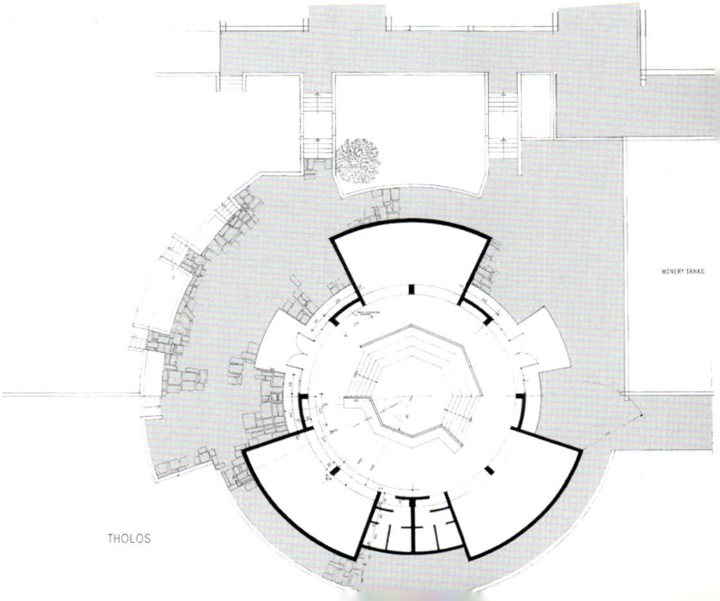
WINERY TANKS
THOLOS

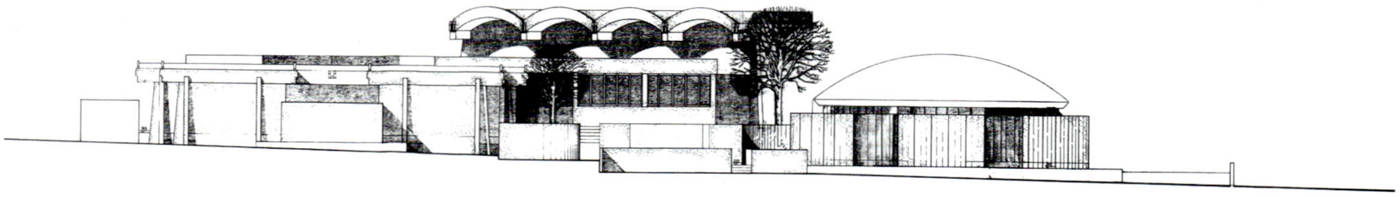

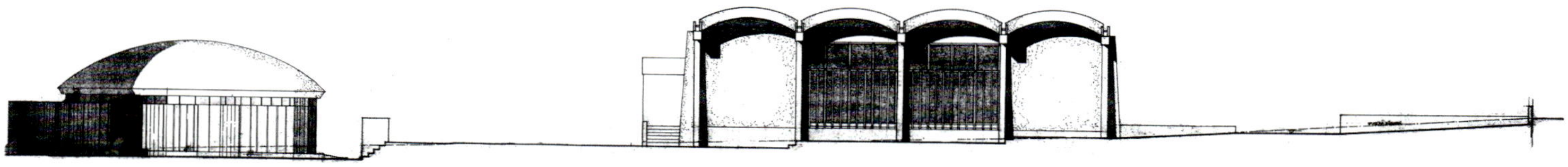

construction of the building repeats a typical Cycladic motif found in the architecture of the island; sometimes white, sometimes dark blue, inspired by the colours of the Aegean. Lastly, the dominant material is white plaster and the flooring used in the courtyard outside the Dome is slate and pebbled tiles which form decorative mosaics in earthy hues.

Architects: Arquitectura X

X HOUSE With Flexible Structure

结构灵活的X住宅

Not having a site when we started design on our house, we set out an elemental scheme that could work both in Quito and the valleys east of the city; this meant distilling our experience into an abstracted form, inspired in the work of Donald Judd, that could be placed in any of the sites we would be likely to find: an open ended box, whose spatial limits would be the eastern and western ranges of the Andes.

As we had no actual place, we looked to the spaces we felt our own, and found the patio as the essential place maker throughout our architectural history.

On the other hand was our fascination for the prototypical glass house and its possibilities in our year round temperate climate.

While the patio creates a sense of place it has to be enclosed in order to work, so the mountains can't become the spatial limit. The glass house is perfect for that unlimited sense of space; the addition of a patio to the glass house gave us the chance to adapt to the different site possibilities.

We separated the private and public spaces defining a patio, the service spaces and circulation could be added as a plug-in as needed depending on site conditions, further defining the patio.

Finally this diagram could be fitted into the open ended box according to specific site conditions that would define orientation, size and proportion.

MATERIALS AND CONSTRUCTION

Construction and materiality were parallel considerations and decided on similar premises: a building system that could be modulated and would allow decision making based on varying budget and site conditions. That meant it had to be as far removed as possible from the outdated, waste generating, inefficient and absolutely limited standard building technique used across the country for every building type and size.

Finally having a site with extraordinary conditions but a very limited budget (cost with garages 155.000 us dollars), the house was ultimately defined:

A light steel structure on a concrete plinth supports the rusted steel and plywood open ended box. Circulation, service spaces, and the elements that make the house function are inserted in white and enclosed in polycarbonate for protection from the stronger western sun. All services run concentrated parallel to circulation, rain water is kept separate from drainage, it is surface collected, and flows down the rusted ends of the box into the ground.

MAIN MATERIALS

Open ended box

3 mm cold rolled steel, rusted and weather coated, for exterior planes.

9 mm varnished plywood for walls and ceiling inner planes and 15 mm varnished plywood for the floor plane (marine plywood for the outside)

Functioning Elements

Concrete floors polished and painted white

Drywall ceiling and partition walls, plastered and painted white

Sandblasted glass and polycarbonate partitions

8mm polycarbonate enclosure, translucent and ice finishes

8mm single pane transparent glass with aluminium bracing and doors

Structure

Reinforced concrete plinth and foundations, reinforced concrete ground floor slabs

Steel columns, 160 mm HEB sections, rusted and coated

Steel beams, 200 mm IPN and rectangular sections, painted, 150 mm rectangular section substructure, painted

First floor and roof, reinforced concrete slabs over steel deck, weatherproof white paint (with polyester reinforcement on roof)

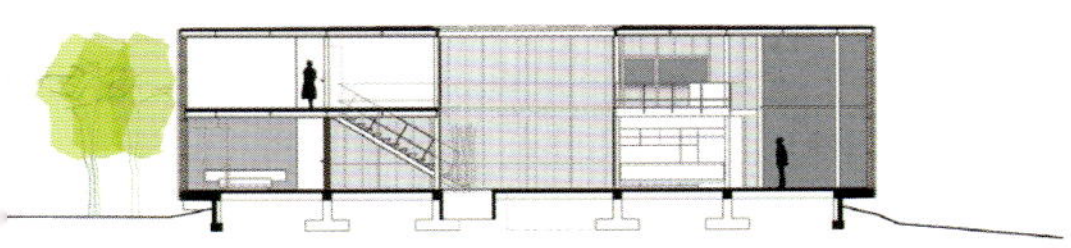

SECTION 2

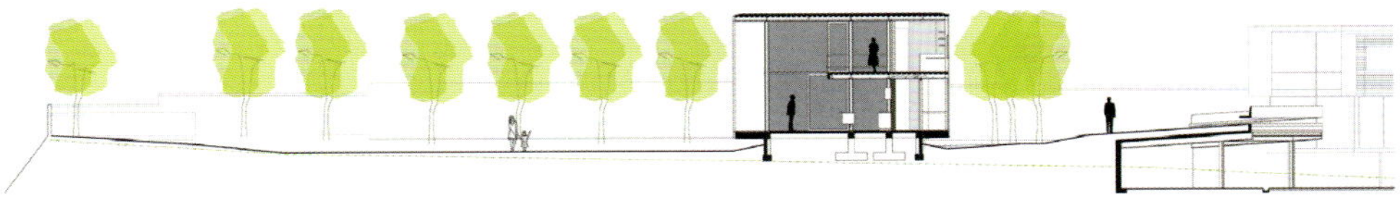

SECTION 1

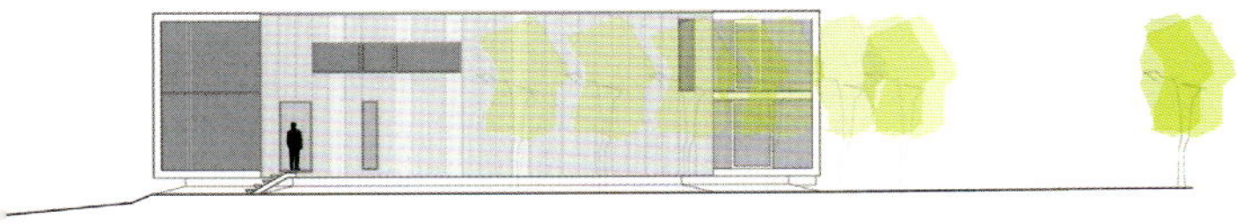

WEST ELEVATION

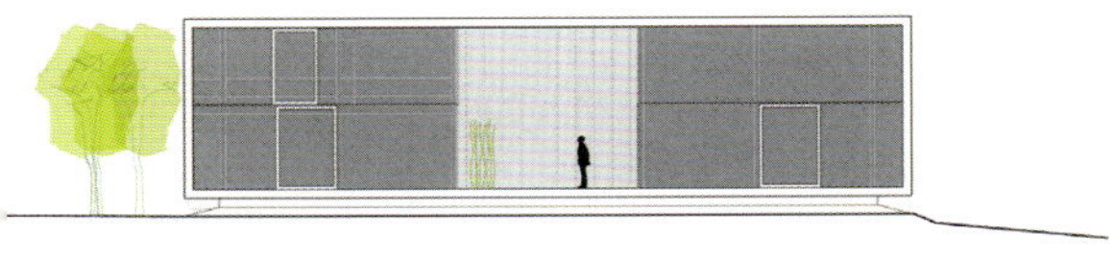

EAST ELEVATION

NORTH ELEVATION

1 HOUSE
2 GARAGE (UNDERGROUND)
3 GARDEN

SITE PLAN

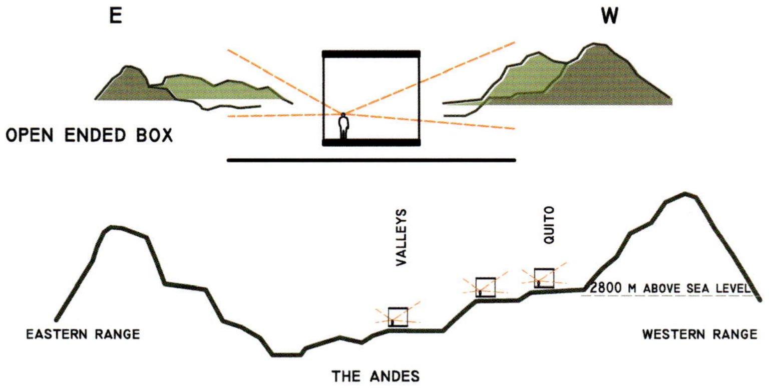

ELEMENTAL SCHEME APPLICABLE FOR QUITO & VALLEYS

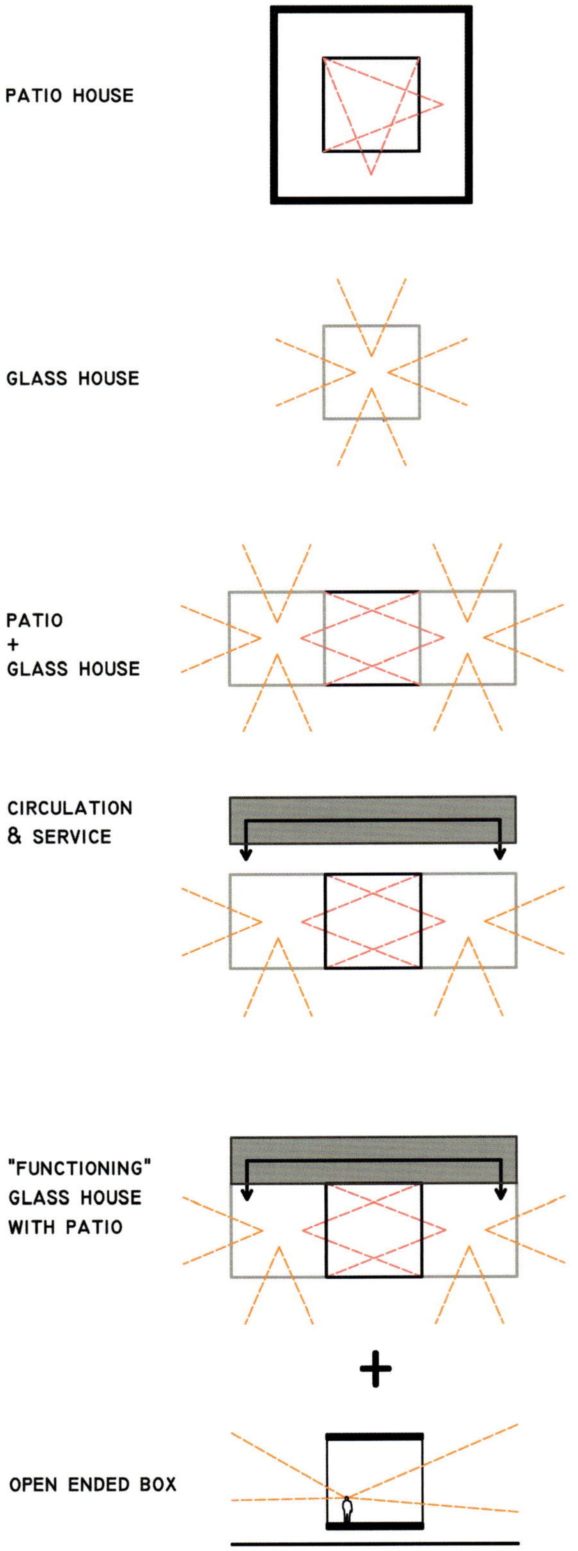

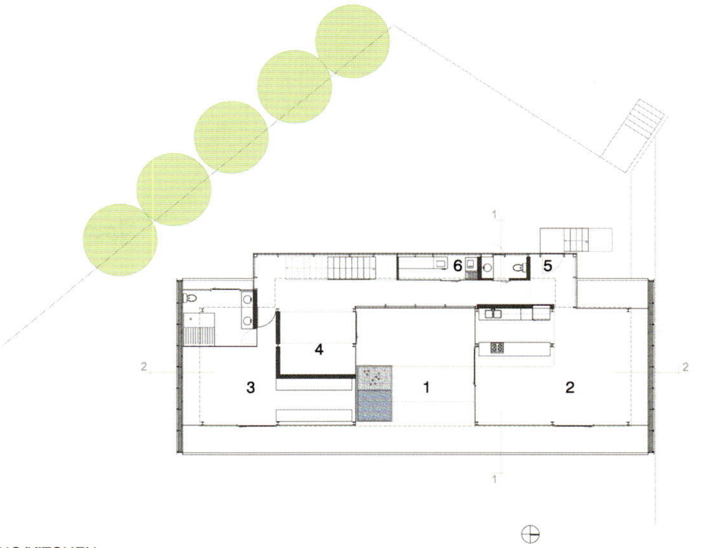

GROUND FLOOR PLAN

1 PATIO
2 LIVING/DINING/KITCHEN
3 BEDROOM
4 FAMILY ROOM
5 ENTRANCE
6 SERVICE/STORAGE
7 LIBRARY/STUDY
8 PLAY ROOM
(GUEST-FUTURE BEDROOM)

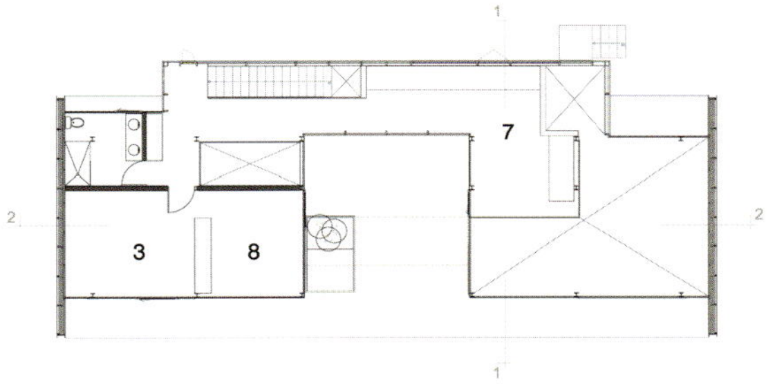

FIRST FLOOR PLAN

Ziranenge/Angel — A Healthcare Center for Children
儿童保健中心

When we downloaded the information regarding the different sites proposed by the competition it was difficult for us to decide which site to pursue as we have limited understanding of the conditions on both sites based on the raw data available. Hence it became critical for us to formulate system/tools of design that could be interpreted for various sites irrespective of boundaries. Keeping this ideology in mind we decided to select Lat. 2°9'"S Long 30° 31' E as our probe site. In order to understand accessibility to the site we began by drawing a circle 1acre in size (depicting the built up allowed) within the site as the loci for he site. We then began to trace paths of people who would arrive at the hospital from distances out of the region as well as those within the periphery of the site which became our point of access into the site. We later moved the foci on site to various locations as to understand how these walking patterns would morph and to challenge the notion of the location of the building within the large site. Our next move was to focus on the acre and develop the overlaps in various fragmented options that could take the form of the program. We selected one of these options to take the form of our design proposal. We have titled our proposal as Ziranenge/ Angel as we feel this project is one that a) will be responsible in bringing relief to the children and pregnant women who have limited access to such facilities b) we also feel that the project in a way would become the regions "child" where everyone around would contribute to it in some for or the other.

Our goal was to create a structure that would offer security, comfort, be cost effective and have the ability to create its own identity. We are proposing the use of Adobe bricks as they are not only sustainable but their ability age naturally and to appear as part of the landscape are essential qualities of design. An important factor in terms of design was also to propose the fortress wall as a design element that rises from the ground around the building and is carved by local artisans in a traditional fashion.

The trellis provided above also rises from the ground at times is to be constructed out of dead wood found and can be built in course

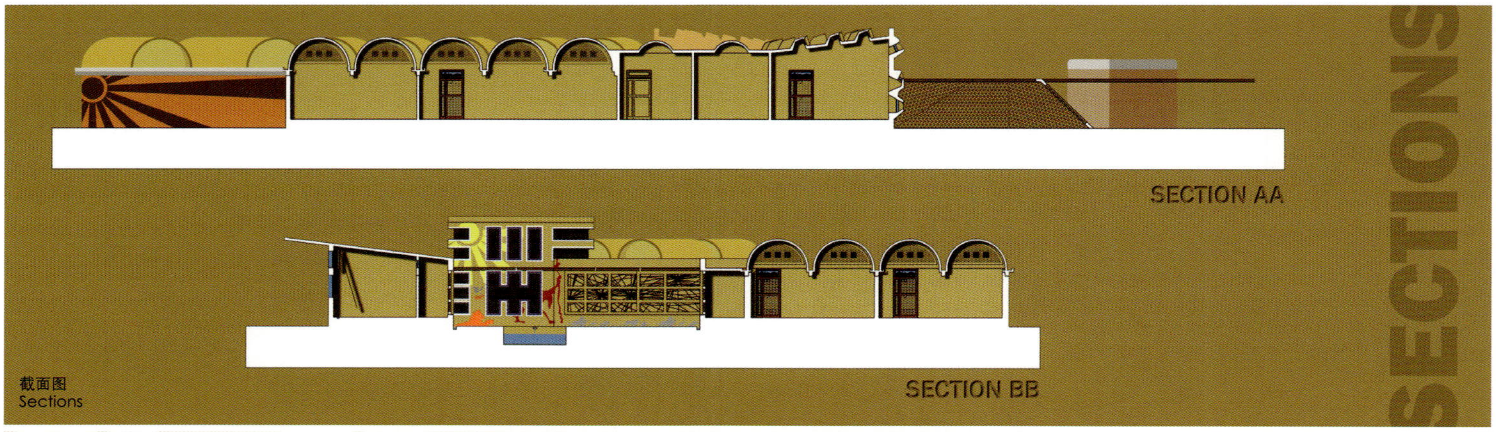

截面图
Sections

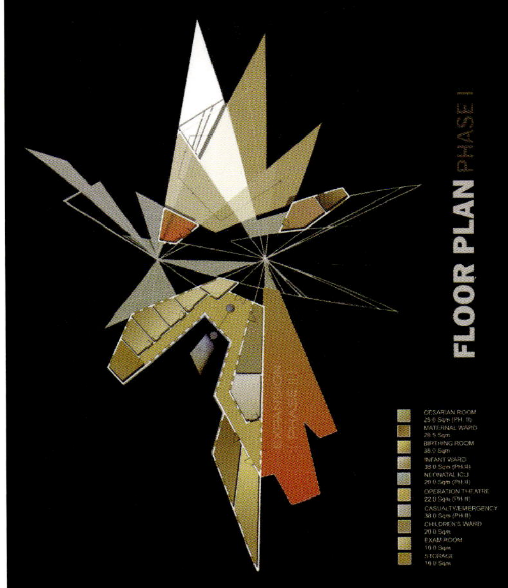

规划一阶段
Phase I Plan

规划二阶段
Phase II

公共住宿区地形图
Diagram Showing Public Stay Zones

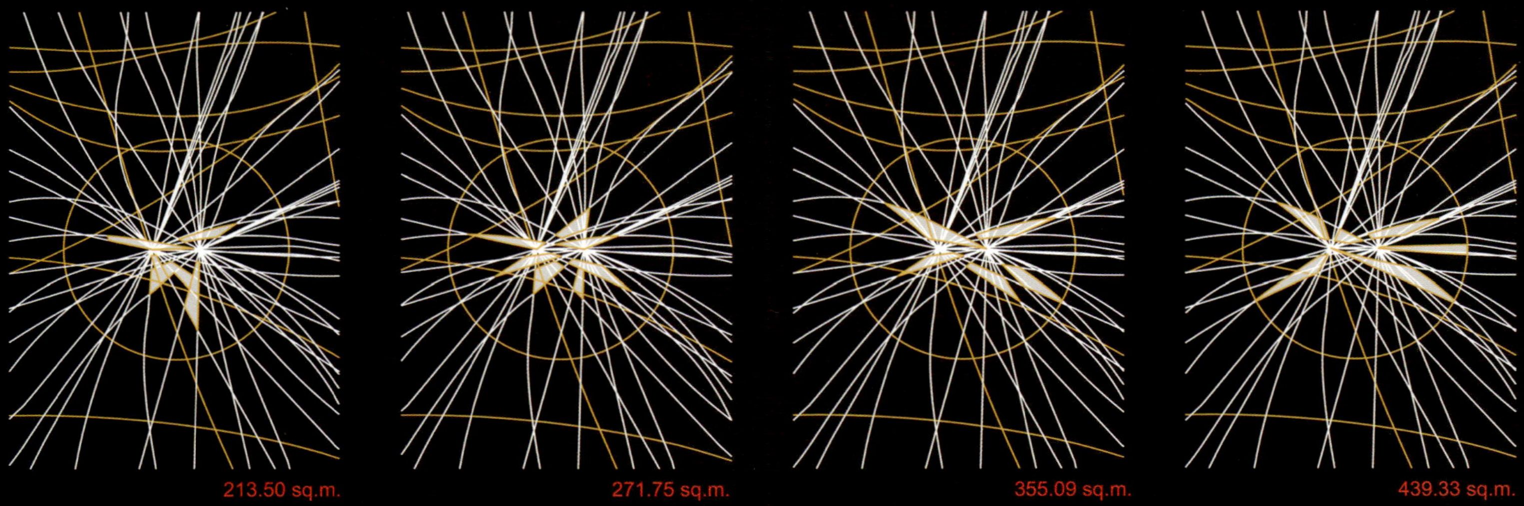

项目各种不同可能性分析
Analyzing the Different Possibilities of the Project

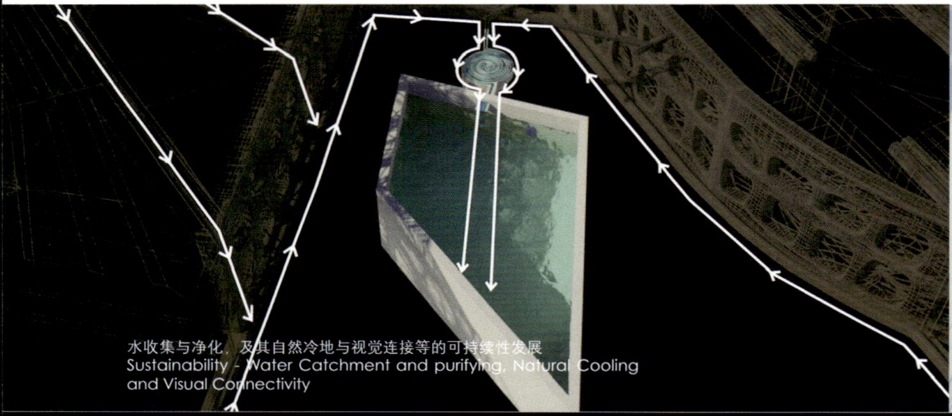

水收集与净化，及其自然冷地与视觉连接等的可持续性发展
Sustainability - Water Catchment and purifying, Natural Cooling and Visual Connectivity

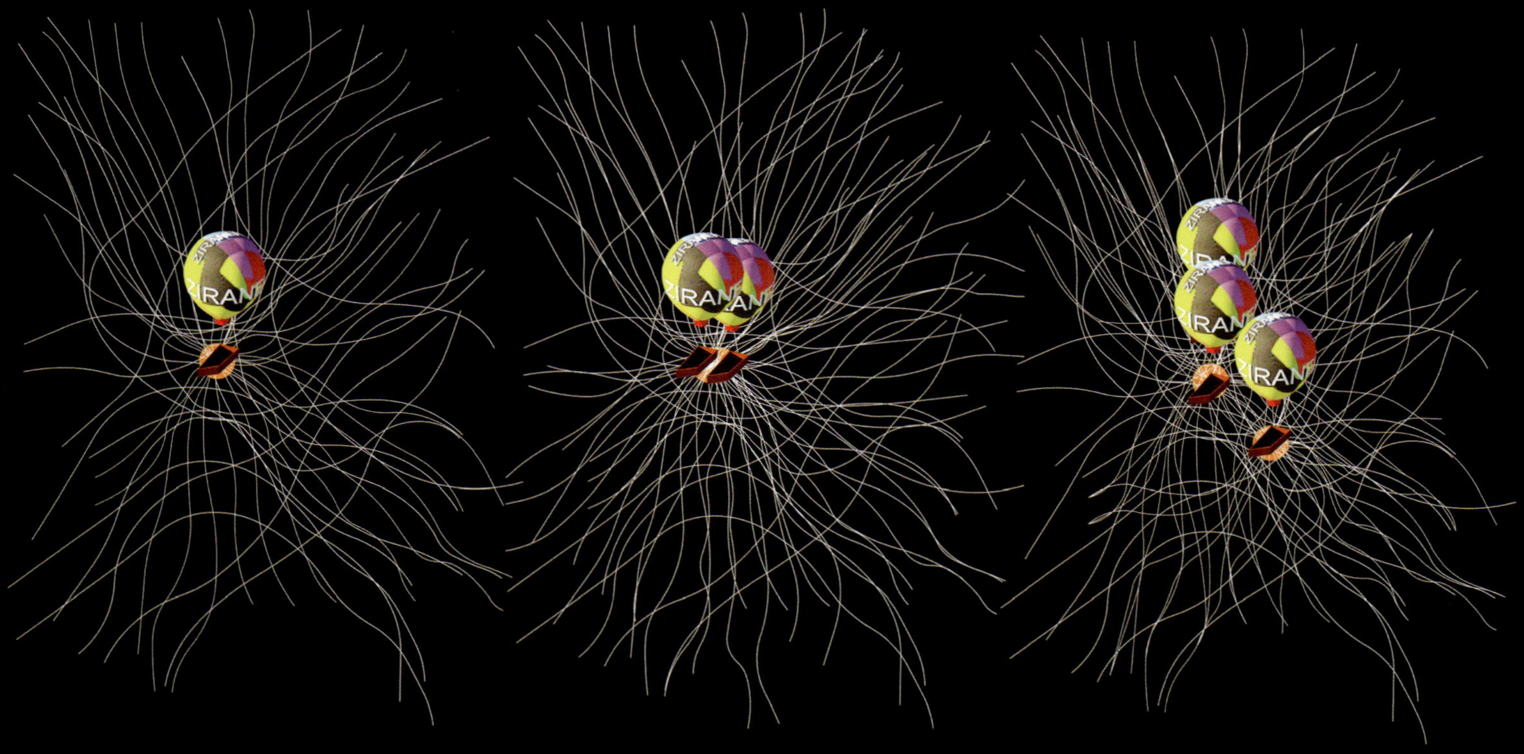

该项目在卢旺达全国的推广行动计划
Strategizing the Creation of Hospitals/angels All Over Rwanda

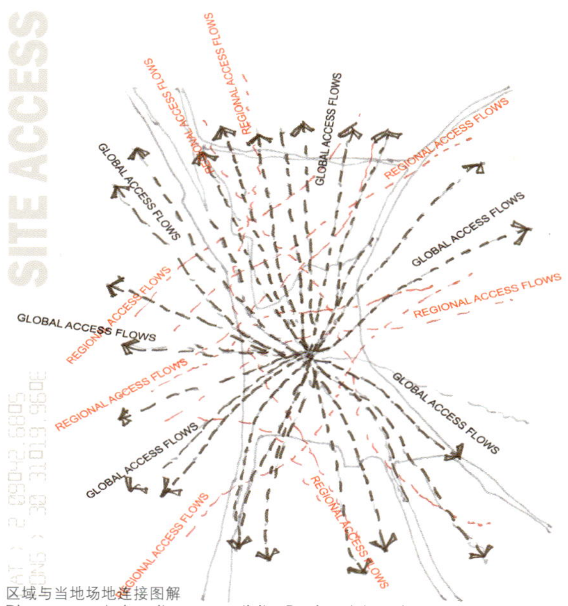

区域与当地场地连接图解
Diagram exploring site connectivity- Regional+Local

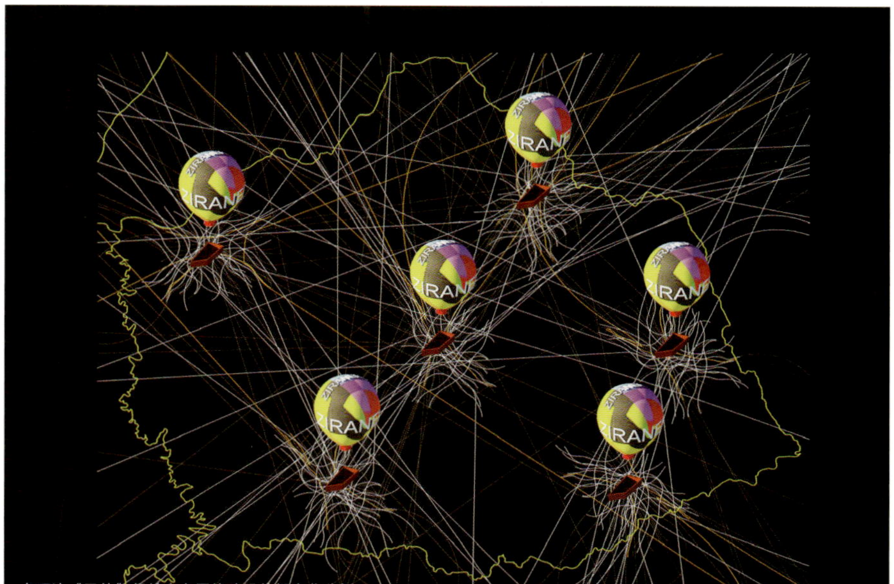
卢旺达“天使”保健中心网络建设的概念化分析
Conceptualizing the creation of Rwanda's Healthcare "Angel" Network

水收集与净化，及其自然冷地与视觉连接等的可持续性发展
Sustainability - Water Catchment and purifying, Natural Cooling and Visual Connectivity

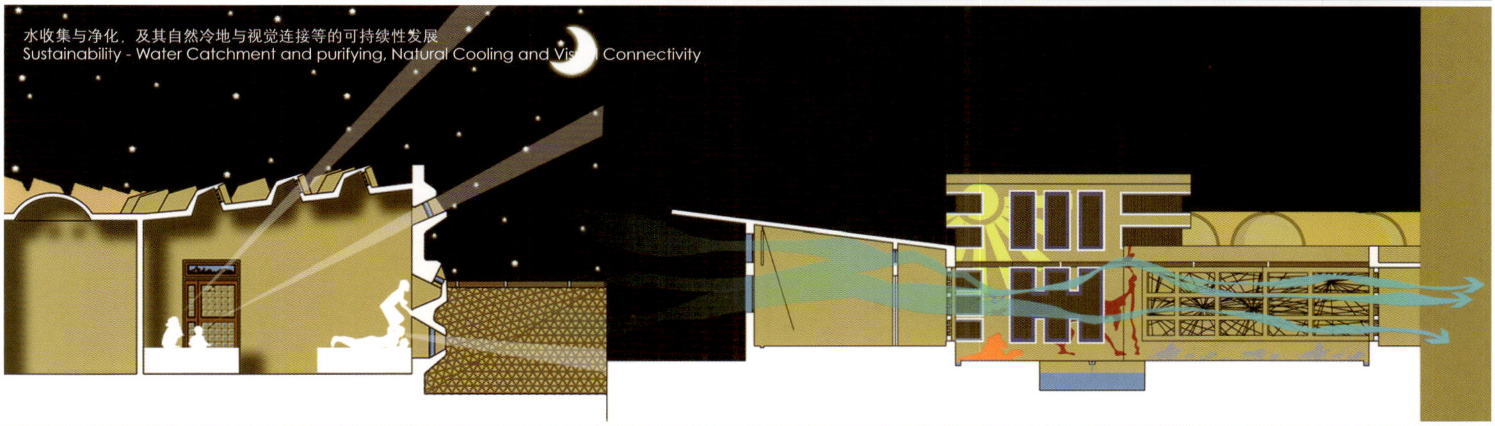

of time. The fabric portions of the proposal are to be considered as bed sheets or canvas to offer protection to people who have travelled for long distances to the hospital. The trellis can also be loaded with solar panels whenever the institution can afford to do so. The roofs of "Angel" are designed to collect water through gutters engraved into the slopes or at the ends of the vaulted roofs. The water is collected from the roofs and processed through an ancient traditional water purifying system designed in Mandu, India which involves the physics of processing water through centripetal and centrifugal forces to rid water of its impurities before use. All the roofs proposed are to be covered in broken china/porcelain in order to reflect heat radiations. For the children's ward and the infant ward we have designed the wall in a way so that they may have an interaction with the outside world even though they are confined to a ward. We have designed the roof of the ward so that the children in bed can look out to the stars through little mud pods which are situated on the roof and the walls. These also add to the insulation of the walls further. We would like to incorporate local children's art on the walls of the structure. We have limited the use of glass and stressed more on the use of intricately locally carved meshes to be used as ventilators for ventilation and cooling purposes.
The room for HIV prevention is incorporated with a niche to place a computer and screen to transmit information to all passers by. The amphitheatre is designed to be in close proximity to the HIV room so it can work as an informal space for conversations or seminars. As part of our escape strategy in case of any catastrophe we are proposing a hot air balloon that may be used to transport the patients from one place to another. The balloon also acts as a marker in the sky indicating the presence of an "Angel".
In time we believe we will be able to create Angels all over Rwanda and they will formulate their own healing constellation which grows from one to two to three to an entire network of angels where each one is designed by the process of understanding the existing conditions with viable solutions that lead to their integration into the fabric of the country.

项目中文

500 Cubic Meters — The Perfect Embodiment of the Architectural Art 006
500立方米——建筑艺术的完美体现

根据北京市新建和改造文化开发区的计划，此设计概念提倡艺术个性化和艺术家个性化的设想，无论他们属于哪一个行业。因此空间达到500立方米的建筑物之间的间隔会很小。这一间隔会形成某种直观而微妙的联系，使人们能短暂地与艺术家的住所及其作品进行互动，这也是处理较大城市规划的一种客观方法。

除了房屋的庭院，还能创造更多的空间，用于更加公开、规模更大的展览会。这样就能将多种作品集中在一个地方。

房屋的内部结构颠覆了个性化的理念。每座房子的房间结构都大同小异，仅在房间长度上有所差别，这是由房屋框架的大小不一所致。尽管都保持了500立方米的空间要求，但在尺寸上却分为三种比例，每种比例分为三种类型，体现在房间大小和窗户大小及其布局上。底层入口处为陈列室，可通过坡道进入，用于陈列艺术家的每件作品。鉴于房屋要低出1米，因而专门修建了这一坡道，这样就能利用房屋平台，遵循房屋正面高度10米的设计规则。

一楼工作室是创作艺术作品的地方。艺术家可在此处的开阔空间中自由创作，利用底层设施也很方便。除浴室和厕所之外，此开阔空间再无其他阻碍。

二楼结构设计微妙，以滑门将过道和客厅隔开。剩余部分则为开阔空间，用于摆放艺术家的个人物品。

所有房屋的顶楼均建有私人泳池，隔空远眺就如一池被房屋间距和混凝土屋面板隔开的湖水一般。

屋内，由于地板采用半透明材质，这些泳池还可起到照明的作用。呆在卧室内，人们可以体味到池水波纹带来的宁静。

不同的艺术理念，不同的艺术家，设计出不同的房屋。

我们设计出高雅的建筑外观来展示艺术家的才华，同时也显示了每座房屋的特色与它所代表的每一种艺术。与此同时，所有房屋的室内设计也是一种艺术的体现。

A House in Jeddah 010
吉达住宅

该住宅位于吉达，项目场地长28米，宽30米。住宅设计旨在探索新的空间形式，在现代的环境中融入根植于当地传统的建筑结构语言。住宅的结构就像是一个凸窗，开放庭院则与立面设计、露台的浅色调融合于一体。

A Kindergarten Pedagogical Building Filled with Naiveness 012
童趣十足的幼儿园教学楼

该项目是一个幼儿园教学楼建筑项目，项目客户要求建筑师设计建造一座拥有6间大型教室的幼儿园教学楼。项目施工分两个阶段进行。该项目的建造应充分利用项目选址周边的自然景观，突出其环保性与可持续性，同时也应考虑到教学实用性。

幼儿园教学楼的设计建造十分注重与周围自然景观之间的自然融合；整个建筑坐北朝南的方位设置，是建筑师们出于对生物气候因素的考虑，这样的方位设置有利于建筑物获取冬暖夏凉的室内舒适度；在教学楼建筑的正立面和与其相对的人造假山之间，有一条通往农场景观的走廊。教学楼的户外空间则作为幼儿园的心理教育基地，为孩子们创造一个舒适、轻松的学习环境。

学校是孩子们除家之外的另一个生活中心，建筑师为了把这一生活中心打造得更加完美，在教学楼正立面的走廊外围铺设了彩虹门帘，这样的设计不仅令整个建筑物显得亮眼夺目，也对激发孩子们的想象力起到了积极的推动作用。

教学楼建筑材料与色彩材料的选用和搭配，在某种程度上是为教学服务的，建筑师在设计建造过程中特别注重这些材料对孩子们的视觉、听觉、触觉、嗅觉等方面的影响，同时也十分注重它们的环保性。

A Novel Carapace House 014
新颖的甲壳住宅

“生活机器”式住宅的构想，早已成为科学家探讨的对象。现在，甲壳式住宅的出现，使得这一设想变成了现实，而这一新颖微型住宅的出现大部分是依据大量的工业原理和运输结构原理来设计的，它取代了传统的建筑材料和设计方法。

这种甲壳住宅，大部分的设计施工都是这些工业及运输结构原理的具体体现，从而使得这一微型可移动住宅，能够适应各种不同的外部环境和内部格局，同时还能符合不同住户的要求。这个项目的开发探究了独立建筑的本质，也以十分环保的形式向人们展示了一种高层次的生活水平以及灵活的生活空间概念。

设计方法

建造甲壳住宅的意义在于，它为一个四口之家在极富挑战性的自然环境中的生活提供了一个独立型可移动住宅。这种住宅由它的三条支柱支撑而起，为主人在房子底部的安装与维修活动留出了一部分空间，而且，由支柱支起住宅高度，可以很好地适应各种不同的地形，也为住宅的一楼提供了很好的接近大自然的机会。现有的地形令三个支柱无法有利地支起一对运输轮。这些结构元素构成了甲壳住宅的主结构。这一住宅的外墙（点）基本不采用钢结构，而是采用隔墙把每一个房间隔开，这些外墙（点）又是由覆盖着鲜亮色彩的纤维玻璃层构成的，它们起到了很好的保护层作用。外墙特质玻璃上的白色漆，有着很强的防紫外线功能，它可以被阳光直射数小时之久，这些特质的墙外玻璃与一种普通的玻璃相结合，组成了一种十分“厚实有效的外墙”系统，这种外墙系统具有冬暖夏凉的环保功能：夏天它能将热气挡在门外，同样，冬天它也可以抵御外界凛冽的寒风，以确保室内一年四季在任何环境中都能拥有适宜的温度。住宅的入口处设在一层；第二层有两间卧室、一间学习就寝两用室和一间桑拿洗浴室；住宅的顶楼是一个可开放型多功能活动室，它可供主人活动、就餐、烹调、观看电视、思考问题等等。在此，还要特别感谢设计师们设计的可配置系统装置，包括：厨房设施、冷风机和液晶电视。

A Secluded House in the Suburb Tokyo 018

东京市郊一座清幽的私宅

这是一座私人住宅，宅内居住着一对夫妇和他们的宠物狗，这座私宅位于距离日本东京市中心30分钟车程的一个典型的卫星小镇内。一直都在迅速扩张城市范围的东京，早在20年前就将这个卫星小镇纳入了它的管辖范围。虽然它成为了东京的一部分，但是，它还依然保留着一片树木林区。这片林区位于这个小镇的城市化边缘地带一个小山丘的南坡上，那里环境清幽、阳光充足，就像一个自然绿色公园。然而，这一地区的地形不佳，呈现出形状不一的梯形状地势。

因而，在此地建造一座住宅，首先应当摆脱这一选址的地形困扰。于是，我们开始从这一选址的中心位置入手，寻找相对平坦的地势。我们在相对理想的地势上建造三个主功能房间，这三个主功能房关系着整个住宅的朝向；同时，也与其他未确定其功能的房间相连接，从而，组成一个同质三枝的立体空间。虽然整个空间是一个连续性的整体，但它的三枝结构能够区分每个房间不同的功能，并且能够使得每个房间多功能化。底楼则形成了一个混合式缓冲区，为屋主提供了自由的活动空间，并且这样的结构设计有利于空气的流通。三个枝架之间的空间还组成了三个风格各异的小花园，形成了室内外景观的平衡点。

由于这一地区的地形特殊，因此，该住宅的地下室与路面平行。我们为主人在最方便的一层楼设置了办公场所，这一办公场所又分为会议室、工作室和档案室。在一楼，有一个通往二楼的入口处，入口处设有一个楼梯，楼梯上去便是二楼。二楼是一个综合性的活动场所，分为起居室、厨房、饭厅、书房、浴室和卧室。一楼与二楼之间还有一个小空间，每个房间都设有一个通往这一空间的入口，以便于大家在这一自由空间内进行各种互动活动。

ACEN — Pilot Project for a Modern Trial Detention Center 022

现代感化中心实验项目

这是一所监狱的设计项目。项目的设计理念基本上分为两个方向，每一种都很好地利用了各自在物质上和概念上的相关联系，但每一种都不可分离地趋集。一种是正式的功能性的，外部世界的真实性在正式的功能空间的辅助下以恰当的形式和程度对监狱进行渗透和影响；另一种是为监狱引入和加强奖励政策，在服刑者积极地改善和转变的情况下，为他们提供较为开放和高质的机会与外界进行交流。后一种必须在许多措施的安全保证下实行。

设计理念体现了如手套功能的结构，所有的服务机构是手掌部分，各狱室则是与手掌相连的各个手指，与之联系并受其牵制。

一个连续的外部路堤状围墙环绕了场地四边中的三边。所有的办公室、连通道路、结构、管理部门和安检部门，以及各狱室都封闭在这一路堤之中，所有的人员必须从唯一的检查点进入监狱。

主建筑包含地面上的三层结构和一层半地下结构。内部每个尺度各异的现代化空间，都通向内部道路节点，体现了视觉上的联系。同时与户外空间、休闲运动设施相连。这些空间的尺寸和大小都很好地提升了南部的入口到北部边界的联系。这也为扩大空间外表和连接两个相邻空间提供了可能性，如此实行的话可以形成一个容纳4人的56平方米的空间，其中包含卫生间、厨房等基本设施。

结构部署通过热质量的研究得到了提升，地热和太阳能这种可再生能源的使用强化了建筑的生态特色。这样，舒适度和节省能源都体现在这项绿色建筑的设计中。

Adharshila Vatika 026

清新的学前教学楼

设计概要

项目设计了5至6座供30个孩子使用的教室，其中包括活动室、音乐舞蹈室和医疗室等，均为2年半至5年级之间的学生设计。管理办公室附近的接待室可以容纳12至15个人同时等待。客户要求房间的设计能够非常贴近自然，并且自然光线充足。经过设计的户外空间包括一些游泳池、滑梯、旱冰道等室外活动区。学校采用了中央空调系统。应客户的要求，设计通过丰富的材料和彩色的应用，呈现出儿童乐园的特点。

设计过程

建筑的设计强调了功能需求和外立面的特点。建筑是两层结构，位于场地的正中间，因此所有的房间都拥有充足的自然光线。设计非常简约，一座走廊将中心庭院周围的所有空间相连接。内部的环境被转换为儿童空间，具有非常清晰的辨识度。孩子们很容易产生归属感，并愿意在这里玩耍。建筑平面设计呈六边形，中心庭院环绕其中成为一处安全的儿童游乐空间。

大楼面向主干道的正面设有一个入口，为了安全起见，供学生出入的大门设在侧面交通较为稀少的道路上。正面的入口通向管理办公室和接待区，另一个儿童出口直接通向地下室和一楼的楼梯。游泳池位于场地后方的角落里。

建筑大型的玻璃窗子为孩子们提供了绝佳的视线。

建筑外立面运用了圆形、三角形和矩形这几种基本形状，但是又与城堡的外形不同。从外面望去，既现代又粗犷。

此外立面的色彩和形式与内部空间的风格相统一，体现了室内外空间的融合。

在大楼的施工和材料色彩的选择上都没有抹杀建筑本身的特质，客户也不希望大楼看起来更像是一座大厅。对于设计而言，景观也发挥了比较大的作用，将不同材料和色彩的空间进行和谐的转换。

每一座教室都有着不同的主题，并依照功能和使用性的不同设定了不同的标志。例如，不同的国家旗帜或者彩色字母等。

作为一个巧妙的连接空间，连接所有教室走廊的墙壁和地板上有着许多有趣的元素。走廊上宽大的窗户也很好地让孩子们与周围的环境进行互动。

色彩选择

在建筑的特点营造上，颜色发挥了很大的作用。色彩和形状都非常悦目，清淡优美的颜色是首选，避免了对红色、绿色和橘色这样浓重色彩的使用。

建筑的色彩和形状相得益彰，整幢建筑像是一个漂亮的水果篮，不同的色调和形状就像是其中缤纷各异的水果。

入口大门采用了简约的设计，长方形面板上绘制了深浅不同的粉色。与大门不同，小门上是各种颜色的圆圈，呈现出卡通感十足的学前班风格。

正立面的圆形元素恰好是正面入口的轴线。

Aeroform — Subsidized Urban Living Module 032

机翼形建筑形态——都市生活空间的补充形式

随着世界人口的增加，土地越来越受到人们的重视，并已经成为了最珍贵的资源之一。尤其在高密度的城市空间，对土地需求的不断增加和有限的土地面积使得许多想在城市中生活的人们的购房历程面临着更大的困难。设计师们必须在城市的空间布局上寻找一种新的方式来满足现代生活的需求。

相较于拆除原有的结构来营造新的空间，在城市中用新的方式充分利用剩余空间显得更为合理。建筑师采取了一种在大多数城市中非常普遍的元素——广告牌。如今，人们普遍认为广告牌是展示两维图画的简单结构，在他们的印象中，这只是一个普通的平面。但如果将广告牌想象成多维的元素，我们将如何利用广告牌背后和周围的空间呢？

作为一种原型，广告牌可以演变为多维体来营造不同的生活空间，满足大城市的人们拥有自己房屋的愿望。同时广告牌和住宅的结合可以为居民提供用广告牌增加收入的机会，由此抵消一部分他们在高消费的大城市的生活费用。广告牌可以做到自给自足。

由于其预装置和轻型材料的特质，广告牌可以组装好再运到项目所在地，还可与现有的广告牌进行组合使用。外表是隔热结构面板制成，并覆以碳渣层，能够减轻结构的重量，并减少对于现有的广告牌的压力。

建筑的形式取决于其位置和项目特点。反映道路和机动车相互关系的外表在户外呈流线造型，符合空气动力学原理和气候特色。项目的名称来源于机翼这种坚固而方向感强烈的飞行物造型。在每个单元尽头，可操控的窗体能够提供良好的空气流通和自然光线。室内的阶段性空气紧缩可以优化总体空间。这个过程可以在空气动力学基础上促进单元之间的空气流通。

对于机翼形建筑来说，可持续发展是必需的。由于土地是珍贵资源，降低对土地的使用是机翼形建筑的重要目标。因此它们都建造在现有的广告板上，与土地的接触面积非常小，对环境的影响也降至最低。由于对原材料需求较小，这种建筑形态可以直接建造在现有广告牌的钢铁结构上。

在一些项目中，可以对两个或更多的广告牌之间的空间进行利用。未来，这种建筑原型将演化成不同的形态，例如V形、三角形等等。正因为其普遍的特色，广告牌可以在世界上任何一个城市或地方实施。

Alley Pond Park 038

艾丽池塘公园

艾丽池塘公园是一座高尔夫公园，里面设有一个18洞的普通球场、9洞的儿童球场，这两个球场都是由Harris Miniatrure公司设计的。公园还包括一个88球座的练习场，214平方米的会所，会所中设有专业用品商店和小吃店。为了与周围的环境融合，并最大化利用太阳光线，会所是平顶的低矮建筑，非常的节能并富有热情的氛围。中庭则是玻璃结构，全天阳光充足。小吃店的高房顶是一大特色，阳光弥漫在中庭之中。当太阳落山的时候，卤素灯营造出优雅的夜晚氛围。

建筑采用了灰色的混凝土和暖色调的西洋杉木作为材料，并使用了粗线条的窗结构勾勒出建筑的形态。会所的形态线条干净而简约，并采用了回收的玻璃材料和可持续性木材，展示了设计者对于环境的兼顾。阳光充足的会所的户外庭院为客户提供了享受花园和儿童球场的舒适空间。建筑师Paul Athineos应客户要求，为会所进行了绿色的环保设计。这座会所不仅适合家庭游客，还适合年轻人进行约会和周末聚会，高端的游客也可以在这里找到自己的享乐空间。会所成为了高尔夫游客寻找快乐的空间。孩子们可以在父母打球时坐在庭院中玩耍。

钢铁结构的会所在材料的颜色上也是精心营造了冷暖色调的对比。外立面的窗体能够为室内引入自然光线。当外立面和中庭的窗子打开的时候，自然空气的流通能够减少冷暖设施对于能源的消耗。

光滑的混凝土外表经久耐用，能够防火防蚊虫，并且一旦需要重建，拆除的材料都可以重复使用。西洋杉木十分适合东北部的气候。室内空间包括形态优美的混凝土地板、灯光、竹木地板以及休息室的陶瓷瓦片。小吃店、休息区和收银台的操作台面则是由回收的玻璃材料制成。

矩形的会所里设有小吃店、厨房、休息室、专业用品商店和员工办公室，总面积达到215平方米。会所建筑由一个宽阔的入口通道分割为两个部分，从南至北穿越整个建筑的玻璃中庭为室内带来充足的光线。

小吃店附近是户外的庭院和玻璃中庭花架，这里可以欣赏到外面的植物和景色。即使在下雨天，这里也是一处舒适的户外休闲区。地面是由碎石铺底上方浇注混凝土制成的，方便排水设施和降水流通的使用。

休息设施空间非常宽阔，这里能够满足ADA要求，并为女士们提供化妆间。考虑到使用的隐私性，休息区设在建筑的中心，入口前树立的幕墙、高背椅的座凳和较高的植栽充分保证了使用的安全性和隐私性。

游客进入会所的时候，右方就是小吃店，非常方便。这里有充足的座椅，可以满足所有客户的需求。早晨提供的咖啡和点心将为人们一天的高尔夫运动赋予活力。除此之外，这里也是享受午餐和冷饮的好地方。

专业用品商店坐落在建筑的西北角，这里设有高尔夫用具的陈列区。专业用品商店的上方能够引入很多的自然光线，从这里可以俯瞰到练习场的景致。

维修店位于建筑的西南角，从用品商店和户外都能直接进入维修点。如果客户需要使用维修业务，他们可以在去会所的路上将物品放下，在离开前可到收银台将修好的物品取回。维修点独特的地理位置为游客提供了极大的方便。另外，作为外立面一部分的遮阳棚在维修店使用时可以很好地屏蔽阳光。

Arezzo Courthouse — The Emblem of Architectural Art 040

阿雷佐市最高法院办公大楼——建筑艺术的象征

意大利阿雷佐市最高法院办公大楼坐落于梅狄奇城堡附近的一座历史公园内，新建的大楼包括主审判楼和法官总部大厅，它与一座新古典风格建筑相连，这座建筑原先是当地一家大医院的所在地，而今，它即将被改造成相关的办公楼和档案馆。

新办公大楼的建筑风格与之四周的阿雷佐中世纪城市结构和草木茂盛的历史公园地理环境十分融洽地结合在一起；办公楼的正面外观由别致的连排贝壳状外墙组合而成，而背面则朝向街区、花园以及周边广阔的土地。

该办公大楼的北面是一堵巨型的黑色花岗岩石墙，它赋予了大楼外观一种暗灰色的优雅建筑格调。大楼南面的墙体是由弯曲折扭成波浪状的透明遮阳板构筑而成；这种波浪状的墙体设计灵感来源于许多生物的表皮形状，例如树叶，而如此复杂的墙体结构仅限应用于直线型建筑中。

因此，新建大楼波浪状的南面银色墙体，在太阳的照射之下出现光影，依靠这样的光影，就能为整个建筑的室内空间起到遮阳作用，并且与周围的公园景观相互映衬，也与周边的新古典风格建筑相互融合。

这一杰出建筑的室内空间设计更是令人惊叹不已。光滑透亮的黑色花岗岩地面与绿色玻璃建筑内墙形成鲜明对比；水平放置的穿孔式灰色铝质隔音板的外围种植着一排枫树。整个大楼的室内空间的格局显得宽敞、肃静。在屋顶灯光和升降式电梯金属拉门反射出的光的共同作用与黑色花岗岩地面的映衬之下，三层楼高的入口大厅显得明亮而庄重。

Argerich Building 044
阿格里奇住宅

“好建筑离不开高效的建造，没有建造就没有建筑。建造的过程就是依据材料的特性使用材料。也就是说，石头在建筑上的运用与金属和混凝土是不一样的。我相信，只要能根据材料的特点正确使用，就可以在当代建筑中使用任何一种材料。如果在某一地区只能找到石头这种材料，那么就用石头来建造吧。”

——Aris konstantinidis, Architecture，建筑师，1964年

这个包含5个住宅的建筑位于布宜诺斯艾利斯del Parque别墅区一个宽8米，深长35米的地块上。其中4个住宅空间为三层，另1个为一层。一层住宅空间主要用于连接它上面的两个露台，并将下面的空间充分利用起来。

建造和结构逻辑都是材料工作的一部分，决定了空间的特点。这个建筑使用了一些非常经济的材料，并显示出它们与以往不同的特性。所有元素组合在一起，架起建筑的结构，使材料拥有了与以往不同的形式、功能和语义。

建筑由两个12米高的部分构成，中间是一个6米高的露台，可作为所有住户的室外休闲空间。

露台边特别栽种了三棵杨树，用于遮挡一些阳光，同时赋予两边建筑一些私密性。露台是聚会空间，但是并没有使用灯光柱来强调它。露台四周围起混凝土墙，整个看起来像个盒子，从不同的角度能得到不同的视觉效果。

Asian Culture Complex in GwangJu 046
光州亚洲文化综合楼

文化舞台

城市舞台为纪念历史性事件及文化活动，使城市广场更富有文化景观特征。城市向现有的城市结构及肌理中延伸，以此来纪念光州悠久的历史。为融入新城市环境中，向外突出的街道被用作柱子。这也意味着现有的城市肌理也将会成为具有历史纪念意义的街道和空间。设计师对旧街道进行了修整，在遵循将城市打造成纪念碑的基础上，界定了城市的结构。城市结构形成的景观片断融入在周围的城市景观中，在满足人流量的同时，还可以举办各种各样的活动。

三维广场中的文化舞台植入了城市结构。01舞台被嵌入街道中成为多功能的场所，与外围的景观元素搭配，成为能举办不同活动的舞台。街道界定的户外空间包括了许多不同主题的公园、商业设施、文化媒体中心等。

街道

通过一系列空间改造，街道中的三维柱子也成为历史矩阵。街道不再是被遗弃的空地，而是积极灵活的框架结构，其中包括很多的活动。重要的历史事件的时间顺序也得以在街道中突出，活动及边界都被用作历史记录。美术馆只将历史事件呈现出来，历史矩阵则主要纪念和缅怀一些具有象征意义的空间。

新文化综合楼在公路交叉口的核心地点将光州市与亚洲连接起来。文化网状结构作为一种设备，将周围地区的功能机制及场地的地形情景联系起来。通过对嵌入网状结构中的线路及方向的重新调整，从现有的人行道中创建出混合结构。

设计师通过文化连接形成一系列历史事件分隔的街道，及以景观形态存在的实践性舞台。高处与低处的景观广场与现有的城市肌理联系起来，街道则融合在整个网状结构中，与所有元素一起形成空间元素。

作为城市广场及景观，宣传街道的制图由一系列的网状元素组成。场地不是单一的个体，更是被用作公共舞台。低处的城市舞台主要是人们的活动空间，而高处举办活动的舞台通过一系列活动的渗透，使城市结构更加多样化。

Atatürk Culture Center and Merinos Park 048
Atatürk文化中心与Merinos公园

该项目原为老美利奴丝绸纺织厂，是20世纪土耳其工业化进程中的重要标志。原先是作为一座公共大楼建于土耳其早期的共和国时期，由M.K. Atatürk于1938年2月2日主持开幕典礼。2000年后整个丝绸厂被废弃，2005年时被移交给布尔萨自治市政府，由其负责将此地开发成文化创意活动中心。在布尔萨工商议院大量的财政支持下，布尔萨自治市政府将此地的大楼及环境进行重新改造，以满足这座正在发展中的城市的当代社会和文化要求。该复兴工程由三个重要部分组成，Merinos公园、Merinos文化中心和Atatürk剧院及音乐厅。

由于时间紧迫，工程在设计阶段便开始施工。具体说来，2006年刚完成招标后就开始着手基础设施及景观工程的建设。Merinos公园及文化中心现已完工，而剧院和音乐厅目前仍在施工中。

Merinos公园

总面积：270,000平方米

景观面积：210,000平方米（其中硬质景观面积为42,000平方米，软质景观面积为168,000平方米）

Merinos公园作为大型的城市区域，将为其周边区域提供服务。园内有成千上万的新老树木、人工湖、慢跑小道、自行车道、座椅区、儿童娱乐区、咖啡吧、快餐吧及其他各项设施，满足了周边地区人们的各种需求。Merinos公园成为布尔萨地区重要的娱乐综合区域。该项景观工程于2006年开始施工，于2008年6月15日竣工。

Merinos文化中心

总建筑面积：61,800平方米（含9,600平方米的停车场）

占地面积：36,420平方米

丝绸厂原来的大楼已失去了实际意义，成为复兴工程的主场地。改造后的文化中心包括以下文化设施：音乐学校、美术展览馆、美术工作室、图书馆、纺织品及丝绸陈列馆。这项工程于2006年6月开始施工，2008年6月15日完成。

Atat ü rk剧院及音乐厅

总建筑面积：44,600平方米

占地面积：5,750平方米

无论在形式上，还是功能上，剧院与音乐厅都与文化中心和公园很好地融合在一起。项目包括一座1,800个座位的音乐厅，专为戏剧、芭蕾和交响乐演出所设计；一个为会议和音乐会所设的、可容纳800人的礼堂，小音乐厅内可容纳300人，还有可容纳1,000人的宴会厅。此外还设有艺术馆及咖啡厅和其他设施。该工程竣工后，必将成为布尔萨地区新的文化中心。

剧院及音乐厅位于文化中心的东面，面临从东面起横跨整个公园的人行道。金属屋顶下的公共广场解决了周边城市肌理问题，并鼓励市民们充分利用文化中心和剧院的各种资源。

Bungalow Olger 052
木质住宅大楼

该项目是设计师在特殊情形下接手设计的一个案例。

设计师当时主要负责TAG建筑公司设计的一个生态旅馆的施工及监理工作。建筑队里面大多是当地的工人，因此，从某种程度上讲，他们需要有人监督他们的日常工作。这也使设计师获得了一年时间的大酒店设计实践经验。

设计师当时住在一个叫做Bijagua的小村里，该地离酒店的施工场地有3000米。整个小村只有800名居民，这为设计师尽快了解并熟悉当地条件提供了难得的机会。一个月之后，设计师便遇到了一位叫做Olger Arrollo的先生，他向设计师讲述了他梦想中的度假别墅。

客户在通向Volcano Tenorio 国家公园的地方拥有一块地，这里有世界著名的旅游胜地。客户对他这块土地的各种情况了如指掌，并且已规划好将他的别墅建造在那里。该地块所处的位置风景秀美，这里可看见Tenorio和Miravalles火山，还可以欣赏到Nicaragua湖。

设计师每天晚上着手该项目的设计，白天则忙于酒店施工。周末大部分时间用于欣赏乡村风景。

客户是一名自行车销售商，拥有森林种植园及锯木厂。因此，该别墅的建造主要使用木质材料，只有在卫生间内采用混凝土来支撑整个结构。该住宅面积为120平方米，两间卧室和独特的浴室扶梯及生活区（如厨房、平台、卫生间和洗衣间）。从二月至五月，该区主要以阴雨天气为主，空气比较潮湿。设计师在窗户上设计了天窗，这样有利于室内的自然通风。悬挂的锌板屋顶起到防止雨水及太阳照射的作用。平台是该住宅内使用频率很高的地方，这样增强了生活区与户外空间之间的互动。尽量减少材料的浪费（尤其是横木），以及合理利用工业制成品，才将房子的尺度控制在这个范围内。

设计师第一次绘制技术平面图，因此住宅的设计与酒店的施工进程密切相关。每周，设计师都能获取不同的灵感，在平面图中加入新的细部、新材料及新的施工手法等。酒店竣工时，设计师也完成了该住宅的设计。该项目是酒店设计理念在木质住宅大楼中应用的典范。

Cash and Carry Food Store　054

C&C食品店

C&C食品店位于布达佩斯的北部，这里原为一处工业用地，后根据城市发展规划，将作为棕地开发进行投资改建。建筑基础面积大约为4400平方米，投资的总价值达300万欧元。

项目场地内污染严重的旧皮革厂需要进行不断的拆除与修复工作。与临近区域的其他场地一样（DIY店、汽车展示厅和服务厅等），城市规划者选定在这块场地内实现商业功能。

设计的重点是这个新的经济功能带必须非常现代，拥有绿色空间，必须保证达成功能性与节约能源。

建筑的设计按照以上原则进行，大厅的结构为钢筋混凝土，夹层板铺面保证了热量不被散发。

项目的设计非常出众，分块的设计从视觉上缩小了建筑尺寸。

每个立方的材料和色彩都不同，强调各自的重要性。建筑入口处的立方则被改成了一个透明的玻璃表面，可将人们的注意力吸引至主要入口。

因为这个项目是这个地带第一个建筑物，所以它有必要为其他规划中的建筑提供足够的建筑基础设计概念，决定未来建筑的风格与形式。

CB30　056

CB30大楼

大楼有三个房间面向街道，后面的两层住宅面朝室内的天井。公共区域被认为是面朝街道或公共室内空间的单一空间。位于大楼前的树木异常壮观，楼上同时还能观看到Polanco公园内的秀美景致。由于设计师想将室内空间延伸到街道和公园中，这些条件便成为设计实现的中心目标。

最后，大楼正面与打破内外区域的限制原则对应起来，使大楼变得更加空旷，保证了各个元素之间的连续性。设计师充分利用结构几何学的设计，采用40厘米的厚木板来支撑15米的横梁，从而避免柱子遮挡人们的景观视线，而后设计师将厚地板与墙体紧密地联系起来，将边缘地带隐藏在大楼正面内。

Cellular Network City — Peculiar Architectural Conception　058

蜂窝式建筑——奇特的建筑构想

建筑规划

该项目的建筑目的是对一系列城市狭小空间进行统一的划分与迁移，从而重新定义“都市生活”的概念。许多古城与历史文化名城都在重新调整自己的城市空间布局，城内的各种建筑都以几何分布形式来划分建筑层次。Am:Pm建筑公司为迎接2010年世博会将规划设计一系列的蜂窝式建筑，使其形成一个城市建筑网络体系。

景观是室内外公共空间的相通媒介，也是文化设施与内部空间的媒介。网络化的建筑格局是各个分散式小空间的基础设施纽带。蜂窝式建筑就是将传统的聚集式空间逐步转变成为一个开放式空间。

这种建筑形式变革的巅峰时刻则从实验型展馆到景观与公共空间的彻底转变中得以充分地体现，因此，博览会的遗产能够得以持久地保留。反之，私人空间则保持其传统建筑外观以及与这种建筑外观同质的转变形式。

在城市空间、建筑规模和与之相对的物质层面上，蜂窝式建筑起着基础作用，同时它还起着集合所有城市狭小空间和多孔式可占用空间的作用，而所有这些空间的集合都试图寻求一个方向不同却又能够成为一个连贯整体的富有条理性的城市建筑结构体系，这种建筑体系之所以称之为一个连贯整体，是因为它必须跨越城市主要河流的两岸，因此，它将会随着时间的推移影响它的形成。这种条理性的城市建筑结构体系具有对不同互交空间、部分移动空间以及不同建筑的规模、密度、高度等方面的统筹组织能力。

2010年世博会的景观建设将跨越浦江两岸，而蜂窝式建筑网络系统的建造目的则在于延伸与强化各建筑间的距离，从而产生高密度建筑组织，并为少数的纵向建筑提供横向视角。

参数化设计系统

一种预设式线型三维建筑网系统在蜂窝式建筑中得以发展，并与各线型元素装配的增长与扩大的特殊原则相匹配。具体的设计行为将安排到阵列式模块的过渡曲面内，同时建立起三维建筑网的设计指导方针，从而改变由立方体转变为曲面模块的传统模式，这种方法进一步适用于不同规模的空间互动，也实现了由简单的地方秩序和地方行为所引发的显而易见的复杂建筑形式。

Centre for Nanoscience　060

纳米科学中心

根据俄罗斯目前支持纳米科技发展的政治体系，莫斯科州立大学将营建一个多功能的纳米科学中心，进行先进的研究、实验性生产和教育，并对信息交换的功能进行总结，对先进技术和成果进行展示。

纳米中心选址位于一系列城市地标的中心，周围的交通和生态环境都非常良好。这幢大厦位于Michurinsky大街和Akademika Holova大街的交叉口，在俄罗斯大学和莫斯科州立大学的区域内。除此之外，新大楼所在地还是两条城市规划轴线的汇集地。

为了保持第一条轴线现有的交通和道路循环，即纵向穿过大学建筑物和植物园的道路，纳米中心的尺度相当的低，并且沿着轴线在拱门处弯曲。

第二条轴线是一条粉丝轴，依据奥运会主会场和麻雀山而定。作为一条重要的城市轴，这条线与大厦主入口的坡道相重合，体现了与纳米中心的融合。

从入口开始，依照许可的程度，整个纳米中心的建筑可以被分成五个区域。

区域1：（对游客和学生开放）主要入口和展览空间；

区域2：（一楼设置次要入口）教育区域；

区域3：（学生、游客和研究人员入口）媒体图书馆、礼堂；

区域4：（底层设有独立入口，出入有所限制）研究办公室、实验室、工作室；

区域5：（只有通过区域4的一个封闭式入口进入）ISO1净化室，楼上楼下均为相关技术部门。主要的净化室和区域1之间有着视觉联系，能够满足所有安全和卫生的需求，同时能够提升中心对于学生、专业人士和游客的吸引力。

在空间的划分和与周围环境的连接上，设计也都做了兼顾：教学区与大学和研究办公室相连，并面向大街；公共区直接面向莫斯科河、麻雀山和奥林匹克会场。区域4位于建筑较高的位置和其他区域的北面，并与南面部分由一个内院所连。这样的布局保证了室内良好的自然光线。

中心的主体形状，能够满足所有的前提和规划目标，形成了独特的表面风格，这就是所谓的克莱因瓶。当外立面的外部长度转变为室内时，一个进入中心的主入口就呈现出来，区域1和5之间的视觉联系也有所体现。克莱因瓶不只是自行交叉，它还需要一个额外的尺度。从区域上来说，这五种尺度通过纳米技术的发展向我们展示了人性的一些特点。

在纳米世界中，六角形栅格非常普遍，例如分子结构。在纳米中心建筑的场地和结构设计中采用了六角形这一独特的几何元素。场地的开发还包括：道路，并以拱门穿过中心大楼，形成六角形点阵。其次，是包含街具和工程的空间布局。

从几何学角度上来说，这座建筑采用的是垂直的蜂窝六角形结构填充了克莱因瓶的外表。因此这一外表可以根据变化的建筑和功能要求来进行调整，例如应对光线的调整等。这一混凝土的蜂窝结构以纳米的六角形结构进行了增强，是一座能够很好地承重的三维结构。通过内部架构的串透和固定，空间得到了很好的组合和划分。根据内部空间的功能需求的不同，其中每个部分的尺度和克莱因外表也各不相同。例如，公共区设有最大的开放口，办公区则是比较小而规则的。分隔元素组成了倾斜的礼堂，净化区的结构则能够在为室内带来充足的光线的同时阻隔来自外部的视线。

另一方面，建筑各个部分间的连贯性使得中心大楼与周围不同高度的物体都有着很好的联系，如建筑、景观环境，以及人们的交流（如主入口）。除此之外，中心大楼可以被看做是环境中抽离出的一个任意尺寸的建筑，没有尺度的限制。

最后，一系列不同部分的大小联系表示了在宇宙中所有物体的尺度的序列，从超大尺度的银河到太阳系，以及各种不同尺度的物体，直至人类所知晓的最小单位——纳米，这些内容都有待于继续研究。

Childhood Evocative Mantle 064

孩童记忆弥漫的建筑覆盖物

该项目意在提出一种与老年人现行生活方式相匹配的新的居住模式。项目提出了解决老年人居住问题的质量方案，除了一系列房子的建造外，项目还力图寻找天然的宜居环境，以此构建一组“套餐”。此“套餐”在空间上、形式上、功能上安置于一个建筑覆盖物下，这样，老年人生活主要方面的问题在不同环境里都得到了解决：住宅、花园苗圃、相邻的公共绿化区。

项目位于智利的比奥比奥大区的康塞普西翁市。康塞普西翁市的气候温暖多雨，每年的雨季持续八个月。这样的环境不利于老年人进行户外活动以及整个的社交生活的发展，因为娱乐是维系老年人心理健康的基础。

项目“孩童记忆弥漫的建筑覆盖物”满足了人们一年365天聚会的需求；空间设计和创建中的可持续变量弥补了这个不利的环境，这成为了覆盖物下温室效应的处理方案。在此建筑覆盖物中，内部和外部空间的温差控制在5℃～10℃，冬天的供暖和夏天的制冷得益于建筑覆盖面上的落叶所产生的太阳能。

项目设计主要考虑了变化因素

社会——金融方面

项目是整合花卉栽培系统的典范，旨在卓有成效地使老年人重新介入社会。其挑战在于实现预防隔离的社会管理和增强老年人的自信心，即通过给予老年人工具使他们再次感到对社会有贡献。就这点而言，“Huerta de Los Ancianos de Tucapel Bajo”在社会标志方面把这个中心变成了城市，并在实践中为老年人获得了额外的收入；此年龄组对资金的需求很强烈，这源于许多原因，特别是该阶段昂贵的医疗费用。

功能设置

“套餐”的理念相当于迷你城市，其中囊括了三个相异的大型区域，而公共场地和花圃显得与众不同。这些区域不仅带来了经济收益，同时也是受此地居民欢迎的舒适环境。“套餐”的中部相当于老人的公共中心，其职责在于满足老人的休闲聚会需求；此外，该公共中心还置有花卉作坊，老人可以自行在此生产产品，接着可以在“套餐”的花店里售卖。右边区域（北边）是“老人指南村”，在此，24个住宅单元围合着一个公共院子。社区左边（南边）是“老人集体公馆”，在此，30间两人间形成一个单元，单元中心是公共场所，有餐厅、客厅和起居室等。

可持续性

建筑覆盖物100%覆盖着这个“套餐”，起着类似温室的作用：从环境里获得热能，减少密闭空间（在冬天）的供热成本，利用建筑覆盖物的落叶表层（在夏天）降低温度，提供大量的庇荫处并营造一个对老人有利的小气候。建筑覆盖物内部的通风设备是通过机械控制的空气交叉系统完成的，同时根据康塞普西翁市的主风设置成南北方向。通风设备与房屋的潮湿区域的直通烟囱互为补充，穿过建筑覆盖物并排出被污染的空气。房屋紧密相拥的目的在于热能的快速传导。雨水聚集在建筑覆盖物的周围，形成植物灌溉区，同时还流至洗盥污水池，从而配给住宅“套餐”用于项目三个区域的厕所。

空间设计

老年人的主要特征之一是感观上的返老还童。这种循环体现在老人生活的许多方面，以至极简单的事物都能刺激他们，在某种程度上形成了一种再生以及对世界的发现或再发现。身体状况发生变化，体态萎缩并接近孩童的高度，视觉角度向下。就此而言，有关唤醒孩童记忆过程的意图便诞生了：花园的特殊内涵、将此处自然地视为孩童时期的内化，这些都使我们想到了游戏、朋友和自由的幸福时刻。考虑到老年人情绪共鸣的影响，营造了一个坐拥自然的开放空间，这也就解释了建筑覆盖物的花卉和植物能富于引发共鸣的色彩、质地和气味的原因。内部空间洋溢着生命的融洽气息，迟钝感顿无，欣喜感渐至，促进了集体的互动，从而把项目打造成潮湿的灰色城市里的一片绿意盎然的景象。

Continuity of Heritage 068

历史遗产的延续

“每个时代都在建筑上书写着自己的故事。”

——路易斯•芒福德

被誉为春天的山丘的特拉维夫拥有以色列五分之二的人口。这座城市诞生于1909年，是由一片片沙丘上的区域连接而成的。显然，这个年轻的城市拥有着现代的风格，从最初的新建筑试验田发展成为建有许多城市剧院和高楼的密集城市。这里也成为了充斥着乐观主义和能源的城市。P.Geddes为特拉维夫进行了初次规划，将这里设想为每个人的公共空间和美丽花园。正是有了这个设计，这里才成为了能够自由呼吸的绿色城市。包豪斯建筑则成为了这里的亮点，被联合国教科文组织列为世界遗产。

每年，有大量的移民涌入特拉维夫。有限的城市面积无法容纳迅速增长的住宅需求，城市开始向周边的郊区扩张。如今，临海的Tayelet散步区成为了最具活力的地方，但是在北面的入口处，仍然存在着拥挤的问题。这里还有一片废弃的半岛。同时，道路没有恰当的规划，河流也已经干涸。但是作为Yarkon河流的入海口，这里却拥有得天独厚的地理位置，能够发展成宜人的生存空间，这与格迪斯思想是一致的。现存和待开发的土地将成为全球性的都市化进程的一部分，能够影响整个半岛的发展，并向海洋推进，最终形成与相邻的雅发城之间的视觉联系。历史的冲击也将赋予这里强烈的意义。道路的开放性规划、对河流的保护、重要的公共空间，这些都是城市开发的重点。同时，对贸易场所的复原和恢复，以及对旧港口和历史连接的重视也是此次规划的重要部分。

为了满足城市发展的需求，设计将发展以下内容：住宅、商店、办公区、花园，城市的延续性。这里也将拥有新的发展方向：海洋，连通欧洲和地中海……

格迪斯中心的设计灵感来源于历史遗产的延续性。在半岛的尽头，这个靠近海滨大道和花园的垂直公共空间就是城市景观发展的结果，并为其发展增添光彩。这里就像是一个地标，展现着特拉维夫的形成过程，连接起众多片段。这里是文化、艺术和城市的融合，是对历史遗产的转换和改变。这座大厦带领我们从喧闹走向静谧、沉思和回顾、发现和惊奇。通过这次历程，我们穿过展览和空间，来到一个适合研究、学习的空间，在其中，我们聆听音乐、阅读智慧，用网络与他人沟通。新的媒体科技为我们提供了学习数据的新方式，其他的文化、语言和景象也相继到来。当我们到达顶端的天台，眺望城市，我们将重新建立起与日常认知的联系。

项目的设计是将特拉维夫作为集体成果，复兴其都市风格和理念。只有回顾过去，展望未来才能有意义。格迪斯中心这座多功能大厦给予了人们表达自我和了解自我的机会。

〝文化就是对知识的连接和丰富。〞

——埃德加•莫林

Contour Museum　070

轮廓博物馆

〝当代艺术博物馆〞（获得竞赛二等奖）计划建在以色列的阿拉伯城镇Umel-Phahem，它利用浮动结构框架漂浮在主街道上，在这个充满戏剧的当地环境里留下了极小的足印。这样，美术馆在空间上形成梯台式，带着对自然和时间的永恒认知成为自然的一部分。博物馆在崖壁和树丛间的土壤和桥梁里诞生；其间，城市和自然、传统和现代、旧的和新的、东方和西方的组合相得益彰。博物馆象征着当地居民以及世界各地旅行者和职业人士的进步、谦逊和乐观，它使这座城市成了艺术讨论领域里的一个重要角色。博物馆有着清澈质朴的形象，然而，其不同角度却都十分引人注意。博物馆是分期建造的，每期都有其自身的审美特征和完整性。

Doğan Medya Center Ankara Cube as a perfect form and more　072

安卡拉DMC中心

DMC中心坐落在安卡拉的Eskişehir大街上，是一座独特的媒体中心，拥有电视频道、新闻报纸集团等媒体，是一座许多品牌媒体的聚集地，也是一处优质的交流中心。

为了体现与正方形场地的关联，建筑设计选择了立方体作为主体结构。在建筑外立面上，一些附加的架构和墙体采用了立方体竖向变形的结构，营造了额外的空间。

建筑清晰动感的外形营造了与当今媒体快速发展的景象相关联的外立面。从各个角度观察，这都是一个具有吸引力的优秀设计。各个不同尺寸的圆孔类似于盲文字母的标志。

从外面望去，建筑外立面上的圆孔单元为人们提供了通透的视线来观察内部的媒体状态。从内部望去，圆孔则展示了优美的城市景观。整个建筑形似一个方盒，上面的圆孔结构则象征着整个公司的多样化业务和氛围。

为了与当地的地形走势相统一，模块的设计进行了细微的调整，南北轴线赋予了整个建筑以活力。

这是一幢通透的智能化建筑，室内外空间得到了强烈的呼应。在人们进入建筑物之前就已经对这座建筑有了深刻的印象。设计还与周围的环境相互融合，与城市景观相协调。

通过回廊的巧妙设置，这座混凝土结构建筑最多达到了7层。中层楼营造了额外的空间，标准为4至4.25层楼。楼层高度为4米，每两层形成一个立方体。这种组合结构为设计提供了柔性的接触。中层楼层是由钢柱和横梁组成的。

主餐厅位于中庭的走廊之后，楼上的露台休息室则作为VIP会议区使用。

在与建筑平行的位置，有一个L形的停车场公园，另外还有两侧地下结构以备使用。第一层地下室是技术设备和储藏室所在地。

另外，科技设备和健身设施为大楼的使用者提供了方便，维护也很简便。

建筑动感的外形与前方的道路形成了互动，成为了周围一处引人瞩目的亮点。工作室和出版室等各个分区有着足够的私密性，同时又组合成这座优秀的建筑。

Dormitory for Staff of NGO Magic Bus — An Expression of Contemporary Sustainable Architecture　078

NGO Magic Bus员工宿舍楼——当代可持续建筑艺术的精彩演绎

NGO Magic Bus将在Karjat大坝下面的一处低凹地带建造一座员工宿舍。由于这一建筑选址位于一个小高原的顶端，因而站立于此能够俯瞰一望无际的稻田和一座座寂静的村庄。宿舍楼建于这样一块特殊地形之上，从不同的视角望去，建筑物所呈现的建筑外观也就各不相同。建筑师利用陡峭的斜坡在靠近森林一边的入口庭院为宿舍楼建造了一个怡人的建筑外观，而宿舍楼另一面则亮眼地矗立着，面对着无边无垠的田野。

宿舍楼的内部结构十分简易，整个建筑由一个过道贯穿始终，一个别致的小楼梯成为了宿舍楼通往一楼地面的主要通道。这个楼梯建在宿舍楼建筑的侧立面上，站在楼梯上，能够观赏到室外的美景，它创造了室内外的互动。经过这个楼梯走下主宿舍楼层，一个比主宿舍楼层高出一倍的大型娱乐空间便映入眼帘。在两层悬浮式空间的底楼建造这样一个大型娱乐空间，为建筑物增添了不少活力。建筑物的顶楼是一个大阳台。在柱栏过道的尽头，有一个遮阴露台，它为楼内的人们观赏树林提供了一个绝佳的视角。站在阴凉的露台之上，远处连绵起伏的Matheran山脉和近处苍翠的树林便尽收于眼底。

这座宿舍楼是一个竹栏环绕的〝笼型〞建筑。楼内的公共流通区域和阳台空间位于建筑外部竹栏和室内空间之间的空隙处。大面积的遮阴阳台与相对狭窄的建筑宽度为楼内的半室内空间提供了良好的通风与遮阴系统。竹栏外墙结构为楼内空间与楼外千变万化的自然景观提供了互赏的媒介。周边的自然景观在不同的季节为人们呈现出不同的美景：雨季中，它们在雨水的滋润中，展示出一片郁郁葱葱充满生机的景象；而在干旱、炎热的夏季，暗淡的棕色灌木荒地又呈现出另一番别样的景色。这座宿舍楼的建造必须迎合这些极端的自然环境，在酷夏中，室内就要求具备良好的通风与遮阴系统；在暴风雨频繁的季节里，楼内就必须安装相应的防水系统。竹栏式建筑外观为整个宿舍楼提供了良好的采光条件，然而，随着天气的变化与昼夜的替换，采光效果也有所不同，因而，这种外墙结构成为了建筑物的光线〝传感器〞。

宿舍楼的竹栏和竹质墙面的淡色处理，天然石材和抛光混凝土组合的地面铺装，雪白的室内粉刷，这些都为建筑物提供了一个幽静简约的室内环境，它是一种冷静而理智的建筑语言的具体体现。竹栏外墙中，每根竹栏的间距为1.5米，这样较大的间距形成了整个建筑物的渗透性，同时这样的外墙结构也能够更为坚固地支撑起整个建筑。竹子是世界上生长速度最快的植物，它是一种高产型可再生物质资源。竹子的高度牢固性和轻便性决定了它显著的抗震与抗风能力。

由于Vilas Gore of GeoScience Services投入大量宝贵的人力与物力资源，创新技术与先进的建筑方法将在该项目中得以充分使用，从而减少整个宿舍楼建造过程中非再生建材的使用量，以达到环保效果。建材与建筑方法的优化选择，减少了建筑物本身的负荷，同时也显示了竹栏外墙结构的优越性。

每根竹栏支柱都由两根直径为90毫米粗的竹子组成，它们被钢绳绑在一块儿组成一根支柱。钢绳不但缠绕在两根竹子的外围，也同时缠绕于两根竹子的间隙处，以免两根竹子之间产生磨损。楼内的预制混凝土桁架钢质压缩电缆横跨于每两根竹栏支柱之间。地板是由削半的竹子、50毫米厚的混凝土以及自然石材等建筑材料混合铺筑而成。楼内的墙壁材料是一种特制的填充壁板，这种壁板的表面由裂竹覆盖，内部则填充着钢筋混凝土材料。
该项目的建筑细部强调结构的轻便性。竹栏支柱的上下两端以及楼内的竹系梁都分别被一根预制混凝土钢板横梁连接固定着。这种预制混凝土钢板贯穿竹栏支柱，从而利用其反向弯曲力，令一系列的竹制竖梁保持垂直结构的稳固性。自然建材与当地建材的完美结合、先进建筑技术的应用，以及木材、混凝土建材与钢材在该项目中的混合使用，必将塑造一种全新的建筑风格，同时，它也将成为当代可持续建筑的一种具体体现。

Drobeta Archaeology Museum 082
迪罗贝塔考古博物馆

背景

在提升国家文化遗产，并与国际文化遗产相结合的需求下，设计师必须对整个博物馆建筑进行改造。原有的考古公园“迪罗贝塔”不仅集合了多种建筑元素，而且也可视为一个保存着罗马文明的博物馆。如今，作为一个博物馆，其建筑设计就需要按照标准进行。这里有令人惊叹的罗马文明和罗马殖民地的文化生活场面，其中一个代表便是古罗马兵营。古罗马兵营是已知的最早的石头堡垒，由罗马的Cacia省人建造。兵营在对抗搭桥进攻方面起着重要的战略作用。1950年，考古学家将其挖掘出土。由于其结构复杂，从很大程度上揭示出当时它作为一些区域重要基础设施的地位。后来，兵营也纳入到了游客游览线路的设计之中。通过普通的测量法，可以很清楚地区分出第一阶段的建筑和第二阶段的建筑。所幸兵营是对称结构，除了东部边缘，其他部分均在测量范围内。
兵营还配有一个独立的建筑单元：罗马浴室。位于西南末端的迪罗贝塔的罗马浴室可能是此类浴室中最复杂的一个，由奥尔特尼亚区域的罗马人建造。20世纪80年代浴室被移至铁路的北部，为铁门水利大坝的建造留出空间。受影响的是一大部分的罗马浴室和角力场废墟，加上缺乏监督，导致大部分的废墟遗址遭到破坏，最终从游览线路中消失。
场地内另一个主要建筑元素是图拉真桥。尽管没有得到良好的管理，且缺乏文物保护法规和一些适当的保护干预措施，拥有2000年历史的图拉真桥依然屹立在此，并为游客展示着这片土地的历史。图拉真桥是遵照图拉真皇帝的命令建造的，由Damasc的Apolodor设计。罗马人将它称为当地的世界八大奇迹之一。图拉真桥的建设只花费了两年时间，正好是达契亚第二次入侵期间，也就是公元前103年至105年。桥长1134.9米，宽14.55米，高18.6米。桥身结构材料为石头，中间混合了一些砖块和天然水泥，入口处宽50米。这个超级结构由橡木支撑。
不幸的是，过去对这个地方的一些改造留下了抹不去的印记。不完全的修复，加上缺少维护，一直影响着遗址的保护与发展。甚至，一直以来长期、间接的影响已经迫使当地政府不得不开始寻求新的解决方案。为此，政府提出了发展规划，其中包括作为整个项目开端的新博物馆。

设计概念——迪罗贝塔考古博物馆

建筑师决定项目的主题和建筑项目的一些基本元素，比如选址。尽管地产文件圈出了场地的范围，但是并没有限制建筑设计理念和设计类型。因此，当要在一个比较敏感的场地内进行建设时，设计师会尽量保留场地的历史，同时通过一系列特别的设计，保证游客能体会得到这种历史。项目中，场地的象征也就变成了博物馆的象征。项目改造的方向并不只是要向罗马式靠拢，而且还要建造一个现代城市的博物馆。建筑设计概念中包括了场地中的各个主要元素。罗马中轴决定了整个场地的走向——并转换为一种设计语言。博物馆与项目设计目的之间是一种互动的关系，已经得到了很好的体现。博物馆的建设已经完成了预期的设计目标，但是设计师却不满足于此。设计师希望将博物馆转变为场地文物保护过程中一个积极、不间断的干预手段——“原址博物馆”（在遗址上直接建设的博物馆）。
和已经出现过的在原址建设的博物馆一样，它最大限度地利用了“游客—游览目的—博物馆”这一框架概念中所拥有的最佳条件，很好地保护了遗址空间，同时也能向游客开放展示。图拉真桥四个桥墩（入口桥墩，一号桥墩和二号桥墩，边墩）当年的视觉冲击效果已不复存在，只是作为对于过去的一种展示。但是，它们的象征意义在整个场地中起到非常重要的作用，无论是从政治因素，还是文化因素考虑，它们在浩瀚的历史中都占有一席之地。设计师对所有的因素进行考量，形成出色的透视图，演绎出图拉真桥失去已久的价值，和它与游客之间的关系。
设计师提出选择使用“原址”的设计方法，将大马士革建筑师阿波罗多拉建造的图拉真桥作为中心。经过兵营，桥的对岸就是罗马尼亚文明。兵营与桥面相连接，是前往罗马尼亚的必经之地。这个防御系统的主要部分已经被设计成了一个用于唤醒某种情感，展示考古公园的元素。同样，兵营的防御布局则服务于游客，让他们对罗马此类建筑的使用情况有一个大致的了解。达成最初的设计要求后，博物馆将与另外一个符号性建筑连接起来——城门上的烽火台。这个烽火台为场地内的垂直元素，同时也能作为一个观景台和场地的突出标志。
在多瑙河岸城区的发展与布局设计概念中，博物馆的位置位于项目场地最底端（现阶段，一些富有工业特色的区域和混乱的布局都是发展进程中不利的一面）。这种布局是考虑到了需要达成的其他设计目的，例如博物馆可作为游客游览的目的地。正如我们所知道的，这种关系并不都具有积极的意义，也可能让这个点将就此成为一个谁也不愿意前往的地方，但是设计师正在努力防止此类事件的发生。迪罗贝塔港口遗址已经完全被破坏，无法纳入博物馆的建设范围。但是，港口仍然属于整个项目的一部分，设计师有责任通过其他的方式将其展现在人们面前。城市广场位于多瑙河岸边，设计形式为当代的奥斯纳布鲁克式，有效地强调了场地的历史，并与公共元素紧密结合。
项目场地和考古公园，以及整个城市海岸线的设计都面临着一个真正的挑战——铁路。铁路所在的位置正好是桥梁废墟（多瑙河边）和罗马兵营/浴室之间，影响了博物馆的整体结构。遗址的游览内容包括遗址和与铁路交叉处形成的一个地下空间，充分利用了自然形成的圆形剧场空间。通过这种运用自然景观的方法，设计师试图保留考古公园现有的一些特色。地下空间则被用作行政区、研究和公共服务区域，内有两个内部庭院。庭院的作用不仅是保证自然通风与自然光线，而且用于分隔地下空间与游客区、考古公园。我们可以很有把握地说，所有这些元素连接起来最后将形成一幅图景，完全以一种现代的方式展示出来。桥梁象征着两种遥远文明之间的联系，玻璃桥的一部分跨过多瑙河，并没有与任何设计实体相连。

EDF Archives Centre 086
法国电力公司档案中心

得到Meuse & Haute Marne对项目的经济支持后，法国电力公司（EDF）决定在布勒建立一个档案中心，集中管理生产中间过程和生产管理档案。
直到现在，这些档案资料还分布存放在那些潮湿闷热的生产车间内，或是工程场地和相关服务区域。它们都分散在各处，处理的方式也各不相同，所以非常有必要将它们集中起来，并最大化地发挥它们的功能：分类、检索、管理和使用。
不仅从以上所讲的功能性方面来考虑有必要管理这些档案，而且这个项目对于地区发展还具有积极的社会与环境价值。
于是设计师与本地的代表一起合作，充分考量项目的经济效益，创造的工作机会等元素。客户要求建筑要完全与景观相协调，达到环境质量标准要求，作为未来EDF大楼的设计基础。
项目的设计目标是要通过一个良好的形态、高效的立面、合适的系统，选择可再生资源产品来创造一个高效的建筑。
项目设计中遵循的一个重要原则就是减少或是最小化对环境的影响。项目设计将得到：

1）减少占地面积；
2）通过每年可以减少档案使用带来的里程数来体现提升的功能性；
3）对景观的影响非常小（从稍微宏观的角度考虑）；
4）最大限度利用建筑地基挖掘出的土壤，用于控制场地中水的管理与恢复；
5）一个不论是从能源角度还是从环境角度来看都非常高效的建筑；
6）创造出一个由Meuse & Haute Marne资助的代表性建筑

这里需解决的问题是如何将轻盈感融入到建筑当中，如何不给自然景观带来影响。

首先，为了打造建筑的轻盈感，设计师建议在混凝土中加入不锈钢立柱，起到模糊建筑限制的效果，反衬出周围自然的色彩与季节变化。由于在混凝土墙面采用了特别的材料与技术（减少热桥），减少了对冷气设备的需求。

项目场地与其他建筑相隔较远，也没有连接到附近的污水处理网络，所以，设计师提出建立一个低成本的自然雨水处理系统，来解决污水、废水和雨水的处理问题。

建筑将采用双流通风系统，回收余热，降低对能源的需求，并保证空气质量。建筑围护的高效结合经济环保的通风与照明系统，减少了对能源的需求。对可再生资源和热泵的使用将保证建筑的高自主性。

Eicher-Volvo India Headquarters, Gurgaon 088

艾彻-沃尔沃印度总部

艾彻汽车公司的商业核心业务反映在大楼的布局中。大楼的设计重点在于其工程部分，外围反映出结构形态。在立方体内，设计师采用以四面体为基础的创新结构。建筑师与Pune市的工程师密切合作，采用既经济又大胆的多层结构来解决实际存在的问题。

公司办公大楼由底层空间、办公区和公司总部办公室三个部分组成。

大楼底层建在线性结构基础上，包括入口大厅、咖啡厅、游戏室、服务区及研发中心。大楼底层两旁是一大一小的两根柱子。大柱子包含公司的整个底层办公空间，小柱子包括艾彻汽车公司高级办公区。这些柱子都相互并列在一起。

植物及防晒保护

以公共平台为例，植物种植在屋顶花园上，柱子便将整个大楼延伸开去。主楼中的攀爬植物慢慢地向两边延伸，减少了大楼吸收的热量，起到遮阴的作用。

大楼西面与南面的天窗减少了大楼吸收的热量。天窗的经度和纬度与艾彻汽车公司的位置及角度统一起来。

设计师试图平衡建筑与当地气候之间的关系。

EKOyapı Research Center 090

EKOyapı科研中心

EKOyapi是一个研究与教育中心，它是伊斯坦布尔技术大学马斯拉克校区自然环境科学学会园区中的一个组成部分。这是与建筑工业中心合作完成的一个项目，而它的总体规划与项目申请等筹备工作则由HAS Achitects承接。该项目所有的设计、技术咨询，以及大部分施工材料、设备都是由赞助商自愿提供的。

EKOyapi处于规划阶段时，该项目的初步讨论议题主要集中于如何定位“绿色设计”建筑典范。该项目总体规划的负责人采用了全美普遍接受的LEED计分制度作为项目规划指南。据LEED计分制度的描述，绿色设计是指，所有的设计和施工方案，都应该本着一个原则：尽可能减少或消除建筑物对环境和建筑用户的负面影响。

领先能源与环境设计制度是根据以下五个主题制定而成的：

可再生选址规划；
水资源保护与水资源合理利用；
能源节省与大气保护；
施工材料与自然资源的合理利用；
内部建筑质量的确保。

因此，EKOyapi的环保达标评估，也应遵循以上五大主题的要求。

EKOyapi的选址

科研大楼地理位置

EKOyapi位于伊斯坦布尔技术大学马斯拉克校区的一个小湖边，它是自然环境科学学会园区项目的一个组成部分。该选址在规划之初，计划整个建筑物的朝向为南面，这样有利于更有效地利用太阳能。考虑到该选址的陡峭地势，该大楼就建于一个斜坡之上，以便减少大量的土地开凿，从而保护周边地质。该项目的选址及其紧凑的建筑形式，有力地减小土地占用面积，同时，不会对周边的自然环境产生不利影响。此外，该大楼依附于小湖畔，接近水体，这恰好符合了领先能源与环境设计评估制度的要求。

水资源利用

感应水流装置

为了有效地控制EKOyapi的用水，设计师们在洗手间内安装了感应式水龙头，同时在卫生间中也安装了感应式冲水装置。

雨水处理方法

另一种节约利用水资源的方法就是充分利用雨水。屋顶上的雨水积水将被储存在一个地下贮水池中，而后再灌注到蓄水库中，以备不时之需。

废水与废物处理

废水将通过一个污水坑自动排出。这一污水坑将设置在一个相对安全的地方，这个地方必须确保废水排出时不会流入旁边的小湖里。EKOyapi的固体废物中，可再生物质将被回收再利用。

能源与大气策略

这是EKOyapi的环保策略之一，这一策略的目的是为了通过一些措施，尽量降低石化燃料温室气体的排放量。当然，这一策略也将同时确保能源节约与当地生产的正常进行。

室温平衡环保设计

在EKOyapi大楼中将不使用锅炉或冷、热空调。为了有效地保护周围的气候环境，设计师们依据德国“Passivhaus研究所”所制定的标准，建造超绝缘建筑物外墙，这种外墙将有效隔绝室外高温与低温。另外，还有一部分是地下室隔热系统的建设。总而言之，该大楼将隔绝一切室外高温，给建筑用户提供一个舒适的办公与科研环境。

能源的替换与产生

校园内放置着的两台风力涡轮机，以及屋顶上面积约25平方米的光伏（PV）电池，都将为EKOyapi提供电能。换言之，大楼本身所能产生的电量将大大超过它目前的用电量，并且还将保持网格内的电量库存。

有效的能源管理

由于EKOyapi是一个研究中心，所以它对能源系统的安装与维护尤其重视，采取了一系列的管理措施，对其进行严格的管理、监测与报告。

建筑材料与自然资源利用

设计师在选择EKOyapi的施工材料时，遵循了下列几项标准：

隐含能耗量较低；
就地取材，并能随时获取；
由可再生或可循环能源制造而成；
耐久且无需维修；

科研大楼内部质量：

挥发性有机化合物的使用。

由于EKOyapi的建筑外墙具有较高的绝缘性，不透水，因此，内部的空气存在着毒气无法释放的隐患。鉴于以上所述的建筑构造，所有的油漆、胶水以及其他的绝缘材料都必须是挥发性较低的有机化合物（VOCs）。同样，在混合型木材的选择上也应该注意到这一点。

ETC n 330 Gutierrez Ruiz, Maronas 092

n 330 Gutierrez Ruiz, Maronas 的全日制学校

这个公立学校在外观和规模上富于变化，形成一个连续的环境，意在与社区相融合。由于场地较小，设计方不得不把学校设计成三层楼，从而留出一个宽阔的庭院。鉴于现存的斜坡，项目将庭院按梯形分成一个露天足球场和三个相异的操场；以此类推，大楼的分层也是根据孩童的尺寸设计的。校舍是一个由不同分区组成的连贯系统。有4组（尺寸、高度、色码定位不同的）教室群，带有形式上的细微扭曲以呈现最佳的视角。这些群由一个流泻在建筑体积中的三倍高的中庭统一，完备了教育社区。教室体积相对于公共场地是透明的，而大多数教室由圆柱支撑着，从而形成一层空地。这种较低楼层比之其余楼层有着细微的差别，其隐透着不同设施的共享（通道、中庭/多功能区、公共餐厅），同时维护其可大可小的灵活性。

Eye Hospital 094

眼科医院大楼

孟加拉国是世界上最大的三角洲，国内有52条河流，这些河流都源于喜马拉雅山脉，它们以一种复杂的方式汇集到孟加拉湾。在雨季到来的时候，这些河流浸没了整个国家三分之二的土地，因此，水便成为了孟加拉国的主要风景元素。当洪水退却之后，留下的是一层肥沃的冲积土壤，整个国家形成了处处稻田与轻风共舞的富饶景象。

在这样的地理条件之下，在孟加拉国内，除议会综合设施之外，其他绝大多数的公共建筑都仅仅只是建筑而已，它们并没有附带的多余公共空间。而该项目的建造，就是要努力创造大一些的建筑公共空间。

Louis I. Kahn的建筑艺术方式已经成功地应用到了议会综合设施和Sangsad Bhaban的建造之中，而它也成为了该项目建造的一个“基础理论”。Sangsad Bhaban建筑项目中的“天空广场”在该项目选址中施行时，已经把原先的一个“过渡空间”转变成为了一个灵活的“转换空间”。基本上整个建筑外墙都采用11.24平方米左右的正方形铁丝网围筑而成，因而，它成为了一堵透明的多孔城市幕墙，由一个绿色庭院把它和“眼科医院旧楼”相连，它的东面与Farmgate相对，西面是Manik Mia大道和Sangsad Bhaban。

面对西面建筑独特的风格，该项目的建筑师着实感到了一种无形的挑战与压力：柔和的建筑线条，如诗如画的视觉效果，这似乎变得不可超越。

孟加拉国是一个气候环境十分独特的国家，它的独特在于它在一年之中拥有6种不同的气候环境，季节变化微妙而多样。因此，建造该项目时，持续的烈日、季风时节的大量降水以及大风等气候因素都应该考虑在内。薄水层的应用不仅仅具有雨水收集器的功能，而且具有三角洲的地域特色与内涵。在该项目的建造过程中，线条、透明度与孔隙度的使用，不仅为新旧两幢楼的通风与连接提供很好的媒介，同时改变了传统的凉亭建造模式。采用绿色调且层次分明的遮阳设施不仅是绿化环境的一种必需，同时也映衬出了建筑物本身的无限魅力。

该项目的主要建筑材料是混凝土。混凝土建材的渗透性和可持续性与坚实的建筑结构相结合，形成了一种非凡的建筑视觉效果。正如Kahn在议会大楼的建造中使用露石混凝土建材一样，该项目也选择露石混凝土，这种建材适应于孟加拉国炎热而潮湿的气候，同时它的使用令整个建筑物显得线条突出，活力无限。

Floating Spaces of the Residence 096

浮动空间住宅

该项目是在一座现有住宅的旁边建造一幢占地约300平方米的新住宅。客户要求将新住宅建造成一个现代化的钢结构建筑，但不能因为新住宅的施工而影响有着30年历史老住宅的完整性。于是建筑师建议修建一个半分离式的建筑连接原有的住宅。经过仔细考量，建筑师认为不宜照搬原有建筑，这样新的创意才能得到充分发挥，而这也是客户所希望的。新建筑将会建在现有住宅的后面，建筑师构想出一个新的设计能使整个建筑焕然一新。其想法倾向于利用现有的车道作为新建筑的主入口。每当夜幕降临，灯光从2、3楼中宽3米的悬空房间里照射出来时，便会呈现一个明亮的浮动空间的视觉效果。此外，新建筑中的客厅和卧室的通风系统良好，并且拥有开阔的视野，能够看到住宅花园全景。由于可使用空间很有限，建筑师将房屋设计成了并排结构。

Folded Bamboo + Paper House 098

“竹制和纸制折叠住宅”

在2008年5月12日我国中部的8级大地震中，6.9万人遇难，数十万人受伤，部分房屋受损；我国政府启动了抗震救灾重建项目，项目包括150多万个临时住房的建造。许多制造帐篷和临时房屋的工厂立即全力投入到生产中。

此项目的主要特点在于临时住房的建造要展现出雨伞和折扇特征的动力结构，并具有顺应各种环境和居住需求而改变的潜力。建筑师们把此结构命名为“竹制和纸制住宅”，即一种即时安装的自我重建结构，此结构能根据变化的内在需求和场所地形构建无限的潜在方案。

竹制住宅

竹子是生长最快的植物，2个月左右便可突长至标准高度。收割竹子时，其根系不受损害且健康发展，还将生长更多的竹笋，就像草地一样。厚壁的竹竿比相同尺寸的木头坚硬二至三倍。竹子7年一收，相较10～50年一收的软木材和硬木材，竹子的产量多达木材的20倍。在相同区域里，相较阔叶树林，竹子能产生更多的生物，产生的氧气比阔叶树林多30%；同时，竹子还能提高分水岭的质量，防侵蚀、修复土壤、生产可食的甜竹笋、去除污染土质的毒素。

与工厂大规模生产统一规格的居所不同，“竹制折叠住宅”采用了简单的竹制动力结构和一种自下而上装配的自我调节的综合适应系统，其与自上而下的普遍原则相反。笔直的竹竿在螺旋表面、双曲线抛物面和回转双曲面上形成平行线条。“竹制折叠住宅”的成品反映了折叠相对于展开、自我结构相对于解构、固定结构相对于活动的逻辑。

“竹制折叠住宅”的特点在于大量地依托雨伞结构的开合角度和骨架之间的空间关系。折叠住宅运送到场地后会根据使用者的社会、经济和文化需求进行改进。围绕使用者的定制系统使住宅在短时间内就可组装，且与外部环境诸如光、风、温度等相融合。

纸制折叠住宅

随着超过50%的垃圾来自纸以及囿于经济原因对垃圾输入降低的渐涨需求，把纸转换成可用的、稳定的、牢固的建筑产品的能力成了较大的优势。

纸的回收利用减弱了对环境的影响，同时还降低了生产新建筑产品和材料的能源成本。自从楼房消耗了人类大部分的自然资源，其中包括40%的天然石块和沙砾及25%的原生木，纸对环境的影响就变得至关重要。

纸用纤维、水、水泥的合成物可以运用于各种各样的建筑物。汲取大量水后的浆被填塞进管子里，其可用作建筑材料。质轻的纸制住宅可以预先在工厂里装配好，并折叠进一个小型包装物，然后装上货车以待运输。

动力结构的角度变化使纸制折叠住宅得以千变万化。由此，这些因地制宜的

方法变化出了住宅单元、社区中心、餐厅及功能各异的商铺。除此之外，时空组织也使建筑资料不断更新，衍生各种各样的组合来表征文化和历史的多样性。

Grafishow Printing Works 102

Grafishow印刷厂房

该建筑位于布宜诺斯艾利斯市内一个宽8.66米、深23米的地块上。印刷厂房建筑面积共220平方米，四周围起金属墙面，一层与二层的层高也因此被提升了起来。

在结构与建造技术方面，建筑使用的都是最基本的资源，设计师用最朴素的设计语言来表达。

一层没有窗户，四周墙壁呈统一的白色，看起来功能性并不强，只留出了门和印刷材料的装卸区域。

二层则是开放式，日光可以透过金属挡片洒进来。设计师使用的材料非常少，使建筑看起来简单结实。

从外部看，二层半透明的结构围合，形成不同的深度。从室内看，金属片与天空之间伸出的常年生树木则提供了一份清凉。

项目侧重建筑结构，为达到预期建筑效果，空间只通过简单的结构与组成元素来构建。建筑室内加上灯光的设计，得到一种强烈的艺术效果，同时，建筑所需要的各种功能要求也得到满足。

狭长的印刷厂房上设计了四个天窗，室内尽量利用自然光。玻璃幕墙也保证了前后光线的通透，强烈的光线洒满了入口处的空地。

Helical Stair Gallery 104

螺旋楼梯美术馆

由匿名建筑师设计的螺旋楼梯在随后的几百年间被用于营造简洁雅致的空间。螺旋楼梯可以说是最赏心悦目的自给结构体系了。项目的灵感源于螺旋楼梯在施工效率方面的结构优势；模块化美化了空间且提高其品质。此外，设计的最终目标在于使螺旋楼梯由空间配角地位转为空间主角。

传统螺旋楼梯的结构中心在结构旋转轴的中心；然而，设计的独特之处在于材料叠加的2个相异中心的错落，这样就整合成空间品质的一个完整结构且富于变化，而不只是一个垂直的通道。就材料组装而言，一大半的材料来自对几个原型尺寸不变地简单重复，而其他的材料则是对这几个原型在尺寸上稍加改动。就空间体验而言，主陈列室沿着螺旋楼梯不间断地绵延至屋顶面，这样，游客就可以同时体验内外结构，还可以欣赏贝辛斯托克运河的风景。

Hortus Officinalis 108

疾病控制中心

建立疾病控制中心（CDC）的目的在于建立一个医学研究中心，通过对建筑周边环境和绿化设施的建设，逐步将其建造成为一个无菌中心。

建筑师们在项目设计的过程中，联想到了一个古老的医学研究中心Hortus Officinalis，这是一个专门从事研究与草药种植的场所。

在建筑师们看来，CDC就是当代的Hortus，唯一的区别就在于实验室内所有的实验设备都是十分先进的当代设施。

因此，该项目的总体设计是基于Hortus Conclusus建筑理念之上的，在这种建筑理念中，各种作用不同、高度不同、规模不同的建筑物都建于一个插入性区域内，以便于建造一个宽敞的中心绿化庭院。

这个绿化庭院是一个步行区：它是一个脱离周边建筑而独立存在的活动空间，它的位置处于公共循环区的下层，与新竹生物医学科学园中央公园的地面高度相当。

而各个建筑楼群又形成了一个室内庭院空间，这一空间又划分为两个部分：第一部分由三幢7米高的小楼组成，这三幢小楼都分布于插入性区域内，以便于创建一个宽敞的连续性底层公共绿色空间。

这三幢小楼有着双重功能：一方面，它们对绿化庭院起到了相应的界定和范围限定作用（这一绿化庭院是CDC工作人员的进出通道）；另一方面，它们的组合创建了一个人工平台，连接着南面的中心公园和北面的公共空间；

第二部分由6幢高度不同的方形建筑组成；这6幢方形建筑建于人工平台之上，突出了庭院极富特色的外围轮廓。

两个部分楼群的楼内空间是根据其特点和具体要求划分的（行政中心、科研与诊疗中心、公共场所、员工活动场所、办公区等）。

因而，集体性多功能活动区，例如，展览区、会议室、礼堂、大厅等，都集中设在地下层，而科研与办公空间（实验室和办公室）都设在6幢方形建筑之中。

这两个建筑楼群既紧密相连又创造出各自不同的功能区：

—— 科研与诊疗中心位于南面的较大区域内；

—— 行政中心则位于北面区域内，并在中心楼前建造了一个临街的主入口。

受到Hortus Officinalis建筑理念的启发，CDC项目的建筑特点与材料选择都十分讲究，从而达到建筑与周边景观的最佳融合。

基于Hortus理念，自然元素是该项目的主要组成部分，但也在一定程度上受到了人为规则的制约。

为此，该项目的3个主要元素以下列三种不同的方式呈现出来：

—— 绿化庭院所处的建筑底层（人工平台）上建有一个连续性的绿色屋顶（这一绿色屋顶上种植着一排排绿树）。建筑底层还在两个楼群（即科研与诊疗中心和行政中心）之间设立了一个分界区域。

在第一部分的建筑楼群区块中，底层的绿色屋顶通过其倾斜坡道与街道相连。而在第二部分的建筑楼群区块中，一系列下倾式阶梯将架高的绿园广场与街道相连。

—— 绿化庭院的主体由两块绿草地及其之间的小水池空间组成。庭院的其余部分则进行有孔铺砖处理，以便种植更多的植物。

—— 该项目中的所有建筑都有双层外墙保护。金属网格外墙的外层还覆盖着一层攀缘植物，这样就可以形成绿色建筑的视觉效果，使其更加能够融入到周围景观之中。

攀缘植物建筑外墙同时也起到了生物气候控制作用，有效地控制了建筑内部空间的温度与空气质量。

Hoshikusa 110

皐草

项目位于古老的〝有马〞温泉度假胜地的一处街角，这处翻新的普通房屋建于二战后，至今已有45年了。它的特色在于铺瓦屋顶、砂浆外墙、黏土墙基、带有厚棉的灰泥墙和带木框的滑动玻璃荧光屏。古老而高大的枫树枝丫从邻家延伸至房屋入口通道处，近在咫尺。夜晚，枫树上空群星闪耀。

此房屋作为私人客房被其主人用来接待异国友人。项目力图保留此房屋以前的日本居民所构建的结构，从而使访客在此放松之余还能体会到历史和自然的氛围。项目保留了木制栏杆、移除了盖板，以此呈现当前的结构，诸如天花板的墙基；同时，外形没有较大的改变，也没有粉刷肮脏皱裂的墙，这样，房屋像是化了淡妆，原汁原味地展露无遗。最后，亚麻制成的网轻柔地反罩在灯下，由此衍生的光影漫射着木制栏杆和滑动纸制荧光屏。

Hotel Hospes Palacio de los Patos in Granada with Luxuriousness and Beauty 114

豪华亮丽的格拉纳达Hospes Palacio de los Patos大酒店

Hospes Palacio de los Patos大酒店原先是由一个19世纪的宫殿改建而成的，该项目是对这一宫殿建筑的扩建项目，酒店的扩建部分内设有22间独立客房。酒店建筑的外观是由一种十分亮眼的雪白玻璃幕墙砌筑而成。

2007年7月，巴塞罗那EQUIP Xavier Claramunt建筑公司设计建造了格拉纳达Hospes Palacio de los Patos大酒店的扩建项目，建筑过程历时18个月。扩建后的Hospes Palacio de los Patos大酒店内共有42间独立客房，其中22间是设于扩建楼内的新客房，其余20间是旧楼内原有的客房。

新建的扩建大楼占地7,100平方米，位于酒店旧楼（19世纪宫殿建筑）的一侧。由于酒店新建大楼的外观是由一种十分亮眼的雪白玻璃幕墙砌筑而成的，因此，整个建筑十分引人注目，给人以脱颖而出的视觉效果。建筑师们为该酒店的扩建项目创造了一种可视化建筑策略。

"该项目不仅仅是Hospes Palacio de los Patos大酒店的扩建项目，它也将成为一个综合性的娱乐场所。建筑师们在楼内也建造了一系列与建筑外观相匹配的白色窗格式玻璃幕墙，这些幕墙围合成一个个独立的功能空间，以便于举行各种活动时使用。扩建大楼之所以不采用实墙作为建筑外观，而采用玻璃幕墙、悬挂物以及白色窗格墙面，是因为建筑师们设想在新楼内为酒店主楼那幢古老的宫殿建筑提供一种视觉上若隐若现的朦胧美感。一侧是曾经发挥着主导作用的历史性建筑，而另一侧则是极富摩登风格的雪白玻璃幕墙外观现代建筑，一个别致的水景花园成为了两个建筑物之间的纽带。"

Hospes Palacio de los Patos大酒店坐落于格拉纳达的中心商务区内。酒店的扩建大楼内有三个截然不同的区域划分：就餐区、温泉沐浴区和客房区。就餐区与水景花园毗邻，两个空间相互映衬、相互融合。温泉沐浴区由若干个浴室组成，这些浴室的照明借助于外界的自然光。22间客房分布于建筑的高层，因此，客人们只要在房间里就可以把宫殿、水景花园这些美丽景致尽收眼底，同时，对建造这座扩建大楼的建筑师们油然而生一种敬仰之情。

House in Carapicuiba 116

Carapicuiba别墅

从地形角度看，该别墅典型的特征便是地势低矮。由于地表突然被峡谷及树林遮挡，人们在街道上看不见地平面。

客户希望将住宅与办公区融为一体，这样便将工作与生活连接起来了。虽然位于同一个空间内，但尽量将住宅与办公区分隔起来。

在设计住宅与办公区域时，建筑师充分地利用了该场地天然的地形优势。

街道与围合空间之间没有联系，是两个不同区域的"底层架空柱"。一个是在靠近街道的地面上，另一个则是垂直的，作为大楼的屋顶平台。桥梁主要使用钢铁材料，将街道与屋顶平台连接起来。进入别墅的唯一入口则是桥梁通向公共空间的钢铁格子地面：楼下通向住宅区，楼上则通向办公区。

住宅区被分成两层，都位于街道水平线下。住宅空间与树木、溪谷、花园融合在一起，而游泳池则位于地面水平线上。住宅将室外的自然景致与室内空间融合起来：滑行玻璃板门朝向生活区平台，形成了一处大型空间。卧室与天井也在低处水平线上连接起来。

从街道上看，办公区位于楼上，为3米宽，25米长，就像是两极的管道。因此，与天井相比，通过窗户能欣赏到更多美丽的景致。

"管道"被两根柱子支撑起来。混凝土柱将整个大楼支撑起来。除材料、混凝土和玻璃外，该别墅以其地形和景观为基础，进行了专门的设计。在施工阶段，不少元素意味着对工作的更多关注。这样便于控制成本，关注施工的具体步骤。

House on Floating Land 118

漂浮大地 漂浮住宅

兼具城市与自然风格的住宅

项目场地位于城市与山脉的交界处。住宅正面朝东，光线透亮，城市中心景观更是一览无遗。住宅的后面就是绵延的山脉和自然景观。

大门到住宅入口的通道

打开透明的住宅大门，就是一条通向住宅的通道。这条长长的通道看起来就像是建筑中的峡谷。

这个地下通道的尽头就是住宅前的庭院，与入口相连。

漂浮土地上的两个盒子

场地底层结构上站立着由复式结构房屋构成的两个木质盒子结构，下面有起居室和餐厅等。其中一个"盒子"里面是主卧与书房，另外一个则是住宅主人两个女儿的卧室与书房。

起居室上方的漂浮花园——特殊空间中光的演绎

在起居室上面是一个9米高的空间，是主卧挑高的一倍，空间内是一个竹园，种满了竹子，十分富有活力。日光可从屋顶投射下来，随着时间的变化和日光的变化，室内空间的氛围也随之慢慢变化。

House TTN — Residence for an "Urban" Extended Family 120

TTN住宅——城市"大家庭"的新式住宅

该住宅是为父母及他们的两个女儿设计的。在外孙子出生后，他们便决定住在一起。于是该住宅设计的主要目的便是满足这个"城市"大家庭所需的功能，使其符合习惯在过独立生活的家庭中长大而现在选择享受大家庭快乐的需要。

该项目首先需要一处能容纳三个家庭的集体住宅，同时又要使三个家庭完全独立。为了最大限度地提供方便、有效合理的使用场地以及享受家人的陪伴，设计师在深思熟虑后决定采用半独立式的公共策略。

为了让这个大家庭很好地融合在一起，有必要在设计上保留一个舒适的距离。因此，每个家庭都有独立的厨房、淋浴间和洗手间，但底层区域与内外的平台将房子连接起来。父母的卧室位于一楼，独立的主卧（这也可作为三家人的公共空间），以及朝向外面的个人房间（日式风格的房间及一张床）。一楼与二楼分成东西两面，作为女儿家的居住空间。在两边的中间，每一层都设有公共平台，大家既可以在这里聚集，也可以作为缓冲区。透明玻璃和日式木制百叶窗将家庭分隔开来，同时也避免了完全分隔。每个家庭都可以独立处理自己的事务，但这些有意建立的公共区域又有利于提高他们的生活品质。适当的距离也有利于家人的团聚。

该住宅的另一个设计重点是结构。在决定不将房子完全分开之后，客户希望能多一个选择，以便将来将房子分成左右两部分。为满足客户的要求，住宅两边的结构都是完全独立的，包括地基在内，以确保房子被分开后的耐久性。当然，如果要修建两栋独立的住宅，则需要遵守各种规章制度。因此，这对住宅的规划及建筑形式产生了重大影响。正由于这一要求，才能够实现一个大胆的设计，若将房子分成户外公共平台，原来的区域将会被移开。客户是否会在未来将大楼分成两部分并不能确定，但这也无疑促进家人之间积极和谐的交流与沟通。

Ilpurla Substance Misuse Rehabilitation Centre 124

Ilpurla药物滥用康复中心

Ilpurla药物滥用康复中心建立的目的是为那些沉溺酒精、毒品和汽油的年轻土著居民提供复原性的护理和教育。

大楼为监督、管理、教育和所有客户的健康恢复营造了一个安全舒适的环境。Ilpurla社区的集体宿舍里有厨房、餐厅、双人间、教室、康乐室和项目文化管理办公室。

大约15年前，巴里神父和他的家人在这个偏僻的土著社区建造了这个中心。那时，他们用极少的资源来帮助年轻的土著居民戒掉其日常生活里难以根除的习惯。这些蚕食身体并毁灭人生的恶习足以破坏这个区域里的偏僻土著社区的社会结构。巴里神父和他的家人继续致力于改革。许多成功的改革项目的关注点在于帮助药物吸食者戒除坏习惯、教授他们生存技能、给他们灌输文化上的自尊、教会他们自己作决定。

大楼的设计考虑到年轻男女在居住期间负责的一系列半私用住宅空间的供给，还考虑到进行正式、非正式的教育和日常活动的公共互动空间的供给。

内部温度的稳定依靠大型屋篷、屋顶的天然通风系统和墙壁的热质，而不是呆板的空调。大楼的发电依靠远处的太阳能发电站，同时大楼还具有高效的限电系统。热水来自太阳能乙二醇的热传导。下水道是通过化粪池的黑水和网状暗水地下系统。

项目由Brendan J Meney Architects in Association with Richard Layton & Associates负责，由来自Probuild (NT)的团队建造。

Innovation within Tradition — Cemetery of Comacchio 128

传统中的创新——科马乔墓地

设计师对现有墓地提出了一个新的设计方案，旨在传递出一种"完整"的概念，尽管目前项目范围只涉及其中的四分之一。墓地初始的设计非常清晰，但是后来的一些扩展与补充使场地看起来有些混乱。设计师的场地设计方案就是要通过强调入口历史悠久的南北走向轴线，恢复最初设计的清晰感。这条轴线将教堂从中间一分为二。

七个相互独立的石墙上壁龛一个接一个排列着，在稍微倾斜的角度上创造出了一种围合感，当然光线也依然可以透进来。这样排列下的石墙投在地面的阴影也随之起伏。

七个石墙都是由混凝土制成，与砖块的颜色略有不同，以示区分。设计师的目标不是要复制现有的砖墙和色彩，而是希望它们与环境在色彩和比例上更加协调。

设计师还选择了树形高大美观的塔形柏树，为场地提供一种自然元素，视觉上拉大这种传统的墙面围合式的意大利墓地纪念空间的尺度。从远处看，树木是景观的一部分，形成了墓地的天际线，仿佛也是在向活着的人说："这里是住着我们祖先的城市"。

整个规划设计的中心就是这一小片柏树林。树林的中心和长轴线的终点相交处是一个大型的十字交叉口——叫做"空气的交错"，是由领近的两块墙面排列组合而成的空间。

设计师希望每个陵墓的设计都有其个性，每个家庭陵墓都可选用不同类型的材料、颜色、建筑造型和风格，即使这样会破坏建筑的统一性。另外，来访者的隐私也会得到保护。

在传统的墓地，壁龛都是垂直平整摆放在一个长长的廊道里，一个完全开放的空间内。而在这个项目中，设计师打破了原来的这种线条，而用小块墙面围成一个个小空间，创造出一种私密性。前来扫墓的人们可以安静坐下来，与逝者说说话，或只是默默地哭泣。

Instant House 132

创新性临时住宅

著名的分析学家詹姆斯·希尔曼（1926-1989）在他的《美貌与政治》一书中曾对被忽视的设计、廉价的色彩、毫无任何意义的声响、结构与空间如何能够赢得生理上的舒适与心理上的平衡进行分析。按照他的这种思维方式，该项目的建筑师们要研究如何在住宿需求与有限的购房能力之间寻求到一个暂时平衡点——此处的住宿需求指的是最低限度的空间需求——二者的平衡能够解决人们生存的基本困难，从而享受快乐的生活。

由于该项目的建筑场地选在一公园内，因而建筑师们要考虑到建筑与周边公园景观的和谐度。在这种情况下，建筑师们选择了Parco Delle Cave作为建筑场地，从建筑策略上看，这里拥有良好的自然景观资源，并且场地周边就是城市未来道路建设的系列项目，其中一部分还包括米兰市中心的规划项目（例如，Via dell′ Acqua），未来这一地区的交通将会十分便利。

该项目中最具影响力的与周边自然景观相互依存的设施之一就是模块集结居住设施。具体而言，整个设施仅由一个总的生活模块构成，这一模块又划分为8个小模块，每个小模块内包含两张床和一个就餐区，而这8个小就餐区由一个廊道似的互通空间串连着，在廊道的两端分别设有卫生间和厨房以及带有控制室的洗浴空间，控制室用于控制整个模块系统的开关功能。

为了尽量做到与周围的自然景观相协调，建筑师们使用天然木材作为建筑主材料，同时建造绿色屋顶，这样既起到了环保作用，又与周边自然景观形成了和谐的统一。此外，这两种建筑材料也充分地体现了这一建筑模块的可持续性：利用木材搭建可以使得该建筑在视觉上不那么突兀，易融于周边景观之中，而且这些木材并非乱砍乱伐的非法木材，而是通过森林管理委员会认证的合法木材，不对自然环境造成任何负面影响。此外，绿色屋顶所产生的隔热作用也对这些木材起到了一定的保护作用。加之，绿色屋顶的纵向倾斜式设计，有效地阻挡了夏季阳光的60°斜射，并且能够充分吸收阳光30°斜射时所产生的太阳能，以供房屋住户们使用。

该项目的建筑规划还同时满足了人体感官知觉上的相关需求。建筑师们有效地控制了约化空间中木材湿度的纵向流变，还可以控制室内的光线，使得有限的空间变得宽敞明亮，为整个住宅的室内空间注入温暖舒适的感觉。另外，每个建筑模块内的残疾人辅助设施也同样是不容忽视的环节：户外的斜坡在必要时可以作为残疾人通道使用，提供了便利的自然场地。

Integrated Parking Construction in the Historic Centre of Thessaloniki 134

塞萨洛尼基历史文化中心停车场综合设施建设

该项目是为了应对塞萨洛尼基历史文化中心不断增加的车流量而在中心周围构建的一个设施完备的地下停车场系统，同时，这一停车场项目也迎合了当地政府未来城市规划的需求。然而，建筑师们说服当地政府构建该停车场项目的最大挑战在于如何证明该项目将对历史中心周边地区现有的严峻交通形势起到积极的疏通作用。因而建筑师们考虑在该项目的建造过程中重新启用传统的停车场建筑类型，以此作为城市交通枢纽的备用装置。通过对该项目功能灵活性的探究，建筑师们认为整个停车场应该具备多层次混合型城市广场的建筑模式，以适应各种不同交通状态之下所产生的车流量。该项目将建成一个集数字电影院、城市广场和巴士客运站于一体的综合型停车场，各活动场所之间紧密相连，形成一个有机的整体；建成后的这一超大规模综合停车场将与周边的城市设施相互串连，使其能够充分发挥催化剂作用，从而大大改善城市拥挤的交通现状。

该项目计划建于希腊塞萨洛尼基的城市边缘地带。项目选址现场虽然被当地人称之为广场，但实际上是一个开放式的露天停车场，停车场内还设有一个巴士客运站。该项目就是在这一开放式露天停车场的基础之上进行改建，使其形成一个集数字电影院、城市广场和巴士客运站于一体，可容纳500辆汽车同时停靠的综合型停车场。建筑师的目的是为自驾车人员提供交通枢纽设施的同时，也为他们提供一系列的生活便利设施，从而为这一地区带来更多的商机，带动经济的发展。出于这一构想，建筑师对传统的停车场建筑模式进行反复推敲，思考如何将该项目建成一个车站式的停车场，而不仅仅只是一个单纯的汽车寄存区。因此，建筑师试图寻求一种既务实又灵活的停车场建筑模式，以便于停靠车辆出入自如。

这一综合型停车场将通过两个渠道实现其多元化的停车模式：一是通过各停车坡道与电影院入口通道的交织相连，使人们停靠好自己的爱车之后，可以十分方便地进入影院观看电影；二是通过坡道自身的空间灵活性来实现的。各坡道之间以Z字形通道作为媒介相互连接，无形之中增强了各自的功能性。每个坡道本身都是一个单独的功能体，在此可以自如地举行各种不同的社会活动，例如，开办各种购物市场、举办各式展览会、就职仪式等等。

Krakow Conference & Concert Centre Krakow, Poland 140

克拉科夫会议中心&音乐厅

该大楼位于克拉科夫市，与维斯瓦河河岸的日本Manggha技术美术博物馆毗邻。大楼的布局是在Grunwaldzkie Roundabout大楼内部平面的一个角落。有机形状的大楼决定了其现代化的形象，一个巨大的玻璃休息厅面向维斯瓦河、老城中心及瓦维尔城堡。

河滨的大楼高度进行了降低以减少其对河岸风景的影响。

会议和音乐会的参与能同时从多功能休息厅中观看到城市景致。从平台放眼望去，整个休息厅和〞五层楼外立面〞的景致便映入眼帘。呈现出大楼及附近的博物馆景致，并与周围的建筑形态互相呼应。

大楼立面主要采用玻璃和钛锌板，辅以个性化设计的玻璃及石头镀层（花岗岩、石灰岩和沙石）。使用的材料是典型的起源于Wawel山的材料，这些材料开创了建筑学的历史。不同材料与镶嵌玻璃的结合在大楼的立面上形成了马赛克风格的外观。

大楼的建筑占地18,000平方米（包括10,000平方米的地下停车库），有三个会堂，各能容纳1,800、600和300人。会议室面积约为500平方米，还有储物室、技术支持室和有320个停车位的地下停车库。

KU School of Architecture 144

美国堪萨斯大学建筑学院教学大楼

该研究性设计建筑项目，是为美国堪萨斯大学建筑学院建造的新型长方体教学大楼。这一研究性建筑项目的建造目的是为了对连接坚固的网格结构及其在实际建筑中的应用方式作可行性研究与探索。事实上这一建筑研究最早是由Frei Otto开始着手的，因而，它已经发展成为一项目研究课题，探究如何使得结构系统适应于主建筑物本身（即结构的有效性研究），同时又能够使其成为建筑物的一种装饰性元素和空间界定元素，当然，它首先必须是一种结构系统，这是先决条件。

上述的这项研究成果在堪萨斯大学建筑学院新教学大楼的特殊建筑设计之中得到了实际应用。堪萨斯大学正在为其建筑系筹建一幢新教学楼。并在筹建过程中寻找一些创新性建筑理念，以便于为项目所需以及项目建造所应遵循的方向做充分的准备。

通过不断的研究与探讨，建筑师们最终提出了一个较为理想的方案，即建造一个有机的建筑结构体系，而该体系又能够创建可支撑整个建筑物的集群空间。尽管该项目的建造本质十分〞不规则〞，但它却能够在建造施工的同时，对这一长方体建筑的内部框架进行清晰地划分：该建筑的底层将设计成一个开放空间，作为各种〞露天〞活动的举办场地，而其他室内楼层则悬浮于这个开放空间之上。该项目的建筑结构系统以碳纤维为主要材料，同时结合特殊的网格结构性能，以便能够支持结构的大跨度，从而最大限度地减少建筑主体与地面的接触，达到最佳的悬空视觉效果。

KUMU — The Main Building of the Art Museum of Estonia 146

KUMU——爱沙尼亚艺术博物馆主楼

这个项目是以一项在国际建筑设计竞赛中的获奖作品为基础进行设计的。在竞赛之后，建筑师就展开了进一步的设计工作，修订后的初步设计方案在1995年初提交给了博物馆。随后的几年，项目一直处于暂停状态，直到1999年资金到位之后设计师才得以开始实际的设计工作。在经历了一系列延迟之后，建筑在2002年夏季获得了建造许可证。接近21.6万立方米的石灰石被移除，整个建造过程在2003年秋季开工，两年后全部完工。博物馆在2006年2月18日正式对公众开放。

项目所在地距塔林市中心约3千米，是一片非常优美的区域。这里位于Kadriorg公园的南端，有一处20米高的石灰石斜坡。为了不破坏公园和周围的皇宫的景致，这座大型博物馆建在整个斜坡之上，形成一处半地下的结构。建筑的圆形造型是为了适应斜坡的地形，并连接整个博物馆中不同高度的地方。一座环绕在雕刻墙之内的庭院形成了室外雕塑展示区。斜坡将博物馆建筑分成两个部分，外部是管理办公室和保护区，内部是大厅和展示厅。停车场和公共汽车站位于整个场地的最高处。游客沿着楼梯隧道走下来，通过室外雕塑展示区到达主入口。旁边是售票厅、博物馆商店和通往临时展览处的入口。博物馆的另一个入口在Kadriorg公园旁边。斜坡带领游客从礼堂到达主入口大厅。大厅区将博物馆分成两个部分，中间以天桥相连。

设计以简洁和清晰为特色。展览厅设计简约而亲切，艺术品则陈列在中心。简洁的几何元素将室内外融合起来。外立面是由石灰石、绿色合金和玻璃建造而成的。

LEGO 150

乐高木建筑

在战后的马歇尔时代，丹麦正在重建，因此国家决定选择预制混凝土作为整个国家建筑的主要材料，以减少开支。由于现在混凝土制造业已经逐渐消失，因而，这一行业能留给社会仅存的记忆就是依靠这些战后预制模块和混凝土建筑，而当代的丹麦建筑，都选择乐高木为主建筑材料。

当乐高木在停车场、零售商店、办公室以及民用住宅等建筑区域高密度使用，以使得乐高木向小而精的简单超密集使用方向发展时，建筑师们也考虑到，这一项目并不适合采用一般的街坊建筑形式。

建筑师们不得不再次冒险，将处在流行巅峰的乐高木材料与现代巨石材料相比较，试图重现早已消失的哥本哈根高层建筑。现在看来，20世纪初的建筑师已经具有丰富的设计想象力，他们在当时，就设计出了教堂或王室的塔式建筑。然而，在这种建筑发展到它的顶峰阶段时，随着公众关注度的渐渐消失，以及设计理念、建筑的功能性和生产技术的逐渐下降，人们的设计灵感也似乎被这些不断重复使用的、死板的长方形石块所禁锢了。其实，人们应当要变被动为主动，充分吸收这些长方形石块的优点，并将其变成为一个创新的平台，而不应该被它的重复使用束缚了自己的想象力与创造力。

与其匆忙地开始设计夸张的建筑外形，而后再去挖空心思、想方设法，将设计好的外形塞进预制的模块中，不如忠实于组合建筑的设计方法。因此，建筑师们决定将该项目设计成组合式建筑。

他们决定，整个项目选址中的建筑都采用现代十分流行的模块组合建筑模式：3.6米×3.6米。他们考虑到由于该选址建筑区域分布不够均衡，因此，应该为这一项目设计弹性模压，而不是将设计与建造简单地组合在一起。而后，就像累积蜘蛛网似的，要慢慢地改变城市中大量的软地形以适应像素化的高峰和低谷景观。塔式建筑就像玛雅林雕塑一样，从一块块梯形石板慢慢堆积成为被人类占据的高耸建筑。于是一种模式化的建筑理念浇灭了建筑师们的创新梦想。

从上面的描述中，可以看出这个项目选址的地形，就像不同高度的托盘组成的一个三维区，是一个很好的现成空间。在塔式建筑的底部，这些托盘能够组成平整的购物街或者一个惬意的露天剧场。灯光和空气渗透进停车场和零售店周围繁忙的步履之中。然而，在稍高一些的露台上就很难看到涌动的人群了，这个露台是专门为外来务工人员提供的休息地。最后，他们又在私人住宅的露台上停留了许久，这些私人住宅的露台，有的都有三到四个出口，可通往建筑物的其他地方。

有一种对高层建筑经常性的批判是，它们的设计不够人性化。于是，一贯模式化的高层建筑周围，往往有着很多人性化的设计元素。

当你要求一位经验丰富的功能建筑师来设计公寓时，他们将会引进一些突出物和休息地，来创造角球窗口，使得室内拥有更为充足的阳光。要充分地利用免费的资源。当你与一个房地产经纪人交谈时，他会告诉你，最吸引人的公寓是角落中的公寓。在这个项目中，所设计的都是角落公寓。实际上，设计角落住宅，已经成为了模式化建筑的增建部分，以达到理想、均衡的建筑面积与配置。这种住宅的设计，给人以城市中的小山丘上的住宅的感觉。为什么不按照常规的平面设计来进行呢？为什么不依照住户或大楼用户的方案来设计呢？

由于丹麦人的收入较高，大量时间的节省就意味着一种巨大的财富节约，或者意味着生活质量的提高。因此，大部分的浴室就都成为了预先密封的吊舱似的封闭空间，加上瓷砖和其他洗浴配置，就组成了现代住宅的浴室。建筑师们决定采用这一设计原理进一步巩固所有的公共设施，使得浴室、厕所、厨房、卧室的墙壁和壁橱形成一个个浓缩的超级可变空间。一旦你已经选择了你的生活领地，你所需要做的就是建立一至两个超级可变空间，这样你的房子才算真正设计完成。

作为最后一步的说服，建筑师们知道，即使私人开发商可能会被他们清晰的设计思路和良好的视觉感受所吸引，他们依然觉得泛泛而谈的设计轮廓不如实际的建筑物来得真实可靠。因此，为了证明这个设计方案技术指标的可行性，建筑师们认为，没有什么比乐高木建筑模型更具说服力。于是，他们决定采用乐高木1:500比例的设计模型，其单位像素的大小与实际建筑的大小恰好形成比例，丝毫不差。而且，开发商们还可以通过乐高的网站下载一个免费软件，建立所有乐高木建筑作品的模型库，以便更为准确地了解设计方案的可行性。当您完成下载时，您按一下提交按钮，就可获取该项目的报价，之后的3天内，您将收到来自LEGO factory的一份文件，里面有您的设计概图和所有内部建筑用砖量的显示。因此，他们的其中一名建筑师仅仅花了2天的时间就完成了为开发商提供的这一网站服务，而另外的两天时间，则是应用于与开发商口头地重复说明这个方案。

所以，在最终的说服中，建筑师们提供给了客户设计方案资料（预先编制的）。通过仔细推敲这些资料，客户把这个项目交由他的儿子负责，并且允诺可以开始进行这一项目。因此，带着一点点运气（和相当的工作速度），在三年时间里，哥本哈根项目终于可以开始着手进行了，这一项目是传统的尖塔形数据的叠加与rigueur功能结构的合理性相结合的产物。应该说，Gaudi满足了Herzberger的要求，或者也可以说，是这些数据迎合了乐高的要求。

Lim Geo Dang 156

森林生活

传统融合现代

“传统融合现代！”这是三十出头的客户的要求，这些客户是作家和出版商。长久以来，把传统成功地融合进现代风格一直是当今韩国建筑师设计的目标。这一直是一个难以解决的复杂题。尽管如此，研究传统遗产是很有必要的，因为传统往往是创造的思源。

空间和别致框架的灵动组合

这个小房子的空间概念是基于灵动的连通，这也是韩国传统建筑式样的特色。古老韩国的连通在空间上无限延伸，即利用参差不齐的互相重叠和渗透来产生和周围自然与楼宇相宜的别致框架。

“马当”的八个场所（传统内庭）

那时，房子必须是外部世界的避风港。与自然的联系性给城市房子带来不同程度的改变（例如，亲眼看花开花落）。“马当”（开放性庭院）是古代韩国的“特质”。

韩国的空间特征

“马当”的八个场所处于不同水平高度，其设计的目的是为了产生吸引人的戏剧性空间效果。这个房子所有的主要空间与“马当”都有直接的联系。入口庭院、地下庭院I&II、野菊花（休息亭）、球场、内庭、二楼庭院I&II都是“马当”，处于不同高度的这些场所在水平和垂直上相互连通。“马当”间的连通性带来了多种多样的有趣改变。

首层虽然窄小，但通过普通的房间布局便拥有了大小相宜的几个外部场所。为了尽可能建造一个与“马当”同宽的区域，首层只设置了卧室和餐厅；其他的在上一层或地下室。此外，二楼的卧室是由底层架空柱建造的；因此，下面的空间也可以用来做“马当”。这个所谓的“RU”型项目使这个房子呈现了建筑环境的许多优点。首先，内庭有了一种安逸的氛围，这归于周围“马当”的重要作用。其次，创造了另一种“马当”并成为内庭的一部分，被屋顶覆盖着。最后，不靠任何围墙（根据区域建筑法规，在这个区域内是不能建造实心围墙的）而把房子和公共道路相分割，同时还解决了内部“隐私”和外部邻区的关系问题。“马当”是房子和公共道路的缓冲地带，还是和邻区相连的地方。

韩国的视觉特征

这个项目使用了许多横飞在路边的木头以期从传统建筑式样中唤醒“飞”的形象。用在吊门和挂门的钢杆上来使木头相互固定的T字形金属绳是传统元素；尽管如此，它们是可现代化的，同时还能唤醒过去。

传统文化是当今创意的起源。这个项目的研究只是一个用现代语言解读遗产的试验。

Living on Display — Harmonious Combination of Private Housing and Public Sales Site 158

私人住宅与公共销售场所的和谐共存

该项目是一个“商店与仓库”结合体的雏形，它是一个集单元住宅与冷冻仓库于一体的大型综合销售场所，销售形式以零售和批发为主。该项目的建造目的是通过借鉴、改造等建筑形式，有效地保护现有的销售产业，使之能够得以平稳发展；同时，它还致力于提高人们对制冷及其相关能源的节能意识。

由于单元住宅内各自的厨房空间都与冷冻仓库的储藏空间相连，因此，这种特殊的销售场所室内格局彻底地将居住于此的居民的私人日常生活空间呈现于购物者的面前。这里的零售与批发活动都采用“厂家直销”的经营模式，这样有利于缩短食品供应链，确保价廉物美。在Living on Display销售场所中，建筑师们还设计安装了一个货品传送带，这样有利于大幅减少商场小范围内产生的服务垃圾，同时，对货品的场内输送、陈列和摆设等方面的观念改善起到很好的实验性作用，为以后的同类建筑设计提供参考。

Low-cost Warm Homestead — School and House for HIV Orphans in Uganda 162

低造价的温馨家园——乌干达艾滋病遗孤校舍

该项目的客户是一个非盈利性组织。

虽然用于项目建设的筹集资金相当有限，仅三万美元，但建筑师的目标还是希望通过采用乌干达国内的低技术（砖墙与木质屋顶结构），为孤儿们建造一个温馨的家园。这样的房屋建造方式，也能让这些孤儿们参与并融入建造过程之中，从中了解房屋建筑的奥秘。

该项目的建筑选址是一块90,000平方米的土地，然而在这片宽广的土地上却没有任何基础设施。在这种情况下，建筑师发现自己原先的规划：几间必要的房间、两间大卧房、三间教室、两间办公室和一个卫生间，在这里是行不通的，因为这样现代化的建筑格局只有在城市里才能行之有效。因此，他改变了自己原先的建筑构想，计划在这片荒芜的土地上建造8间简陋的小屋，这8间小屋组成一个屋群，形成一个小型村庄；这8间小屋有着各自的功能，但又是相辅相成的。

建筑师希望为孩子们营造一个轻松愉悦的上课环境，而他理想中的课堂应该是孩子们围坐在一棵大树下上课，所以，他计划在这个8间小屋组成的小型村庄的院落中央种植一棵大树，这样孩子们就能在大自然的怀抱中获取新知。小屋与小屋之间由屋顶的檐篷连接起来，这样8间小屋就组成了一个有机的整体，并且在屋距间形成了一个个三角形带顶篷的室外空间，因此，在规划中形成一个由褶状中央庭院与周边若干个三角空间组合而成的室外活动场地。同时，这些三角空间上的檐篷还可以为在庭院中玩耍的孩子们遮阴挡雨；并且，这样的室外空间设计很容易激发孩子们户外活动的热情，即便在乌干达的烈日之下，他们依然可以在这片小天地中努力地学习，尽情地玩耍，享受生活的无尽乐趣。

在庭院的中央还有一口水井，房屋的顶上都安装了太阳能电池板发电系统，孩子们在上课之余还可以学习如何耕作。

将来，孩子们长大了，他们可以自己动手建造更多的小屋，使这里成为一个真正的村庄。建筑师希望，这8间小屋能够成为这个孤儿村的起点，在今后的日子里，这里的小屋将会越来越多，让更多的孤儿感受到家的温暖。

由于该项目的建造没有任何特别的技术含量，因而，这种低造价的建筑理念适用于发展中国家，值得推广。

Luxurious House on the Water 164

水上豪华住宅

该项目位于希腊扎金索斯的赞特岛上，是一座豪华的水上住宅。

然而，将这一豪华住宅建成一座坚固的独立水上住宅并非易事，它的特殊建筑选址给建筑师们带来了极大的挑战。

住宅主人原先的构想是将其建造成为一座单一的家庭住宅。但建筑师们却计划将该住宅结构建造成游艇式的建筑外观，这样的设计是考虑到住宅可以随着周边海水水位的增减而进行灵活地上下浮动。

建筑师们为了充分利用这一特殊选址的美丽海景，赋予了这座豪华的水上住宅最佳的观海视角，在住宅的每一个房间里都能够观赏到无边无垠的大海。试想一下，在水上住宅的室内花园或顶楼甲板上，欣赏着四周蔚蓝的海水，悬壁、陡石与海水俏皮地嬉戏着，迎面吹来阵阵温柔的海风，这样的生活多么惬意！

住宅的室内外结构和谐相依，与周围迷人的海景相互映衬。

该项目在其设计过程中，涉及到了海水淡化、能量积累、通风方法、废水的回收利用、热能与其他能源的消耗、潮汐规律以及太阳能利用等多方面的调查研究，因此，它完全是一个可持续性的住宅设计方案。

M House (Module House) 168

模块住宅

建筑师从活动住屋获得了灵感，设计了模块住宅的理念。这种活动住屋在美国非常盛行，是低收入家庭经常居住的一种建筑形式。

设计面临的挑战是如何将这种住宅形式变换为经济的、尺度灵活、易于移动的建筑，并且能够吸引各种收入群体。通过这样的方式，人们将逐渐改变从住宅外表来判定一个家庭的经济情况的看法。

模块住宅的概念是以临时展览设施的功能需求为基础的。临时展览设施通常是附加在地面上的独立结构，由一些元素连接而成，能够提供必要的水电连接等。墙体与结构进行连接，并且易于调节和替换，通常是由多种设计和功能丰富的材料制成的。为了使这些材料经济可行，需要许多不同类型的生产商和设计师加入其中。正如开放软件资源那样，这种开放建筑资源也将对许多非建筑行业的公司开放，诸如电力和家具公司。这将提供一种新的方式来设计、供应、使用和循环使用住宅建筑。

模块住宅最大的特点就是灵活性，在任何时间都可以对墙体结构进行增加、缩减或是替换。如果居住其中的家庭成员数量发生了改变，就有必要对居住空间进行扩充或是缩小。收入的变化也将提高或降低住宅的开支。此外，模块住宅的方式还将促进住宅与节能标准和个人舒适度保持一致。破损的部分也可以进行替换或是修补。

模块住宅不同于其他组件建筑的是，这种方式汇集了许多不同的公司和机构加入其中。可以想象，如果公司和赞助商能够提供各种模块部分，这些模块还可以重复售卖及使用。

MACRO Museum of Contemporary Art in Rome 170

MACRO罗马当代艺术博物馆

走入一个古老的工业区，面对一种历史遗产，一个新的博物馆的建成旨在利用空间的流动性、运动性，以及景观实物，唤醒现存的历史静物，赋予其生命力。

在建筑作品中，它的断面图往往向人们展示了无法直观的建筑物内部构造。在罗马当代艺术博物馆的断面图中，从横向到纵向、从内到外、从馆内大厅到屋顶景观花园，清晰地向人们展示了整个建筑物的总体布局。

崭新的艺术作品与年代久远的艺术作品搭配摆放，展览空间与其他的活动区域也相互交错，这样的设计不会引起视觉的疲劳，令参观者在这种交错设计中，得到耳目一新的视觉感受。在这个交错设计系统中，新鲜事物无处不在，任何地方都有一种难以言表的清晰感，这种清晰感赋予了整个博物馆一种活跃的氛围。

博物馆的顶楼是一个屋顶景观花园。它是一种艺术足迹的投影，创造了一种抽象的延续。

从这一意义上看，它似乎是一个抽象的艺术花园，这一景观露台是为博物馆的参观者设置的，在参观艺术展品之余，来到这里享受一番自然景观艺术之美，一定会令参观者们感到赏心悦目；同时，它也是城市绿化中的一部分。这一屋顶花园还为参观者们设计了极富脚感的网状结构地板，踩上去，粗糙而结实，平滑而柔软，织网密集而富有亮泽……在花园的中央有一根自来水管，用于净化周边空气，以助于园内植物的生长。砂粒、碎石、光滑的石灰石、清澈的水面以及精美的植物……这些组成了一幅美丽和谐的露台景观图，给人以宁静、清新的感观。

博物馆展厅，由于性质、场地，以及人流量的不同，向参观者们展示的又是另一番景象。整个博物馆就像是一个流动的空间，从公共空间到私人空间再到公共空间，就这样循环交替着，不断给予参观者视觉刺激。

与整个博物馆灵活、流动的大空间相比，那些展厅就显得不那么具有流动性。它们为艺术家提供一个不规则但十分简约的展示空间，展厅内还提供了各种艺术品的展示方式：吊挂、悬挂、浮挂、置于地板上或安于地板内等等，为艺术家充分展示其精妙绝伦的艺术作品，提供了一个极佳的平台。

Manav Sadhna Activity Centre — Ahmedabad 172

艾哈迈达巴德市的Manav Sadhna活动中心

印度的城市中心每日可产生大约2,740万公吨的废弃物。像艾哈迈达巴德这样的城市就可产生2,750公吨。很遗憾，人们从未在真正意义上对这些废弃物进行过加工处理。这些废弃物只是简单地倾倒在垃圾掩埋场，消耗了大量的化石燃料，形成一个污染严重、危及健康的异化景观。所幸的是，印度有着废物回收的优良传统，在人们平日的生活起居里这种传统清晰可辨。通过把废弃品和盈余食物给乞讨者和动物的方式，残羹冷炙等废弃物便不被列入其主循环的过程。食物及许多其他东西凭借其用途多样的属性被赋予了高附加值。难道建筑业没有从中学得些经验么？位于艾哈迈达巴德市Rama Pir Tekra，Wadaj的一个活动中心正是沿此方向的一小步尝试：回收市政或家庭废弃物并用作建筑材料。

活动中心位于艾哈迈达巴德最大的违章建筑区，由社会非政府组织Manav Sadhna主持建造。活动中心的功能众多：幼童的非正式学堂、成人的夜校、培训中心、妇女和老人的手工作坊。校园里设有集体宿舍、行政室和宗教冥想室。校园的建筑材料来自市政或家庭废弃物的回收利用。这同时解决了环境、经济及经济适用房的问题。利用家庭单元产生的市政废弃物来生产建筑材料大大降低了废弃物的污染。废弃物回收工序的附加值为穷人带来经济效益，也为建筑获得了某种授权感。最后，由于回收利用的建筑材料比传统材料便宜且质优，城市贫民能获得价低且优质的房屋。

关注焦点

环保、经济效益的授权和价廉质优的结构是此项举措的三个维度。项目是Vastu Shilpa Foundation基金会在环境设计方面三年多实验研究的成果，此研究旨在把来自家庭单元的市政废弃物转变为建筑材料。利用废弃物、简单的手动工具、当地资源和行内知识对建筑材料进行的创新得益于第一手实验和现场勘察。项目表明了房屋的建造也可以带来经济效益，还可以使贫民获得某种授权。项目体现了经济自维的家庭手工业的潜力，以及利用价廉的另类建筑材料来提高房屋品质的可能性。

建筑施工

校园生动地展示了废弃物的回收并用作建筑材料的价廉质优和审美趣味。项目中的墙壁、屋顶/厚板、门和窗集合了市政或家庭废弃物，由简单的手动工具建造。墙壁的建造使用了六种材料和技术，即粉煤灰水泥黏合砖、来自垃圾掩埋场的废弃物浇铸砖、加固土块、回收的玻璃瓶、满是灰和废弃物的塑料瓶、蔬菜包装箱。包装箱的木条用作内部隔板墙。

同样的，活动中心的地板和屋顶厚板的应用材料包括：填充木板的玻璃瓶、塑料瓶和砖、石板、覆有黏土瓦的水泥板、黏土瓦屋顶的捆有光导管的白铁皮。门板使用碎包装皮和废弃的铜版纸作为增强材料，从而替代玻璃钢（FRP）。厕所的通风百叶窗用蔬菜包装箱的木条做框，用油罐做叶片。管理楼办公室厕所的木板门也是采用蔬菜包装箱的木条做框，油罐做填充。内嵌陶瓷工业废品的粉煤灰和废弃物瓷砖也作为镶嵌物（工序发生在瓷砖铸造过程）零星地点缀着瓷器。

所有这些产品都是第一手发明并制造的。产品是在其工程性能通过实验鉴定后才生产的，结果证明了产品的环保性、节约性、观赏性及其作为当代建筑另类材料的可执行性。

Meditation House — A Tranquil and Classic Spiritual Homestead 174

冥想住宅——恬静而典雅的心灵家园

这是一个倾注了建筑师许多心血和情感的建筑项目，它的建筑风格具有丰富的元素。建筑师必须了解住宅主人的心情，此刻，主人也许不会关注住宅的艺术感，但对于建筑师而言，他们需要为客户创造一个恬静而典雅的心灵家园。为此，建筑师们以埃及“墓室”建筑或帕伦克的一些玛雅建筑为参照物，建造这座极富古代建筑艺术风格的住宅。

建筑师们通过对建筑材料和照明设备的抽象性处理，达到了创造住宅的冥想性与视觉和谐性的效果。这个住宅坐落于一个住宅区内，带有一个室内照明庭院，建筑师们应主人的要求，计划将其改建成一个与众不同的个性住宅建筑。整个住宅的建筑外墙是由Grissal flamed花岗岩砌筑而成的。

住宅的迎宾门是一扇高约9米，宽约2米的三角形木门，进入这扇高大的迎宾门，映入眼帘的便是一条相同形状的迎宾道，建筑师们赋予入口处一种庄严的氛围。这种庄严的氛围止步于一个大厅入口过道，走过这一过道，一个宽敞明亮的双倍高度花岗岩大厅呈现于眼前，给人一种豁然开朗的视觉震撼。

大厅的光线来自于北面的室内庭院，在这个雅致的室内庭院中央，种植着一棵高大的千年蕉，旁边摆放着艺术家Saul Kaminer的象征性抽象雕塑作品，这是住宅内唯一的一件艺术装饰品。

室内没有任何家具摆设，只有一张环绕室内一圈的波浪状木长凳，整个住宅建筑的木质装饰都是使用这种与木长凳相同的木材建造而成的。这种木材还可用于隐藏住宅内所有的空调、音箱和手感极好的嵌壁式照明系统，使得整个室内空间显得整洁、清爽。室内虽然没有任何的装饰，但多样的灯光变化效果反射在花岗岩墙面上，为室内空间增添了几分活力。悬挂式天花板是用黑色龙凤檀木建造的，与四周的墙面并无任何衔接，在天花板上还有一个别致的天窗，整个天花板组合就像一个精美的艺术品。

Meghna Residence — Living in Detail 176

位于三角洲的梅克纳住宅

由于缺乏适当的总体规划、建筑规则和建筑意识，世界上人口密度最大城市之一的达卡，已经成为了一座市区建设杂乱无章的城市。400多平方米的达卡市区，人口不断增长，所拥有的绿化面积却不到5%，加之其他公共事业发展困难、缓慢，因而，达卡正在迅速降低它的城市生活质量。

达卡是一座人口稠密的城市，因此，大多数带有开放绿色空间的单层住宅楼正在迅速消失，取而代之的则是多层（6层）公寓楼。

梅克纳住宅是位于Dhanmondi的一座单一的大家庭式住宅，它在六十年代初期，是这一地区的规划住宅之一。这座住宅的设计考虑到了三个方面——

周边环境、当地气候，以及建筑物本身的传统风格。

周边环境

梅克纳住宅坐落于通往东北方向和西北方向两条道路和连接处的一个角落中。西北公路两旁种植着高大美丽的桃花树，它们就像一个屏障，保护着梅克纳住宅的隐私。这一住宅朝向一个开放式的大型游乐场和妇女体育联合会的一个游泳池，这个游泳池在东北公路的对面。在它的东南面，有一座6层楼高的公寓，这座公寓的正面设有49个窗口，这使得公寓内的一切活动，一目了然；而在它的西南面，是一座两层楼的住宅，这一两层住宅也可以被认为是今后达卡市区高层建筑的雏形。

当地气候

达卡市区位于北回归线附近纬度24.5度的地方。每年夏季，东风、东南风和南风转变为西南风时，给达卡市区带来了约1523毫米的年降水量。由于它所处的纬度位置，所以夏至对达卡的影响并不大，因为太阳在夏天的正午，位于东北极点上，而到了下午，太阳的位置又转向了西北方向。夏季太阳的照射比较平均，但到了冬天，太阳光倾向于对南面的照射，而其他地方就缺少阳光，因此在冬季就需要采取一些人工措施，使得各个方向都能得到充足的阳光。

传统风格

在孟加拉国，住宅内的庭院往往是一个具有浓郁的当地传统风格的创造型私人生活空间，庭院内都设有良好的通风、日照系统，同时，主人也会种植一些自己喜爱的植物。

为了保护梅克纳住宅主人的隐私，尤其是保护主人在五楼泳池内的活动不受外界干扰，同时，也为了确保住宅西北角和西南角的住宅生活主空间冬暖夏凉，住宅主人要在有着49扇窗户的住宅东南面建造一堵围墙，然而这却是一个极大的挑战。这堵围墙的建成，将利用其暖房层进一步营造一个传统的生活氛围；同时利用水流将整个住宅建造成一个三角洲。

基于以上设计考虑，我们可以将这堵围墙设计成一个“U”字形结构。

同时，我们还采用了各个不同楼层的“叠加”设计来扩大地面和其他楼面的使用面积。一个大部分面积与地面连接的地下室空间被充分利用，作为污水处理系统、水压容器系统和发电设备的安装空间，它也可作为停车场和电梯房等等。二楼作为一个正式生活空间用于会客、就餐，另外，在二楼还设有专门的会客室、住宅主人的家居办公室和整个住宅的主厨房。住宅的三楼则是主人的私人空间，所有的卧室都设置在三楼，三楼还设有主人私用的休息室、餐厅、书房、浴室和几个别致的微型小花园。主人在四楼安装了泳池的隔底装置和其他的泳池机电配套设施。梅克纳住宅的顶楼，也就是五楼，设有桑拿房、健身房、泳池（这个泳池的一部分被屋顶遮盖）、一个与池边阶梯相连的露天平台，在这个露天平台的旁边，有一个大花园。此外，在泳池边，远处一大片树林的美景尽收眼底。

Memorial of Flightno.655　180

655次航班纪念馆

伊朗航空公司655次航班，也被称为IR655，是一架在两伊战争期间被美军导弹击落的民用飞机；它于1988年7月3日在霍尔木兹海峡上空坠毁。这架伊朗航空公司IR655次客机是一架A300B2型空中客车，坠毁当日，该次航班从伊朗阿巴斯港飞往迪拜，机上乘客与机组人员共290人，其中包括66名儿童。此次事件被认为是伊郎第七大致命坠机事件。美国海军CG-49文森号巡洋舰在穿越伊朗领海内的霍尔木兹海峡时，攻击了正在伊朗领空航行的IR655次航班，致使坠机事件发生。

该项目是为伊朗航空公司655次航班坠机事件中的殉难者而建造的纪念馆。该纪念馆的建筑设计与任何其他相关图片的展览序列空间的设计有着很大的区别，它必须全面地反映出这场灾难性事件的连续性后果以及它对海洋的破坏性影响，并且展示出海洋在搜救过程中的客观意义。它还集中反映出了战争给整个国家带来的贫穷、孤立与阻力。蔚蓝色的大海，在这样的阴霾笼罩之下，失去了往日的美丽风采，它仿佛也在为死难者们哀悼，海面上不断扬起的汹涌波涛，似乎都在怒斥这场无情的战争。人类社会的进程将记录下这一惨痛的时刻。为了追忆和证实这个血淋淋的历史事件，建筑师以当地宗教建筑形式为雏形建造了这一规模庞大的海洋式纪念馆。纪念馆建筑分为两个部分：纪念碑和碑座。碑座由两层序列空间组成，这两层序列空间内都设有瞭望设施，可以通过它们观察655次航班遇难的那片海域，这两层序列空间在其两个尾端各自连接起来。入口通道上设有明显的“肃静”标志，以示对逝者的缅怀。这一规模宏大的建筑在坦尼斯日落余辉的照射之下显得格外壮观，这种壮观似乎在向人们呼吁一种和平，呼吁战争远离这个国家。尖木桩的设计显示了国人的使命感与对战争压迫的一种反抗精神。

Milhundos House　184

Milhundos住宅

该地块位于佩纳费尔市郊区Cavalum河的山谷上，与Quinta das Lages公园毗邻，公园的秀美景致也纳入住宅中。

设计师对原来地块改造的设计方案进行了修改，并提出了不在L8和L9地块中寻找住宅的场地，而是将其与最近的柱子联系起来。考虑到地块与施工图之间的差别，设计师提出了新的设计手法与策略。

住宅由相互补充、互相独立的实体组合起来，功能性较强，并与两根独特的柱子融合起来。每根柱子与外围空间都有着不同的联系：白色石灰石柱与远处的景观遥相对应，而另一根黑石柱子却与室内地面的联系更为紧密。

Moon Project — Villa　188

月球度假区

月球度假区是建筑师在2007年设计的一个概念性项目。在研究了月球的环境和各种限制条件之后，建筑师对在月球上的设计项目进行了论证，最终确定了一个以光和影为主题的方案。

月球是一片没有任何生命迹象的区域。建筑师的设计旨在打破这样的状况，营造与周边环境相融合的空间。

Musee Des Conflluences, Lyon, France (2001 - 2010)　190

法国里昂博物馆——现在与未来的完美结合

知识的结晶

未来社会将是知识的社会。尽管如此，将知识划分到某个特定的领域却不是一件容易的事情。

创新将随着室内空间、模糊空间及重叠空间而发展。未来社会的问题将会出现在技术、生物及道德等方面的过渡区域，这三个方面也是该项目的主题。

形式、穿透、变形、故障和变异的突变将会对建筑产生影响。由此产生的建筑将被赋予由不同实体所产生的互动、融合和突变的特点，并构成一个新的形状。

该项目本身并未被界定成无所不包的知识领域，但却是通向当今知识时代的公共场地。

其本身不仅仅是博物馆场地，同时也是小镇中的集会场所。它刺激了当地人的直接需求，鼓励他们积极参与进来。其建筑形态将博物馆及城市娱乐空间这两种类型很好地结合起来。

硬质空间——软质空间

两个连接复杂的建筑实体的概念是建筑场地相互联系而产生的结果。

结晶状的建筑物面朝小镇的一边，被设计成城市论坛和入口大厅。其建筑形态象征了日常生活的世界。与之形成鲜明对比的是，白云掩盖了关于未来的认知。软质空间内包括隐藏的溪流及数不尽的过渡区域。

该项目中所呈现出的现在与未来、已知和未知被看做是“引发公众好奇心”的空间排列。南部岛屿顶端的公园向外延伸形成了新的城市空间。斜坡及地面景观将内外空间结合起来，引发了一系列空间活动。

展览厅内的空间结构紧随其后。利用两层高的双人房，封闭的黑盒子及免费展览区域可以轮换使用。整个结构如项目所希望的那样呈现出不同的变化形态。

重现城市活动的理念使得里昂完美地将自身融入到该区，乃至整个世界中。

Museum of Ivan Marchuk 192

伊万马尔丘克博物馆

该项目的主要任务是将斜坡与广场这两处公共空间连接起来。人行道位于大楼结构周围，大楼大部分地区主要是公共的圆形剧场和户外博物馆（延伸部分为内部展览厅）。

城市边界划定了隐藏的建筑：垂直表面是由玻璃制成的，水平面则是天然草皮。不同的视觉角度在公园内反映出来，大楼内部的树木都得以保留。

假设草皮附近是办公区，博物馆及图片展览室则位于柱子“内部”。

室内及地面的装饰与伊万马尔丘克油画风格保持一致。

My House 196

我的小屋

我的小屋是一座很小的观景房，供我观察自然。结构没有将我的视线和周围的环境阻隔开来。

这座房屋位于山上，恰好能够欣赏到优美的山景。在这里，每一天都能接触到自然景观。

这是一个小型的设计和小型的空间。

生活还在不断延续……

Nashr Yadavaran Administrative Building 198

Nashr Yadavaran行政大楼

该项目的主要设计理念可分为几个层面进行阐述。建筑师们对大楼空间进行科学合理的布局，使其既具备完善的社会功能，又能够保证楼内工作人员拥有灵活的个人活动空间。

由于大楼的入口处采用了半公开式设计，人们可以透过入口处看到行政大楼建筑的整体外观，加之大楼外墙使用大量的玻璃建材，因而整座大楼有着极强的透明度。这座大楼拥有着所有行政大楼所应该拥有的各种大型空间元素：入口处大型2米×2米的半掩式旋转门由一根单独的铁柱支撑着，它是大楼周边公共空间与门内大楼私密空间的连接媒介，创造出一个具有独特风格的“共享空间”；位于一楼阶梯上方的木制屋顶平台，为大楼举办各种书展或团体阅览等图书活动创造了理想的活动空间。

行政大楼建筑外墙以红色石灰石为主要建材，令整个建筑外观醒目而庄重；它的使用也体现了当地社区建筑风格的主基调。由于整体外观显得庄重大方，因此，这座大楼在若干年之后，极有可能成为当地一个地标性建筑。

在大楼的空间设计上，建筑师遇到了一定的阻碍，设计受限于伊朗城市发展规定中的“四六分”原则，所谓的“四六分”是指建筑物本身的占地面积只能占总建筑面积的60%，其余的40%必须用于建造与建筑物相匹配的庭院。这样的强制规定严重束缚了建筑师们的设计灵感，从而无法创造出更高质量的建筑空间。

大楼的室内空间并没有十分明显的间隔，各个小空间融合成一个有机的空间整体，但从建筑外观的整体设计角度看，却容易令人产生室内空间间隔分明的错觉。这也正是这一建筑设计的巧妙之处，既加强了室内空间的灵活度，又确保了棱角鲜明的建筑外观。

大楼各层的日照量取决于各层楼的不同功能：楼内工作人员办公室内能够接收最为充足的阳光照射，其次是会议室，而顶楼阁楼空间接收到的日照量则是全楼最小的。

National Museum of Pakistan 200

巴基斯坦国家博物馆

建筑理念

一座历史博物院是一个国家数百年历史的最终展示。因而，博物馆的建造应该是气势宏伟而又形态万千的。博物馆往往也是一个国家未来历史的储藏处。

下列是博物馆设计的几个基本理念：

a. 博物馆建造设计中最重要的是要建立一个人性化的友好建筑外观。

b. 整体的建筑风格应当是愉悦而不单调。建筑形式应当是经受得住岁月的考验。同时，博物馆建筑还应当拥有醒目的建筑轮廓和建筑背景。此外，这一国家级博物馆应该成为夏克巴里安山森林周边最耀眼的建筑。

c. 博物馆内各个流动形长廊空间的大小并不局限，也不需要统一的风格限制，但必须与其相应的展览空间形成一种自然的融合。这种流动空间的设计灵感可以借助于当地那些川流不息的河水。

d. 室内的景观摆设要注重干净、整洁，与整个室内空间的线条式简约格调保持一致。展示区的室内空间强调的应该是展示艺术，而非建筑艺术。

e. 悠久而博大的历史画卷，一个博物馆是不可能以一个时代的反映把它完整地展示出来的，哪怕只是象征性的展示。因此，可供参考的展示形式更多地应该借助于土地本身。棱角渐变墙的设计是受到锯齿状Potohar Plataue的启发。使用各种不同土黄色调的石砖砌筑博物馆的外墙，这样的建筑设计理念，突出表现出这一独特建筑的历史见证者特殊身份。

建筑规划

该建筑分为南北两个不同的区域，由一个宽敞的人行过道将这两个区域连接起来。

博物馆的北面由长廊以及相关的活动空间与行政办公场所组成，南面是图书馆、礼堂和公共餐厅。

博物馆大厅本身就分为多个层楼，厅内还设有多处采光井和室内庭院。在大厅的东北角有一个半开放式的文物展厅，厅内带有一个幽长的走廊。

博物馆东面的长廊设有多个不同的工作区域。走出博物馆工作人员的办公楼层，就可见一个悬于长廊上的夹层空间，在上面可以俯瞰大厅全景，并与它下层的长廊空间直接相通。

经过办公楼层的下层楼梯，再经过一个坡道，便是大礼堂和图书馆。餐厅设在博物馆的高层。餐厅的下方是一个半开放式的空中人行广场。

在博物馆的建筑用地中，南面沿街的一部分土地（约12,140平方米）将被划为临时展区。

建筑入口与人员流通

建筑入口与人员的流通取决于建筑功能的可行性和建筑用户对建筑物的了解程度。

1. 游客方面。空间人行广场能够通往博物馆大楼的任何地方。广场上设有一个信息布告廊，这个布告廊可通往博物馆主楼、图书馆以及大礼堂。地下停车场中有一个通往人行广场的独立楼梯。公共汽车和旅游巴士的停靠处在博物馆主入口的左侧。在这些公共交通工具的停靠处也有一个较宽的人行通道可供游客自由行走。

2. 管理方面。博物馆的工作人员可通过下层人行广场或与他们办公室指定的地下停车场相连的楼梯进入办公区域。

3. 服务方面。博物馆设立三个服务领域：

a. 机电空间设于底楼（半地下室中），机电空间的对面是公共停车场。

b. 经过登记区的博物馆运输区位于北楼的后侧，纪念品小作坊也设在此地。

c. 餐厅运输区和固体垃圾处理区位于南楼的后侧，礼堂的后备供应区也设在这里。

上述所有区域都可经由人行广场底下的一个地下通道相互连通，因此，行人与车辆并无直接冲突。

环境因素

博物馆一向被认为是一种绿色建筑。它的建筑设计与建筑定位都决定了它能够避免夏季的烈日暴晒。厚厚的空心墙以及朝向西面与西南面的环保设施的应用，使得整个建筑天然能量的储存得以实现。北楼与它的百叶窗屋面能够吸收大量的光能。东面的办公室窗户则设有避光保护系统，因而有着极佳的遮阳效果。建筑物的南面和西南面的建筑外墙几乎不采用玻璃外墙结构。

结论

显然，该项目中的许多施工细节和小的改动都必须纳入到实际建造阶段之中。所提供的图片也仅仅只能显示出建筑师的设计意图。但最值得一提的是，这座国家级博物馆是顺应周边自然要素，且迎合用户需求的现代建筑。

Nautilus 204

鹦鹉螺大厦

作为多功能商住楼，鹦鹉螺大厦是环保理念与现代可持续性建筑的完美结合体。这座建筑与它的环境一同成为了住户的第二层皮肤。这也是一个兼顾环境保护和使用功能的优秀建筑设计。对于居住在这座大厦的人们来说，他们的现代生活不仅保护了环境，更同时服务于自己的生活。在当今社会，我们如何应对这些担忧，并在项目设计中很好地处理呢？雨水的收集、过滤和处理；具有休闲、景观和渗透功能的绿色屋顶；太阳能的利用；利用风能的垂直轴；环保砖；舒适的双层隔热玻璃；迷你污水处理站；有机的废弃物堆肥；废弃物分类；用住户处回收的烹饪油来生产肥皂、清洁剂、消毒剂的小型生产区，满足住户使用需求；地下室和停车场的圆顶能够更好地利用阳光。大厦周围有四条道路，主体建筑建在一个大型的平台之上。为了更好地实现循环通风，主体建筑被分成三个部分，各自设有独立的垂直通风设备和三个储水池。这三座独立的塔楼都建在一个平台之上。五楼以下的公寓的户型平面设计都是相同的。从五楼到二十七楼的公寓则交错排列，以满足不同的需求：露台的渗透性、东西风向的自由流通，这样就带来了新鲜和充足的空气循环。每个公寓在前后立面都是开放的，营造了更多的视野。这种布局方式也能很好地阻隔附近道路上的噪声，并提供更好的自然光线。平台设计：地下为商场的停车库；一楼为商场，与周围区域的水平高度相统一，位于建筑正立面的较低部分则进行了提升；二楼为住户单元的专用车库。住宅楼设计：底层是整个区域的最高处，与周围的道路直接相连；1至5楼是标准型公寓；6至27楼为有楼台的双居室。整幢大楼一共有59套1至6居室的公寓单元。住户车库为120个车位，商业停车库分为16个单元，共1202个停车位。设计概念和几何学设计：项目的设计理念和几何学设计灵感来源于鹦鹉螺的几何形状和结构。这种软体类动物以微生物形式开始它们的生命，然后在钙化的贝壳内部继续生长。随着它们不断长大，它们离开旧的贝壳并在其背部钙化出新的贝壳。鹦鹉螺经常占据着最后一个螺旋空间，这样其他的部分都被空气充满，使它能够在其中漂浮。据说鹦鹉螺就是用这种螺旋形的生存方式呈现了黄金比例，也被称为黄金螺旋。这些成功钙化的贝壳将称为鹦鹉螺的新家，也形成了1:1.618的黄金比例。这一相同的比例也应用于建筑的公寓设计之中。同时，这种比例关系也经常用在文艺复兴的绘画上，例如达芬奇和其他大师的作品。这种黄金比例与生长的本质有着联系。在贝壳的空间分配（如鹦鹉螺），人体结构（趾骨和手指），甚至是世界各地的男女关系上。由于它与生长的特质相联系，因此这一数字比例变得如此普遍。正是由于这种频率，黄金比例变得极为特别，对研究人员、艺术家、建筑师和作家而言都是非常实用的。除了这个特点之外，黄金比例最重要的地方在于它嵌入的背景。事实上正是数学的发展过程赋予其意义。这也是我们在这个项目中将其融入设计的原因。

New Sicilian Deli Corner — Food Experience 210

城市饮食广场中的建筑哲学

一座城市是由不同的空间元素组合而成的。城市在各种光的作用之下，创造出活力四射的瞬间动态元素，在这些动态元素的映衬之下，城市建筑更具魅力，城市空间也变得更加五彩缤纷。该项目的建筑理念主要是为了展示现代都市人的一种生活情趣，它融入了意大利城市广场建筑中最为经典的传统元素，同时也表达了建筑师对意大利建筑史的深刻理解和对建筑艺术的无限热情。意大利著名建筑师伦佐•皮亚诺在美国接受查理•露丝的一次采访中曾经说过这样一席话——

伦佐•皮亚诺：透明度和光的结合能够组成一种奇妙的元素，这种元素可以建造出一个公共空间，一个封闭空间内的小广场。

查理•露丝：这是您的意大利建筑精神吗？

伦佐•皮亚诺：广场就好比一个图标；我梦想中的城市应当是这样，街道与桥梁之上充满着光和各种透明元素，例如，水，道路两旁种植着各种绿树。城市广场是都市人约会的场所，人们在这里相约、相见，而后匆匆离别……因而，城市广场的建造，应当与其周边的街道、建筑等城市元素相融合，使其成为城市空间的一部分，而并非对一个孤立的空间体的建造……这并不完全是指所谓的风格，更确切地说，它是对空间创造和探种的一种强烈的欲望……“探种”，在此处是指对一个广场建设场地的深刻理解，用心去聆听它的故事……每一块土地的背后都有一个动人的故事……

从上述的对话中，不难得知建筑师们是如何定义一座城市广场的主题风格的，建筑场地固有的历史韵味和建筑师们独到的选材风格赋予了这一城市饮食广场一种意大利西西里美食文化体验。一个城市广场的建设并不仅仅是简单地对不真实的几张设计图纸的加工和改造。试图赋予一个城市广场一种浓郁的当地特色，仅凭设计图纸是行不通的，建筑师们必须探究如何在一个小小的城市广场内融入更多的生活与体验元素，以吸引更多都市人的眼球。在这样的城市广场中，各种动态设计显得尤为重要，它们可以赋予广场空间以充实的视觉效果，让在此就餐、约会的人们充分感受到生活中丰富多彩的一面。

Olympic House and Park, Nicosia, Cyprus 212

塞浦路斯奥林匹克之家及公园

奥林匹克之家及公园的建筑构想要充分表达奥林匹克理念。以下三点是体现这一理念的基本要素：

•在全球层面上，鼓励并教育人们崇尚奥林匹克精神，发扬手足之情，培养高尚情操，维护世界和平。

•在历史层面上，保留和追忆伟大而古老的运动。

•在运动层面上，使人类在必要的活动中追求身心的协调发展。

从个人到集体，从古代到现代的转变可以通过对建筑形态的改变来体现，通过调整规划中的建筑物的间距和截面图表达出来。在总体规划图中建筑物与非建筑物、主体建筑和公园景观的和谐统一寓意着身心的协调发展。

项目的平面图中，展现出了古代体育场的整体概貌。建筑物是围绕着中心的空地修建的，出入口处面朝着城镇，使外界可以看到建筑内部的概况，也方便游客出入。
在截面图中，从古代到现代的转换是通过抬高建筑物、架空建筑物底层来表现的，并使这里变成一个展示奥林匹克运动历史的博物馆。
在总体规划图中，建筑物不与公园并置，也并非背对而建。图中，建筑物单独矗立于空地之上，但实际上，未建造区域则穿过建筑物中心并与建筑物融合在了一起。建筑物和未建造区域、主体建筑和公园相互依存表达了“体质与心智”的相依相存。
中心的空隙为人们集会提供了巨大的空间，使人们能够通过背阴的天井从公园到达大厅的入口，而后从多功能厅折回公园。从外面看办公室是环形的，所以在天井任意一侧的空间都是成线形的，让人仿佛置身于奥运赛事之中。

Orphanage at Chandpur, Bangladesh 214

孟加拉孤儿院

孤儿院位于孟加拉郊区，周围是低地和不同的水体。该项目对24名孤儿的宿舍、计算机室、室内及室外娱乐设备的综合大楼进行扩建工作。孟加拉郊区住宅在规划、材料的使用及地理位置等方面都有着自己独特的特征。与城市建筑相似，郊区建筑风格也发生着变化，但孟加拉郊区建筑风格依然保持着传统的风貌。数百年来，郊区建筑一直以本地材料及施工手法为主。综合楼也是基于对文化及传统的敬仰与崇敬，调整设计策略是该项目面临的最大挑战，这不仅要采用较高的技术手法，还需要获得当地人们的社会认同。除此之外，自然照明及通风成为衡量对环境产生影响的重要因素。场地四周被农田包围着，并沿东西方向进行延伸。运河位于大楼西边，主要用于灌溉。
平面图中的大楼呈L形，与公共空间相互对应起来。宿舍及教室位于大楼的两侧，宿舍向东南方向微微倾斜，这样便于交叉通风，防止将过多的热量集中在宿舍内。设计师采用当地竹子设计的屏风有效地阻挡了太阳光的直接照射。屏风后的窗户被用作挡光架子，将照射进屋子的日光分散开来。教室位于南北方向，该项目的目的是将室内空间与周围的自然景观融合起来，营造出室内的娱乐空间。

Pachacamac House 218

秘鲁巴恰卡马隐藏别墅

该别墅位于离利马南面40公里的巴恰卡马市的小山内，客户是一对退休的哲学家夫妇。设计师对场地条件做出的反映就是将别墅隐藏在山中，试图在建筑与景观之间形成一种平衡的对话，这里室内、室外都成为对黑暗与光明这两种对比的完美阐释。别墅的玻璃外立面沿着小山向外延伸，也象征着建筑对大自然的影响与干预。

Passive Solar Habitat 220

被动式太阳能住宅

为使可持续生活方式更加深入人心，实现与自然和谐相处的目标，设计师通过利用当地的资源和调查可行性方案，提出了采用被动式太阳能住宅设计手法的建议。
设计师受到自然栖息地的启发，与被动式太阳能设计保持一致的同时，直接利用当地的太阳能进行加热和制冷，收集大量废弃的自然原料，重新探索修建被动式太阳能住宅的方法，从而实现可持续发展策略。该项目主要利用可回收材料，如从当地工业区、木材厂等地收集的钢铁，以及从采石场和矿渣中收集的材料。建筑结构主要采用预制的建筑部件，从而有效地控制了成本，降低了人力成本，减少了对场地造成的影响。该住宅典型地突出了使用的材料，成为可持续发展策略的典范。
与自然环境相关的是项目中采用的灵活部件，尽管在许多被动式太阳能住宅中通常不采用可拆分部件，设计师努力寻求可行性方案，将自然风与阳光融入到住宅结构中，以抵御恶劣的天气、大风及西面日照。建筑构造上采用的屋顶通道、墙体及天窗等都是对通风条件及日光等方面的探索性创造。设计师采用预制正面及屋顶板使室内空间变得更加舒适。玻璃窗朝向南面和东面的阳光，通过以地板与循环中心为主的地热系统，将太阳光的热量收集起来。
地面是室内生活空间广场的主要入口。设计师通过提升原来的居住、餐饮及休息空间，在各种资源不变的情况下，使功能空间在视觉上增加了一倍，减少了对场地的影响，加强了与周围景致之间的联系。屋顶与绿色屋顶系统的建造对住宅及户外娱乐区起到了制冷的作用。

Plan[T]is 224

未来城市住宅新模型之典范

设计公司提出将该住宅项目作为适应建筑环境的模型。设计团队在尊重生态系统的基础上，积极探索城市发展能促进自然生长，协调和适应现有城市肌理的模式。设计师还从自然界中引用成长和叶序等法则。为满足不同的城市需求，该项目被看做是阶段发展目标的衡量指标。
按照项目发展阶段及场地的住宅承受力，设计师提出修建2,500～4,500个住户单元。努力将该住宅打造成能适应不同的城市条件，满足不同的城市需求的城市公共系统。设计师在充分理解伦敦五个社区的游牧生活方式的基础上，通过建筑设计营造出游牧式互动氛围。
设计师采用代理生成系统和叶序自然模型的分发系统，以时间作为阶段限制进行电脑处理，利用反身系统来进行城市设计的开发。该项目位于伦敦机场和泰晤士河之间，港口区与千禧穹顶协调着整个边界与交通网状系统的各项资源。由于伦敦是世界著名的游牧式生活城市，设计师旨在满足时代的要求。如今的节奏营造了新的城市生活模式，使城市生活更加灵活，瞬息万变，人们得做好充分的准备来适应这些变化。该项目试图寻找一种新的生活方式，为不同的人群打造不同的居住方式，以满足城市发展的需求。该项目中对塔进行了深入的研究，设计公司试图挑战现有的时间逻辑，提出新问题，打造新的住宅类型。住宅作为一种新的模型和居住方式必将带来新的开发潜能，为未来住宅带来深远的影响。

PUMA CITY — A Nimble Dismountable Container Building with Convenience 230

PUMA CITY——便利灵巧的可拆卸式集装箱建筑

PUMA CITY是一个集装箱建筑，它是由24个海运集装箱改造组合而成的可运输型零售活动中心。它随着长达21米Puma号帆船参加2008年沃尔沃海洋赛，周游世界一年之久，于2008年10月初停靠在西班牙的阿利坎特。
该建筑是完全可拆卸式建筑，它可在拆卸后装于货船之上，与帆船一同远航；它适合在任何一个国际港口停泊，组装与拆卸都十分便利，且结实耐用，可经受多次拆装。
建筑主体由三层集装箱堆叠而成，并实现室内外空间、大型悬垂空间以及露天平台之间的相互转换。在建筑物的正立面，也就是在三层堆叠集装箱的正面，印有PUMA巨幅字样和一只巨狮，这是PUMA公司的徽标；由于集装箱的交替堆叠，这个徽标被分散成几个部分，这样的徽标印制布局给人一种零乱美的视觉感受。

Puma City底层是两个完整的零售商店，其层高是一般集装箱高度的两倍，底层还带有一个比普通集装箱宽4倍的室外空间，这样的设计布局与集装箱室内的模块化布局形成鲜明的对比；二层是办公室、媒体采访区和储藏室；在Puma City的三层，也就是顶层，有一间酒吧、一间休息室和一个带有开放式阳台的活动空间。

该建筑使用约12米长的统一规格的集装箱，同时，采用一些现有的集装箱连接器将这些集装箱个体连接成一个有机的整体，以增强整个建筑水平与垂直方向的安全系数。在这座特殊的建筑中，每一个建筑模块的设计都是为了通过一个结构覆盖控制板系统实现常规船舶集装箱的运输，而在这一结构覆盖控制板上的所有大的孔隙都经过焊接处理，以便与大型开放式室内空间实现牢固的衔接。

Puma City是一座真正的实验型建筑，它充分利用了全球海运网络进行广泛的自我宣传。它是首个真正可移动式集装箱建筑，占地约1,022平方米，它的设计旨在回应建筑设计中的各种挑战，包括国际建筑法规的相关规定、恶劣的气候变化、电源插件和高压交流电系统，以及安装与操作的舒适程度等等。

Reclaiming the City — Urban Nexus 2020　232

改造城市——城市枢纽2020

密集城市的“可持续基础设施和发展模式”，例如孟买。站点的复兴涉及一系列的商业项目，包括零售商铺、一个酒店以及拥有5个屏幕的影城等文化设施（两个500个座位的观众席和一个1000个座位的音乐厅）。项目包括在混凝土板上建造绿色屋顶作为下面车站降温的吸热设备、利用高楼朝向的自我设荫策略、东西朝向办公楼的生态与气候设计原理、利用窄楼板（平均宽度：12～14米）使昼光穿透和自然通风最大化、办公楼南立面铺陶砖和玻璃增强复合jaalis、在向阳面栽种台地（空中花园）。

问题引述

这个未来的世界城面临着层层危机。无地点性的物体正以新时代建筑为名义慢慢摧毁着社会的内聚结构感及每个地方或社区的特有规模——这座城市是否失去了对层次感的控制，取而代之的是无方位的整形意象？此外，还有许多关乎这座城市存亡的重要问题亟待解决，例如可持续发展模式和区域建筑杂乱无章的问题；更甚的是，在工业废弃地和灰色地带充足的背景下，为了发展而胡乱破坏绿地和自然生态系统。

超级网络化的实体城市

这座城市作为一系列网点的功能是这个项目实施的前提。网点里每个火车站四通八达的交通使其成为所在郊区的中心。在这个连通性最强的地方继续着密度最大的真理。在孟买的城市结构里，每个郊区作为更大城市系统的自维部分，与一张大众移动系统的非线性网连接着，从而形成了超级网络化的实体。这促进了这座城市的平衡发展，也确保了城市未来的可持续远景。

嵌入式建筑设计

安德里联合运输项目

项目的实施大部分是由本地居民完成的。项目的建筑物是城市的延伸，其鳞次栉比地牢牢嵌入周围多样化的环境里，目的是使其每部分都能与城市里特定变换的环境相呼应，即与城市风景相融合，而不赋予雕塑式的色彩。

每天大约150万人徘徊在安德里车站附近。安德里不仅是小型工业中心和著名的集体宿舍郊区，它还是这座城市服务业的非官方枢纽。这得益于安德里的重要地理位置，其近旁就是国际机场和一个正在发展的酒店商务地区。

项目的发展预期和市政府的官方立场相矛盾。项目提出通过密集的空间层次组织使车站的城市化更加明显——使其天衣无缝地整合。

把车站场地想象成城市的交叉点或“枢纽”，其按传统被分成不相通的东部和西部，由宽宽的铁轨横亘其间。此项目需要把这些深深的切口缝合起来，即复原城市风光并改造这块宝地。

就概念而言，首先把连通性的表层拉伸，及至形成一个流动的公共空间，匀出视觉的清晰度和极好的朝向。这样的表层就成了十字路口或枢纽，能容纳众多人工建筑形状徘徊其上。

这些建筑形状大致分为两类：一类是沿着院脊南北朝向的多功能低矮线性建筑物，能使昼光和自然通风最大化；另一类是东西朝向的窄楼板办公楼，下面是能举行活动的大型实体地下室。

这两种朝向表达了该场所内在的两种重要流型。铁路的南北动向残酷地截断了与东西朝向所建部分相结合的愿景尝试。

沿东西轴的高楼挤聚一团，俨然一幅表征该场地主交叉点的视觉晴雨表。同时，基地的线性楼房作为入口点。这种朝向有益于太阳能吸热的弱化以及自然通风和昼光的最大化，同时每层的开放式后阳台上栽种的足量植物能降低内部的温度。从南边看这些高楼，外立面的陶砖和玻璃增强复合材料的纹理现出鲜活的土壤色调，由此赋予了项目强烈的色彩感。这些鳞片似的大块jaalis能调节大楼内部的光质。

枢纽的创意延展成一片波浪式的风景——浅浅“土丘”的线条由缓缓凸起的屋顶或后阳台板组成，因大楼的立轴而凸显。这带来了岩层和失却地质宝藏的记忆——特有的玄武岩崖壁，其六角岩层在91米高处飞翔。这些六角岩层早在1900年就被人们漫不经心地挖掘走了，它们曾经是安德里风景的标志。

空间上形成了由大型斜坡和稀松的表面组成的模糊城市地貌——它模糊了层次间的分界线和不同处，并顺应了该场地的不规则性。

从屋顶台面往下看，透过稀松表面的斑驳缝隙就看到了车站下的风景。空间戏剧性地向下铺展并延伸进车站的内里（月台的水平面）。月台上部空间的层次交错感体现在漫不经心的一瞥里，还体现在不时的光线和视觉里。

个体沉浸在紧张的都市体验里，这总归是不完整的——还包括沉浸在重复使用的演变过程里。

项目一直试图表达的是：孟买并不像大多数人和规划家认为的那样，它并非平坦（2005年7月的洪灾使大家从睡梦中醒来并意识到这一点）；其次，使其失却的地貌感在都市环境里复原是当前急需做的。因为人们早已忘了自己是从来都不会把孟买的地面设计成平坦的。

Recreation Center　238

康乐中心——建筑与光的完美结合

该项目是一个康乐中心建造项目，其设计理念是对Trenta地区的主要地形、气候（山脉、浓雾、水蒸气、烟雾、村庄等）进行总体分析与整合。通过不同元素的使用及其组合，为该地区保持瞬间美景提供了适应性的环境系统。

正如人们已了解到的，迄今为止许多建筑借助光、影的作用来表现其优美的线条，比如该项目。建筑师们的设计想法也与这种作用有着一定的关联。如果建筑师们计划设计一种烟雾缭绕的建筑效果，那么所有的设计理念都应当围绕着球面投影这一概念，并且烟雾流体边界须与建筑的周边环境保持一种持续的互动。那么这一烟雾缭绕的建筑空间将会给予游人们怎样的视觉体验呢？建筑师们认为这一建筑一定可以带给人们一种朦胧感和无限的幻想。烟雾将成为该项目的一个亮点，它将引领人们进入一个奇妙的空间世界。

该项目将成为一个室内外互动的建筑系统，虽然它以烟雾作为空间的假定边界，但它几乎具备与普通建筑相同的功能结构。

—— 由于有着烟雾环绕的视觉效果，建筑内部空间的边界时虚时实，因而整个内部空间就仿佛一个无边界的整体。

—— 波浪空间中没有任何的空间角度之分，整个空间由一条回旋曲线连接着（在空间的中央位置与每个空间节点之间各平面曲线的曲率与其长度成正比）因而，整个空间无形之中就形成了一个视觉上的未知空间。

—— 在这一波浪空间内不存在水平空间。

—— 这一互动空间系统由于坐落在游客集中的历史古迹旅游线路上，加之独特的波浪外形，十分引人注目。

此外，特别的曲面外墙将数个大小不同椭圆空间连接成为一个空间整体，这也为整个建筑提供了一个行走通道，当受到大风天气影响时，室内就会产生一些雾状体。因而，这一特别的建筑物，几乎无内外之分，但它却是一个十分独特的康乐中心空间，同时也是周边游客歇脚的佳所。

由于这是一个康乐中心空间，为了重新建立起周边民众的社区意识，建筑师们考虑对现有的空间设施进行更新、改造，以腾出空间建造一个社区中心，以供人们交流之用。

该项目的庞大规模与强大功能，以及它独特的波浪状建筑外形，将吸引无数的当地民众与外地游客。

Rising Core House 242

增长型核心住宅

在城市中，任何一件有益的事情都具有其阴暗面。正如快速的城市化进程，发展的同时也带来了一系列问题，贫穷、高失业率、政治不稳定、社会动荡以及环境的恶化。在发展中国家，这些情况更加严重，促使了许多人士寻求解决的方法。

菲律宾建筑师Albert Zambrano试图用增长型核心建筑这一项目提议来解决这些问题。增长型核心建筑（RCH），是一种住宅传送系统和建筑形态。在菲律宾这样的发展中国家的城市区域内可以营造更多的可持续性动态居住空间。

作为专门为低收入人群设计的项目，增长型核心住宅的发展过程有两大策略。

一、提供能够满足未来需求，并可以根据情况变化而灵活调整的居住空间；

二、为社区和居民创造生存或收入再增值的机会。进而，促进社会的安定和政局的稳定。

项目的中心论点是，贫穷不应该一直延续。增长型核心住宅就是一个变化加速器，将安全、创意、能量和资源组合起来，帮助贫穷的人们不断改善生存的环境。

Zambrano对于马尼拉穷人的专注源于他对低收入人群住宅的学习、研究和规划设计，他将这种专注根植于项目之中。增长型核心建筑就是Zambrano早期关于营造低收入居住空间的研究衍生的作品。

增长型核心建筑将本地的建筑风格与贫民住宅、可持续建筑方式相融合。每个单元面积由12平方米至150平方米不等，还可以作为商铺使用。低楼层可以开展创收入的活动，高层则用作居住。如果需要，高层的居住空间还可作租赁之用。

通过建造、扩建、修复和维修工程为社区带来更多的工作机会，居民的收入也将随之增加。这些建设工程还将刺激区域内的经济的增长和中小企业的创收，因而为城市的贫民区创造增加收入的机会。

在理论陈述中，建筑师展示了一个居民如何通过家门前的商店生存并进而获得积蓄来加盖一层房屋的过程。这种建筑形式还将促使其他类型的商业发展，并以多米诺效应来影响企业的增长，同时促进社区中居民之间的和谐相处。

通过这种建筑形式的应用，生活和工作空间的隔阂也将消除。虽然这两者的结合已不新鲜，但是在当今城市条件上的应用将为我们生活、工作和与社区交流的方式展现新的前景。

Zambrano设计的核心住宅将操作费用降至最低，保证了安全的同时，能够很好地支持当地的经济和中小型及家庭型企业的发展。这是一项有助于消灭贫困和城市复兴的整体分析。

RM Library 246

造型独特的RM图书馆

项目场地

该图书馆坐落于里贾纳玛丽亚大道上（大道连接着Unirii广场与Rahova街道），大道周边都是布加勒斯特中部地区最具文化特色的建筑，在大道街角的一小块三角形土地上，就坐落着许多宝贵的历史性建筑，例如，著名的6.11教堂和罗马尼亚东正教教堂。

研究目标

—— 重新建立一个功能型图书馆，并建造相关设施以供举行各类以信息科技为主题的研讨活动

——在图书馆中专门建立一个有效的文化活动空间建筑，以唤起人们对渐渐淡出日常生活的民族文化的感知与认识

—— 利用现代化的手段对腐朽的城市结构进行整合

项目场地的主要特点

—— 这一地区的集约型住宅较多，它们是商业与服务相结合的具体体现

——这一地区的住宅房形多为火车形，狭长的庭院空间为住宅与外界的公共空间提供了良好的过渡，既保护了住宅内部的私密性，又确保了与外界的互通性。这样的过渡空间为人们提供了一个全新的连续视角，就仿佛一幅油墨叠加的浓彩画一般，划定了城市的肌理。

建造新建筑的主要目的在于整合现有杂乱无章的城市休憩用地，从而创造一个治理有序、环境优美的城市开放空间。

该图书馆项目是一个生态设计项目，设计理念受到了南非闪蝶身形和马蹄莲花形的启发，馆内的开放式公共空间呈现出渐变状态，逐步向封闭的室内公共空间发展。这种设计理念也就是所谓的火车型建筑理念，从某种程度上说，它近似于平房建筑，这符合这一场地的地形条件。

在这一建筑结构中，两边仿佛闪蝶翅膀的建筑空间逐渐向两旁的街道处延伸开去，而这两部分的建筑空间中包含图书馆的阅览室、餐厅和附属书店。北面建筑空间起到了缓冲区的作用，它的存在实现了里贾纳玛丽亚大道和公共广场（开放式公共空间）的顺利连接。而夹于两个翅膀空间中间的开放空间显得尤为重要，它扮演着一个城市广场的角色，同时它也是图书馆的一个户外阅览室，为整个图书馆建筑提供了一个被动式的环境调制空间。基于这些建筑设施，图书馆一楼就可开设多媒体间和图书馆行政办公空间。

该图书馆的设计建造过程由计算机全程参与其中，主要的三角形空间建造都经过建筑师们精确的结构分析，此外，其空间划分、功能设置以及环境参数与人文参数也都是经过准确计算的。

Rossignol International Headquarters 250

全球总部——建筑对山的赞礼

Rossignol是一个历史悠久，享有国际声誉的滑雪装备品牌，它的形象总是与山脉和雪有关。因此，其全球总部的设计也是与其他典型的办公建筑有着完全不同的风格，表达了对自然和山脉的敬意，同样还拥有运动与科技等不可分割的元素。

屋顶景观

建筑是专门按照Rossignol公司的要求特别设计，通过细节体现出了滑雪运动的灵动，雪山顶的风景与冰川的棱角等元素。屋顶在建筑上起伏，与自然景观融为一体。有机木质铺装模式很好地将周围山脉的形状勾勒出来。

从屋顶风格可以区分出下面三个不同的区域：

——项目场地边高速公路一侧的产品车间与技术空间；

——建筑间的道路空间，两侧用灯光点亮，非常宜人；

——办公室。

高速公路一侧的建筑立面为运动风格，充满活力，装饰着规律排列的标志。曲线形的立面慢慢向屋顶延伸，下面是产品车间，然后在抬起处开始再向下一直伸展至办公楼。这两者之间又插入了一个平台庭院，两旁种满桦树，直冲并穿过屋顶。

屋顶的不规则和办公楼的立面都使得在需要的时候扩展建筑空间成为可能，而且不会打破现有的和谐，破坏建筑各自的特点。从一开始，建筑就一直在不断地增加。

屋顶的最高处处于建筑间道路的最上方。道路两侧是玻璃立面，最高层是一

个餐馆，与建筑处于滑雪坡道的顶端成为对应：都拥有宽阔的景观。在从屋顶向外伸出的露台上就餐，壮丽的山景做背景，可远离高速公路的嘈杂。

Rossignol"家族"

Rossignol集团计划将旗下不同的公司实体都集合在一起，虽然它们现在还分布在不同的地方，但是它们都是这个"家族"的一员，拥有家族印记。

从内部来看，建筑就像一个蜂箱，划分不同的活动空间，通过道路连接，人们在其中享受工作与社交活动。这个项目的初衷是要沿行政区域建立起其他的各种功能空间：

——滑雪比赛产品车间，可作为产品科技含量的展示区；

——研究与开发、设计部门位于建筑的中心；

——两个用于展示Rossignol和Quiksilver成品的展示区。

将这些不同的元素放在一起是为了激发整体的协同合作，避免设计风格的偏差，还有行政和科技使用的混乱，为此每个不同功能——工程、设计、技术、行政、销售等空间都相互交织在一起。为了鼓励内部的交流，社交空间被设计分布在了建筑的各个部分。顶层的餐厅，正位于道路的顶部。在这里人们可以悠然地工作生活，两个大型立面玻璃外面就是宽阔的天空与山脉风景。无论是在露台，还是在壁炉边，在这里午餐休息绝对是独一无二的体验。

对环境的尊重

建筑结构的设计都将对环境的影响降到了最低。建筑设计采用了最先进的技术，高效节能。夏天，天然的木制屋顶可以将阳光阻挡在外。所有设备都进行了最优化的设计，车间机器产生的热量将由暖气系统收集使用。

设计师还格外做了水的管理设计：将雨水收集起来，工业用水与马路上流下的雨水一起进行进化与循环使用。

Royal Botanic Garden in Jordan 254

约旦皇家植物园

游客活动中心与冬季花园

约旦皇家植物园游客活动中心坐落在园内人工小山的山顶之上。该活动中心现有的主入口位于中心大楼的东南角，建筑师们计划在此处建立一个迎宾缓冲区，使游客们进入这一缓冲区就产生一种特别的感觉，感觉自己即将进入一个全新的世界去感受一种全新的体验。活动中心的正门前有一条颈巷，设置这条颈巷的目的在于控制来往的人流量与车流量。建筑师们将在植物园中设计建造一座人工小山，这座人工小山将建筑技术与园内现有的自然景观融合于一体，它成为了游客与植物园的纽带，因此，游客活动中心就设于山顶，以便游客们欣赏植物园的全景。实际上，这一人工小山功能繁多，集露天阶梯看台、展览厅、购物中心、图书馆等功能于一体，主要为参观、集会、体验美景提供良好的活动场所。停车场将设在地下室，当游客们停放好自己的爱车，走出停车场的刹那间，就会感受到置身于植物园之中的美妙体验。

俯瞰整个游客活动中心建筑，就像植物园中自然流动的线条一般，动感十足。这是一个层次感极强的上升式旋涡状建筑，建筑构造十分奇特，仿佛一棵从地里生长出来的植物一般，慢慢壮大成为一座规模宏大的建筑物。该建筑中的旋涡式过渡是依靠整个建筑物的主基准墙面来实现的，事实上，这个主基准墙面就是位于建筑中间层的一座连接桥。当整座建筑开始发挥它的功能的时候，它的建筑魅力也就得到了极至的展示，迷人的建筑外观与动态十足的线性美感，为游客们提供了丰富的视觉享受。该中心还为游客提供形式多样的娱乐设施，如，艺术馆、灯光展示馆、剧院等，通过这些娱乐设施与周围植物景观的互动，人们可以更加深入地了解植物的特性、气味及其生长变化的过程，从而增长植物科普知识，增强植物保护意识。

总体规划

建造皇家植物园的主要目的在于：为植物提供良好的养护场所；对人们进行普及植物常识科普教育；为植物学家提供科研场地；供游客休闲游览。因此，它是一座集植物养护、科普教育、植物科研、休闲娱乐于一体的综合性植物园，它不仅吸取了本地区各个园林的建造优点，同时还借鉴了全世界众多园林建筑的精髓。这座皇家植物园将成为约旦国内又一迷人的景观公园，在这优美的环境之中，游客们可以边欣赏美景，边了解约旦的植物遗产，还能够观赏到精心养护的稀有植物。它是一个国家级公共景点，在约旦，这样的公共景点屈指可数。

由于约旦是一个适合发展旅游业的国家，因而，这座皇家植物园的建造也将对当地旅游业的发展产生积极的影响。

虽然该项目的主园区位于Tal Al Rumman，但在约旦全国都将建立它的分园区。皇家植物园建成后将采取一系列措施对这些约旦稀有植物物种进行有效的保护，它将成为约旦国内重要植物科普基地。在这里，人们可以更好地了解和欣赏约旦的植物遗产。

该植物园还将实施和促进生物多样性保护，并对这一保护行为进行有效的管理，其重点放在本土植物的合理利用、科学研究、起源追溯、科普教育以及本国传统药用与烹饪植物历史的展示等方面。

植物园远景展望

约旦环境部于1992年编写了生物多样性公约，1993年约旦政府批准了这一公约。该公约要求施行生物多样性战略与行动计划，而皇家植物园的建立是这一计划的重要组成部分。

因此，该项目的建设受到了约旦国际合作与规划部、农业部和水源灌溉部的大力支持。

皇家植物园将成为约旦生物多样性养护中心；它将在主园区和全国的各个分园区建立自然栖息地、主题花园和国家植物标本馆。这将促进约旦植物遗产知识的推广与普及，并将提高人们的植物鉴赏能力。

此外，皇家植物园还将建立一个全世界最权威的先进植物科研中心，它将为世界植物研究的发展作出巨大贡献。

皇家植物园在Tal Al Rumman地区的具体任务：

——Tal Al Rumman地区生态系统的修复；

——提高对当地植物和生物品种重要性的认知度；

——建立有效的濒危植物收藏与保护系统，并促进珍稀濒危植物的易地保护工作；

——建立国家植物标本馆。

皇家植物园在约旦的全国性任务：

——维系现有存活植物的收集工作；

——建立一个植物标本收藏室；

——有效培养珍稀濒危植物的遗传因子，以确保大多数珍稀濒危植物物种的安全留存；

——承办一些植物发源地参观活动或生态系统修复宣传活动，以便于人们对生态系统功能动态的及时了解，同时，对本地植物种群的恢复和管理起着积极的促进作用；

——在植物种子库中安装一个珍稀濒危植物种子备份系统；

——繁殖珍稀濒危植物，为物种的数量增长、回归引种以及植物发源地的修复提供有力的参考材料。

Shanghai Expo 2010: Italian Pavilion International Contest, 3rd prize winner 258

上海世博会 2010：意大利馆国际竞赛，三等奖

"更好的城市，更好的生活"，项目的设计灵感来源于意大利城市，展现了独特的意大利风格——多元化的历史名城。

21世纪都市化进程面临着由不当的发展和对环境的忽视造成的持续危机。由于之前对历史和地理环境的充分重视，意大利建立的都市系统仍然能够满足当今发展的需求，并能提供高效的生活方式。

这种多元化的城市概念仍然反对现代社会人与人之间的疏远，更多地兼顾人

们的心理需求，强化人们感受这一空间实体的感觉。这一概念连接着历史，经过时代的更迭和发展能够很好地与居民的生活相关联。

意大利广场是这个实体的重点，也是无法替代的核心内容。它能够存在并且停留，是一项非常重要的研究。同时，艺术也表达在这个独特的空间环境中。

针对上海世博会设计的这个意大利馆是一个创作优秀、生产优秀的场所，是表现社会和都市意大利灵魂的原型，并与历史保持一致性和连贯性。

这种连贯性是解决现代发展和科技与可持续要求的完美方式。

这一项目设计的主导思想是分为数个层面的时间轴，能够展现意大利都市化进程。

像意大利部门一样多的二十个层面，由三个意大利制造的窗体进行分割。它们组成了底层的主要结构。

室内的展示空间受到了这种空间组合的影响，在不同的层面间形成了不同的风格。在里面参观如同经历历史的变更。

建筑围绕中心广场进行建设，游客们走进去才会发现外部难以发现的各个层面。徜徉其中，如同探索某个古老的意大利城镇。

建筑内部到处都是材料和灯光，保护并营造出当地的心理学空间。所有层面都是对环境开放的，由可视的太阳能面板组成。

内部广场是一个供游客放松和聚会的场所。

Silhouette Istanbul　262

伊斯坦布尔城市轮廓

该项目将建成伊斯坦布尔市的地标性建筑，它的建造将为伊斯坦布尔的城市发展带来一个全新的面貌，也为公共空间的建设提供了参考。

由于伊斯坦布尔地处欧亚大陆相交处，加上博斯普鲁斯海峡和伊斯坦布尔的地形特征，因此，建筑师们规划将该项目建筑结构定位为城市轮廓对话。

此外，该项目建筑结构将打破传统的垂直建筑物常规模式，成为Buyukdere大道上的另类建筑。

该项目的建造目的是为现代都市打造一场全新的视觉盛宴，为都市人提供不同生活方式的参考。其中，概念性基础结构建造，将分为以下几个关键点进行阐述：关于降低能源消耗的敏感建筑系统的建设；室内循环系统的设计，如，半开放式走道、举行文化与公共活动的半开放式公共广场和花园。楼内大部分公共文化区域的绿化设施都十分完备，营造一种和谐、自然、轻松而又高贵的伊斯坦布尔都市生活氛围。

这样的室内空间布局，是依据普通公共区域转换为花园与活动区域的基本设计理念来施行的。其规模大小以d′Etoile广场、罗马Collesium、托普卡帕宫等大型知名公共空间的建筑方法为主要依据。

该项目选址位于一个小山坡上，坡下有一个面向博斯普鲁斯海岸的小山谷，项目的建设重点之一在于保护现有绿化带，并在此基础上扩大绿化范围，以恢复山坡原有的绿化规模。

建筑师在该项目的建筑规划中，将购物中心设置在一楼和地下一楼。由于整个建筑整体的跨度相当大，因此，在山体中心就形成了一种与城市地形特征的自然连接。建筑师们考虑到该建筑项目的整体外围空间都是绿化带，所以他们决定在外围空间建立一个庞大的公共活动区域。

该项目中的私人活动空间与公共活动空间基本上是完全分离的，这样能够确保个人的隐私得到有效的保护。但建筑师们依据当地的地形与建筑的布局，将私人活动空间和公共活动空间与周边的自然环境巧妙结合，形成一个有机的整体。

该项目的建筑师们把Zorlu地产公司宣布的竞争条例上榜上有名的一些不同的功能型建筑视为独立的建筑元素。在整个概念性设计的过程中，建筑师们一直在思考这样一个问题：“如果将每一座功能型建筑都设计成一个完全独立的建筑，那将会出现怎样的一种景象？”于是他们制作了实际占地面积为625平方米的正方体建筑群落模型，以便更好地观察其三维效果。然而，这时第二个问题又产生了：“如何在不改变这些功能型建筑的本质功能的基础上，对它们进行横向建造？”

如果留心观察Buyukdere大道上的城市建筑结构，就会发现，其实这些建筑都有一个共同的特点：大部分的建筑都与这一街区的原始建筑风格相去甚远；此外，伊斯坦布尔这座古老的历史文化名城，它的历史与传统在这些枯涩无味的直挺挺的建筑中得不到丝毫的体现。而该项目的建造恰恰就是要打破这种单调的城市建筑格局，充分展现伊斯坦布尔的城市文明，同时从建筑物的各个不同角度建立起博斯普鲁斯海峡与伊斯坦布尔城的和谐关系。另外，由于该项目的建造没有对这一区域内原有的绿化带造成破坏，而是完整地保留了原有绿化带，因此，它为伊斯坦布尔留下了宝贵的环保财富。

建筑师们还计划将地面与地下交通以及人行通道进行有效的连接，为该项目提供便利的交通环境。另外，在博斯普鲁斯大桥的中轴线上，除现有的公共汽车停靠站外，还增设一些站点和一个路边公园。

住宅区、酒店和办公区域都水平设置于建筑物的高层中。购物中心和文化活动中心分别位于建筑物的各层与地下层中。它们都与绿化带相连，以展示博斯普鲁斯的自然地形特征。

该项目的室内循环系统是由一些半开放街区和半公共区域组成，还有一些半开放公园与室外景观相连，以展示现代化的生活方式。

该建筑的最底层是购物娱乐中心，其间的公共空间由音乐厅、会议大厅和露天电影院所组成，露天电影院还可承办一系列的户外活动。

SKMS Institute　270

SK管理系统研究所

缓坡下是繁枝错节的栗树和高丽参农场，形成另一番地貌。稀疏的栗树林（产在质而不在量）给建筑师们留下了神秘的初次印象。土地在晨露的滋润中焕然一新。嫩叶和树杈在和风中沙沙作响。这些都刺激着建筑师们的感官。为了今日栗树果园能如此迷人，长期以来人们辛苦劳作地重塑这个崎岖小山。曾经那些重视恶劣环境的人们的精神和刺激建筑师们感官的大自然交流：那时，人们相信前者诠释了进取精神，而后者则释放了建筑师们的哲学思维。建筑师们的工程旨在寻求潜在土地来建造一个极小型的建筑物。这个项目与栗树林和新建场所之间的长长分界线并列。基于附近高丽参农场建造的简单方法，建筑师们将这个建筑物融入这片风景，而不是成为一个刺眼的大建筑物。建筑师们将创造这样一种大自然和建筑物相融合的环境：在宽松的构架里漫步并体验一种不同的时间感。因此，这个建筑物将成为环境的一部分、土地的一部分，也会在和自然的融合中为思想提供源泉。建筑师们希望留下的建筑物会随着时间的推移渐渐成为这片土地的一部分。

工程目标

项目旨在为研究员研发这个公司的管理系统创建一个驻扎地。栗树林的繁茂枝丫和高丽参把昔日被稀疏的老栗树簇拥的缓坡点缀成另一番景象。项目以探寻潜在土地为开始，接着是在这个基础上设计节约型的建筑布局。通过利用景观的整体性，即栗树林和稀疏土地的广阔分界线，建筑师们就能使建筑物和自然并置。建筑师们从高丽参农场的宽松构架获得灵感，并最终想出这个方法：使建筑物以景观的形式存在，而不是一个碍眼的物体。穿梭在农场里呼吸着不同时刻的空气，感受一种把大自然带入建筑物里的氛围。那些把这片恶劣环境开垦成繁茂栗树林的人们的辛苦劳作依然像记忆一样留存在这片土地里。建筑师们相信这记忆能够帮助客户诠释其企业远景，同时研究员能够在与大自然的感官接触中驰骋思想。于是，这个建筑物是这片土地不可分割的一部分，也是景观的延伸，还是与自然交流的认知构架。因此，建筑师们耐心地等待他们留下的痕迹在若干年后融入植物景观里。

选取材料

建筑项目的主要材料来自场地周围的自然环境。栗树林里有为质不为量而栽种的稀疏栗树和晨露滋润的土壤，还有和风中沙沙作响的栗树新叶和枝丫。建筑师们设法赋予这样的栗树林某种建筑风格。通过以最中性的表达方式呈现赤裸裸的混凝土，建筑师们把这个建筑物设计成与自然相协调的背景。此外，在整个项目中，建筑师们深入考虑到大自然的可循环材料的理念。建筑师们在这个久远的栗树林里小心翼翼地收集一小部分木材并从场地周围收集土壤和石块。除了给宿舍楼的挡土墙增添非结构性的土石材料，收集来的土壤和石头还用来做景观土丘。在花园里移植了一些栗树林里的栗树。土石挡土墙的完成像是沉积层的隐喻。尺寸垂直向上扩大的石头代替了土地过去的底层，形成了一个显眼的非结构性墙。因为比喻的元素存在于建筑师们干预的痕迹里，收集并使用的材料将展现为一个介于自然和建筑物之间的物质性、时间性、记忆性的媒介。

选取技术结构

建筑师们致力于不断地使土地呈水平方向，同时承认周围景观的重要性。为了把栗树农场和稀疏土地边缘的土压力转换成自重量，建筑师们统一把挡土墙的基脚做成50厘米厚的席式基脚。这样，土地和建筑物就在物质环境里和在土地潮流方面毗连成一体。

利用背景材料

土壤一直留存着栗树林和风景小道的记忆。因为大自然渐渐地辐射这个建筑物，建筑师们可以把长长的风景小道的记忆延伸到建筑物的内部空间。作为大自然和建筑物的模糊界限，即户外和户内，这个建筑物就是大自然，同时大自然也是建筑物。因此，源于这个基本原理，板模系统使用中性材料。此外，这也使人们相信建筑物可以再次和自然共存。

Skyscrapers City 276

摩天大楼建筑群

世界上最早的摩天大楼建筑出现在19世纪末的芝加哥、纽约等城市。如今，一个世纪过去了，摩天大楼成为了世界各大城市中心的标准建筑模式。

现代城市中，城市居民的数量正在形成逐年递增的态势。根据2008年的调研数据显示，当年世界城市人口在全球总人口中所占的比例首次超过50%，据科学家预测，到2050年这一比例将上升至75%。50年前，世界上只有8个大城市的居民人口超过500万；而今，世界上有64个城市的居民人口超过了500万，其中26个城市超过了1000万，东京3400万的城市人口数量一直以来居高不下。城市人口数量的不断增长与严重不足的城市居住空间之间的矛盾已日趋明显，而解决这一矛盾成为当下不容延缓的当务之急。多年来，城市居民们在对自己的公寓住宅空间布局的设计上都绞尽脑汁，竭尽全力扩大自己的居住空间，以缓解城市居住空间严重不足所带来的压力。在这种情况之下，高层建筑模型的建造成为了唯一的解决途径，这种建筑模式甚至可以改变整座城市的总体结构。

建筑师们在大都市建造高层建筑模型的构想灵感主要来源于美国纽约的曼哈顿，这里是世界上人口密度最高的地方。假设在这样人口密集的大都市内安置一种特殊材质的巨型气球，这种气球能够吸收人口过剩城市所排出的暖湿气流以使其自身的体积得以膨胀，而后使它们有足够的承受能力可以承载更多的高楼，如果将这样的巨型气球安放在整个中央公园中，将会是怎样一幅壮观的场面？！

城市的区域划分决定着摩天大楼的建筑形式。随着社会的发展和科技的不断进步，城市中心高层建筑模型的建造将成为一种城市建筑的主流。这种建筑模型一般分为三个楼层：第一层为住宅区，由众多的单元公寓组成，这种单元公寓在许多大城市的郊区随处可见；第二层是高级住宅区与塔式大楼，第三层则是文化与商业中心。

Small Library in the Centre of Thessaloniki 278

萨洛尼卡市中心小型图书馆

位于市中心的这座建筑，靠近全市最热闹的广场。项目地块非常狭小，环绕在一片高耸的住宅楼之中。地块只有一面临街，另外两面均临近其他建筑，后面面向市区的开放空间。

建筑包括一座三层的图书馆、一间小型会议室、一处小型展览空间和一座咖啡厅。建筑的主要理念是将众多功能空间汇集在一起，并赋予每个空间不同的形态特点。一个重要元素就是建筑中分辟出的公共区，强调了建筑原址的古迹。入口设置在稍里面的位置，这样能够为古迹区引入更好的光线，并将其与公共区和谐过渡。小型会议室的外观富有动感，座位的设计与整体风格相统一。沿着入口的基本轴线设计的会议室在结构上为一层的古迹区提供了自由贯通的光线和气体，非常巧妙。建筑的中庭与三层结构结合，三层中只有第三层是与外立面一致的，其他两层结构均较为后退，为一层的古迹提供了更好的光线。第三层采用了木质百叶窗，缓和了强烈的日光。建筑的顶部是一座咖啡馆，这里可以欣赏到广场的景致和金属的流线外形。作为公众建筑，图书馆面向市民开放。混凝土的外表和多样化的楼层尺度、精致的外形等都为建筑吸引了众多目光。室内外空间、开放空间及封闭空间也和谐地融合。项目是由萨洛尼卡亚里士多德大学的希腊建筑系学生设计的。

Solar Tube — A Symbiosis between Nature and Architecture 282

太阳能管式住宅——自然与建筑的共生体

太阳能管式住宅位于维也纳西北部郊区的Döbling，这是一个环境清幽的富豪级居住区，区内大都是独门独户的别墅住宅。该项目独特的太阳能管式建筑风格，在周边传统住宅建筑的包围之中，显示出十足的个性，虽然它的个性十足也引起一些周围居民的不满。“一位居住在该住宅对面的妇人抱怨说，这样的住宅是对整个居住区建筑风格的破坏。”建筑师Driendl说道。该项目的建筑灵感来源于一个简易的管状天窗，这种天窗是一个通过反射管、屋顶集热器和散光器向室内输送阳光的特殊设备。当初的建造想法是利用整个住宅作为太阳能集热器，它仅仅需要少量的加热、冷却或电力照明。这是建造这座住宅的出发点，也是确定建筑形式、结构与选材的基本依据。

客户购买的这块住宅建筑用地，面积约1,300平方米，地形窄而长，周边绿树成荫，是一块朝北的沿街坡地。建筑师将住宅建于坡地的最低处，这样就在街面的基础上提升了居住空间的建筑高度。这样的建筑地形设计可以充分利用自然通风系统和四周绿树所提供的自然遮荫环境，同时周边的自然生态环境也不会因此而受到破坏。

该住宅中有一个全玻璃制造的天井前厅，它将整个住宅连接成一个有机的整体。住宅的沿街入口与屋顶阳台处于同一水平线上，经过这一沿街入口，再走过一座悬空的木桥，就可直接到达客厅。另一个通往该住宅的方式是：沿着住宅另一端的阶梯往下走，这一阶梯在低于街面10米处。

在着手实际设计之前，所有的细节都需经过再三考虑。所谓的考虑，也就是对住宅所处的位置特点、空地的合理利用以及住宅的景观设计等做最终的确定。另外，太阳照射点、日照时长以及其他一系列的天气因素也都应该纳入到着手实际设计之前的考虑因素范畴之内。

得名原因

一般说来，“太阳能管”是一种安装于住宅屋顶上的小型光热收集器。因此，该住宅就是这样一个从四面八方集取光能和热能的收集器。

所以，整个住宅的建筑结构也是模仿了光热收集器的外型结构。

由于该项目选址是一块树木茂盛的阴凉之地，因此，建筑师可以大胆地采用玻璃作为建筑外墙的主要材料，还可在住宅的各个不同位置对玻璃外墙进行倾斜或弯曲处理，从而令整个建筑外观显得另类、个性。屋顶和地板也部分采用了玻璃材料，这样整个住宅的核心部分就形成了一个综合型的天井前厅。至于在住宅的最高层上安装的玻璃管装置，它有利于节省冬季的取暖费用。玻璃建材除了有节能的优点之外，还能够为住宅提供更多亲近大自然的机会。由于建筑外墙绝大部分采用的是玻璃建材，所以室内外的界限也就不复全在了。特别是在夏季，整个住宅被郁郁葱葱的树木包裹着，住宅主人将会感受到一种特别的清凉体验。这座住宅充分地体现了自然与建筑之间相互依存的合作关系，同时，它也是一种开放与透明的建筑理念的具体体现。

低能源概念

夏季的室外高温并不会影响到住宅，因为它有着四周茂盛树木的荫庇，同时它还拥有一个烟囱似的特别通风系统——天井前厅能够向上散发室内的热量，使之通过屋顶上一扇敞开的天窗向室外驱散开去。在冬季里，取暖费用可降至最低，这是由于住宅外墙上有一层结实的保护层，其作用在于，它可以在某种程度上吸收和反射大量的太阳能。当然，另一方面，冬季里，室外树木上的枝叶都凋零了，因此，阳光可透过玻璃外墙与玻璃屋顶，直射进室内空间，这也是取暖费用降低的原因之一。

设计与功能

从街面看这一住宅，别有一番趣味体验。路人们都会被它夺目的建筑结构和亮眼的建筑外观所吸引，尽管面对街面的部分并不那么透明（相对于住宅两侧的花园而言），毕竟还是得给住宅主人留有一定的私密空间。

建于住宅中心的天井前厅能够通往住宅的每一个角落。即使在一楼，也可以观赏到不停晃动的屋顶，这是因为，一方面，每层楼的地板都是部分透明的；另一方面，顶楼的卧室建于一个回廊的上方。整个室内布局也贯彻了开放型居住空间的建造精神。宅内所有的家具（厨具、书房用具、杂物柜、浴室、一个便于浏览的大型壁式书架等）都是由Driendl建筑公司统一设计的，因此，它们与整个住宅建筑完美地结合在一起。

Stair House 288
楼梯住宅

这是一幢某设计师为其父母和叔叔一家设计的住宅。最初的规划是在占地120平方米的土地上设计两幢住宅和一个小工作室。前期已经竣工的房子正在使用中。由于设计师家的造房预算较低，所以要选择一块相对便宜的地皮。于是，设计师鼓励家人买下现在这块长30米、宽3.9米，占地面积约119平方米的土地造房，由于这块地长而窄，不宜建造住宅，所以它比镇上其他面积相当的地皮要便宜一半。但这对设计师来说，却是一个展示建筑设计才能的绝佳机会。然而，由于资金有限，建筑施工的质量低下，设计师只能省略一些设计方面的细节，但总的设计理念并没有受到影响。设计师的初步构想得到了全家人的一致认可，但是依照德黑兰市对私人住宅建筑高度、面积、层数的严格规定，设计师不得不对最初的施工方案进行调整。这个住宅原计划在一年半时间内完成，但是依然由于资金问题，可能需要更多的时间来完成。设计师的哥哥负责设计整个住宅的钢筋混凝土结构的建造，而设计师的父亲则身兼二职——承包商与机械工程师。

材料、结构与施工

这一住宅以钢筋混凝土为主结构。墙壁由表面有绝缘层的砖块砌成。建筑外墙由石灰华筑成，石灰华是一种当地极易获取的建筑材料。室内墙面则由石膏、小块砖（它们都是既便宜又容易获取的建材）和粗糙的石灰华共同构砌而成。很有意思的是，这些砖块都是一些石材厂的剩余材料，所以价格十分便宜，并且由这些石材厂中没有受过正规训练的工人们砌割成设计师想要的形状。每一层的地板大多由掺杂着少许木料的石灰岩覆盖而成。住宅内部楼梯由木材和钢材组合而成，这些木材虽然有些粗糙，但体现出了自然的风格。

意义与影响

设计师的这一住宅楼位于镇上人口密集地区，四周都是四层楼的住宅区。狭窄的宽度（3.90米）、狭小的面积、被规定的建筑高度和住户们的表面设计总面积等都是影响设计构造的重要因素。最终，我决定在这块狭长的土地上建造两幢内部相连的三重住宅楼，这两幢住宅和书房由一个公共楼梯连接，其他位于两幢住宅中间的公共部分是天井。由于这一连幢住宅建在一个斜坡上，因此，一幢楼比另一幢楼多出了一层半，而这多出的一层半空间在它的北面可以为整个连幢住宅提供充足的阳光。整座连幢住宅以三维立体设计理念为主，以便于人们把注意力集中到房屋的高度上，避免了对其狭窄的宽度的注意，这一三维立体设计，将最终达到取长补短的视觉效果，把所有的优点都展示出来，把所有的缺点都隐藏起来。两幢楼都分为五层，这五层都有独立的楼梯和空间，这样的空间处理方式，使得整个住宅从原本狭长的占地面积的缺陷中摆脱了出来，从而，展示出通风良好的立体视觉效果，这样的设计就显得比它们原来的地表面积宽敞了许多。基于各个不同的视角，各个楼层的互动还有许多值得开发和探究的地方，而楼层之间的空间和楼梯是解决各楼层互动的最佳选择，在这些空间和楼梯上，可以创建出各式各样灵活的楼层互动区域，也使得整个连幢住宅富有生机。功能图对整个住宅，从公共空间到私人空间，从底楼到顶楼，都做了详细、统筹的安排。谨慎地使用住宅中那些自然、粗糙的建筑材料，它们会让你感受到来自大自然的舒适与惬意。

这一地区寒冷的山区气候，要求我们必须设计一个坡形屋顶，以抵御凛冽的寒风；同样，两幢楼之间的公共楼梯也需要这样的坡形设计，并且增加1.20米的坡道，以巩固两楼间的连接。这样的设计会产生额外的面积，这些额外的面积可以用于建造厨房和卧室。

虽然，这一住宅的内部结构十分复杂，但它的外观却体现了简约的风格，给人以一目了然的视觉体验。

Stone Terrace 292
石头平台

稻田平台用来将森林中的水分散到水平地表中。人工系统有效地利用了光照、水资源及风能等自然资源，有利于农作物的生长，也大大影响该地的景观形态及景观材料。该项目的目标是将现有的人工系统转换成更为人性化的新系统。设计师考虑到石墙形态的连续性，在稻田平台中间置入小柱子。与周围景观形态互相对应，简洁的方形室内空间营造出各种各样的景致。两个闭合花园的朝阳面与室内空间直接联系起来，并成为夏日生活空间。浴室内可以欣赏到森林及天空的别样风景。水园及大面积的田园景观位于北面窗户不远处，雨水落在屋顶及北面墙体上，石墙上生长出美丽的苔藓。来自稻田平台的雨水汇聚起来形成水园。水园不仅是北面窗户一道亮丽的风景线，同时还起到将夏日北面窗户吹进来的热风进行冷却的作用。夏日，冷风从北面窗户吹过来，室内的热风沿着屋顶斜坡慢慢升起，冷热风在南面窗户顶端相遇。夏日，倾斜的屋顶为室内提供了阴凉处，冬天，充分地利用太阳光照，使屋子变得暖和起来。从这个角度上说，石头平台成为由水、空气及光组合而成的这个自然循环系统中的一部分。

The Consulate — An Ingenious Building 296
别出心裁的领事馆建筑

该项目是一座领事馆建筑。垂直的混凝土墙体与软质聚四氟乙烯膜构成了这座领事馆独特的建筑外观。两种建筑材料相互衬托，创造出一种简约而壮观的视觉效果，充分展示了领事馆建筑独特的外交功能。这一建筑外观体现了宁静美与奔放美的和谐统一，它似乎在告诉人们，馆内的外交人员虽然对生活有着无限的热情，但他们必须对自己的特殊工作谨慎行事。

上述两种建筑材料的和谐共存，不仅赋予整个建筑外观以充分的视觉美感，而且使其拥有很强的可持续性。在多层半透明软膜墙面与室内灯光的共同作用之下，整个领事馆就仿佛一个巨型灯笼一般，向室外透出一种朦胧的光，而室内却是光亮一片。建筑外墙软膜层与室内环保玻璃墙面之间有一个夹层，这一夹层为该建筑提供了良好的通风空间系统。由于混凝土建材具有较好的散热性，建筑外墙的混凝土部分能够将室内囤积的热量渐渐排出，从而确保室内温度保持良好状态。

这种建筑外墙的另一个优点是，通过在各个层面间放置光线投影器作为信息的载体。

The Lightbox 298

光线的璀璨交错

这个竞赛项目是一个新型的社区文化中心，为附近工作的人们提供休闲的场所。作为社区的衍生建筑，这个多功能的空间是一处会议场所，并且是不同群体间交流的重要场地，也为人们在参观美术馆之余提供了一个去处。项目于2007年9月正式对公众开放，并获得了公众信任奖和艺术基金奖。

项目位于城中心，坐落在贝辛斯托克运河与一个六车道的高速公路之间。建筑面向道路的外立面使用了金色和银色的材料，其他立面则使用了木质材料，显得更为田园和怡人。

The Mashrabiya House 300

Mashrabiya住宅

该地位于耶路撒冷与伯利恒附近的阿拉伯巴勒斯坦村庄。住宅将阿拉伯传统元素与当地特征重新融合起来，同时也采用新的设计手法与村子的文化景观进行转换。

平台景观

斜坡住宅中采用传统的阿拉伯景观元素，如平台及墙体，形成主要的"景观表面"。地下室被雕刻在山内，形成石头外环的平台，与下沉庭院中的平台采光井对应起来。平台中还包括工厂、工作室及美术展览馆。因此，将生活区内的工作及公共功能分开。场地后边，平台内石头堆砌而成的墙体内有几套公寓。

大楼的地面景观形成一个大舞台。地面的厚重感与垂直结构的轻盈形成鲜明的对比。通过对阿拉伯风情的重新演绎，营造出漂浮的效果。木制边框窗户将私人空间与公共空间分隔开来，用大尺度的石块将大楼包围起来，格子幕墙窗户还与木制边框及石头联系起来。设计师通过将石头分开，形成不规则空间，使该空间灵活多变，并且渗透性良好，营造出典型的半透明效果。公寓后面石头外围被分成狭窄的小缝。公共空间内的娱乐元素在保护私密性的同时，为人们提供了室内的景致。

阿拉伯传统元素

设计师通过对传统阿拉伯元素的积极应对，在坚持当代建筑布局及使用当代材料的同时，发展了住宅的公共空间理念。后墙突起的庭院使人回忆起传统元素，而屋顶上的木制边框便成为屋顶花园墙体，在遵循布尔斯坦设计理念的基础上，花园成为人们沉思的良好场所，同时还有效地保护了私密性。

阿拉伯的可持续性风情演绎

设计师在遵循传统阿拉伯建筑布局的基础上，保护当地的历史文化遗产，探索并发展了可持续发展理念。木制边框是将公共空间与私人空间分隔开来的门，有效地控制了当地的气候条件。大楼外围堆砌的石头起到缓冲作用，白天便于吸收太阳光中的热量，在寒冷的耶路撒冷夜晚，将热量释放出来。这同时也使大楼免受太阳辐射及风雨的侵蚀。与烟囱的功能相似，石缝便于空气的自由流动，还能够吸收屋顶的热量。热量从下往上窜，新鲜空气从石缝下流出来。

与大多数耶路撒冷的村子一样，Beit Safafa正面临着重大的社会及文化挑战，尤其是将村庄发展成城市社区。在这种背景下，该住宅提供了将传统及现代城市公寓住宅完美结合的设计新理念。该住宅对村子的传统元素做出了积极的应对，同时又将工作与生活连接起来。新美术馆代表了巴勒斯坦文化，该住宅也成为美术馆的新家园。

The Museum of Modern Art 304

现代艺术博物馆

该博物馆建于一个倾斜式的地下救济所中。所有的建筑材料都是轻质材料，它们都是用吊挂工具从地面吊至这一地下空间之中的。馆内还设置了自动扶梯，以便参观者进出。整个博物馆是一个单一的参观系统空间，因此，参观者们仅需通过一个通道就可以到达博物馆的每一个展区，所有展品一目了然。

The New Venice Biennale Australian Pavilion 306

澳大利亚威尼斯风情阁

"南方沃土"与"水上城市"

罗马时代的澳大利亚素有"未知的南方沃土"称号。这样的描绘现在仍然是澳大利亚的象征，也隐喻了它巨型多样的自然环境。人烟稀少的内陆、大堡礁、奥古斯塔斯山及哥斯山的大分水岭只是众多你所能体验和感受澳大利亚独特自然风情中的很小一部分。而被誉为"水上城市"的威尼斯，则是建在由150条运河所贯穿的118个岛屿中。自然风貌对威尼斯悠久的文化产生了深远的影响，形成了独特的建筑环境。对比鲜明的情景也深深地影响着该地的建筑风格。

澳大利亚新楼阁

该项目力图塑造一个既具有澳大利亚的自然风情，同时也将"雕刻"出具有威尼斯风情建筑环境的新澳大利亚楼阁。整个大楼布局与周围的环境紧密相连，入口处的斜坡镶嵌在大楼中，直接与周围的运河连接起来。大楼正面将澳大利亚的景观形态及色泽完美地展现出来，微风吹来时，外面天窗的颜色便会发生不同的变化。设计师采用分层贝壳散发出的微光，使整个室内空间光线明亮起来。材料使整个空间富有艺术化气息，可作为艺术品展览场地。

The Rainbow Tower and Poljane Masterplan 310

彩虹大厦及Poljane总体规划

我们把竞赛方案看做是一个经典的分区，赋予不同场地不同的功能。我们发现这样有一点刻板。然后我们为Poljane区域引入了各功能相互作用的系统理念。我们设计了一项开放而坚固的系统，并设置了模糊的边界。我们的目标是积聚多种功能，为公众提供开放空间。我们将Poljane看做是一个先进的综合区域，第三产业、文化和居住功能并存的空间。大型的绿色空间、真正意义上的生态居住环境也是其中的特色。

这一系统使我们在规划水门和城市的边角地区时更加地细致和小心，并兼顾不同区域间的地域连接、结构组成和综合性绿色走廊以及流通性方面的相互作用。
这似乎是一种综合系统，并与各个区域紧密相连。这也是一个开放的、自给自足的、延续性和自行更新发展的系统。
这一系统以浓缩的方式应用在彩虹大厦中。作为一处新的地标，大厦象征着自然法则，是活力的标志。它内部充满了生活的气息，外部则是卢布尔雅那能引起人们无限遐想的优美天际线。彩虹大厦名字的来源有几个不同的原因：首先，我们能够在这里领会大自然活动的内涵。它阶段性的灯光和棱镜反射效果将唤起我们对太阳能的意识，如何利用太阳能，以及与色彩和景色上的联系。作为一种建筑形态，彩虹的拱形结构象征着已经改变的建筑结构。这是一项综合了象征性、庆典作用、沉思空间和功能设计的结构系统。矗立在这里的这座拱形结构也提示我们这是一个对社会、政治、功能和能量都有兼顾的概念。它是一个综合物、一座商店，也是一处魅力空间。在这种新的理念中，所有古老的标志都混合在一起并融合成一种全新的符号。同时，这是一个综合性的、垂直水平兼容的独特元素。这项设计如同包含着不同活力的结构元素的容器，如今转变成一个新的魅力和能量地标，也为斯洛文尼亚民主政治建立了一个全新的生态标志，全新的内容和理念也开启了现代的多样化综合时代。

The Special Container Building in the South of SANLITUN 316

三里屯南面特色集装箱建筑

该项目是一个新成立的零售集团综合建筑，项目的总体规划由香港欧华尔顾问公司设计完成。整个综合建筑呈现出一种中世纪村庄式的建筑风格，建筑内有许多相互交织的狭窄小道和几座层数不多的小楼，这些小楼由高架人行通道和人行天桥连接着。LOT-EK建筑公司将对这一综合建筑东北部进行设计建造，使其成为一个集零售、饮食、活动于一体的生活空间。而其他三家建筑公司——KKAA/Kengo Kuma & Associates、SHoP Architects、Sako Architects——则负责该综合建筑其他楼层的设计建造。
LOT-EK的建筑理念侧重于将建筑物内的小道改建成多层次、开放式空气流通走道。在每一个小道内，都有一个棚架式的金属框架将两幢相邻的小楼连接在一起。这种金属框架适用于各种宽度不同的小道截面，它们由建筑中水平放置的一系列金属棒串连于建筑外墙的周边，充当建筑物的栏杆和遮阳板，从而在建筑物的高层形成一个凉廊和一个四通八达的隧道。
由于这一建筑主体是由许多通过ISO认证的2米多高的海运集装箱叠放组合而成的，这些集装箱被随机塞置于各个小道内，突出的部分恰好与建筑外墙形成极强的层次视觉效果。在一层楼，这些集装箱就充当各零售商店入口和房屋展示的檐篷，也为室内空间提供一些相应的小功能；而在上层空间中，这些集装箱则穿梭于水平通道上，作为各零售商店的入口以及凉廊沿边的商品橱窗。
这种棚式的建筑风格延伸至整个三里屯主街区，吸引了许多路人驻足观赏。从建筑外墙上看，置于各个小道内的集装箱，它们的突出部分，大大增强了整个建筑的立体感。橙色的网格式外墙，加上特色集装箱的点缀，给这个综合建筑的东北面增添了几分活力，这样的建筑外墙也为楼内人员提供了良好的隐私系统和遮阳功能。

The Waterfall House 318

建筑与水的灵动组合——瀑住宅

建筑师应邀在这处密集的社区中为一位年轻的商人设计了这座住宅，客户的特点、需求和场地的特色都融入了设计的过程。
通过构思，建筑师用水与设计结合的方式将场地的边界隐藏在水流之中。设计还兼顾了周围环境和日照轨迹的特点，场地位置和周围景致也结合其中。场地后方，水面位于与地面保持一致的高度。因此观景点需得到提高，才能观赏到优美的水面。建筑师将活动区布局在二层，这样即使在白天也可以欣赏到周围的全景。
在这里，水扮演了非常重要的角色。建筑包围在水中，流动的水与建筑的倒影相互掩映，使得居住空间也生动了起来。
从居住者的角度观赏过去，熠熠生辉的水波和水流丰富了人们的视觉和听觉享受。
浅水池将建筑环抱其中，赋予其无限的空间感。水池柔和了建筑与自然的分隔，将室内外空间完美过渡。
瀑布的高度进行了提升，降落在低处的游泳池中，为周围的环境注入一丝新鲜。同时，瀑布还保证了住宅内部的私密性。水流还减缓了北面强烈的光线，使环境变得柔和起来。
瀑布成为了建筑的亮点，这也是建筑师设计它的重要原因。
建筑师为住宅设计了丰富的功能空间，不仅能满足客户目前的单身社交活动，还兼顾到未来家庭的使用需求。二楼和一楼的公共区域都与私人空间进行了有效的分隔。
建筑还包括一个多功能空间，将住宅同室外的街道和空间相联系。这里可以作为车库使用，也可作为游乐厅或聚会场所。
三层垂直的外立面为室内带来了优越的光线环境，并将每一层都连接起来。
建筑的北立面有较大面积的开放，方便人们与水面亲密接触。
整体设计疏密适中，避免了大型窗体结构与隔热作用间的冲突。
服务区设置在住宅的前方，为室内提供了合适的私密性。

The Women's Health Centre 322

布基纳法索女性健康中心

2005至2007年间，一个名为AIDOS的意大利国际民间组织为保障发展中国家妇女权利建立了位于布基纳法索国的女性健康中心（CBF）。
这个国际项目在经济上得到了意大利某政党的支持，欧盟也予以了大力支援，主要致力于对抗这一区域针对女性的割礼行为。
在瓦加杜古市郊一个农村人口集中区域，AIDOS与当地的合作伙伴一起建立了这一社会/健康服务项目。其主要目标是提供教育性服务，提升女性对于性与生育权利的意识。
这个社会性项目也需要一个能够在恶劣环境下容纳多种功能的建筑，以应对并不乐观的条件。
这个建筑在FARE建筑设计工作室的带领下，由当地的建筑承包商共花费了15个月建造完成。建筑不仅在成本与功能性方面都达到了AIDOS的要求，也成为了整个区域的中心与标志所在。
建筑的科技与结构运用都符合这个项目的社会性，代表了对当地传统建筑方法的一种创新，呈现出一种自然的变化，达到AIDOS的预期。
项目首先从整体设计着手，建筑空间与当地气候条件（建筑的朝向、布局，对于能源使用的控制，自然植被的使用，技术的选择）之间的互动都是基于可持续性与适合性的考量。
项目空间参考CBF的主要职能活动，共划分为连接在一起的两个部分：培训中心和咨询中心。前者用于CBF的行政与管理职能；后者则用于医疗、法律援助和心理咨询。
这两个建筑空间都位于同一个升起的平台上。平台为人工建造，可服务于不同用途的建筑。平台需高出地面，不同于当地文化和建筑类型，是出于对卫生与气候条件的考量：防尘、防泥、防潮。
建筑上方是一个轻型的可回收防水PVC天棚，可遮挡雨水和阳光，下方由金属杆支撑。倾斜的防水布屋顶还可收集和储存雨水，用于花园的灌溉。
建筑的空间体量可根据需要，自由在地上平台设置，与屋顶结构相互独立。
这个项目中，建筑围合在一起，形成一个天井，保证了室内的隐蔽性。房屋的这种结构还方便将来建筑的扩展，而并不破坏现有建筑格局。墙面材料使用的是大块干燥的黏土砖，由场地内的混合土壤加入黏合剂与水制成。砖块完全是天然晒干，不涉及任何能源消耗，对环境不会带来消极的影响。

选择使用这种砖块来做墙面是出于对天气的考虑，因为可以起到很好的防潮效果。

材料与技术的采用也是源自对可替代与可持续性科技的追求，与科技融入场地达成标准的愿望，而不是一味地追求最好的建筑设计与依赖国外材料的进口。

建筑四周装置了一些波纹铝片制成的半通透通风口，用于自然采光，减少对人工照明的需求。

建筑屋顶与防水层之间的空间、地面平台下的空间，还有室外开放空间都安装了可开关的玻璃肋板，用于帮助室内完成自然通风，减少对空调等制冷设备的需求。

室外空间与室内空间的设计原则一致，作为一个开放空间，供整个社区的民众使用。人们在这里分享、交流——这正是CBF所倡导的。CBF还希望人们在一些小型活动与公开讨论上也能有间接的与非正式的交流。建筑围合而成的花园构成一个微环境，充分利用了建筑围合与树木产生的阴影，以及植物产生的水分。

花园的地面覆盖着一层草，帮助减少水土流失，同时，还植栽了两排来自西方和撒哈拉沙漠以南的多个不同品种，用以创造更多的阴凉，保留住更多的本土植被。

温度控制也许是最主要的天气问题，对整个项目的设计也有着非常大的影响。针对这个方面，设计师采用的策略是：

——仔细研究建筑的朝向，利用建筑与植物的背阴，降低热风的影响；

——选择一些大型材料阻挡阳光；

——广泛使用可开合的窗户；

——围合空间由传统形式空间分割，例如走廊或是天井。

由于该区域无法供应水力与电力，中心必须要自给自足，融入对能源消耗的控制、可再生能源的利用设计。

水来自一个新钻的水井。

为了能够达到不依赖外部能源的目标，中心外围墙面装置了光电设备，以补充电能。

所有这些因素都影响到了健康中心内个人的行为与集体的责任感：项目在环境可持续方面取得的最大成果或许就是不使用空调。

外墙没有任何直接对外的开口，表面覆上了一层塑料，上面涂满鲜艳的色彩，标上了被翻译成五种语言的当地NGO的标语。每面墙都成为了一个画布，以非正式的方式传达着一种社会信息。

建筑内部装饰有多彩的马赛克瓷砖，传递出一种活跃与欢愉的感觉，其建设过程也很简单，都是当地材料的另外一种方式的运用。

The Yoga Studio 324

瑜珈室

客户

业主是一对虔诚的佛教徒夫妻，同时也是现代主义者。他们希望在华盛顿特区之外的谢南多厄河谷建立一座周末度假屋。项目所在地是岩石地表，可以远眺到优美的山景。业主是环保主义者，他们希望将住宅设计成能源节约型建筑，同时将生活、运输、建造和材料维护方面对环境造成的影响减至最低。为了保证原汁原味的当地特色，业主和设计师都同意选择本地工匠进行建造，这也是环保理念的一种体现。项目由Carter + Burton Architecture设计。作为LEED实验项目的一部分，设计还获得了金质认证。

设计

考虑到克拉克郡对于工作室面积的相关规定，建筑设计尽可能地扩充空间感、强化景致，使其光线充足和具有较强的利用率。建筑有机的形状在保证设计简约风格的同时引入了足够的朝南光线。房屋现代的外形与周围环境的和谐融合在这种郊野地区非常少见。结构的循环特色在门道的每一端都有体现。房屋东面是小平台入口，西面则是一处类似树屋的观景台。墙壁和天花板的设计能够很好地节能，满足了空间布局的要求。简洁的设计在细节和材料上非常细致，许多定制元素也为项目增添了光彩。

场地

项目所在的山地为设计和建造带来了挑战。项目占地约公顷20,000平方米，位于一处粗犷的山壁之后，距华盛顿特区仅1个小时车程。工作室距业主住于山中的住宅仅30米，这样的选址能够在功能和系统上与住宅相互协调和补充，减少对周围环境的破坏。项目设计对周围的景观进行了保护，使得这处自然的森林美景很快恢复了天然状态。

系统

交换系统能够提供有效的空间温度调节和瑜珈室所有的热水需求。这一系统由双层的热量转换圈和相连的液体-液体加热泵和液体-气体加热泵组成。转换圈进行垂直和水平的水管内流动，能够满足工作室和住宅的调温需求。

加热泵通过内部的热水储存来调节和保持温度。机械工程师在设计系统的时候还考虑到了水泵中生物繁衍的影响。因此，在热量储存系统中应用了热量转换装置。

液体-气体加热泵可以通过空气压缩系统进行温度调节。热量恢复系统与净化系统能够保证室内空气的质量。

材料

经过测试表面，项目的建造技术符合所在地区和交易商的权限。以下内容是营造独特空间、细节和功能属性方面的关键：

设计过程中，团队考虑了各种因素，诸如结构强度、高R阻值、低空气渗透等，并使用了回收材料制造的复合板材和膨胀性聚苯乙烯。这样，瑜珈室就具有了独一无二的外形和风格，客户可以带领练习者们畅享这个节能而节约材料的空间。

项目还采用了许多得到绿色建筑认证推荐的环保材料，满足了客户对于健康的室内环境的要求。包括：

附近木材厂购买的白杨木作为混凝土基墙的面板使用。作为白杨木的自然隔离剂使用的豆油还重复使用在弯曲的墙壁面板和地板上。房屋使用了不含挥发性有机化合物的地毯和涂料。室内和门廊都采用了不锈钢的天花板面板，这种材料不直接反射光线，同时还扩充了空间感。浴室的墙壁和天花板采用了彩色玻璃材料，地板则是混凝土楼板和水磨石材料，门厅也采用了这样的材料搭配，营造更有宽阔感的空间。顶楼采用了订制的钢制扶手栏杆，与树脂玻璃面板的搭配营造了舒适安全并艺术感强烈的氛围。订制的木窗是建造商从16公里外组合好运来的，能够为室内引入足够的光线，还具有很好的隔热和热量储存功能。

文化

建筑的小型结构非常简洁，细节和材料部分就需要格外的精致来营造宽阔的空间感。以下是几项对项目效果作用重大的元素：

建造商从5公里之外的木材厂购买了约55厘米（22英寸）宽的白杨木作为基础模架使用。

建造商团队还在其工作室为房屋打造了木质门窗。

门采用了定制的钢制按钮，尺寸设计符合人们的使用习惯。这为紧凑的房屋节省了空间。

瑜珈室的另一个特色是它可以当作临时居所。白天，地板上的活门将床隐藏起来。顶楼的双层钢架床和定制床垫都能为人们休息提供方便。

一座钢木结构的楼梯为底楼节省了空间的同时，避免了对自然光线的阻挡。

瑜珈室的所有细木家具都是由当地的工匠打造的。

所有材料都是来自项目所在地。

项目的所有承包商都来自当地，路程不超过50公里。与采用环保材料相同，启用当地工作人员也是一种环保而有创意的方式，能够方便建筑的使用，营造长远的效益。

Tokyo Matsuya Unity 330
东京松屋单元住宅

现代日本建筑，旨在重新应用江户时代的传统建筑技术与装饰文化。
东京松屋公司这一住宅设计，正是重新应用江户时代的传统建筑技术与墙纸装饰艺术的体现。厚墙纸（最早是从我国进口的）上的图案是由碾磨的牡蛎壳粉和云母粉混制而成的，通常，这种厚墙纸用于装饰滑动门。"东京松屋"是一家信誉卓著的建材企业，该企业继承日本传统工艺，制造日本式传统手工隔扇用纸与一般日本传统建材用纸。该公司从江户时代就开始雕刻木桩制造，并且进行批发销售。东京松屋在这一单元住宅的底层为住户提供一个可"观赏、了解和购买"建材产品的平台，同时，在住宅楼的顶层为住户提供各种住宅的"生活、使用和体验"的现场体验区。用于装饰滑动门与墙壁的江户时代传统建材用纸，加之现代工艺元素，使得这些传统纸张也有了些许现代气息。采用江户时代传统建材用纸装饰的滑动门赋予整个室内空间以美感。障子（一种纸质滑动门）普遍应用于日本住宅中，它可以让房间吸收户外充足的自然光线。
东京松屋还在中间楼层，为住户提供空中花园、照明系统以及通风系统。一种传统看法认为："亲近大自然是日本传统建筑的一大特点"，这些日本传统建筑往往都与四周的公共区域紧密相连，构筑了和谐的居住环境。因此，在这种传统高层建筑中，即使在室内，也能感受到充足的阳光、徐徐的微风和户外的自然美景。
日本现代建筑的墙面粉刷依然延用了日本传统墙面色彩haijiro（灰白色）。即使现代建筑的外观极富现代气息，但还是应用了日本的传统图案作为装饰，例如在现代百叶窗上增添sudare（芦苇图案）加以装饰。

Touching Water 334
水上建筑

科特卡以木材及造纸产业而闻名于世。整个城市结构也随着Kotkansaari和Karhula两个中心的变化而不断发展。与海洋及港口相关的活动已成为该地区的重要特征。
场地中空旷的区域为将城市打造成符合当今生活理念的设计提供了难得的机会。
该项目提出根据城市内在的"系统化"尺度，将复杂的空间和建筑联系起来，而不是一味地将单一物体或建筑片段进行简单的叠加。
在这种"系统化"尺度理念的指导下，不同的建筑及景观元素以适当的方式结合起来，使城市变化更加明显。
该项目有利于人们进入各个空间去体验大楼的不同功能，在不同位置及城市其他地方形成联系，更重要的是形成了陆地与海洋融合的新城市空间。
土地与水这两个相互限制的现实条件，是该项目中主要关注的两个方面。海岸线被当做是临时划分城市起点与终点的重要因素。
广袤的自然与人类组成的有序画面也进一步增强了该地的特质。
城市部分有效地协调了自然与人类需求之间的关系。
欧洲风情的9个主题与项目结构
竞标区内，设计师对建筑内涵及形态作了详细的研究，满足了该项目的要求，在遵循现有的基本线路基础上，创建与散步码头相对应的专门区域。
场地场景
竞标场地的具体位置将原来城市场地作为中心区域，波兰码头作为东边的竞赛场地，西边为住宅区。
根据竞标要求，可移动房子的主体不仅仅是简单数量上的满足，而是修建一个能向海洋延伸的小渔村，给人一种永恒的感觉。
滨水
滨水地区的主题与持续的功能需求及海岸线不同的空间变化紧密相连。在西边，住宅区前是绿色的公共空间、小型平台与桥梁。游泳区内的小亭子及停车区域有序地排列。
广场是城市西部文化的集合地，这里物质文化与象征主义、宗教习俗与社会习俗融合在一起。广场中的绿色元素也宣告城市公共空间内自然复兴的宣言，绿色空间在"人性化"的当代空间中变得尤为重要。
滨水系统包括用于船只停靠及提供娱乐活动的固定及移动码头。
两种自然元素
四个互相连接的系统突出了该项目的可持续性，这些系统可概括为水和土地两大类。
流动的住宅系统
停泊码头上的桥梁突出了海洋村的特殊地形，使该空间与水元素更加紧密地联系起来。漂浮住宅之间的缝隙空间被用作水花园，与住宅周围的小水池联系起来。
精良的码头系统：餐馆、咖啡店、商铺、工厂及与小码头相关的活动使人想起现有港口设备的几何形状及排列。
土地
滑道：长长的住宅区将住宅区联系起来，与其他的住宅正面相比，一些住宅区内的滑道更加突出了长柱子与小空间及人类活动的联系。
天井：住宅区内的住宅单元结合起来形成小型的天井花园。尺度为6x3的柱子突出了天井花园周围的墙体。花园的木质平台，狭小的街道通向住宅区。
路线、街道、自行车道及人行道这四个元素互相交织，公共空间成为连接这四组元素的枢纽。
公共空间也被专门设计成绿色空间，将现有的绿色生态系统在周围住宅区或更大的尺度范围内连接起来，增强该区人们的主人翁意识。
该项目的设计理念建立在技术-空间结构的组织上，满足了客户灵活多样的需求，成为了将高层建筑与特殊情景结合的典型实例。

Tracing Perception 340
认知研究

这个项目主要目标是开发一种能够分析现代技术如何影响都市空间的方法。更细致地说，它着重于研究手机在公共空间中的使用。首先，它通过对人们行为和运动方式的分析，尝试呈现人们认知的变化。
伦敦的利物浦大街是这次研究的观察区域，这里每天的人流和特点都将被记录和研究。通过这片区域的选择，研究所有使用手机的路人的行动。公共区域的组合和多样化是选择这条大街的重要标准。随着映射技术的发展，这一分析研究提炼出短期个人动作的可视化结果。
研究的下一步是对映射过程之外的空间品质进行提炼。这是通过转换过程的逐步发展而慢慢实现的。转换过程可自行提取公共空间的研究内容。提取内容和空间关系的研究就是此次设计师所致力的内容。
设计师最终提取出转换结果并将其转变为一个三维的钢制结构。这个结构具有独特的几何特征，能够营造多样的空间关系。通过此项研究，可以用多媒体的手段来体验城市公园与这些空间关系的互动。木材这种软质材料体现了空间关系较轻的特质。木材与钢铁结构得到了巧妙组合，营造了不同的空间和内在关系。通过媒体方式的应用，设计最终呈现了一个由空间定义的半公共场所。

Tranquil Multifamily Residence on the Hillside 346

宁静惬意的山间群居住宅

该项目选址位于危地马拉城山区边缘一个陡峭的斜坡上，这里地处地震带，因此经济发展缓慢，但四周的风景却十分优美。

这一群居住宅采用长跨度式建筑结构，同时借助植入陡坡深处的混凝土桩支撑起这个特殊的住宅结构，从而稳固住宅内的架空悬臂。

住宅内所有的房间都使用大跨度的落地玻璃窗，所有的墙壁都是由轻型钢夹板砌筑而成的，屋顶以木质Prodema rainscreen与铝质复合板为主要建筑材料。

为了使室内空间在视觉上给人以宽敞舒适的感觉，建筑师对室内的钢结构布局并没有做过多的修饰。住宅内的各个卧室是为一群30岁左右的高收入专业人士设计的，因此，卧室空间的设计显得现代而前卫。

Tulipan House 350

Tulipan住宅

漫步于华沙Mokotow商业中心的街道上，映入眼帘的大概都是现代化的高楼大厦，但有一栋建筑如同林立的高楼中出现了一个缺口，这便是Tulipan House，它的低小会让你大吃一惊。要想领略这一建筑的魅力，还待你亲自去看看。

Tulipan House的存在证明，预算的节约、设计的灵活性、结构的高效性以及项目执行的效率不一定和建筑的高质量相冲突。

走进Tulipan House你就会发现，越往里，建筑的那些细微的特征就会体现得越明显。引人注目的前庭中，开放式的廊台将四层连为一体，形成一个大的公用空间，通过透明的玻璃隔墙与建筑前面的庭院融为一体。然而，这一建筑的设计精华却并不在于其前立面，而是体现在将被租出去的那一部分。这些未来将被用作办公室的空间，高度与深度都是经过仔细权衡的，从而实现自然光线的照射以及工作空间的最佳流通。合适的形态、合乎逻辑的维度及比例使得整个建筑显得环保并符合人的尺度。

从现实的角度来讲，建筑是即将被出售的产品，然而，建筑师在设计中，同时要实现建筑的商业目的并展现自己的设计技巧，让建筑在坚固和实用的同时显得美观，是有很大难度的。

Tulipan House的成功设计就是一个惊喜。那些曾经不相信Tulipan House的设计会在这样一个特定的时期和地点成功的人，现在不得不承认，一切规则皆有例外。

Universidade Agostinho Neto Campus Plan 354

安哥拉Agostinho Neto大学校园规划

安哥拉Agostinho Neto大学校园的面积为2000万平方米，在校学生约17,000人。校园的整体规划旨在通过一些设计原则提升城市中没有得到良好维护的生态环境。

规划中采用的第一个原则就是集中发展项目的一部分，保留原有的植被和水域，利用景观提升整体环境。校园的主要区域被一个椭圆形车道包围，位于两个水道之间，发展受限。这条环路中间是不同的空间层次和向中心伸展的道路系统，这些道路在中心交汇处就形成了一个庭院，有图书馆和广场。

第二个设计原则是建筑定向，排列的方向要利于自然通风，夏天凉爽。所有建筑都通过最佳的方位和单排教室的设计达到自然通风的效果。屋顶采用了遮阴系统，将日照面积降到最低，同时通过加压，实现教学楼的前后通风。

Villa in Ojai 358

Ojai别墅

该别墅位于洛杉矶郊外一片空旷的山坡上，满足了客户对新家的精神需求。设计师采用灵活的设计策略，既保护住宅的私密性，又不影响整体视觉景观。项目中采用的线条丰富了全景与围合景观。空间内外的景致宛如自然世界，富有生机活力的公共空间给人一种亲密感。

客户大部分时间穿梭在欧洲两个城市之间。而加利福尼亚居所不仅仅是住宅，还是两个家庭与他们三个孩子居住的温馨的家。家就像聚集在一起的凉亭，将走廊空间联系起来。场地中的斜坡有利于屋顶线条与墙壁之间的联系，在特定的平面内，使建筑结构更具有动感。

Water — Touristic Complex in Kaliningrad Center 362

城市中心旅游开发区建设

该项目的主要建筑理念是根据战前划定的城市区块和相关历史资料建立一个城市中心旅游开发区。建筑师们打算在这一开发区项目中添加进新的建筑技术，即新型建材与战前大多数建筑所采用的玻璃材料相结合。建筑场地的北面是Karbushev河堤，西面是Oktiabriaskaia大街，南面是Solnechnui林荫道，东面是Pavlov大街。新型的开发区布局中包括小型私人旅馆群落、生活区和艺术家活动空间。一楼将设有咖啡屋、餐厅、酒吧、纪念品商店、旅游者服务中心、小型博物馆和画廊等。

在Pregol河畔将建造一个大型海产品交易市场，这里曾经是加里宁格勒人的骄傲。此外，当地的相关基础设施建设不仅给加里宁格勒当地居民的生活带来了极大的便利，也为来此旅游的国外游客带来了极大的乐趣。在该项目的规划过程中，建筑师们还打算对现有的建于20世纪70年代的旧旅游区及其周边地区进行相应的改造，从而更好地展现加里宁格勒的新面貌。滨海地区的改造将与Pregol河的水上交通改造挂钩，这样就将形成一个良好的水上游览交通系统，游客们可以尽情地在游艇和其他水上交通工具上欣赏Pregol河两岸的美丽景致。

Pregol河沿岸地区还将建造9家系列滨水酒店，以便为游客们提供更加开放的观景视野，这样游客们只要透过酒店的窗户，四周的灵动水景都将尽收眼底。这些滨水酒店有一个统一的地下室，内设公共活动空间、餐厅和管理室。滨水区步行林荫道与Knaiphov岛上的大教堂相连，而Evangelicheskaia教堂则成为了酒店区和当地居民的生活区的界线。

在每个居住区内都有一个内部庭院，但仅供居民们使用。之所以要限制使用人群，是因为室内庭院的造价比其他公共空间的造价要高，因而有必要限制其人流量，以确保居民的权益。在居住区的一层设有商店、图书馆、运动场和小型电影院等设施。居住区的地下室将计划修建一个仅供居民使用的停车场；五、六两层将建造系列双层公寓。这一项目的建成，将展现出一个全新的城市面貌，由于项目是根据战前划定的城市区块和相关历史资料而建的，因而，它向游客们充分展示了当地的文化与历史，在时尚中透出一股怀旧的韵味。

Winery in Santorini　368

圣托里尼葡萄酒厂

基拉克迪群岛建筑

希腊的基拉克迪群岛位于爱琴海的中部、东部和西南部，这里的规划体系对岛上建筑的形态进行了统一规定。岛上的建筑均为单层或两层结构，围绕在庭院周围，与周围的天然地形相互融合。在炎热的夏天，这样的布局既能够保护居民的隐私，又可为他们提供相互交流的机会。

当地的法律还规定，建筑必须使用白色作为统一色调。门窗的颜色有三到四种色彩可以选择：淡蓝色、灰色、绿色和赤陶色。群岛的地形粗糙、植被稀少，小型的建筑结构在视觉上对周边的环境进行了缓和。这些建筑组成的村庄和其中不超过两米宽的小巷以及入口台阶都成为了群岛的独特风格。

Boutaris葡萄酒厂综合大楼

Boutaris葡萄酒厂坐落在群岛中最著名的圣托里尼岛上，靠近阿科罗提利。酒厂有一座设计于1987年并于1989年建成的圆顶建筑，在岛上的酿酒生产中发挥重要作用。建筑对公众和游客完全开放。

建筑布局

建筑环绕在葡萄园之中，紧邻火山口。这里包括一座酿造中心、一个储存区、一座管理大楼、销售部门和店铺。著名的圆顶建筑——一座室内的圆形阶梯剧场也坐落于此。

这处角落区域沿着道路从Megalohori一直延伸到阿科罗提利。客车和卡车自北面进入，从西面离开。公众使用的通向行政楼的主入口也位于此地。所有的建筑围绕着中心庭院分布，酿酒厂位于北面和东北面，储藏区位于东面。南面是圆顶建筑，待建的住宿设施则位于西面。

整个布局以庭院为中心，与所有功能区相连。这处区域不受北面和东北方向的狂风侵袭。庭院成为了连接所有区域的重要元素，并再现了当地的传统建筑布局。

行政楼和储藏区是屋顶平坦的单层建筑，酿酒厂和住宿设施为圆顶结构。如此一来，建筑平坦的外部结构和室内景致都与中心庭院形成了巧妙的对比。圆顶的几何元素在设计中发挥了重要作用，将周围的区域有力地组合起来。游客们在参观完周围的设施之后，可以在这里观看有声影像，从而了解圣托里尼最著名的Boutaris葡萄酒厂。建筑的圆顶结构反映了基克拉迪岛统一的建筑特色，白色和深蓝色的组合与美丽的爱琴海相互辉映。

设计中使用最多的材料是白色石膏，庭院和圆顶建筑周围的地板采用了鹅卵石块组成的蓝色马赛克。

X HOUSE With Flexible Structure　372

结构灵活的X住宅

建筑师们在着手设计这一住宅之前，还未寻找到理想的建筑场地。他们只是根据基多及其城市东部的山谷地形，为该住宅设计出了一个概念性的基本建筑方案，使其能够适应各种地形环境。这就意味着这是一个抽象化的设计过程，建筑师们只能从唐纳德•贾德的成功设计作品中获取一定的灵感，以此达到设计一座结构灵活的住宅的目的，住宅的东西两侧的空间变化不受局限，可随地形或环境的需求随时改变空间大小。

由于没有实际的项目场地，建筑师们只能凭借着自己的设计经验来判断建筑空间的实际尺寸。在设计过程中，建筑师们意外地发现，原来住宅天井的建造在人类住宅建筑史上起到了至关重要的作用。该项目的建筑师们对典型玻璃住宅情有独钟，该项目的建筑主风格由此形成，同时，这种玻璃住宅也适合于当地长年的温和气候。

虽然住宅天井的设计为建筑师们提供了一定的灵感发挥空间，然而它毕竟是住宅的一个组成部分，它必须与住宅的功能相匹配，因此，建筑师们的灵感发挥还是存在着一些局限性。这一玻璃住宅的魅力之处在于它开阔的视野，无论处于住宅的什么方位，户外美景都能一览无余。在这一玻璃住宅中，天井的设置，还能够使得住宅适应不同的地形环境。

建筑师们在住宅中的私密空间和公共空间之间设计天井的目的在于，在具体建筑场地确定之后，可根据场地的实际情况，对天井进行临时改造，在住宅中灵活添加适当的活动空间及空气流通空间。这一设计理念进一步证明了天井在住宅建筑中的重要性。

总而言之，这一设计方案适用于任何场地，并且可以根据场地的需求，灵活改变住宅的朝向、规模和相关比例。

材料与施工

住宅的建筑结构与选材是两个同等重要的考虑因素，除此之外还应考虑到与该项目局限条件相关的几个前提：该住宅建筑结构必须能够随时进行灵活调整，并能够随着预算和场地结构的不同而对建造策略做及时的调整。这就意味着，为了成功地设计该项目，建筑师们必须尽可能地收集全国范围内的各种不同风格、不同规模的旧式住宅、废弃住宅、低使用率住宅和数量有限的标准建筑的相关资料，从中提取其设计技术的精华，进而应用于该项目之中。

虽然该项目的场地十分理想，但由于建筑预算十分有限（加上车库的建造，一共仅限155,000美元），因此，建筑师们确定下来的最终住宅结构是：

在一个混凝土基脚上架起一个轻钢结构住宅，主要由不锈钢和胶合板搭建而成。所有的空气流通空间、活动空间以及其他住宅空间的内部墙面都以白色为主要色调，而外墙则采用聚碳酸脂材料，以防止西面阳光的暴晒。所有的活动空间与空气流动空间的建造呈平行状，天井的雨水处理系统与住宅的其他排水系统分离：雨水落在住宅的外墙上，而后向下流入不锈钢槽中，进而流向地面，以灌溉周边的草地。

Ziranenge/Angel — A Healthcare Center for Children　378

儿童保健中心

在建筑师们下载了该项目的两个投标选址的相关资料之后，他们仍然对该项目的场地选择举棋不定，因为即使通过这些网上原始资料，建筑师们对两个场地具体情况的了解也相当有限。因此，制定适合于判断建筑场地选址可行性的多制式系统及其相关的工具程序成为了建筑师们迫在眉睫的任务。在这种思维的引导下，建筑师们选择了一块南纬2°9'东经30°31'的土地作为他们的实验地。为了促进系统制定工作顺利准确地开展，建筑师们在选定的实验地中划定了一块面积约4,047平方米的圆形土地作为实验的轨迹。而后他们沿着这一圆形轨迹，追踪人们从该场地的外围地区以及原先所设定的建筑场地到医院的距离。从而，建筑师们将他们的研究焦点转移至场地中其他不同的位置，以便了解这些步行模式如何通过自身的变形而影响整个场地的概念性建设的。接下来的研究重点将放置在那块面积约4,047平方米的圆形土地上，并同时开发各个可执行相关工作程序的零散土地。建筑师们选择了其中一块零散土地作为我们建筑设计的目标场地。

建筑师们之所以为该项目取名为Ziranenge /天使是出于两个方面的原因：一是该项目的建造将改善当地的医疗条件，为当地的孩子和孕妇提供有效的医疗设施；二是该项目建成后可以成为当地的救济中心，任何需要救济的人都可以在此得到相关的救济，而所有有能力救助他人的人们也都可以通过这里奉献出自己的一片爱心。

建筑师们的目标是将该项目建造成一座安全、舒适、具有较高性价比和自身特性的救济医疗场所。他们打算选用风干砖坯作为该项目的主要建材，不但因为这种建材具有很强的可持续性，还因为用这种建材砌筑的建筑外观与周围的自然景观十分搭衬。该项目建造的另一个特点是建筑物一楼的堡垒式建筑外墙设计，这种外墙造型是由当地工匠用传统方法雕刻而成的。

建于堡垒式建筑外墙上方的格子墙是采用当地的枯木建造而成的。这种格子墙围合而成的室内空间，是为那些长途跋涉来医院治病的人们提供的休息场所。这种格子墙还有助于院方随时安装太阳能电池板。“天使”楼中特殊的屋顶设计，是为了通过屋顶斜槽或拱形屋顶的两端来收集雨水，屋顶收集而来的雨水经过印度慢杜人设计的一种古老而传统的水净化系统的物理净化处理，排除水中的杂质之后，便可成为日常生活中的饮用水。建筑师们建议所有的屋顶都采用陶瓷碎片覆盖其上，以阻挡室外强烈的太阳光照射。儿童病房与婴儿监护房的墙面设计必须注重其通风性和光线渗透性。同样，儿童病房的屋顶设计也应注重其通风性与可透视性，以便于孩子们在满天星辰的夜晚，可以躺在病床上仰望星空。这些设施也为墙壁提供了进一步的绝缘系统。建筑师们将当地的儿童艺术特点与儿童病房的墙体设计相结合。该项目的建造中，尽可能不使用玻璃作为窗户建材，以避免孩子们被玻璃划伤，建筑师们大量使用当地雕刻的网格窗，这样的窗户结构有利于室内的通风和降温。

艾滋病预防中心房间内放置着一台电脑，电脑显示屏上不断显示艾滋病预防的措施，以便路过的人们参考阅览。在艾滋病预防中心的旁边是一个圆形露天广场，可以作为一个非正式的会谈或研讨空间。建筑师们在该建筑的上空放置了一个巨型热气球，以便病人在灾难性紧急情况来临的时候逃生。热气球在空中飘动，就仿佛随时替人们排忧解难的天使一般。

建筑师们坚信，随着时间的推移，他们将把这样的建筑推广到整个卢旺达，使得整个国家的各个地区都拥有同类设施完善的救济医疗中心。

图书在版编目（CIP）数据

A+C建筑+方案 / 凤凰空间·上海编. -- 南京 : 江苏人民出版社, 2011.10
ISBN 978-7-214-07465-2

Ⅰ. ①A… Ⅱ. ①凤… Ⅲ. ①建筑设计—世界—现代—图集 Ⅳ. ①TU206

中国版本图书馆CIP数据核字(2011)第194140号

A+C 建筑+方案 凤凰空间·上海 编

策划编辑：石 莹
责任编辑：刘 焱 石 莹
责任监印：彭李君
出 版：江苏人民出版社（南京湖南路1号A楼 邮编：210009）
发 行：天津凤凰空间文化传媒有限公司
销售电话：022-87893668
网 址：http://www.ifengspace.com
集团地址：凤凰出版传媒集团（南京湖南路1号A楼 邮编：210009）
经 销：全国新华书店
印 刷：天津市豪迈印务有限公司
开 本：965毫米×1270毫米 1/16
印 张：26
字 数：208千字
版 次：2011年10月第1版
印 次：2011年10月第1次印刷
书 号：ISBN 978-7-214-07465-2
定 价：328.00元（USD 65.00）